Materials Processing during Casting

Materials Processing during Casting

Hasse Fredriksson
Ulla Åkerlind

John Wiley & Sons, Ltd

Other Wiley Editorial Offices

John Wiley & Sons Inc., 111 River Street, Hoboken, NJ 07030, USA

Jossey-Bass, 989 Market Street, San Francisco, CA 94103-1741, USA

Wiley-VCH Verlag GmbH, Boschstr. 12, D-69469 Weinheim, Germany

John Wiley & Sons Australia Ltd, 42 McDougall Street, Queensland 4064, Australia

John Wiley & Sons (Asia) Pte Ltd, 2 Clementi Loop #02-01, Jin Xing Distripark, Singapore 129809

John Wiley & Sons Canada Ltd, 22 Worcester Road, Etobicoke, Ontario, Canada M9W 1L1

Wiley also publishes its books in a variety of electronic formats. Some content that appears in print may not be available in electronic books.

Library of Congress Cataloging-in-Publication Data

Fredriksson, Hasse.
   Materials processing during casting/Hasse Fredriksson, Ulla Akerlind.
      p. cm.
Includes bibliographical references and index.
ISBN-13: 978-0-470-01513-1 (cloth :alk. paper)
ISBN-10: 0-470-01513-6 (cloth :alk. paper)
ISBN-13: 978-0-470-01514-8 (pbk :alk. paper)
ISBN-10: 0-470-01514-4 (pbk :alk. paper)
1. Founding--Textbooks. 2. Metal castings--Textbooks. I. Akerlind, Ulla. II. Title.
   TS230.F74 2006
      671.2--dc22
                                    2005023378

British Library Cataloguing in Publication Data

A catalogue record for this book is available from the British Library

ISBN-13 9-78-0-470-01513-1 (Cloth)   9-78-0-470-01514-8 (Paper)
ISBN-10 0-470-01513-6 (Cloth)   0-470-01514-4 (Paper)

Typeset in 10/12pt Times by Thomson Press, New Delhi, India

Contents

Preface

To cast molten metal in a mould and let the melt solidify and cool there does not seem to be a very complicated process. It would be reasonable to assume that casting is a fairly simple business. However, this conclusion does not at all agree with reality. Casting processes have to be carefully supervized and controlled to give useful final products that correspond to modern demands.

The present book covers most aspects of casting. It deals with the principles behind modern casting with applications incorporated in both non-ferrous and ferrous metals throughout the theory. It consists of 11 chapters.

- Chapters 1–2 give a short introductory survey of casting methods and equipment.
- Chapters 3–6 give an account of the theoretical basis behind the modern casting processes.
- Chapters 7–11 have a more practical approach to casting of various metals. The principles derived in Chapters 3–6 are applied to teach the reader how various problems can be mastered and how the casting processes can be designed in an optimal way, according to the present knowledge of casting science. Due to the importance of steel- and iron-based alloys a good deal of the text is devoted to these alloys but the principles and phenomena are general.

Most chapters contain several solved examples in the text. At the end of each of Chapters 3–11 a number of exercises are given. The answers to the exercises are listed at the end of the book. However, the step from passive studies of the solved examples in the text to active problem solving alone may initially be too high. The aim of the separate Guide to Exercises* is to overcome this difficulty.

- The guide to exercises gives complete solutions to them all, step by step. The solutions are far from the direct and complete solutions of the solved examples in the text. The solutions in the Guide are designed in a way which encourages the student to his or her own activity. The idea is *help to self-help* to achieve increased understanding.

This extensive book offers material for several courses on different aspects of the subject. It is also possible to some extent to vary the level of the courses. Some of the mathe-matical derivations are independent of the text and included in special boxes. They can be omitted without loss of continuity of the text. The choice belongs to the users.

There are differences in American and British nomenclature. We have chosen to use the word *cast house* for a place where large quantities of semiproducts are manufactured and save the word *foundry* for a place for component manufacture.

It has been our ambition to follow and utilize the development of international research in the field of casting of metals and alloys to make this book up-to-date and in agreement with the present state of scientific knowledge. Many of the research results during the last 30 years in the shape of papers published in international journals and/or dissertations at the Division of Casting of Metals at the Royal Institute of Technology in Stockholm (KTH) have also been included. We refer particularly to the works performed by Johan Stjerndal, Anders Olsson, Bo Rogberg, Roger West, Carl M. Raihle, Nils Jacobson, Per-Olov Mellberg and Jenny Kron. We want to express our gratitude to them for the privilege of including their contributions to new knowledge in various fields of casting into this book. We also want to thank Mats Hillert for fruitful discussions on microsegregation in the old days.

We would also like to thank Jonas Åberg and Tomas Antonsson at the Division of Casting of Metals (KTH), Thomas Bergström, Erik Fredriksson and Gunnar Edvinsson (University of Stockholm) for their valuable support concerning practical matters like annoying computer trouble and application of some special computer programs. We are most grateful to Alrik Östberg who has been our consultant concerning modern foundry terms in Chapters 1 and 2. We also want to express our warm gratitude to Elisabeth Lampén. Her faithful support and never-failing care has been a great safety and comfort.

We are grateful for financial support from The Iron Masters Association in Sweden and 'Stiftelsen Sven och Astrid Toressons Fond'.

At last and in particular we want to express our sincere gratitudes to Karin Fredriksson and Lars Åkerlind. Without their constant support and great patience through the years this book would never have been written.

Hasse Fredriksson
Ulla Åkerlind
Stockholm, September 2005

*Available on www.wiley.com

1 Component Casting

1.1 INTRODUCTION

1.1.1 History of Casting

As early as 4000 years BC the art of forming metals by casting was known. The process of casting has not really changed during the following millennia, for example during the Bronze Age (from about 2000 BC to 400–500 BC), during the Iron Age (from about 1100–400 BC to the Viking Age 800–1050 AD), during the entire Middle Ages and the Renaissance up until the middle of the Nineteenth century. Complete castings were prepared and used directly without any further plastic forming.

Figures 1.1, 1.2 and 1.3 show some very old castings.

In addition to improving the known methods of production and refining of cast metals, new casting methods were invented during the Nineteenth century. Not only were components produced but also raw materials, such as billets, blooms and slabs. The material qualities were improved by plastic forming, forging and rolling. An inferior primary casting result cannot be compensated for or repaired later in the production process.

Steel billets, blooms and slabs were initially produced by the aid of *ingot casting* and, from the middle of the Twentieth century onwards, also by the aid of *continuous casting*. Development has now gone on for more than 150 years and this trend is likely to continue. New methods are currently being developed, which involve the production of cast components that are in size as close to the final dimensions as possible.

1.1.2 Industrial Component Casting Processes

As a preparation for a casting process the metal is initially rendered molten in an oven. The melt is transferred to a so-called *ladle*, which is a metal container lined on the inside with fireproof brick. The melt will then solidify for further refining in the production chain. This is performed by transferring the melt from the ladle into a mould of sand or a water-chilled, so-called *chill-mould* of metal. The metal melt is then allowed to solidify in the mould or chill-mould.

This chapter is a review of the most common and most important industrial processes of component casting. The problems associated with the various methods are discussed briefly when the methods are described. These problems are general and will be extensively analysed in later chapters.

In Chapter 2 the methods used in cast houses will be described. The methods used in foundries to produce components will be discussed below.

1.2 CASTING OF COMPONENTS

1.2.1 Production of Moulds

A cast-metal component or a *casting* is an object that has been produced by solidification of a melt in a *mould*. The mould contains a hollow space, the *mould cavity*, which in every detail has a shape identical to that of the component.

In order to produce the planned component, a reproduction of it is made of wood, plastic, metal or other suitable material. This reproduction is called a *pattern*. During the production of the mould, the pattern is usually placed in a mould frame, which is called a *flask* or *moulding box*. The flask is then filled with a moulding mixture which is compacted (by machine) or rammed (with a hand tool). The moulding mixture normally consists of sand, a binder and water.

When the compaction of the flask is finished the pattern is stripped (removed) from it. The procedure is illustrated in Figures 1.4 (a–d). The component to be produced is, in this case, a tube.

Stage 1: Production of a Mould for the Manufacture of a Steel Tube

The cavity between the flask wall and the pattern is then filled with the mould paste and rammed by hand or

Materials Processing during Casting   H. Fredriksson and U. Åkerlind
Copyright © 2006 John Wiley & Sons, Ltd.

Figure 1.1 Stone mould for casting of axes, dating from 3000 BC.

Figure 1.2 A knife and two axes of pure copper, cast in stone moulds of the type illustrated in Figure 1.1.

compacted in a machine. The excess mould paste is removed from the upper surface, and the lower part of the future mould is ready. The upper one is made in the same way.

Components due to be cast are seldom solid. They normally contain cavities, which must influence the design of the mould. The cavities in the component correspond to sand bodies, so-called *cores*, of the same shape as the cavities. The sand bodies are prepared in a special *core box*, the inside of which has the form of the core. The core box, which is filled and rammed with fireproof so-called core sand, is divided into two halves to facilitate the stripping. The cores normally obtain enough strength during the baking process in an oven or hardening of a plastic binder. Figures 1.4 (e) and 1.4 (f) illustrate the production process of a core, corresponding to the cavity of a tube.

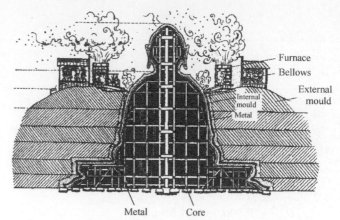

Figure 1.3 Picture of a cast bronze Buddha statue, which is more than 20 m high. The statue was cast in the Eighth century AD. Its weight is 780×10^3 kg. A very special foundry technique was used in which the mould production and casting occurred simultaneously. The mould was built and the statue was cast in eight stages, starting from the base. The mould was built around a framework of wood and bamboo canes. Each furnace could melt 1×10^3 kg bronze per hour. Reproduced with permission from Giesserei-Verlag GmbH.

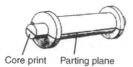

Figure 1.4 (a) A pattern, normally made of wood, is prepared as two halves. It is equipped with a so-called core print at the ends as dowels. Reproduced with permission from Gjuterihistoriska Sallskapet.

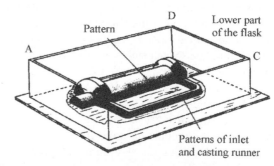

Figure 1.4 (b) Half the pattern and the patterns of inlet and casting runner are placed on a wooden plate in half a flask. A fine-grained powder, for example lycopodium powder or talc, is distributed over the pattern to facilitate the future stripping of the pattern [see Figure 1.4 (d)]. Reproduced with permission from Gjuterihistoriska Sallskapet.

Stage 2: Production of the Core in what will become the Steel Tube

When the mould is ready for casting the complete cores are placed in their proper positions. Since the fireproof sand of

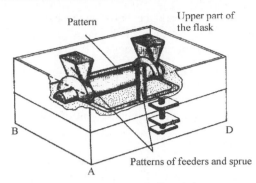

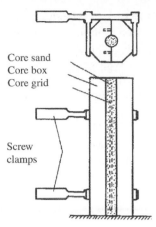

Figure 1.4 (c) The upper pattern half and the upper part of the mould flask are placed on the corresponding lower parts. A thin layer of fine-grained dry sand, so-called parting sand, covers the contact surfaces. Special patterns of the future sprue and the feeders are placed exactly over the inlets in the lower parts of the tube flanges. Reproduced with permission from Gjuterihistoriska Sallskapet.

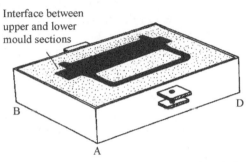

Figure 1.4 (e) The cavity in what will become the tube is formed by a sand core, produced in a core box. The two halves of the core box are kept together by screw clamps while the sand is rammed into the mould. A cylindrical steel bar is placed as core grid in the lengthwise direction of the core to strengthen the future core. Reproduced with permission from Gjuterihistoriska Sallskapet.

Figure 1.4 (d) The mould parts are separated and the pattern parts are stripped, i.e. lifted off the upper and lower parts of the mould. The patterns of the inlet, sprue and feeders are also removed. The figure shows the lower part of the mould after the pattern stripping. Reproduced with permission from Gjuterihistoriska Sallskapet.

Figure 1.4 (f) The core is lifted out of the parted core halves. The core is often baked in an oven to achieve satisfactory strength. Reproduced with permission from Gjuterihistoriska Sallskapet.

the cores has a somewhat different composition than that of the mould, one can usually distinguish between core sand and mould sand.

A necessary condition for a successful mould is that it must contain not only cavities, which exactly correspond to the shape of the desired cast-metal component, but also channels for supply of the metal melt. These are called *casting gates* or *gating system* [Figure 1.4 (c)]. Other cavities, so-called *feeders*, which serve as reservoirs for the melt during the casting process, are also required

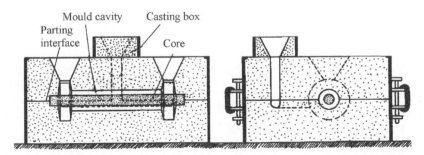

Figure 1.4 (g) The core is placed in the lower part of the mould. The parts of the mould are joined with the parting surfaces towards each other. The dowels through the holes in the outer walls of the flask parts guarantee the exact fit of the corresponding cavities in the upper and lower halves of the mould. A so-called casting box, which insulates the upper surface of the melt and prevents it from solidifying too early, is placed exactly above the sprue and the parts of the mould are kept together by screw clamps. The mould is ready for use. Reproduced with permission from Gjuterihistoriska Sallskapet.

[Figures 1.4 (c) and 1.4 (g)]. Their purpose is to compensate for the solidification shrinkage in the metal. Without feeders the complete cast-metal component would contain undesired pores or cavities, so-called *pipes*. This phenomenon will be discussed in Chapter 10. When the casting gate and feeders have been added to the mould, it is ready for use.

Stage 3: Casting of a Steel Tube
The casting process is illustrated in Figures 1.4 (g), 1.4 (h), and 1.4 (i).

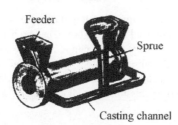

Feeder

Sprue

Casting channel

Figure 1.4 (h) When the casting has solidified and cooled after casting, the mould is knocked out. The casting is cleaned from remaining sand. The feeders and inlet are removed through cutting or oxygen shearing. The section surfaces are ground smoothed. Reproduced with permission from Gjuterihistoriska Sallskapet.

Figure 1.4 (i) The complete steel tube.

1.2.2 Metal Melt Pressure on Moulds and Cores

During casting, moulds and cores are exposed to vigorous strain due to the *high temperature* of the melt and the *pressure* that the melt exerts on the surfaces of the mould and cores.

To prevent a break-through, calculations of the expected pressure on the mould walls, the lifting capacity of the upper part of the mould and the buoyancy forces on cores, which are completely or partly surrounded by melt, must be performed. These calculations are the basis for different strengthening procedures such as varying compaction weighting in different parts of the mould, locking of the cores in the mould and compaction weighting on or cramping of the upper part of the mould.

The laws, which are the basis of the calculations, are given below. The wording of the laws has been adapted to the special casting applications.

> ### *The Law of Connected Vessels*
> If two or more cavities are connected to each other, the height of the melt will be equal in all of them.
>
> ### *Pascal's Principle*
> A pressure that is exerted on a melt in a closed cavity, is transferred unchanged to all parts of the mould wall.
>
> ### *Liquid Pressure and Strain*
> $$p = \rho g h \qquad\qquad F = pA$$
>
> ### *The Hydrostatic Paradox*
> The pressure on a surface element is universally perpendicular to the element and equal to $\rho g h$ where h is the depth of the surface element under the free surface of the melt, independent of the direction of the element.
>
> The pressure on a lateral surface = the weight of a column with the surface as a basis and a height equal to the depth of its centre of mass.
>
> ### *Archimedes' Principle*
> An immersed body (core) seemingly loses an amount of weight equal to the weight of the melt displaced by the body.

The laws given on above are valid for static systems. During casting the melt is moving and dynamic forces have to be added. These forces are difficult to estimate. The solution of the problem is usually practical. The calculations are made as if the system were static and the resulting values are increased by 25–50 %.

An example will illustrate the procedure. The pressure forces are comparatively large and the moulds have to be designed in such a way that they can resist these forces without appreciable deformation.

Example 1.1
A cavity consists of a horizontal cylindrical tube. Its length is L and its outer and inner diameters are D and d, respectively. The interior of the cylinder is filled with a sand core. The density of the sand core is ρ_s. The axis of the cylinder is placed at the depth h below the free surface of the melt. The density of the melt is ρ_L.

Calculate

(a) the buoyancy force on the upper part of the mould
(b) the total buoyancy force on the sand core when the cavity is filled with melt, and

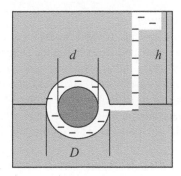

(c) the mass one has to place on the upper part of the mould, to compensate for the buoyancy forces, if $d = 50$ mm, $D = 100$ mm, $h = 200$ mm and $L = 300$ mm.

The densities of the melt and the sand are 6.90×10^3 kg/m^3 and 1.40×10^3 kg/m^3 respectively.

Solution:

(a): Via the sprue the melt exerts outward pressure forces acting on the surface elements of the upper part of the mould. These are equal to the pressure forces that act on each surface element in the figure but are opposite in direction because the forces in this case act from the melt towards the mould. The desired buoyancy force is thus equal to the resultant in the latter case

(b): The lifting force is equal to the weight of the melt, displaced by the sand core, minus the weight of the core. This force acts on the core prints [Figure 1.4 (a) on page 2].

$$F_{\text{lift}} = \left(\frac{\pi d^2}{4}\right) L \rho_L g - \left(\frac{\pi d^2}{4}\right) L \rho_s g = \left(\frac{\pi d^2}{4}\right) L g (\rho_L - \rho_s) \tag{2'}$$

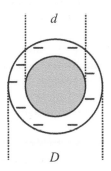

(c): The forces, directed upwards and acting on the upper part of the mould, are equal to the sum of the forces in equations (1') and (2') because the lifting force on the core, via the core prints, also acts on the upper part of the mould.

$$F_{\text{total}} = LD\rho_L g \left(h - \frac{\pi D}{8}\right) + \left(\frac{\pi d^2}{4}\right) L g (\rho_L - \rho_s) \tag{3'}$$

$$F_{\text{total}} = 0.300 \times 0.100 \times (6.90 \times 10^3) g \left(0.200 - \frac{\pi \times 0.100}{8}\right) + \left(\frac{\pi \times 0.050^2}{4}\right) 0.300 \times g (6.90 - 1.40) \times 10^3 \, \text{N}$$

$$= 40.57 \, g + 3.24 \, g = 43.81 \, g$$

but has an opposite direction. We will calculate the resultant.

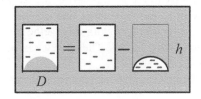

$$F_{\text{total}} = F_{\text{box}} - F_{\text{cylinder}}$$

The pressure on the mould varies with its height. For this reason it is difficult to calculate the resulting force directly. We choose to calculate it as the difference between two pressure forces, which are easy to find.

$$F = LD \, h \rho_L g - \frac{1}{2} \left(\frac{\pi D^2}{4}\right) L \rho_L g = LD\rho_L g \left(h - \frac{\pi D}{8}\right) \tag{1'}$$

This force, directed upwards, is equal to the weight of the mass M and we get:

$$M = \frac{F_{\text{total}}}{g} = 40.57 \, \text{kg} + 3.24 \, \text{kg} = 43.81 \, \text{kg}$$

Answer:

(a) The pressure force on the upper surface of the mould is $LD\rho_L g \left(h - \frac{\pi D}{8}\right)$.
(b) The lifting force is $\left(\frac{\pi d^2}{4}\right) L g (\rho_L - \rho_s)$.
(c) 44 kg.

1.2.3 Casting in Nonrecurrent Moulds

Sand Mould Casting
Sand moulding is the most common of all casting methods. It can be used to make castings with masses of the

magnitude 0.1 kg up to 10^5 kg or more. It can be used for single castings as well as for large-scale casting. In the latter case moulding machines are used. A good example is in the manufacture of engine blocks.

In sand moulding an impression is made of a pattern of the component to be cast. There are two alternative sand-moulding methods, namely *hand moulding* and *large-scale machine moulding*.

Hand moulding is the old proven method where the mould is built up by hand with the aid of wooden patterns as described earlier. This method has been transformed into a large-scale machine method where the mould halves are shaken and pressed together in machines. It needs to be possible to divide the mould into two or several parts. Large-scale moulds get a more homogeneous hardness, and thus also a better dimensional accuracy, than do hand-made moulds.

The advantages and disadvantages of sand mould casting are listed in Table 1.1.

TABLE 1.1   The sand mould casting method.

Advantages	Disadvantages
Most metals can be cast	Relatively poor dimensional accuracy
Relatively complicated components can be cast	Poor surface smoothness
Single components can be produced without too high initial costs	

The disadvantages of the sand moulding method have been minimized lately by use of high-pressure forming, i.e. the sand is compacted under the influence of a high pressure. The method can be regarded as a development of the machine moulding method of sand moulds. Normal mould machines work at pressures up to 4×10^6 Pa (4 kp/cm^2) while the high-pressure machines work at pressures up to $(10\text{–}20) \times 10^6$ Pa (10–20 kp/cm^2). The higher pressure offers a better mould stability, which results in a better measure of precision than that given by the low-pressure machines. Development in sand foundries proceeds more and more towards the use of high-pressure technologies.

Shell Mould Casting

The shell mould casting method implies that a dry mixture of fine-grained sand and a resin binder is spread out over a hot so-called *brim plate,* which covers half the mould. The resin binder melts and sticks to the sand grains, forming a shell of 6–10 mm thickness close to the pattern. The shell is hardened in an oven before it is removed from the plate with the pattern. The method is illustrated in Figure 1.5.

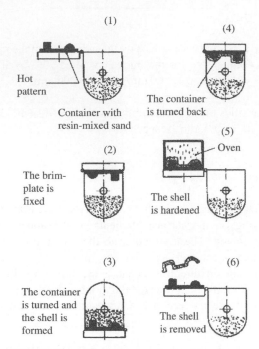

Figure 1.5   Shell mould casting stages 1 to 6.

Two shell halves are made. After hardening they are glued together. Before casting, the mould is placed into a container filled with sand, gravel or other material, which gives increased stability to the mould during the casting process.

A shell with a smooth surface and a good transmission ability for gases is obtained with this method, which can be used for most casting metals. The advantages and disadvantages are given in Table 1.2.

TABLE 1.2   The shell mould casting method.

Advantages	Disadvantages
High dimensional accuracy	High initial cost for the model equipment, which must be made of cast iron
Good reproduction of the shape of the component	
Good surface smoothness	Profitable series size for masses between 0.1–1 kg must be at least 50 000–100 000 components
Easy finishing of the surfaces of the component	
No burnt sand sticking to the surface	Small maximum mass of components, due to fragility of the mould
No sand inclusions	Maximum mass is 60–70 kg.
Components with thin walls can be cast	
Complicated core systems are possible	

Precision Casting or Shaw Process

In the *Shaw process*, a parted mould is made of fireproof material with silicic acid as a binding agent. The mould is heated in a furnace to about 1000 °C. The method gives roughly the same measure of precision as the shell mould casting method but is profitable to use for small series and single castings because the pattern of the mould can be made of wood or gypsum. The Shaw process is especially convenient for steel.

Investment Casting

Investment casting is also a precision method for component casting. In this method, a mould of refractory material is built on a wax copy of the component to be cast. An older name of the method is the 'lost wax melting casting' process. The method is illustrated in Figures 1.6 (a–f).

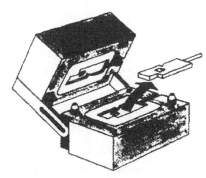

Figure 1.6 (a) Wax patterns are cast in a special tool made for the purpose. Reproduced with permission from TPC Components AB.

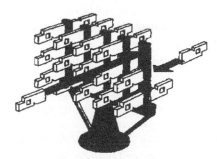

Figure 1.6 (b) A so-called cluster is built of the wax patterns. The trunk and the branches are inlets. Reproduced with permission from TPC Components AB.

In investment casting a wax pattern of the component has to be made. The wax pattern is then dipped in a mixture of a ceramic material and silicic acid, which serves as a binding agent. When the mould shell is thick enough it is dried and the wax is melted or burnt away. Then the mould is burnt and the casting can be performed.

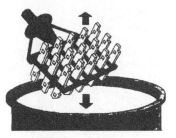

Figure 1.6 (c) The cluster is dipped into ceramic slurry, powdered with ceramic powder and dipped again. The procedure is repeated until the desired thickness has been achieved. Reproduced with permission from TPC Components AB.

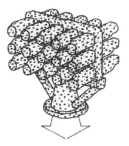

Figure 1.6 (d) The wax is melted away and the mould is burnt in an oven. The wax can only be used once. Reproduced with permission from TPC Components AB.

Figure 1.6 (e) Casting is performed directly into the hot mould. Reproduced with permission from TPC Components AB.

Investment casting can be used for all casting metals. The mass of the casting is generally 1–300 g with maximum masses up to 100 kg or more. The advantages and disadvantages are listed in Table 1.3.

Investment casting offers very good dimensional accuracy. With the proper heat treatment after casting the component acquires the same strength values for stretch and fracture limits as do forged or rolled materials.

The investment casting method and the Shaw method are complementary to each other in a way. The Shaw method is used when the casting is too big for investment casting or

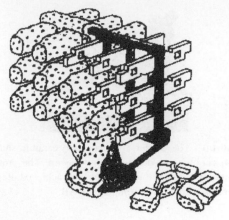

Figure 1.6 (f) The ceramic mould is knocked out after casting and solidification and the complete component is revealed. Reproduced with permission from TPC Components AB.

TABLE 1.3 The investment casting method.

Advantages	Disadvantages
Good accuracy	High mould cost
Good mechanical properties	Size limitations
Good surface finish	
Thin components can be cast	
No shape limitations	
Can be used for all casting metals	

when the series is too small to be profitable with the investment casting method.

1.2.4 Casting in Permanent Moulds

Gravity Die Casting
In gravity die casting permanent moulds are used. Such a mould is made of cast iron or some special steel alloy with a good resistance to high temperatures (the opposite property is called *thermal fatigue*).

The gravity die casting method is often used for casting zinc and aluminium alloys. It is difficult to cast metals with high melting points due to the wear and tear on the mould, which is caused by thermal fatigue.

Cores of steel or sand can be used. It is also possible to introduce details of materials other than the cast metal, for example, bearing bushings and magnets. The advantages and disadvantages of the method are listed in Table 1.4.

Due to the high mould cost, series of less than 1000 components are not profitable. In these cases, another casting method must be chosen. There is also an upper limit, which is set by the thermal fatigue of the mould. In alumi-

TABLE 1.4 The gravity die casting method.

Advantages	Disadvantages
Good mechanical properties	High mould cost
High dimensional accuracy	Materials with low melting point only
High surface smoothness	

nium casting the maximum number of components is around 40 000.

High-Pressure Die Casting
The molten metal is forced into the mould at high pressure as indicated by the name of the process. The method is described in Figure 1.7.

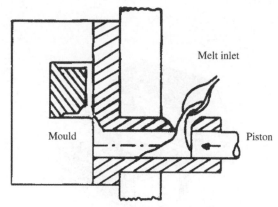

Figure 1.7 High-pressure die casting machine. During casting, the molten metal is transferred into the shot cylinder. The piston is then pushed inwards and forces the melt into the mould.

The permanent mould is made of steel and the mould halves are kept together by a strong hydraulic press. The method can only be used for metals with low melting points, for example zinc, aluminium and magnesium alloys.

The mechanical properties of the components are good with this method, better than with the gravity die casting method. However, weak zones may occur in the material due to turbulence in the melt during the mould filling.

Due to high machine and mould costs, the high-pressure die casting method will be profitable only if the number of cast components exceeds 5000 to 10 000. The method is useful for production of large series of components, for example in the car industry.

The 'life time' of the high-pressure die casting machine varies from about 8000 castings for brass to 800 000 castings for zinc alloys.

The advantages and disadvantages of the method are listed in Table 1.5.

TABLE 1.5 **The high-pressure die casting method.**

Advantages	Disadvantages
The process is rapid	Very high workshop costs due to high pressure and high thermal fatigue
Very thin and complicated components can be cast	
Very high precision compared with conventional casting	The rapid filling process is very turbulent and the melt absorbs large amounts of gas
Little work remains after casting	Components containing cores are normally impossible to cast
Inset parts, for example bearings and bolts, can be introduced from the beginning	Only metals with low melting points can be cast

Low-Pressure Die Casting

The principle of this method is illustrated in Figure 1.8. Contrary to the high-pressure die casting machine, the low-pressure casting machine contains no pushing device and no piston. Nor is it necessary to apply the high pressure, required in high-pressure die casting, at the end of the casting process.

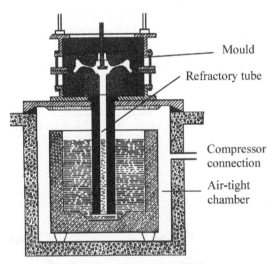

Figure 1.8 Low-pressure die casting machine. The melt is included in an air-tight chamber connected to a compressor. When the pressure is increased in the chamber, melt is pressed upwards through the refractory tube into the mould. Reproduced with permission from Addison-Wesley Publishing Co. Inc., Pearson.

Mould
Refractory tube
Compressor connection
Air-tight chamber

Air, or another gas, is introduced into the space above the melt. The gas exerts pressure on the melt and causes it to rise comparatively slowly in the central channel and move into the mould. The mould is kept heated to prevent solidification too early in the process. This is a great advantage when small components, with tiny protruding parts, are to be cast. In this way it is possible to prevent them from solidifying earlier than other parts of the mould. This is one of the most important advantages of this casting method.

The walls of the component to be cast can be made rather thin. The low melt flow gives little turbulence in the melt during the mould filling and very little entrapment of air and oxides. When the casting has solidified the pressure is lowered and the remaining melt in the central channel sinks back into the oven.

A list of the advantages and disadvantages of the method is given in Table 1.6.

Squeeze Casting

Squeeze casting is a casting method that is a combination of casting and forging. It is described in Figures 1.9 (a–c).

When the mould has been filled the melt is exposed to a high pressure and starts to solidify. The pressure is present during the whole solidification process so that pore formation, which causes plastic deformation, is prevented and the mechanical properties of the castings are strongly improved as compared to conventional casting.

TABLE 1.6 **The low-pressure die casting method.**

Advantages	Disadvantages
High metal yields are obtained	Lower productivity than in high-pressure die casting.
Little work remains after casting. Use of cores is possible	More expensive moulds than in conventional sand casting
Easy to automate	
Dense structure of the component, compared to chill-mould casting and high-pressure die casting	Only metals with low melting points can be cast
Lower workshop costs than in high-pressure die casting	
Better mechanical properties than in conventional sand casting	

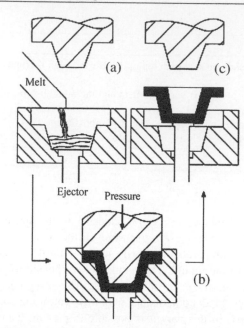

Centrifugal Casting

In centrifugal casting centrifugal force is used in addition to gravitational force. The former is used partly to transport the melt to the mould cavity and exert a condensing pressure on it and partly, in certain cases, to increase the pressure, thus allowing thinner details to be cast and making surface details of the metal-cast components more prominent.

There are three types of centrifugal casting, distinguished by the appearance of the mould and its construction and the purpose of the casting method:

- true centrifugal casting;
- semicentrifugal casting or centrifugal mould casting;
- centrifugal die casting.

The principal differences between the three methods are described in Table 1.7.

True Centrifugal Casting

This method is characterized by a simple mould with no cores. The inner shape of the casting is thus formed entirely by the mould and centrifugal force. Typical products produced in this way are tubes and ring-shaped components.

Figure 1.9 Squeeze casting. (a) The melt is poured into the lower mould section; (b) the melt is exposed to a high pressure from the upper mould section; (c) after solidification the upper mould section is removed and the casting is ejected by the aid of the ejector. Reproduced with permission from The Metals Society.

TABLE 1.7 **Schematic description of various centrifugal casting methods.**

Characteristics	True centrifugal casting	Semicentrifugal casting	Centrifugal die casting
Horizontal axis of rotation			
Vertical axis of rotation			
Inclined axis of rotation			

The dominating product, with respect to mass, is cast iron tubes.

The permanent mould, i.e. the mould, is normally cylindrical and rotates around its central axis, which can be horizontal, vertical or inclined. Figure 1.10 shows a sketch of the most common tube casting machine.

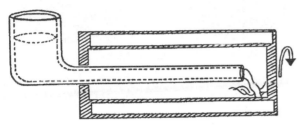

Figure 1.10 Sketch of a tube casting machine which works according to the principle of true centrifugal casting.

Semicentrifugal Casting

During semicentrifugal casting (Figure 1.11) the mould is rotated around its symmetry axis. The mould is complicated in most cases and may contain cores. The detailed shape of the casting is given by the shape of the rotating mould. The centrifugal force is utilized for slag separation, refilling of melt and increase of the filling power in order to cast components with thin sections. Cogwheels are an example of components that can be cast using this method.

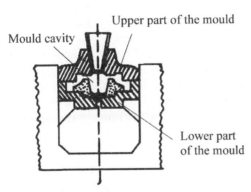

Upper part of the mould

Mould cavity

Lower part of the mould

Figure 1.11 Machine designed for semicentrifugal casting of double cogwheels. Reproduced with permission from Karlebo.

Centrifugal Die Casting

The principle of centrifugal die casting is illustrated in Figure 1.12. The mould cavities are symmetrically grouped in a ring around a central inlet. From this the metal melt is forced outwards under pressure into the mould cavity and efficiently fills all its contours. The centrifugal force supplies

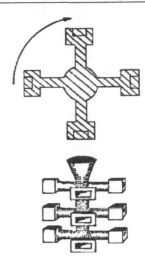

Figure 1.12 Mould for centrifugal die casting. Reproduced with permission from Karlebo.

the necessary pressure to transport the melt to all parts of the mould.

The method is extensively used for casting in moulds prepared by the investment casting method. It is often used in the dental industry to cast gold crowns for teeth.

1.2.5 Thixomoulding

Thixomoulding should logically have been treated under the heading 'Casting in Permanent Moulds' in Section 1.2.4. The reason why it has been extracted from its proper position is that it differs radically from all other mould casting methods and deserves special attention.

Thixomoulding is a very promising method for casting components of various sizes. Fleming originally developed it in 1976 at MIT in the US and introduced it under the name *semi-solid metal processing* (SSM). During the late 1990s the method was primarily used for the casting of magnesium on an industrial scale.

Thixomoulding can be used for casting many alloys. It is a very promising new method, which will probably develop rapidly and is expected to obtain a wide and successful application in industry. Zinc and aluminium alloys followed magnesium as suitable for thixomoulding on an industrial scale. Thixomoulding of Zn and Al alloys has been commercialized and other types of alloys will follow.

Principle of Thixomoulding

In ordinary mould casting it is necessary to melt the alloy and superheat the liquid to above its melting point. Sufficient heat must be provided to retard the crystallization process, i.e. reduce the formation of so-called dendrites, and maintain sufficient fluidity of the melt until the mould has

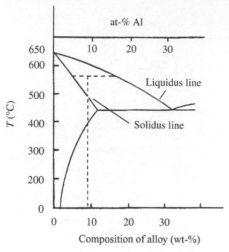

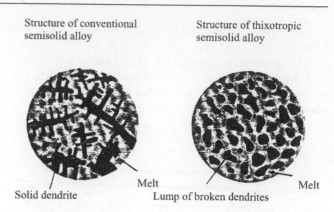

Figure 1.14 The structures of a partly molten MgAl alloy before and after mechanical treatment. Courtesy of Thixomat Inc., Ann Arbor, MI, USA.

Figure 1.13 Phase diagram of the Mg–Al system. An alloy with ~90–89 % Mg and ~9–10 % Al at a temperature of 560–580 °C represents a suitable mixture for thixomoulding die casting. Reproduced with permission from the American Society for Metals.

been completely filled. Dendrites and dendrite growth will be discussed in Chapter 6.

In contrast to the superheating required for conventional mould casting, thixomoulding is carried out at a temperature between the liquidus and solidus temperatures of the alloy (Figure 1.13). At this temperature the alloy consists of a viscous mixture of a solid phase with growing dendrites and a liquid phase. This can be seen from the phase diagram, which shows the composition of the alloy as a function of temperature (Figure 1.13).

The solidifying metal is exposed to shearing forces, which break the dendrites into pieces and a fairly homogeneous mixture is formed. The solidifying metal consists of spherical solid particles in a matrix of melt. The appearances of the structures of the partly molten alloy

before and after the mechanical treatment are seen in Figure 1.14.

The homogenous viscous solidifying metal is the material used for casting.

Thixomoulding Equipment

Figure 1.15 shows a machine designed for thixomoulding of Mg alloys.

Room-temperature pellets of the alloy are fed into the rear end of the machine. An atmosphere of argon is used to prevent oxidation at high temperature. The pellets are forwarded into the barrel section where they are heated to the optimal temperature below the melting point of the alloy. The solidifying metal is carried forward and simultaneously exposed to strong shear forces when a powerful screw is rotated around its axis.

The solidifying metal is forced through a nonreturn valve into the accumulation zone. When the required amount of the solidifying metal is in front of the nonreturn valve, the screw forces it into the preheated metal mould and a product of desired shape is formed. The injection of the

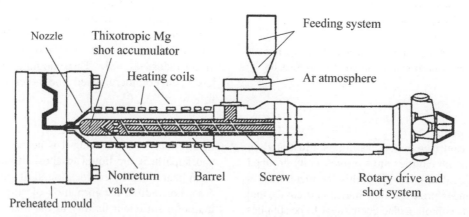

Figure 1.15 Thixomoulding machine designed for moulding Mg alloys. Courtesy of Thixomat Inc., Ann Arbor, MI, USA.

TABLE 1.8 **Method of thixomoulding compared with conventional mould casting.**

Advantages	Disadvantages
No transfer operations required	High equipment cost
Good dimensional stability, i.e. good dimensional accuracy of products	Two-step process
Low temperature, which reduces the melt costs, gives low corrosion and low porosity of products	Risk of formation of oxides and other inclusions
Good mechanical properties of products in most cases	In some cases coarser structure and less favourable mechanical properties
The Ar atmosphere results in little oxidation, which contributes to low corrosion	
No secondary machining and heat treatment of the cast components are required	
Environment-friendly process	

solidifying metal into the mould occurs under pressure. The process is reminiscent of the squeeze casting process, described on pages 9–10.

The advantages and disadvantages of thixomoulding are listed in Table 1.8.

2 Cast House Processes

2.1 INTRODUCTION

A short survey of component casting methods is given in Chapter 1. This second chapter is also devoted to casting methods, but here we will concentrate on casting processes for the production of plates, strips, wires and other basic products. The processes, which are used for casting of the most common metals, i.e. steel, iron-based alloys, copper and aluminium, will be particularly discussed.

The production of steel and iron is incomparably more extensive than that of other metals. In addition, the pro-blems encountered in the process of casting iron and steel are markedly more severe than those involving other common metals such as copper and aluminium. The reason for this is the higher melting point and the poorer thermal conductivity of steel compared with other metals. It is therefore reasonable to devote more space to the processes in hammer mills and steelworks in this section and this book than to the casting processes of other metals. However, *the majority of what is written is valid not only for iron and steel alloys but also for most other metals.*

2.2 INGOT CASTING

In ingot casting, the metal melt is teemed or poured from the ladle into one or several moulds where it solidifies. In steel casting, the mould is usually made of cast iron. In this section, ingot casting of steel will be chosen as an example of this process. It is also valid for other metals.

The shape of the chill-mould can vary from a rectangular to a circular shape, with or without corrugated sides. Two very common types of mould are seen below in Figures 2.1 (a) and (b).

The teeming can be performed in two different ways. Either the melt is teemed from above into the mould, *downhill casting* [Figure 2.2 (a)], or the melt is supplied from below, *uphill casting* [Figure 2.2 (b)].

On top of the cast ingot one normally introduces an insulating layer, called a *hot top,* in order to reduce the size of the pipe, i.e. the pore or cavity, which is formed due to the solidification shrinkage. This topic will be discussed in Chapter 10.

2.2.1 Downhill Casting

One problem with downhill casting is producing a good ingot surface. The quality of the surface depends on how well the mould has been prepared before the casting. In order to improve the surface quality it is common to coat the ingot mould with mould black, which mainly consists of tar.

A common surface defect when using downhill casting arises when the melt is teemed into the mould and drops of molten metal splash from the bath and stick to the

Materials Processing during Casting H. Fredriksson and U. Åkerlind
Copyright © 2006 John Wiley & Sons, Ltd.

Figure 2.1 (a)   Chill-mould with its wider end upwards.

Figure 2.1 (b)   Chill-mould with its wider end downwards.

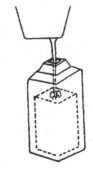

Figure 2.2 (a)   Downhill casting.

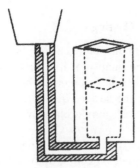

Figure 2.2 (b)   Uphill casting.

mould wall in the form of small droplets. These solidify and oxidize there before the melt has reached the corresponding level. It is important that the jet stream is well centred, as this minimizes these types of fault as well as ripples.

The stream of melt has a large area exposed to the air. This exposure greatly increases the risk of oxygen and nitrogen absorption by the melt. These gases react chemically with alloying metals in the melt, for example Al, Ti and Ce, which may occur in steel, and lead to the formation of oxides. The use of mould black reduces the chemical reactions between the metal melt and the air. The black is partly evaporated by the heat from the metal melt and displaces and insulates the mould from the air or reacts with oxygen to form an inert atmosphere.

2.2.2   Uphill Casting

In uphill casting, the melt is cast into a gating system and enters the mould via a stool in the bottom (Figure 2.3).

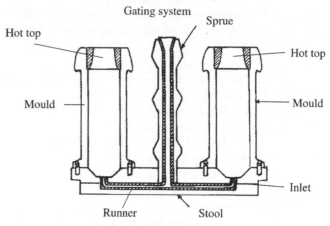

Figure 2.3   Equipment for uphill casting.

Normally there are four or six moulds on a stool. The purpose of uphill casting is to eliminate scab and ripple formation and thus obtain a better surface quality than with downhill casting.

Uphill casting has long been used for surface-sensitive steel qualities. The result was initially rather poor and it was doubtful whether the improved surface quality compensated for the higher cast house costs of uphill casting.

Casting powder was introduced at the beginning of the 1960s. This improved uphill technology considerably. The common disadvantage with uphill casting without casting powder is that the upper surface becomes covered with an oxidized and viscous film. Such a film easily sticks to the steel meniscus or to the mould wall. It may remain there and be passed by the rising melt surface. When this occurs, the solidifying ingot shell gets a weak zone where cracks may easily arise.

When the casting powder on top of the melt in the mould melts, a layer of molten slag is formed. This layer provides three major advantages:

- It reduces the heat loss at the upper surface and prevents the steel meniscus from solidifying.
- It protects the steel against oxidation and serves as a solvent for the oxides precipitated or present in the steel melt.
- It forms a thin, insulating layer between mould and ingot. This results in a reduced number of surface cracks because the shrinkage strain decreases for two reasons. Firstly the cooling rate is decreased and secondly the motion of the ingot surface relative to the mould is reduced, which is important if surface cracks are to be avoided.

Casting powder which successively melts during the casting process is normally used. Initially fly ashes from heat power plants were used, but today the powder is produced by mechanical mixing of different powdered components, such as graphite, calcium fluoride, aluminium oxide and sodium silicate. Figure 2.4 illustrates what actually happens in uphill casting.

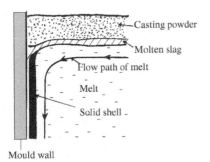

Figure 2.4 A cross-section of the upper melt surface in uphill casting in a mould.

Figure 2.4 shows both the flow pattern and the presence of a thin layer of molten casting powder next to the steel surface. The solidification of the steel melt starts at the vertical steel surface, close to the mould wall, where the heat extraction is at its maximum. Due to the ferrostatic pressure, i.e. the pressure that the melt exerts on the solidifying layer, the latter is straightened and adapts itself to the mould wall. The wall is covered with a layer of casting powder of about 0.5 mm thickness.

The qualitative demands on the casting powder are that its melting point shall be a couple of hundred degrees lower than the temperature of the melt and that it is not changed during the casting process. In addition the casting powder shall have a low viscosity and contain a small percentage of graphite. The graphite forms a layer between the molten slag layer and the casting powder, and controls the melting rate of the casting powder.

2.2.3 Comparison between Downhill and Uphill Casting

Both downhill and uphill casting are used in industry. The choice of alternative depends on the balance between quality advantages and the higher cost of uphill casting as compared to downhill casting. In Table 2.1 the advantages and disadvantages of the two methods are listed.

2.2.4 Ingot Casting of Nonferrous Metals

So far, the treatment of ingot casting has mainly concerned steel casting. Ingot casting of metals other than steel also occurs. These types of ingot casting are not as complicated as steel casting. If the metal, for example Al, Cu, Mg, Pb, Sn or Zn, is to be remelted later, it is first cast in a chest-shaped mould, made of copper, iron or sand [Figure 2.5 (a)]. Big copper ingots for the rolling of copper sheets and copper ingots for wire drawing can also be cast in a similar way. For such copper castings the type of mould shown in Figure 2.5 (b) is used.

Such a mould *cannot* be used if the ingot is to be machined immediately because the upper surface will become very ugly due to slag and pipe, which have to be removed by grinding. In such cases it is preferable to use moulds with long vertical axes in order to make the upper surface, which is exposed to the air, as small as possible.

TABLE 2.1 Advantages and disadvantages of uphill and downhill casting of steel.

Uphill casting with casting powder	Downhill casting
− Additional consumption of material to channel stones, gating systems and stools.	+ Cheaper casting.
− Additional stages in the work process such as stool masonry, raising and cleaning of moulds.	
+ Better surface quality. Internal cracks become clogged with molten casting powder. Oxide particles flow upwards during the casting process and become trapped by the molten casting powder.s	− Scabs and ripples cause surface weaknesses.
+ Better quality of the steel, i.e. fewer micro- and macro-slag inclusions (Chapters 7 and 11).	− Poorer quality of the steel.
+ Casting is more rapid. Several ingots can be cast at the same time.	− Slower casting.
+ The problem with the centring of the melt stream is eliminated.	− More wear and tear (cracks) on the mould.

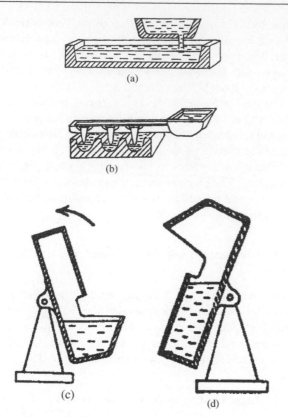

(a)

(b)

(c)   (d)

Figure 2.5 Casting in chest moulds (a) and (b). The Durville process. The melt is teemed into the cup in position (c). The cup is then tilted slowly to position (d). Reproduced with permission from M.I.T.

Metals and alloys that are easily oxidized and/or show large segregation effects upon solidification (uneven distribution of alloying elements, Chapters 7 and 11) often cause problems during the casting process. A special type of chill-mould has been developed for these metals and alloys in order to reduce the turbulence during teeming and the contact with the air as much as possible. This can be done by slowly tilting a cup like the one in Figure 2.5 (c) to about 180° around a horizontal axis into position 2.5 (d). The melt then slowly flows into the other end of the mould, which has a long vertical axis. Another advantage with the slow tilting of the mould is that the slag inclusions remain in the cup when the cup is slowly tilted.

Attempts are made to keep the casting temperature as low as possible to make the melt solidify in the shortest possible time. To promote the solidification process the mould with a long vertical axis may be designed as a chill-mould, like the one in Figure 2.6, and be cooled with running water.

2.3 CONTINUOUS CASTING

The method of ingot casting has its limitations. It is impossible to increase the production capacity beyond a certain

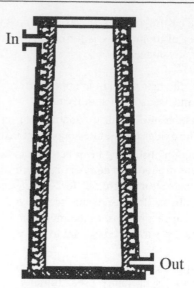

In

Out

Figure 2.6 Water-cooled chill-mould. Reproduced with permission from M.I.T.

limit. If the amount of melt in an ingot is increased, the cooling rate will decrease. This leads to poorer material properties and difficulties in handling the final, bigger ingots in most cases.

For this reason, scientists and engineers have devoted a great deal of time and effort to developing successful methods of *continuous casting*, and have succeeded in spite of considerable initial difficulties. This casting method consists of continuously drawing out a strand of solid metal from the chill-mould while it is continuously fed with new melt from above. There is a distinction between completely continuous and semicontinuous casting. In the latter, the length of the strand is limited.

2.3.1 Development of Continuous Casting

As early as 1856 Henry Bessemer suggested a continuous casting method, which is illustrated in Figure 2.7 on page 19. Bessemer's idea was to cast molten metal between two water-chilled rolls. The process turned out to be very difficult to control and he only succeeded in producing some small plates of very poor quality. The process did, however, prove to be important for future development. At that time the metallurgical knowledge had not yet reached the stage where the practical problems could be overcome.

At the end of the 1800s, Bessemer's idea was again taken into consideration and several new processes, applicable to nonferrous metals, were developed. After 20 to 30 years of experimental work these methods had advanced so much that they could be used on an industrial scale. During

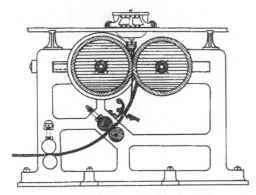

Figure 2.7 Continuous casting machine according to Bessemer, 1856. Reproduced with permission from Pergamon Press, Elsevier Science.

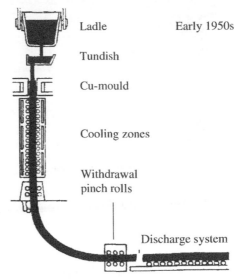

Ladle · Early 1950s

Tundish

Cu-mould

Cooling zones

Withdrawal pinch rolls

Discharge system

Figure 2.8 Old model of a continuous casting machine. Reproduced with permission from Pergamon Press, Elsevier Science.

the 1930s and the 1940s continuous casting of nonferrous metals became a common production method.

Inspired by this success, new experiments were made with continuous steel casting. In 1943 the German scientist Junghans demonstrated a newly developed process of continuous steel casting on a pilot scale using a vertical machine. An extensive development started and several pilot plants were set up within a decade of Junghans's demonstration.

Today continuous steel casting is a well-established method. The development has led to the fact that continuous casting has become the dominating casting process for steel. It corresponds to more than 90 % of world production per year. Today a great number of steel qualities are cast in a very wide variety of dimensions.

2.3.2 Continuous Casting

The principle of the continuous casting method is simple and is illustrated in Figure 2.8, which shows one of the earliest casting machines.

Steel flows down from a ladle into a so-called *tundish*, which is an intermediate container for the steel melt. Metal melt is continuously teemed from the tundish into a vertical *water-cooled copper chill-mould* while a solid shell of metal, which contains metal melt in the centre, is simultaneously extracted from the bottom of the chill-mould. The chill-mould is made of copper because of its excellent thermal conductivity. During the passage through the chill-mould, a solid metal shell must be formed that is stable enough to keep its shape unchanged while the metal is continuously extracted vertically from the chill-mould.

When the metal shell leaves the chill-mould it enters a series of *cooling zones*. The cooling medium is water, which is sprayed directly on the entire periphery of the string casting. Close to and below the chill-mould there is

a short zone with very heavy water spraying. Additional cooling zones follow the first one. The amount of water per unit time in the cooling zones often decreases with the distance from the chill-mould.

After passing the cooling zones, the strand reaches the driving and guide rollers, called *pinch rolls*, which control the exit rate. The strand is then cut into desired lengths in a *discharge system*. After cutting, the cast pieces are transported away for subsequent treatment.

One of the most difficult problems with a vertical continuous casting machine is that the casting rate and the distance between the chill-mould and the discharge system are limited. When the strand is to be cut, no liquid steel can be allowed to remain in its interior. The length of the column of liquid steel inside the strand increases with increasing casting rate. The necessity of making the distance between chill-mould and discharge system short depends on increased difficulties of constructing vertical machines with high centres of gravity and high costs for tall buildings. Besides, it is difficult to transport large amounts of molten steel in a safe way. In order to reduce the risks of severe accidents, big holes were dug in the ground where the casting machines were located and the casting could be performed at ground level. However, this alternative turned out to be even more expensive than using tall buildings.

The casting rate in a vertical casting machine is low, which results in a low production rate, making it difficult to justify high investment costs. To avoid these complications, *bent* casting machines were built. Figures 2.8, 2.9 and 2.10 illustrate three different stages of development. In addition to the advantage of a reduced machine height,

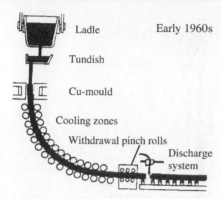

Figure 2.9 Modern bent casting machine with its main curvature below the chill-mould. Reproduced with permission from Pergamon Press, Elsevier Science.

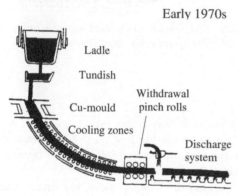

Figure 2.10 Modern bent casting machine of low height. Reproduced with permission from Pergamon Press, Elsevier Science.

it is preferable for the castings to be extracted horizontally from the machine.

To start the continuous casting, a chain arrangement, a so-called *starting bar*, is used. It is moved between the pinch rolls up to the chill-mould. At one end of the starting bar a *starting head* is located, which serves as the base of the chill-mould and initially extracts the casting from the chill-mould. It is easier to bring the starting bar into its proper position and prepare it for use in a bent machine than in a straight one.

2.3.3 Comparison between Ingot Casting and Continuous Casting

The advantages of continuous casting compared to ingot casting are:

- the metal yield is often high because only part of the strand has to be cut off and remelted due to pipe formation and other factors;
- small casting dimensions can be cast directly with no need for a process like ingot rolling;

- a larger part of the casting work can be mechanized;
- the castings have a more even composition than with ingot casting.

2.3.4 Semicontinuous Casting of Al and Cu

In order to cast copper and aluminium alloys a semicontinuous casting process, with casting lengths of 10–15 m, is used in most cases. The principle of this casting method is illustrated in Figure 2.11.

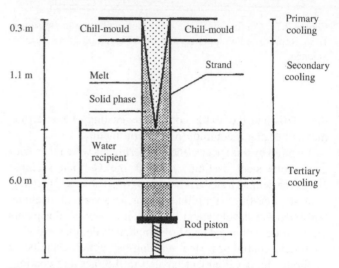

Figure 2.11 Semi-continuous casting machine for Al and Cu alloys. Reproduced with permission from P. Sivertsson.

The casting machine in Figure 2.11 has a 0.3-m water-cooled chill-mould. It is completely straight. When the casting has passed the chill-mould the secondary cooling begins, i.e. water is sprayed directly onto the strand. The length of the cooling zone is somewhat more than 1 m. After secondary cooling the strand is carried downwards into the water basin. A piston bar in the water carries the whole strand. The speed of the piston bar controls the casting rate. It normally amounts to a couple of decimetres per minute.

2.4 FROM LADLE TO CHILL-MOULD IN CONTINUOUS CASTING OF STEEL

Following the review of the ingot and continuous casting methods, we will discuss the transfer of melt from furnace to mould in the casting machines and the optimizing of this part of the casting process, which precedes the casting itself. It is especially important to control this subprocess in continuous casting. For this reason we concentrate on the conditions relevant to continuous casting.

The principle of continuous casting was described in Section 2.3.2 on page 19–20. Figures 2.8 to 2.10 on pages 19–20 show how the melt flows from a *ladle* in a ladle stand down into a *tundish*. From the tundish the melt flows down into the *chill-mould.*

2.4.1 Construction of the Ladle

When the melt is ready after melting and teeming, it is transported to the chill-mould or the casting machine in a *ladle*. The ladle is equivalent to a big bucket, lined with a ceramic material in order to withstand high temperatures. The ladle is usually emptied or teemed through a hole in the bottom. Figure 2.12 shows the opening and closing mechanisms of such a ladle.

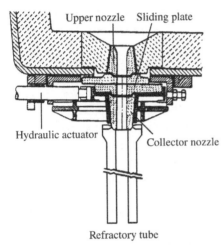

Upper nozzle   Sliding plate

Hydraulic actuator   Collector nozzle

Refractory tube

Figure 2.12 Bottom emptying ladle. Detail of ladle sliding gate valve. The ladle sliding gate valve including the collector nozzle, the sliding plate, the hydraulic actuator and the refractory tube can be moved sideways. In the figure it is in its teeming position. Reproduced with permission from the Institute of Materials.

During teeming the ladle is placed in a so-called ladle stand. The teeming of the ladle is controlled by a slide disc system, which consists of two discs. The relative position of the two discs is adjusted by a hydraulic or pneumatic system. Because the lower disc is placed on the outside of the ladle it can easily be replaced when necessary.

2.4.2 Design and Purpose of the Tundish

Figure 2.13 shows how the tundish is designed. It is an important link between the ladle and the chill-mould in continuous casting.

The tasks of the tundish are to:

- control the flow (amount of melt per time unit) into the chill-mould;
- force the casting beam into a proper position in the chill-mould;

- distribute the melt between different machines in a multi-strand machine;
- function as a slag separator.

In order to obtain satisfactory slag separation, the melt must be kept in the tundish for as long as possible. The bigger the tundish, the better will the slag separation be. However, it is not realistic to build too big a tundish. The size of the tundish is chosen in such a way that the melt is kept in the tundish for about 4 minutes, which is a convenient time for a good result.

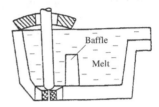

Baffle

Melt

Figure 2.13 Tundish with a baffle. Reproduced with permission from Pergamon Press, Elsevier Science.

Experiments show that big tundishes give very good slag separation. The use of thresholds and baffles gives no improvement in this respect. In small tundishes the slag separation is greatly improved if a baffle is used. The result is especially good if the baffle is located in such a way that the outlet space gets a comparatively large volume.

The level of the melt in the tundish is also of great importance. The time of residence increases with increasing bath depth, although it will not increase the amount of rest melt in the tundish. The amount of melt left in the tundish at the end of the casting process will be the same. With a small bath depth there is a considerable risk that the slag, which is already separated, will run back into the bath. This is especially so when a *vortex* (compare the outflow of water in a hand-wash basin) is formed above the outlet. Such a vortex appears in the flowing melt when the bath depth is less than four slide disc diameters. It is desirable for tundishes, especially small ones, to have a shape such that the melt cannot move directly from inlet to outlet but is forced to divert and use the whole bath. In the latter case, the slag separation is better than in the former for two reasons: (i) because the outlet part of the bath becomes less turbulent, and (ii) because the effective contact area between the walls and the bath increases. Slag separation occurs at these contact areas.

The tundish is equipped with a slide disc in the outlet hole. When the tundish is to be used in combination with big casting machines a pipe is attached to the slide disc. It leads the melt from the tundish to the chill-mould. The refractory materials in the slide disc and the pipe are exposed to considerable erosion from the liquid steel during the casting. Under some circumstances the opposite may occur. Oxide inclusions may precipitate onto the slide disc and the casting tube to such an extent that they become blocked, which makes casting impossible.

The factors that influence the erosion and the blocking, respectively, are:

- the refractory material;
- the type of steel;
- the type of slag inclusions.

Various refractory materials show different tendencies to bind inclusions. In continuous casting a constant steel flux is desired, which can be achieved by control of the melt level in the tundish. This control is facilitated by constant diameters of the slide disc and the casting tube or pipe during the casting operation. Erosion as well as blocking are thus avoided.

2.4.3 Heat Loss

Temperature control during continuous casting is very important for production free from disturbances, and high quality of the castings. A successful casting result depends to a great extent on the casting temperature of the steel. During the casting process the temperature must be kept very even and as low as possible. It must be no more than 30 °C above the liquidus temperature of the steel.

Too *high* a temperature results in castings of poor quality. The consumption of refractory material and the number of breakdowns also increase. Too *low* a temperature causes blocking in the slide discs because the steel solidifies on them. In the worst case the casting process has to be interrupted.

The temperature of the ladle is adjusted to be a little higher than the calculated value and is brought down to the optimal level by gas flushing. Problems caused by too low a temperature often appear, especially at the beginning and at the end of the casting process. It is preferable that the temperature is somewhat too high rather than too low during the whole casting period.

Heat loss of the steel melt occurs by thermal conduction to the lining of the ladle and the tundish and by convection and radiation from the upper surface and the mantle of the steel bath. Figure 2.14 shows the heat loss to the surroundings from the ladle, the tundish and the casting beam.

The factors that influence the temperature of the steel in the chill-mould are:

- temperature of the steel after furnace and gas flushing;
- temperature gradient of the lining after gas flushing;
- heat loss into wall lining by thermal conduction;
- heat loss from the melt surface of the ladle;
- temperature of the melt at teeming into the tundish;

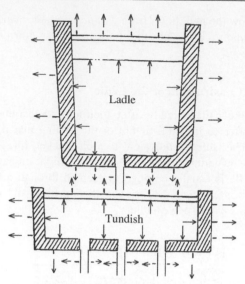

Figure 2.14 Heat loss in ladle and tundish.

- temperature gradient in the ladle;
- casting rate, i.e. how rapidly the melt flows through the tundish;
- heat loss by thermal conduction into the lining of the tundish and heat loss by convection and radiation from the lining;
- heat loss from the melt surface in the tundish.

In the casting beam, heat loss occurs mainly through radiation. The heat loss through the lining is influenced by the wear and tear of the lining and the degree of preheating. The convection and radiation heat loss of the tundish depend on the use of casting powder and/or lid.

The influence of the casting powder on the steel temperature is strong. By adding plenty of casting powder it is possible to prevent heat loss from the surface almost completely. If covering powder is used in the tundish the temperature becomes about 10–15 °C higher than without powder. The cover has a maximal influence at the beginning of the casting process.

A lid on top of the tundish causes the heat flow from the bath surface to be distributed between the walls and the lid. Part of the heat flow is absorbed by the lining and is conducted away through it. The rest is reflected back into the melt. The material of the lid influences the magnitude of the heat loss extensively. The lowest possible heat loss is obtained if the lid is made of an insulating ceramic material.

Example 2.1

At a steel plant for slab casting, the temperature of the steel in the chill-mould has been measured as a function of time.

The result is the function given in the figure. T_L is the liquidus temperature of the steel.

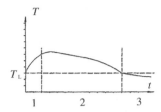

The time axis has been divided into three regions. Explain the appearance of the temperature curve for each of these regions and also discuss the casting parameters that influence their extensions in time.

Solution and Answer:

Region 1
The reason for the temperature increase is the following: in the tundish there is either a so-called warm or cold lining. As is evident from the name, preheating has to be performed in the first case. In both cases the lining is colder than the steel melt. At the beginning of the casting process heat is removed from the melt, which initially reaches a lower temperature than later in the casting period.

The volume and the temperature of the preheated tundish determine the time interval for region 1.

Region 2
The reason for the even temperature is that the casting occurs under relatively stable conditions. The slowly falling temperature depends on heat loss to the surroundings.

The size of the ladle, the length of the casting tube or pipe between the ladle and the tundish, and the volume of the tundish determine the time interval.

The temperature continues to fall slowly, due to heat loss to the surroundings until the liquidus temperature is reached.

Region 3
When the temperature has fallen to the liquidus temperature the curve suddenly changes. The temperature falls considerably more slowly in region 3 than in region 2. This is explained by the following: the measured temperature lies slightly below the liquidus temperature, which means that free crystals are formed within the melt. Solidification heat is released, which reduces the further fall in temperature.

The time interval is determined by the quality of the temperature control in ladle and tundish and by the agitation in the chill-mould. In some cases artificial agitation is performed in the chill-mould, which gives a finer crystal structure than normal.

2.4.4 Casting Powder

The use of casting powder has been discussed earlier on pages 16–17 in this chapter in connection with uphill casting. Casting powder is also used in continuous casting. Its function is to *insulate the surface of the melt* in the chill-mould; *absorb the nonmetal particles*, for example oxides, which rise to the surface; and to *function as lubrication* between the chill-mould and the casting. Casting powder that is to be used in continuous casting must have certain special properties.

It is very important that the strand does not stick to the chill-mould during the casting process. The slag between the chill-mould and the steel casting gives a lubricating effect. It is thus imperative that the powder, which forms the slag, surrounds the solidifying strand completely. The slag must fill the space between the chill-mould and the casting at all times. The slag must wet the steel substrate well in order to give a layer of even thickness.

The casting powder must melt homogeneously. For this reason granulated powder is used. The viscosity of the slag must be adapted to the casting process in order to give a sufficient post-feeding of slag between chill-mould and casting. The powder must not be too free-flowing because this would result in an unnecessarily high rate of consumption.

Furthermore, the powder must have a good insulating capacity in order to keep the slag fluid. The whole layer of the powder must not melt because this would remove the insulating effect and several layers are needed during the melting. Besides, the powder must melt at a temperature that is sufficiently low, and have a wide melting range. In its solid form the powder must have a good fluidity (see Section 3.4 in next chapter) and must not react chemically with the steel.

Casting powder may consist of fly ash from heating plants or a mechanical mixture of various fine-grained components, such as, for instance, graphite, SiO_2, CaF_2, Al_2O_3 and Na_2SiO_3. These components are often parts of fly ash. The proportion of graphite is around 5 wt-%. Graphite determines the melting rate and the fluidity of the slag layer.

2.5 NEAR NET SHAPE CASTING

Near net shape casting is defined as a variant of continuous casting, where the products are shaped as closely as possible to the final products, for example thin strips and wires.

In today's society it has become more and more important to use raw materials and energy in an optimal way. Methods used for manufacturing products with the minimum number of necessary intermediate stages are of special interest. The use of near net shape casting methods, used optimally, can save much material and energy.

The conventional method of producing plates, strips, wires and tubes is continuous casting or ingot casting. To assign the desired shape and properties to the product it is necessary to let it undergo subsequent treatment after solidification and cooling. This procedure requires much time, work and energy.

A considerable energy and material saving has already been achieved if continuous casting instead of ingot casting of steel is used. The former method gives a better yield of useful castings per charged kilogram of steel than conventional ingot casting. In addition, continuous casting gives a better shape of the casting before further treatment. Both these properties result in energy savings. However, subsequent treatment is still necessary in both cases.

In near net shape casting, subsequent treatment can be omitted. A necessary condition for omission of the subsequent treatment is that the casting obtains the desired material properties during the casting process. This condition can only be fulfilled at *very high cooling rates* of the cast metal melt (*quenching*).

Due to the composition of the alloy and the cooling rate, various structure modifications are possible. It is possible, for example, to suppress segregation, refine the solidification structure, significantly increase the solubilities of the alloying elements in the solid phase and form new metastable phases.

If the cooling rate is high enough, it is possible to suppress crystallization completely and open the possibility of formation of a noncrystalline metallic phase, a so-called *amorphous phase*. Due to all these structure modifications it has been possible to develop a new material class with quite unique properties.

2.5.1 Methods of Near Net Shape Casting

Two main methods of thin plate casting, i.e. casting on one and two rollers, respectively, are used. The solidification is different in the two cases. It means that both the casting rate and the metal structure that is formed are different.

The principle of the *single-roller process* is illustrated in Figure 2.15. In the single-roller process it is difficult to increase the period of contact between the cast strip and the roller to control the solidification process. The strip solidifies from the concave side, which causes surface problems on the convex side, such as irregularities in structure, coarseness and anisotropy (properties vary with direction) along the strip.

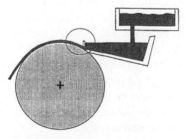

Figure 2.15 Principle sketch of a single-roller casting machine. Reproduced with permission from the Scandinavian Journal of Metallurgy, Blackwell.

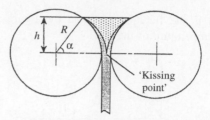

Figure 2.16 Principle sketch of a double-roller casting machine: h = height of melt; R = radius of roller; α = contact angle. Reproduced with permission from Scandinavian Journal of Metallurgy, Blackwell.

The types of problem mentioned above are reduced in a *double-roller process* (Figure 2.16). This process does, however, lead to other types of problem. It is necessary that the solidification fronts meet at the 'kissing point'. Otherwise internal cavities may appear or the shell may burst and release the melt.

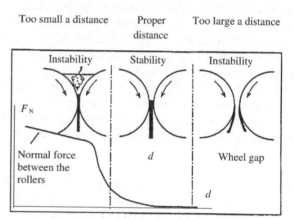

Figure 2.17 Functional conditions of a double-roller machine for near net shape casting. The curve under the rollers represents the normal force between the rollers as a function of the distance between them. Reproduced with permission from Jean-Pierre Birat.

The distance between the parallel rollers must be carefully adjusted in order to achieve stability during the casting process (Figure 2.17).

2.5.2 Thin Strip Casting

For the production of plates and wires a number of different methods are used, where the casting is simultaneously cast and rolled. The surface of the casting must maintain such a quality that hot rolling can be performed directly after the casting without grinding. Several different types of machine have been developed over the years. There are, for instance, machines that produce castings with thicknesses of the magnitude of a couple of centimetres. Two typical examples of such machines are Hazelett's and Hunter-Douglas's machines.

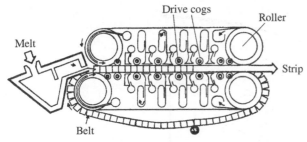

Figure 2.18 Cross-section of Hazelett's casting machine. Two metal belt loops are kept stretched and are driven by two pairs of parallel rollers. The water-cooled belts are parallel and form the chill-mould walls for the metal melt, which is fed between them. Reproduced with permission from Pergamon Press, Elsevier Science.

The idea behind Hazelett's method is to cast the melt between two water-cooled driving belts of steel, which serve as chill-mould walls and follow the metal on its way from the liquid to the solid state (Figure 2.18). This arrangement gives the best and most even cooling of the solidifying material. In this way, crack formation can be avoided.

In order to improve the fit of the cooled chill-mould wall, attempts have been made to use belts made of textile materials instead of steel belts. When the cast materials shrink at solidification, the textile belts are able to follow during the shrinkage. Their fitting properties are consequently better than those of the steel belts. Thus textile belts give better cooling properties than steel belts do.

Hunter-Douglas's and Hazelett's methods are based on the same principle. The only difference is that steel box sections are used in the Hunter-Douglas process instead of steel belts. Hunter-Douglas's 'belt' is reminiscent of a caterpillar track.

There are also machines for production of thicker castings. In these cases the melt is brought into contact with water-cooled rolls and solidifies between them. Examples of processes of this kind are the Hunter Engineering and the Harvey processes. Cooled rolls are used in the Hunter Engineering machine. The melt is fed between them under pressure.

The last two processes show considerable similarities to the near net shape casting methods mentioned above and described below. The castings, produced by the aid of the Hunter Engineering and the Harvey processes, are thicker than near net shape castings and need subsequent treatment.

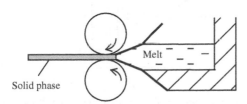

Figure 2.19 The Harvey process for horizontal steel casting.

The Harvey casting method (Figure 2.19) works principally in the same way as the Hunter Engineering method. The casting is performed horizontally.

2.5.3 Thin Billet Casting

Metal wire production is the field where near net shape casting is the dominating method. The most common products are wires of aluminium, copper and alloys of these metals.

There are several methods of wire production. The two most common methods are the Properzi process and the Southwire process.

The casting machine consists essentially of a moving steel band and a rotating water-cooled wheel. In its periphery one or several grooves have been milled. The steel belt runs over the grooves. The melt is fed between the belt and the grooves and solidifies there (Figure 2.20).

Both types of machine work in roughly the same way. The difference is mainly in the design of the steel belt loop.

2.5.4 Spray Casting

In order to reduce the risk of macrosegregation (uneven alloy composition due to flowing melt, Chapter 11) a method of making castings of powdered alloy metals was developed in the middle of the Twentieth century. The melt is atomized and the droplets brought to rapid solidification. The powder is then compressed at high pressure and high temperature (HIP-treatment).

In order to reduce the cost, methods have been developed where the powder is directly compressed during the powder production process. Such a process has been developed by the English company Osprey Metals Ltd. This process starts by the melting of scrap metal goods in an induction furnace. The melt is atomized into particles with diameters of roughly 150 μm using nitrogen gas. The stream of hot particles is then sprayed into a chill-mould where the casting is formed (Figure 2.21). The flat particles are then welded together in the chill-mould and a very dense and compact material is obtained.

The casting is used as a preformed forging, which is heat treated in an oven and is then transported to a forging press and finally to a shearing device. During these treatment processes part of the material is separated as scrap and carried back to the melt together with external scrap and the spray which does not stick to the chill-mould. Thanks to the nitrogen atmosphere the spray particles do not become oxidized. The nitrogen atmosphere also protects the material from inclusion formation in the preformed castings.

Production of rolled metal strips is a process where continuous operation is very important. A. R. E. Singer at the University of Swansea in Wales has developed a process of continuous production of metal strips from metal spray. The

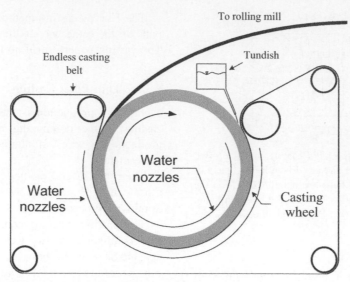

Figure 2.20 Principle sketch of the Properzi and the Southwire machines for production of thin wires. Rollers with parallel axes drive the steel belt. The small circles indicate schematically the water-cooling.

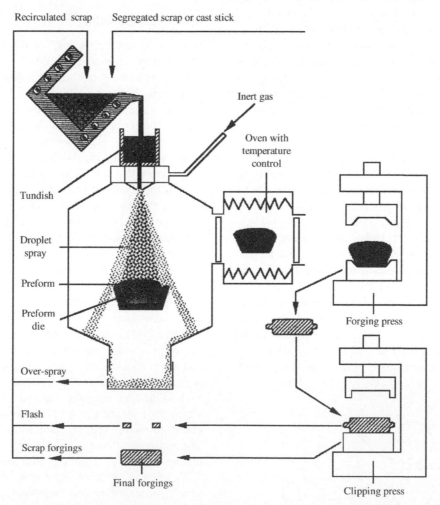

Figure 2.21 Sketch of the Osprey process. The figure shows the complete process chain from scrap supply to final forging. Reproduced with permission from John Wiley & Sons, Ltd.

process is partly reminiscent of a metal powder process. Instead of the comparatively expensive method using powder, which has to be compressed before sintering, so-called *spray rolling* (Figure 2.22) is used.

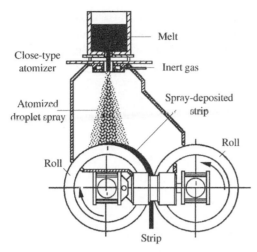

Figure 2.22 The figure shows an example of 'spray rolling', i.e. a sketch of an experimental set-up for the manufacture of plate metal products. Reproduced with permission from John Wiley & Sons, Ltd.

In spray rolling the melt is atomized by use of nitrogen gas and is then directly sprayed onto a substrate. In immediate rolling the existing thermal energy of the sprayed material is used optimally. The atomized particles, with a diameter of about 100 μm, hit the substrate at high speed. They flatten and form overlapping thin plates (Figure 2.23).

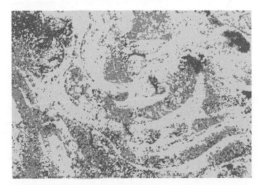

Figure 2.23 Structural picture of a plate, manufactured by 'spray rolling'. In the lower part of the figure the cross sections of some thin wave-shaped plates are seen. They consist of flattened droplets, which solidified when they hit the substrate. The black lines between the plates are associated with the solidification structure.

2.6 THE ESR PROCESS

Electro-slag refining or ESR as it is always denoted, is, strictly speaking, not a casting process but a refinement process, a remelting process to improve the material properties of ingots. The quality of the castings increases, which justifies a higher selling price and/or an improved competitive edge.

As early as the 1930s the American metallurgist Hopkins transformed a method, which was primarily a welding process, into a steel refining method. He used direct current, which in some respects turned out to be inconvenient. For this reason the method did not spread until during the 1950s the Russians began to use alternating current. In this way it was possible to eliminate the unfavourable direct current effects and ESR achieved a breakthrough as an industrial process.

2.6.1 Description of the ESR Process

The principle of ESR is very simple. An alternating voltage is connected between an electrode and the chill-mould which is in contact with an ingot (Figure 2.24) and a slag bath. Because the molten slag has considerably lower density than the melt, it forms as a layer on top of the metal melt.

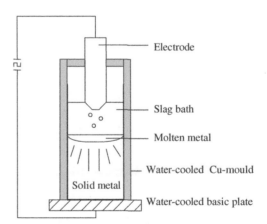

Figure 2.24 The principle of the ESR process. The lines below the molten metal indicate the directions of crystal growth.

The slag has approximately 10^{16} times greater resistivity than the metal. The slag bath works as a large resistor when the alternating current passes the closed circuit. The generated heat will thus concentrate in the slag. Consequently it reaches a temperature that exceeds the melting point of the metal by 300–400 °C and the metal electrode in contact with the slag starts to melt. The molten metal forms a thin film on the electrode tip. Drops of molten metal sink slowly through the slag bath to the metal bath, which solidifies from below. The molten zone in the metal bath moves slowly upwards and a refined ingot is successively formed.

The chill-mould is usually insulated from the mould by a very thin slag film that solidifies between the chill-mould and the ingot. The thin slag layer gives the ingot an extremely smooth surface. It is easy to remove the slag layer by so-called stripping of the ingot.

2.6.2 Function and Importance of Slag

The slag has two functions. It will (i) serve as a heat source, and (ii) refine the ingot.

The most important properties of the slag are its *resistance* and its *concentration of* FeO (oxygen content). The melting point of the slag must be lower than that of the metal to give stable melting, i.e. for steel alloys less than 1500 °C in most cases. A fluoride–oxide complex with some addition of Al_2O_3 on CaF_2/CaO basis is normally used as slag material.

The presence of slag, the favourable ratio between the interface slag/metal, and the volume of molten metal provide good possibilities of refining. Above all, an excellent separation of slag inclusions, especially large ones, is obtained.

3 Casting Hydrodynamics

3.1 INTRODUCTION

The motion that is generated in a metal melt at casting and solidification is important both for the material properties of the casting and for the reduction of serious casting defects. The flow in the melt is generated at the teeming processes from ladle to mould when the melt is forced to pass a so-called gating system. An incorrect design of the outlet hole in the ladle or of the gating system may cause serious defects such as cold shuts and numerous macroslag inclusions.

In addition, flow of melt is also developed during the solidification process in the form of natural convection, which arises due to temperature or concentration differences in the melt. This flow influences the structure of the material and its physical properties. In this chapter we will treat filling processes of moulds and chill-moulds, describe different types and designs of gating systems, and discuss the concept of fluidity extensively.

3.2 BASIC HYDRODYNAMICS

Experiments have shown that the same laws of hydromechanics are as valid for molten metals as for other fluids, for example water. The result is that the laws of hydrodynamics can be applied to the casting of metals. From fluid mechanics we borrow two very important laws, the principle of continuity and Bernoulli's equation.

3.2.1 Principle of Continuity

The principle of continuity is valid for an incompressible liquid. It means that no fluid appears or disappears during the flow. Consider a part of the flowing liquid limited by the end surfaces A_1 and A_2 (Figure 3.1). After the time interval dt the liquid element has moved to the dotted position.

If the velocity of the flowing liquid is v_1 at the A_1 surface and v_2 at the A_2 surface we get according to the continuity principle $A_1 v_1 \, dt = A_2 v_2 \, dt$. After division with dt we get:

$$A_1 v_1 = A_2 v_2 \tag{3.1}$$

This is the principle of continuity.

3.2.2 Bernoulli's Equation

The flow in the liquid is driven by the liquid pressure p. By considering the total energy at any point – here two arbitrary points 1 and 2 – of the flowing liquid, it can be proved that:

$$p_1 + \rho g h_1 + \left(\frac{\rho v_1^2}{2}\right) = p_2 + \rho g h_2 + \left(\frac{\rho v_2^2}{2}\right) \tag{3.2}$$

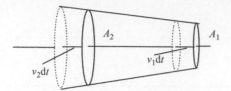

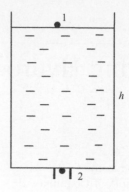

Figure 3.1 Illustration of the principle of continuity on a liquid flow with variable cross-section.

where

p_1 = pressure of the liquid at point 1
p_2 = pressure of the liquid at point 2
ρ = density of the liquid
h_1 = height of point 1 related to a certain zero level
h_2 = height of point 2 related to a certain zero level
v_1 = flow rate at point 1
v_2 = flow rate at point 2.

Equation (3.2) is Bernoulli's equation. It is valid if the flow is laminar and not turbulent. The derivation of Bernoulli's equation is given in the box below for those who want to penetrate its origin.

Figure 3.2 Teeming of a metal melt out of a ladle.

and the outlet velocity becomes (index is dropped):

$$v = \sqrt{2gh} \qquad (3.3)$$

Derivation of Bernoulli's Equation

We assume that the flow is laminar and consider the two small volume elements 1 and 2 in Figure 3.1. The flow is driven by the pressure difference – without a pressure difference there will be no flow.

The pressure of the liquid does a pressure work on the surface A_1. According to the energy law this work at point 1 must be equal to the sum of the corresponding pressure work on the surface A_2 and the changes of the kinetic and potential energies:

$$p_1 A_1 v_1\, dt = p_2 A_2 v_2\, dt + \left(\rho A_2 v_2\, dt \times gh_2 - \rho A_1 v_1\, dt \times gh_1\right) + \left[\rho A_2 v_2\, dt\left(\frac{v_2^2}{2}\right) - \rho A_1 v_1\, dt\left(\frac{v_1^2}{2}\right)\right]$$

$$\underbrace{\qquad}_{\text{pressure work}} \qquad \underbrace{\qquad}_{\text{increase of potential energy}} \qquad \underbrace{\qquad}_{\text{increase of kinetic energy}}$$

According to the principle of continuity we have $A_1 v_1 = A_2 v_2 = Av$. These expressions are inserted into the equation above and it can be divided by the factor $Av\,dt$. If we separate the terms with indices 1 and 2 we get:

$$p_1 + \rho gh_1 + \left(\frac{\rho v_1^2}{2}\right) = p_2 + \rho gh_2 + \left(\frac{\rho v_2^2}{2}\right) = \text{const}$$

Bernoulli's equation can be applied to the teeming of a metal melt out of a ladle. A ladle can be assumed to be an open container from which the melt flows through a hole in the bottom (Figure 3.2).

We will compare two points 1 and 2 in the flowing melt. Point 1 is a point at the free surface of the metal melt and point 2 is chosen at the exit hole. The hole is small compared to the area of the free surface of the melt, whose velocity v_1 can be neglected in comparison with the exhaust velocity v_2.

We choose the lower surface of the container as zero level and get $h_2 = 0$ and $h_1 = h$. We also have $p_1 = p_2 = p$ where p is the atmospheric pressure. These values are inserted into Bernoulli's equation and we get for $v_1 \approx 0$:

$$p + \rho gh + 0 = p + 0 + \left(\frac{\rho v_2^2}{2}\right)$$

The outlet velocity decreases when the height of the melt in the ladle decreases.

Example 3.1

A ladle with a circular cross-section (diameter 2.0 m) contains 2.7×10^3 kg molten steel. The steel is teemed through a circular hole in the bottom of the ladle. The diameter of the hole is 3.0 cm. Calculate the time required to empty the ladle.

Solution:

According to the continuity principle we have $A_1 v_1 = A_2 v_2$ or

$$v_1 = v_2\left(\frac{A_2}{A_1}\right) = v_2\left(\frac{\pi \times 0.015^2}{\pi \times 1.0^2}\right) = 2.25 \times 10^{-4} v_2 \qquad (1')$$

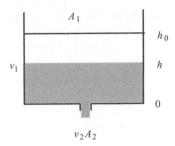

$v_2 A_2$

Obviously the velocity v_1 can be neglected in comparison with v_2 in Bernoulli's equation in this case. Equation (3.3) above can be applied. v_2 is solved from equation (1') and inserted into Equation (3.3):

$$v_2 = \left(\frac{A_1}{A_2}\right) v_1 = \sqrt{2gh} \qquad (2')$$

The velocity v_1 is equal to the decrease of the height h per unit time:

$$v_1 = -\frac{dh}{dt} \qquad (3')$$

If we combine Equations (2') and (3') we get:

$$\left(\frac{A_1}{A_2}\right) \times \frac{-dh}{dt} = \sqrt{2gh} \qquad (4')$$

The variables h and t are separated and Equation (4') is integrated:

$$\frac{A_1}{A_2} \int_{h_0}^{h} \frac{-dh}{\sqrt{2gh}} = \int_0^t dt \qquad (5')$$

and we get the time t as a function of h:

$$t = \frac{A_1}{A_2} \times \left[\frac{2\left(\sqrt{h_0} - \sqrt{h}\right)}{\sqrt{2g}}\right] \qquad (6')$$

The time to empty the ladle is obtained if we introduce the value $h = 0$ in Equation (6'). To calculate this time we must know the original height h_0 of the melt in the ladle.

$$h_0 = \frac{m}{\rho A_1} = \frac{70 \times 10^3}{(7.8 \times 10^3)(\pi \times 1.0^2)} = 2.86 \text{ m} \qquad (7')$$

The value of h_0 is inserted into Equation (6') which gives:

$$t = \left[\frac{\pi \times 1.0^2}{\pi (1.5 \times 10^{-2})^2}\right]\left[\frac{\sqrt{2}(\sqrt{2.86} - 0)}{\sqrt{g}}\right] \approx 3391 \ s = 56.5 \text{ min}$$

Answer:
The time to empty the ladle is approximately 1 hour.

Example 3.2
In order to accentuate the importance of a tundish between ladle and chill-mould in continuous casting we carry out the following virtual experiment.

Calculate how the extraction rate u must vary as a function of the height h in the ladle of a continuous casting machine, which has no tundish. There is no device to control the outlet velocity v_2 of the melt from the ladle.

The extraction rate u is expressed as a function of h, the area A of the bottom hole in the ladle and the cross-section dimensions a and b of the casting.

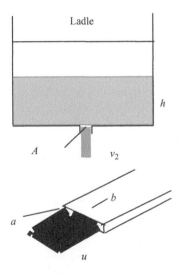

Solution:
No melt stays in the chill-mould. The mass per unit time of the input melt must be equal to the mass per unit time of the output melt. This condition gives the relationship:

$$v_2 A \rho_m = u \, ab \rho_s \qquad (1')$$

where ρ_m and ρ_s are the densities of the melt and of the casting respectively.

According to Equation (3.3) on page 30 we have:

$$v_2 = \sqrt{2gh} \qquad (2')$$

Equations (1') and (2') give:

$$u = \left(\frac{A\rho_m}{ab\rho_s}\right) v_2 = \left(\frac{A\rho_m}{ab\rho_s}\right)\sqrt{2gh}$$

Answer:
The desired function is:

$$u = \left(\frac{A\rho_m}{ab\rho_s}\right)\sqrt{2gh} \quad \text{or} \quad u = \text{const}\,\sqrt{h}$$

Example 3.2 confirms the statement in Chapter 2 that it is difficult to achieve the constant casting rate that is a necessary condition at continuous casting, without a tundish.

3.3 GATING SYSTEMS IN COMPONENT CASTING

In Chapter 2 we discussed the construction of the ladle and the purpose of the tundish at casting. In this section we will examine various designs of the gating system and the mathematical conditions that are valid in different cases.

The purpose of a gating system is to provide the mould with melt at the proper rate without unnecessary temperature losses and without unwanted gas or slag inclusions. It is essential to make an optimal choice of gating system for a given combination of metal, mould material and casting process on the basis of available theoretical and experimental data. Figure 3.3 is a sketch of a common gating system for component casting. Important parameters for the design of gating system are the *casting time* and the *area of the gating system*.

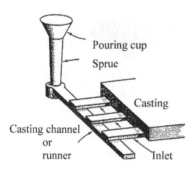

Figure 3.3 Gating system for component casting. Reproduced with permission from M.I.T.

Casting Time
A wrongly chosen casting time causes several faults. Too long a casting time risks misruns, gas pores, slag inclusions and in sand moulds also faults due to expansion of the sand.

Too short a casting time causes too strong a mechanical strain of the mould material. During rapid casting the mould material must be able to withstand the increasing erosion of the melt and in addition allow gases to pass easily.

The casting time changes with the viscosity of the melt. The viscosity varies strongly with the temperature,

the solidification pattern and the composition of the melt. Measurement of the so-called maximum fluidity length may be used as a tool to find the optimal casting time. Maximum fluidity length will be discussed in Section 3.7.

It is very difficult to find the optimal casting time. From experience one knows that there is an empirical relationship between the mass of the casting and the casting time of the type:

$$t = Am_{casting}^n \tag{3.4}$$

where A and n are constants, t = time, and $m_{casting}$ = mass of the casting.

The constants A and n are determined experimentally. A and n vary with the chosen casting process, for example with the binder, the sand coarseness and the packing degree of the sand in sand moulds.

Area of the Gating System
Molten metal can be regarded as an incompressible liquid. In spite of turbulence and heat losses due to possible friction, the equations for an incompressible liquid are valid as long as the system is filled and the walls impenetrable. The total flux in any point of the system will then be given by the principle of continuity [Equation (3.1)]. Unfortunately these assumptions are not always valid. Sand moulds are penetrable, which means that the laws of hydromechanics in these cases are not fully applicable.

In spite of the difficulties of analysing gating systems theoretically, we will discuss the matter below with the aid of the basic hydrodynamic laws. There are many existing computer programs that can be used to supply accurate information on the flow through a gating system. However, one must keep in mind that, even in this case, the calculations can only be used as guidelines.

3.3.1 Gating System in Downhill Casting – Impenetrable and Penetrable Moulds

Impenetrable Mould
The simplest case corresponds to an impenetrable mould and a straight sprue, which is part of the mould (Figure 3.4). The mould is assumed to be impenetrable to gases.

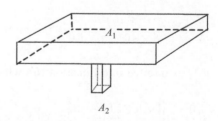

Figure 3.4 Gating system with impenetrable walls.

p = 1 atm at points 1 and 3

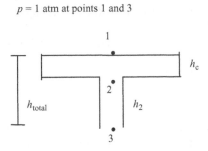

Figure 3.5 Cross-section of the gating system in Figure 3.4.

For simplicity we assume that the height h_c is kept constant during the whole casting time. We apply Bernoulli's equation at points 1 and 3 in Figure 3.5 and choose level 3 as zero level. According to Equation (3.3) on page 30 the exhaust velocity v_3 will be:

$$v_3 = \sqrt{2gh_{\text{total}}} \qquad (3.5)$$

We also apply Bernoulli's equation at the points 2 and 3 in Figure 3.5.

$$p_2 + \rho g h_2 + \left(\frac{\rho v_2^2}{2}\right) = p_3 + \rho g h_3 + \left(\frac{\rho v_3^2}{2}\right) = \text{const}$$

According to the principle of continuity, the velocities v_2 and v_3 must be equal because the cross-section areas A_2 and A_3 are equal. If we choose the zero level in such a way that h_3 is equal to zero we get:

$$p_2 = p_3 - \rho g h_2 \qquad (3.6)$$

where p_3 is the atmospheric pressure p. According to hydrodynamics the pressure at point 2 is consequently *lower* than the atmospheric pressure. This is true if the mould is impenetrable.

However, there are other effects that contribute to a higher pressure than the hydrostatic pressure p_2. Different types of mould gases may be generated by chemical reactions between the melt and the mould, close to walls. The mould gases are heated by the melt and their pressures increase because the walls are impenetrable. The gases dissolve partly in the melt, and the amount of dissolved gas depends on the partial pressures of the mould gases and whether they react with the melt or not.

If the mould gases do *not react* with the melt their total partial pressures contribute to a higher total pressure than p_2. In such cases gas pores may be trapped in the casting. In order to reduce this problem the mould must be equipped with channels to let the gases escape during the casting operation. If the mould gases *react* with the melt oxygen, hydrogen, or carbon dissolve in the metal melt and oxides, pores, or carbides precipitate during cooling and solidification.

All these effects are undesirable but can be tolerated more in some metals than in others.

Penetrable Mould

If a mould that is penetrable to gases (in practice made of sand) replaces the impenetrable mould above, the situation will be different. As the mould is penetrable, air is sucked into the melt to compensate for the pressure p_2 being less than 1 atm. The complications with mould gases will be the same as above. The resulting total pressure including air and mould gases, close to the walls everywhere in the gating system, will be equal to the atmospheric pressure.

In order to find out if the sprue of a penetrable mould can be designed in a better way than above, we assume that the gating system walls are completely penetrable to gases. We will examine the shape of the metal beam with the aid of Figure 3.6.

p = 1 atm in the points 1, 2 and 3

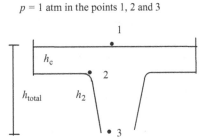

Figure 3.6 Gating system of a mould with permeable walls and a sprue with variable cross-section.

We choose two points 2 and 3 on the envelope surface of the beam where the pressures in both cases are equal to the atmospheric pressure p.

If we apply Bernoulli's equation at points 1 and 2, and 1 and 3 respectively, the same assumptions as on page 30 are valid and we get:

$$v_2 = \sqrt{2gh_c} \qquad (3.7)$$
$$v_3 = \sqrt{2gh_{\text{total}}} \qquad (3.5)$$

According to the principle of continuity the flow is the same:

$$A_2 v_2 = A_3 v_3 \qquad (3.8)$$

By combining Equations (3.5), (3.7) and (3.8) we get:

$$\frac{A_3}{A_2} = \frac{v_2}{v_3} = \sqrt{\frac{h_c}{h_{\text{total}}}} \qquad (3.9)$$

The gating system of a mould with penetrable walls should be designed in such a way that the cross-section area of the sprue decreases in accordance with Equation (3.9).

The ratio of the areas for points between 2 and 3 also follows Equation (3.9) but one can use approximate linear interpolation between the points 2 and 3 and design the sprue in accordance with these calculations.

3.3.2 Gating Systems in Uphill Casting – Penetrable Mould

So far we have only considered vertical gating systems with atmospheric pressure at the base. In many cases it is favourable to locate the gate at the bottom of the mould. This method is called uphill casting (Chapter 1). In this case the calculation of the casting time has to be modified.

We will consider a sprue that does not absorb gas, as in Figure 3.7. We assume further that the runner has parallel walls and that friction losses can be neglected. We introduce the following designations:

t = time interval after the start of casting

h_{total} = total casting height

h = height of the melt in the mould

A_{mould} = area of the upper surface of the melt

A_{runner} = cross-section area of the runner.

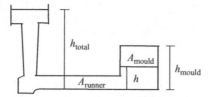

Figure 3.7 Gating system located at the bottom of the mould.

During the time interval dt the height h increases by the amount dh and the cast volume by $A_{mould}\,dh$. The volume of the melt that has passed the runner during the time interval dt is $A_{runner}\,v\,dt$, where v is the velocity of the melt in the runner at time t. As the mould is penetrable the velocity at the runner is given by the expression:

$$v = \sqrt{2g(h_{total} - h)} \qquad (3.10)$$

The increase of the ingot volume during the time dt is equal to the flow through the gate during the same time interval:

$$A_{mould}dh = A_{sprue}\sqrt{2g(h_{total} - h)}dt \qquad (3.11)$$

The variables t and h are separated and the equation is integrated:

$$\left(\frac{A_{sprue}}{A_{mould}}\right)\int_0^{t_{fill}} dt = \left(\frac{1}{\sqrt{2g}}\right)\int_0^{h_{mould}}\left(\frac{dh}{\sqrt{(h_{total} - h)}}\right) \qquad (3.12)$$

where t_{fill} = time required to fill the mould, and h_{mould} = height of the mould.

The filling time can be calculated from Equation (3.13):

$$t_{fill} = \left(\frac{2A_{mould}}{A_{sprue}\sqrt{2g}}\right)\left(\sqrt{h_{total}} - \sqrt{h_{total} - h_{mould}}\right) \qquad (3.13)$$

Uphill component casting is often used for metals that easily form skins of oxide. Aluminium alloys are common examples.

The advantages and disadvantages of the downhill and uphill casting methods are discussed on page 17 in Chapter 2. The methods are used both in component casting and ingot casting. Ingot casting is discussed in Section 3.4.

3.3.3 Gating System with Lateral Sprue

The introduction of a gating system with a lateral sprue is an attempt to combine the advantages of the uphill and downhill casting methods.

A small casting requires only one sprue, one casting channel and one inlet. With increasing dimensions of the castings, the mechanical stress on the mould will be strong and the risk of overheating and bursting of the mould material, close to the sprue, increases when larger amounts of melt are to be transported. The solution of the problem is to distribute the melt in casting channels with several inlets. The inlets are located either in the same horizontal plane or on successively higher levels while the mould is filled.

In such a branched system it is essential to control the distribution of melt between the different inlets and to adjust the flow in order to achieve an even distribution of the melt. Examples are given in Figures 3.8 (a)–(c). The flow has a tendency to be lower at larger distances from the sprue, due to friction forces in the melt.

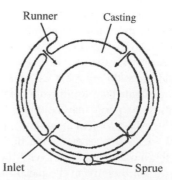

Figure 3.8 (a) Design of a mould with multiple inlets. Reproduced with permission from the Butterworth Group, Elsevier Science.

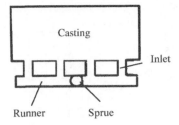

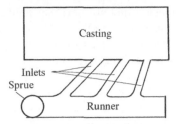

Figure 3.8 (b) Position of the sprue in case 3.8 (a). Reproduced with permission from the Butterworth Group, Elsevier Science.

Figure 3.10 Gating system with inclined inlets. Reproduced with permission from the Butterworth Group, Elsevier Science.

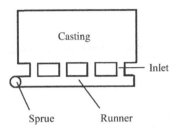

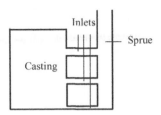

Figure 3.8 (c) Alternative position of the sprue in case 3.8 (a). Reproduced with permission from the Butterworth Group, Elsevier Science.

Figure 3.11 Gating system for tall components. Reproduced with permission from the Butterworth Group, Elsevier Science.

Measures to adjust the pressure in the casting channel to make it as uniform as possible are:

- The inlet areas can be different. The areas close to the sprue are designed to be larger than the ones further away from the sprue.
- The inlet areas are contracted relative to the casting channel to achieve a flow rate in the channel that will be comparable to the inlet rate and a pressure at every inlet that will be about the same.
- The casting channel is made more and more narrow after each inlet (Figure 3.9).

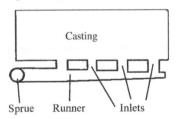

Figure 3.9 Gating system with a decreasing area of the runner. Reproduced with permission from the Butterworth Group, Elsevier Science.

- The inlets are designed to form different angles with the casting channel. The angles are chosen in such a way that the decreasing flow rate along the channel from the sprue results in equal flows at all the inlets (Figure 3.10).

Combinations of the measures suggested above are used in practice.

For tall castings (Figure 3.11) several inlets at different levels are required to combine the advantages of optimal temperature distribution, characteristic for downhill casting, and of calmer and splash-free inflow, which is typical for uphill casting. Even in this case measures are required to get the desired flow of melt, for example more and more narrow inlets at increasing heights.

3.3.4 Gating Systems in Pressure Die Casting

The gating systems, discussed in Sections 3.3.1, 3.3.2 and 3.3.3, are all adapted for flowing melts and the driving force of the casting process is the gravity force on the melt. They fit the majority of the methods for component casting that are listed in Chapter 1.

However, the gating systems of pressure die casting machines are entirely different. This type of component casting machine allows high production rates. As the name indicates the driving force of metal melt transfer into the mould is high pressure. The mould is always made of a metal with a higher melting point than the alloy to be cast in order to provide rapid cooling. Such a mould is called a *die*.

The high-pressure die casting and low-pressure die casting methods are briefly discussed (pages 8–9 in Chapter 1). As a complement we will here later discuss the gating system and the ejection and opening mechanisms of machines of the high-pressure die casting type.

The great majority of pressure die castings are produced in zinc- or aluminium-base alloys. Two alternative types of injection system are used: the so-called hot and cold chamber systems. Either system permits production rates of one casting in a few seconds.

Gating System of a Cold Chamber Machine

In a hot chamber machine the alloy is kept in a reservoir of molten metal at a temperature well above its melting point.

The melt is injected directly into the mould with the aid of a high air pressure ($> 20 \times 10^6 \, N/m^2$). Hot chamber machines are extensively used for zinc-base alloys but are unsuitable and useless for aluminium and other alloys of high melting points. They give contamination with iron during the contact between the molten alloy and the walls of the chamber. In addition there arises a considerable inclusion of air in the metal during the injection phase.

Aluminium-base alloys are cast in a *cold chamber machine*. The principle of such a machine is shown in Figure 3.12. The molten metal in a cold chamber machine is kept at a constant temperature in an adjacent holding furnace, often by electrical heating. Metal for a single shot is loaded into a cylindrical chamber, injection cylinder A.

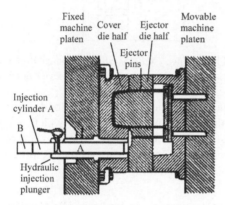

Figure 3.12 Cold chamber pressure die casting machine. Reproduced with permission from the Butterworth Group, Elsevier Science.

The metal is forced into the die by a piston B at a high pressure (magnitude $(70 - 140) \times 10^6 \, N/m^2$). The whole injection operation is completed in a few seconds and so the risk of dissolving iron from the mould is greatly reduced.

Opening and Ejection Mechanisms
Figure 3.13 (a) illustrates the injection of metal into the die at high pressure. The pressure is not constant. It varies during

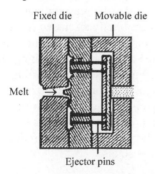

Figure 3.13 (a) Closed die, ready for injection of molten metal. Reproduced with permission from the Butterworth Group, Elsevier Science.

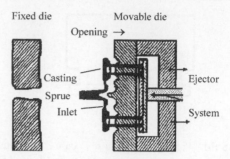

Figure 3.13 (b) Open die, indicating the ejection mechanism of the casting machine. Reproduced with permission from the Butterworth Group, Elsevier Science.

the process, as will be discussed later. Figure 3.13 (b) shows the open die after casting and how the ejector pins push the casting out of the die.

Injection Technique
Modern cold chamber machines are designed for close control and variation of the pressure and the speed of the plunger. A pressure and speed cycle is illustrated in Figure 3.14.

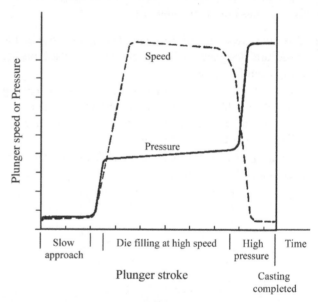

Figure 3.14 Plunger speed and pressure during one metal injection cycle in pressure die casting as functions of time. Courtesy of British Foundrymen, IBF Publications, *Foundryman*.

The gating system is primarily filled at a relatively low speed, which minimizes the capture of air in the molten metal. During the filling of the main cavity the speed is high. At the end of the plunger stroke the pressure is raised to its maximum level to make sure that the die cavity is completely filled. The high pressure is maintained during the solidification in order to minimize porosity from solidification shrinkage and air bubbles.

Filling of the Die Cavity

The injection of molten metal has been extensively studied and a review of pressure die casting was published by Barton in the 1990s. It is generally agreed that the filling process mainly occurs in two steps at pressure die casting (Figure 3.15).

Step 1 corresponds to Figure 3.15 (a) and (b). When the jet flow reaches the inlet after passage through the runner it strikes the die cavity at a point opposite to the gate and spreads laterally along the walls of the cavity. This moment is of utmost importance for surface quality of the casting.

Step 2 consists of turbulent filling of the cavity in a way illustrated in Figures 3.15 (c) and (d). The cavity is filled from the top mainly in the direction back to the gate.

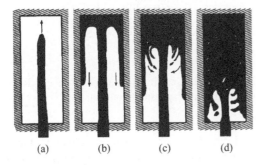

| (a) | (b) | (c) | (d) |

Figure 3.15 Melt flow during the casting process in pressure die casting. Reproduced with permission from the Butterworth Group, Elsevier Science.

Vents have to be installed in order to remove as much trapped air as possible. Venting is accomplished by use of narrow channels, 1–2 mm in diameter. In addition small cavities are placed along the symmetry axis to provide escape for the air and to facilitate the cavity filling. Complete removal of the air inclusions cannot be achieved. Minor porosity, caused by air bubbles, has to be accepted as a normal phenomenon of pressure die castings so far.

Various methods are used to reduce the porosity. One method is to suck out air from the mould cavity during the filling process. Another method is to replace the air in the mould cavity with nitrogen. Nitrogen is dissolved into the cast metal or forms nitrides and air pores have thus been avoided.

3.4 GATING SYSTEM IN INGOT CASTING

Ingot casting resembles component casting very much. The difference between ingots and components is merely a matter of size. The mass of an ingot is of the magnitude of 10^3 kg. Thus all ladles, gating systems and moulds are much bigger in ingot casting than are those of component casting. The solidification time during ingot casting is much longer than that in component casting, being of the order of hours instead of seconds or minutes.

The gating systems, used in component casting, are discussed extensively in Section 3.3. The principles, formulas and location of gating systems are the same for component casting and ingot casting, with allowance for modifications due to differences in size.

Thus the discussion of the gating systems in ingot casting can be rather short with a reference to Section 3.3. Only a minor addition is given below.

Dimensioning of the Gating System in Uphill Casting

For ingot casting both downhill and uphill casting are used. These two methods are described in Chapter 2 where their advantages and disadvantages are listed (Chapter 2 page 17).

As is seen from Example 3.3 below, Equation (3.13) on page 34 can be used for calculation of the time required to fill the mould in uphill casting, or of the design of the sprues in order to achieve a certain casting time.

The level of the melt in the runner is controlled by the outflow from the ladle. The shortest possible casting time is obtained if the sprue and the runner are filled with melt during the whole casting process.

***

Example 3.3

A steel company formerly used downhill casting for ingot production but has decided to change to uphill casting. They want to use a central sprue and locate the six moulds symmetrically around it. The dimension of the sprue determines, in principle, the casting time for the six ingots. The casting time is not to exceed 10 minutes. The sprue and the runners are kept filled during the whole casting time.

Suggest an appropriate design for the sprue and calculate the smallest possible diameter for the sprue. The ingot height is 100 cm and the ingot areas are squares of 20 cm × 20 cm. The free surface of the sprue lies 10 cm higher than the ingot height.

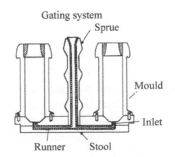

Solution:

At uphill casting the melt is cast in a central gating system and enters the bottoms of the moulds via a step plane. Normally two to six moulds are placed at the same step level. In the present case we have six moulds.

The diameter d of the sprue can be calculated from Equation (3.13) on page 34:

$$t_{\text{fill}} = \left(\frac{2A_{\text{mould}}}{A_{\text{sprue}}\sqrt{2g}} \right) \left(\sqrt{h_{\text{total}}} - \sqrt{h_{\text{total}} - h_{\text{mould}}} \right)$$

where

$t_{\text{fill}} = 600\,\text{s}$
$A_{\text{mould}} = 6 \times 0.2 \times 0.2 = 0.24\,\text{m}^2$
$A_{\text{sprue}} = \pi d^2/4$
$h_{\text{total}} = 1.10\,\text{m}$
$h_{\text{mould}} = 1.00\,\text{m}.$

We insert the values above into Equation (3.13) and solve the diameter d:

$$d^2 = \left[\frac{2A_{\text{mould}}}{\left(\frac{\pi}{4}\right) t_{\text{fill}} \sqrt{2g}} \right] \left(\sqrt{h_{\text{total}}} - \sqrt{h_{\text{total}} - h_{\text{mould}}} \right)$$

$$= \left[\frac{2 \times 0.24}{\left(\frac{\pi}{4}\right) 600 \times \sqrt{2g}} \right] \left[\sqrt{1.10} - \sqrt{(1.10 - 1.00)} \right]$$

which gives:

$$d \geq 2.2 \times 10^{-2}\,\text{m}$$

Answer:
The diameter of the sprue must be at least 22 mm.

3.5 GATING SYSTEM IN CONTINUOUS CASTING

The principles of continuous casting are treated in Chapter 2. The major differences between ingot and component casting on one hand and continuous casting on the other are the following features. With continuous casting:

- A water-cooled metal mould, the so-called *chill-mould*, is used to provide strong and effective cooling of the strand. The solidified shell of the strand must have enough stability and strength before leaving the chill-mould.
- An intermediate casting box, the so-called *tundish*, is located between the ladle and the chill-mould. The aim of the tundish is to provide a constant casting speed. The negative effects of absence of a tundish are illustrated by Example 3.2 on page 31.

3.5.1 Submerge Entry Nozzle

In continuous casting it is desirable to have the most even velocity of the casting that can be obtained. By use of a *tundish* as an intermediate container (Chapter 2

Section 2.4.2, page 19–20) between the ladle and the chill-mould, the disadvantage of the velocity of the melt varying with the height of the melt in the ladle (Example 3.2) is eliminated. The tundish is kept filled and the distance between the tundish and the chill-mould is constant. A sliding gate or a *stopper rod* controls the flow from the tundish (Figure 3.16). Often flushing with argon gas is used to prevent air intrusion or clogging of the nozzle.

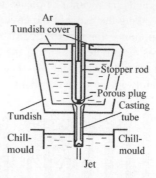

Figure 3.16 Tundish, casting tube and chill-mould. Reproduced with permission from the Institute of Materials.

Another problem is the risk of *slag inclusions*. To decrease the problem of macroslag inclusions during continuous casting, the upper surface of the melt is covered with casting powder and the melt transferred from the tundish to the mould through a *submerged entry nozzle* (SEN) or *casting tube*.

A casting tube cannot be used for castings with small dimensions, for example small billets. In all other cases the device can be used.

The flow depends on the geometry, depth and dimensions of the nozzle exit and the flow rate of the jet (Figure 3.16). The jet causes forced convection in the chill-mould in addition to natural convection that is always present during the solidification process. To avoid inclusion of trapped slag particles at the solidification front, the penetration depth should be moderate. Therefore the nozzles are often designed with exits on the sides at some angle between 0° and 90° relative to the vertical axis (Figure 3.16).

The casting tube must be made of a material that resists chemical attacks from steel alloying elements, such as aluminium, sulfur and manganese. Casting tubes are often made of a mixture of Al_2O_3 and graphite.

Flow of Melt

The melt flow in the mould is greatly influenced by the jets from the casting tube. When the melt, which is superheated by about 30 °C, leaves the nozzle it impinges laterally into the mould, where it is split into two strongly circulating flows, one directed upwards and the other one directed downwards. Figure 3.17 shows the flow pattern in the mould. The violent motion in the melt contributes strongly to the formation of a homogeneous fluid.

Figure 3.17 Flow pattern of the melt around a casting tube in the mould.

3.5.2 Sequence Casting

A wide range of steel products is used today and there is often need of products with slightly different compositions. The simplest way to cast two such products of qualities, which we will call grade 1 and grade 2 below, is to cast them separately and stop the casting machine, when grade 1 is finished, and then restart it to cast grade 2.

The productivity in continuous casting processes is improved by increasing the casting speed. However, an increase in production efficiency also requires long sequences of ladles without stopping and restarting the casting machine. It is thus desirable to cast ladles with the same composition in sequence, and also two ladles with slightly different compositions in sequence without stopping the casting process.

One method, and the most common way of casting steel batches with slightly different compositions in sequence, is *ladle exchange online*. The two steel qualities mix in the tundish and move to the chill-mould and mix there as well. The main task is to minimize the mixed region, which has a slightly different composition in comparison with the two batches. The ladle exchange is associated with costs due to wasted material.

Another method is *flying tundish exchange* when both the ladle and the tundish are exchanged simultaneously. In this case the mixing occurs only in the final product, the strand. A 'grade separator' plate can be inserted into the chill-mould in order to minimize the mixed region. It is a plate, which is impenetrable for the melt and follows the melt through the chillmould and divides the strand into two parts.

Composition Change of Cast Steel during a Grade Transition in the Tundish
During a typical ladle exchange and grade transition in sequence casting, the casting rate, the total tundish volume and the flow rate into the tundish vary with time as shown in Figure 3.18 (a)–(c). Mixing begins when the new ladle is opened and the new grade of steel starts to flow into the tundish, which defines $t = 0$. The curves at times <0 represent events before the opening of the new ladle.

As a preparation for the exchange of the ladle, the casting rate is generally decreased. This lowers the flow rates temporarily. Simultaneously the volume of the melt in the tundish decreases. The tundish volume at $t = 0$, when the new ladle is opened, almost always corresponds to the minimum tundish level.

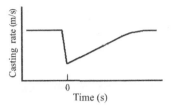

Figure 3.18 (a) Casting rate as a function of time. $t = 0$ is chosen as the time when the new ladle is opened.

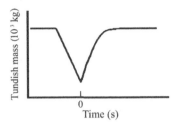

Figure 3.18 (b) Tundish mass as a function of time. $t = 0$ is chosen as the time when the new ladle is opened.

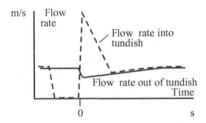

Figure 3.18 (c) Flow rates into and out of the tundish as functions of time. $t = 0$ is chosen as the time when the new ladle is opened.

When the new ladle is opened, the tundish refills to the desired operation level, in equilibrium with the inlet flow rate and the casting rate. The mixing of the two alloys starts at $t = 0$ and occurs in the tundish. It depends on the filling rate and the casting rate, which both increase and reach their stationary values.

When the melt has left the tundish it partly solidifies in the chill-mould. In the mixed region the composition of the casting changes gradually from grade 1 to grade 2.

Example 3.4 below shows how the concentration of the alloying element of the cast steel varies with time in sequence casting in a simplified case. The casting rate has *not* been slowed down [Figure 3.18 (a)] and the volume of the melt in the tundish is *not* reduced at the ladle exchange

[Figure 3.18 (b)]. The calculations below illustrate clearly how necessary it is to take these steps.

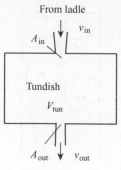

From ladle

Tundish V_{tun}

To chill-mould

Example 3.4

Two stainless steel alloys are cast in sequence by ladle exchange during continuous casting. A grade separator plate is inserted into the chill-mould at the same time as the ladle exchange occurs. The casting rate is assumed to be constant during the rapid shift of the ladle.

Alloy 1 contains 2.00 wt-% Mo while the Mo concentration in alloy 2 is 0.50 wt-% Mo. The mixing of grade 1 and grade 2 alloys in the tundish is assumed to be instant and complete. The volume of the tundish is 1.0 m³. The dimensions of the slab are 0.20 m × 0.90 m and the casting rate is 0.60 m/min. The sprue from tundish to chill-mould is assumed to be straight. The volume of melt in the tundish is constant.

(a) How do you get the distance z from the chill-mould outlet as a function of time t if you know the casting rate v_{cast}?

(b) Calculate the Mo concentration in the strand, which leaves the chill-mould, as a function of distance from the chill-mould outlet and illustrate the function graphically. Define $t = 0$ as the time when the new ladle is opened.

(c) What measures can be taken to reduce the intermixing region of the strand?

(d) The same question as in (b) with the only difference that no grade separator plate is used. The mixing of grade 1 and grade 2 alloys in both the tundish and the chill-mould are assumed to be instant and complete. The height of the chill-mould is 0.70 m.

Solution:

(a) The casting rate is equal to the velocity of the strand.

$$z = v_{strand}\, t = v_{cast}\, t \qquad (1')$$

where $v_{cast} = 0.60$ m/min $= 0.010$ m/s

(b) Mo balance in the tundish:

If a grade separator plate is used the composition of the melt in the tundish and the chill-mould will be the same. No mixing with the old melt in the chill-mould is possible. We analyse the Mo balance in the tundish at time t, due to the flow of melt during the time dt. The designations are given in the figure.

At time t the homogenous melt in the tundish has the composition c_{out}.

$$V_{tun}\, dc_{out} = A_{in} v_{in}\, dt\, c_{in} - A_{out} v_{out}\, dt\, c_{out} \qquad (2')$$

Increase of the amount Mo in the tundish during the time dt	Added amount of Mo coming from the ladle during time dt	Lost amount of Mo due to the outflow from the tundish during time dt

where $c_{in} = c_2$ for $t \geq 0$.

The principle of continuity gives:

$$A_{in} v_{in} = A_{out} v_{out} \qquad (3')$$

Equations (2') and (3') give:

$$V_{tun}\, dc_{out} = A_{out} v_{out} \left(c_{in} - c_{out} \right) dt \qquad (4')$$

Integration gives:

$$\int_{c_1}^{c_{out}} \frac{dc_{out}}{(c_{in} - c_{out})} = \frac{A_{out} v_{out}}{V_{tun}} \int_0^t dt \qquad (5')$$

The lower limit of the integral corresponds to $c_{out} = c_1$ at $t = 0$.

The solution of integral (5') is:

$$-\ln \frac{(c_{in} - c_{out})}{(c_{in} - c_1)} = \frac{A_{out} v_{out}}{V_{tun}} t$$

or

$$\ln \frac{(c_{out} - c_{in})}{(c_1 - c_{in})} = -\left(\frac{A_{out} v_{out}}{V_{tun}} \right) t \qquad (6')$$

At $t \geq 0$ $c_{in} = c_2$ which is inserted into Equation (6'):

$$c_{out} = c_2 + (c_1 - c_2) \exp\left(-\frac{A_{out} v_{out}}{V_{tun}} \right) t \qquad (7')$$

c_{out} is also the Mo concentration at the inlet and outlet of the chill-mould in presence of the grade separator plate.

The principle of continuity $v_{out} A_{out} = v_{strand} A_{strand} = v_{cast} A_{strand}$ and the relationship $z = v_{cast} t$ are applied to Equation (7') which gives:

$$c_{out} = c_2 + (c_1 - c_2) \exp\left(-\frac{A_{strand} v_{cast} t}{V_{tun}}\right)$$

$$= c_2 + (c_1 - c_2) \exp\left(-\frac{A_{strand} z}{V_{tun}}\right) \qquad (8')$$

Introduction of known values gives the function:

$$c_{out} = [0.50 + 1.50 \exp(-0.18 z)] \text{ wt-\%}$$

This function is illustrated in the answer on page 42.

(c) Equation (8') shows that the mixed zone is infinite as ($c_{out} = c_2$ at $t = \infty$). In practice it is shortened if V_{tun} is *small* and v_{out} is *large*. The latter one is constant. The volume of the melt in the tundish shall be decreased as much as possible before the ladle exchange, when the tundish is filled with melt of a new composition.

(d) Mo-balance in the chill-mould:
If no grade separator plate is present, Equation (8') does not describe the composition in the strand any longer. The mixed melt from the tundish with composition c_{out} mixes with the melt in the chill-mould to a new composition c_{strand}.

The differential equation for the Mo concentration in the chill-mould as a function of time can be obtained from Equation (4') by replacing:

V_{tun} in Equation (4') by V_{mould};

A_{out} in Equation (4') by A_{strand};

v_{out} in Equation (4') by v_{cast};

c_{in} in Equation (4') by c_{out};

c_{out} in Equation (4') by c_{strand}.

According to the principle of continuity we also have:

$$v_{out} A_{out} = v_{strand} A_{strand} = v_{cast} A_{strand}$$

All these substitutions result in the relationship:

$$V_{mould} dc_{strand} = A_{strand}(c_{out} - c_{strand}) v_{cast} dt \qquad (9')$$

c_{out} is a function of z according to Equation (8'). This function is introduced into the differential Equation (9'), which gives:

$$V_{mould} dc_{strand}$$
$$= A_{strand}\left(c_2 + (c_1 - c_2) \exp\left(-\frac{A_{strand} z}{V_{tun}}\right) - c_{strand}\right) dz \qquad (10')$$

By use of the relationship $V_{mould} = A_{strand} H_{mould}$ it can be written as:

$$\frac{dc_{strand}}{dz} = \frac{1}{H_{mould}}\left(c_2 + (c_1 - c_2) \exp\left(-\frac{A_{strand} z}{V_{tun}}\right) - c_{strand}\right) \qquad (11')$$

$$c_{strand} = c_2 + \frac{(c_1 - c_2) \exp\left(\dfrac{A_{strand} z}{V_{tun}}\right)}{1 - \left(\dfrac{A_{strand} H_{mould}}{V_{tun}}\right)} + \left((c_1 - c_2) - \frac{c_1 - c_2}{1 - \left(\dfrac{A_{strand} H_{mould}}{V_{tun}}\right)}\right) \exp\left(-\frac{z}{H_{mould}}\right) \qquad (12')$$

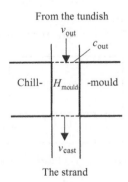

From the tundish
v_{out}
c_{out}

Chill- H_{mould} -mould

v_{cast}

The strand

The solution to Equation (11') is:

Boundary condition: $c_{strand} = 2.0$ wt-% at $z = 0$

Answer:

(a) $z = v_{cast} t$

(b) $c_{out} = c_2 + (c_1 - c_2) \exp\left(-\dfrac{A_{strand} z}{V_{tun}}\right)$

(c) The volume of the melt in the tundish shall be as small as possible at the time for the ladle exchange. It is important to decrease the volume of melt in the tundish before changing other qualities of the melt.

(d) c_{strand} as a function of z is obtained by inserting the known numerical values into Equation (12′) above. It is plotted in the diagram below. The Mo concentration becomes slightly higher when the grade separator plate is excluded.

Lower curve:

$$c_{out} = [0.50 + 1.50 \exp(-0.18z)]\,wt\text{-}\%$$

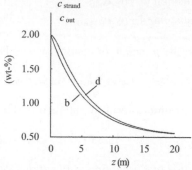

Upper curve:

$$c_{strand} = [0.50 + 1.716\exp(-0.18z) - 2.16\exp(-1.43z)]\,wt\text{-}\%$$

3.6 INCLUSION CONTROL IN GATING SYSTEMS – CERAMIC FILTERS

During casting it is necessary to separate slag inclusions, mould particles and other nondesirable particles from the melt before the melt enters the mould cavity. In case of there being sufficient density difference between the melt and the impurities, the separation can simply be done in an upper chamber at the inlet, where light impurity particles may float and be separated from the melt (Figure 3.19). Another method is separation by centrifugation.

For metals where the density differences between the impurities and the melt are small, filtering can be used.

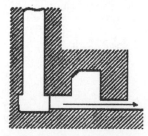

Figure 3.19 Mechanical impurity trap. Reproduced with permission from the Butterworth Group, Elsevier Science.

Swirl Trap
A fundamental claim in the production of quality castings is to minimize the presence of slag inclusions and dross defects. This can be achieved by two different methods:

- The gating system can be designed in such a way that the metal flow in all parts is as laminar as possible. A laminar

flow keeps the possible reactions between metal and mould/air at the lowest possible level and consequently minimizes slag formation.

- Some kind of additional equipment can be installed into the gating system to prevent all kinds of nonmetallic inclusion entering the mould.

An example of the latter method is a so-called 'swirl trap'. This can be used to separate inclusions from the melt in metallic systems. A swirl trap is designed to centrifuge inclusions lighter than the metal melt. The particles are forced to the centre of the trap, where they aggregate and float. Figure 3.20 (a) and (b) illustrate the principle of a swirl trap.

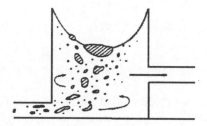

Figure 3.20 (a) Swirl trap in action. The light slag and dross are aggregated in the centre of the trap. Reproduced with permission from Elsevier Science.

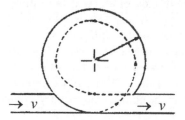

Figure 3.20 (b) The melt moves forwards after a cleaning turn in the swirl trap. Reproduced with permission from Elsevier Science.

Swirl trap technology can evidently *not* be used in cases such as aluminium and magnesium, where the oxides have a higher density than the metals themselves. In such cases ceramic filters can be used (page 43–44).

Runner Extension
The casting channel or *runner* often extends a little bit further than the last inlet (Figure 3.21). This is called a *runner extension* and has the purpose of separating the

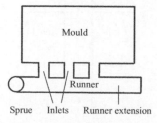

Figure 3.21 Gating system with extended runner.

first batch of melt that enters the gating system from the rest. The first batch is often the worst part of the metal melt, which contains and carries impurities floating on the surface of the pouring ladle. Additionally, it simultaneously diverts away the first cool metal melt. It helps to fill the mould only with melt that has flown through a preheated gating system.

The runner extension is designed in such a way that no circulation occurs within the extension, making sure that the first melt does not enter the mould.

Ceramic Filters

Ceramic filters are extensively used in the foundry industry to improve casting purity and reduce the cost of casting production. Ceramic filters are included in the gating system and remove slag inclusions, dross and other nonmetallic particles, which can seriously damage the physical properties and appearance of the casting. The particles often include:

- oxides, formed during melting, metal transfer and pouring;
- refractory particles from furnace and ladle;
- refractory particles present in the gating system;
- reaction products from metallurgical operations;
- dissoluble metallic and/or nonmetallic particles, added to the molten metal for microstructural modifications.

These particles or inclusions act as discontinuities in the metal matrix of a casting and have a variety of undesired effects, for example:

- reduction of tensile strength;
- problems during machining;
- decrease of surface finish;
- affects on subsequent surface treatments.

The conventional method is to design the gating system to separate the impurities from the metal by a swirl trap or by centrifugation, as has been discussed above. Ceramic filters are a good alternative to other separation methods and the only one that works in case of minor density differences between the impurities and the metal melt.

Advantages

Correctly installed ceramic filters are reliable devices for trapping undesired particles before they enter the casting cavity. They contribute to:

- reduction of inclusions in the melt;
- improvement in the ability of machining the casting;
- improvement in the physical properties of the casting;
- higher reliability of the casting process.

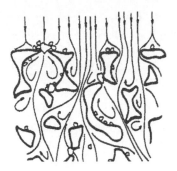

Figure 3.22 Sketch of the metal flow in a ceramic filter.

Filtering (Figure 3.22) is based on two mechanisms: (i) physical screening, and (ii) chemical attraction.

Properly installed ceramic filters do *not* restrict the metal flow significantly. Gating systems, specifically designed to include ceramic filters, are more effective than conventional separation systems.

Filter Types

Ceramic materials are the only filter materials that can stand the temperatures of molten metals. They are available in a wide variety of materials and in many different shapes (Figure 3.23). The open frontal areas of most ceramic filters cover 60–85 % of the cross-section area. The flow through a filter [Figure 3.23 (c) and (d)] is much less turbulent than through a strainer core [Figure 3.23 (a)].

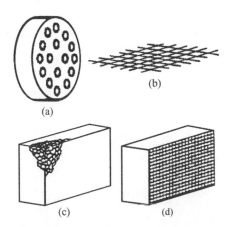

Figure 3.23 Filter types: (a) strainer core; (b) woven cloth; (c) ceramic foam; (d) extruded blocks. Reproduced with permission from Elsevier Science.

The structures of ceramic filters are of two kinds: (i) extruded forms with long, straight and parallel holes, and (ii) open cell foams.

The open cell foams have an average pore size of approximately 0.5–2 mm. They are made by impregnation of plastic foams with ceramic slurry, squeezing out the excess slurry and firing out the plastic in order to develop strength of the ceramic.

Installation of Ceramic Filters in Gating Systems
Design of optimal gating systems with included ceramic filters follows the principles listed below:

- the optimal position of the filter must be accomplished;
- mould-filling time must be constant and unaffected by the presence of the filter;
- the filter type must be optimal for the application;
- the gating system design must provide minimum metal turbulence through the filter and in the casting cavity;
- gating system size must be kept as small as possible.

If we want to keep a constant filling time both with and without a filter we have two options: increase the runner area or increase of the height of the sprue. The last option can be a bad solution because of increased wear and higher demands on the sand in the gating system.

The first option is favourable for removal of slag inclusions. The runner area is frequently enlarged by a factor of $\sqrt{2}$ when a filter is introduced in the gating system. An example of a proper arrangement is seen in Figure 3.24. If a coarse filter is used, the runner area must be enlarged by less than the factor of $\sqrt{2}$. A fine filter should be combined with a larger runner area, which is enlarged by more than the factor of $\sqrt{2}$.

Figure 3.24 Proper position and design of a ceramic filter in the runner.

Ceramic filters are usually not wetted by liquid metals. If too fine a filter is used, there may be problems for the first melt to pass, because of the surface tension between the melt and the nonwetting filter. Coarse filters are less effective for removal of small inclusions, but they reduce the amount of large oxide films.

3.7 MAXIMUM FLUIDITY LENGTH

3.7.1 Definition of Maximum Fluidity Length
During casting on an industrial scale it is of utmost importance that the process is designed in an optimal way. The melt must fill the mould rapidly and safely and, in addition, solidify as soon as possible.

As an aid to choosing an adequate temperature for the melt, knowledge of the ability of the melt to fill the mould completely and rapidly is useful. As a practical measure of this property the concept of *maximum fluidity length* has been introduced.

This concept is by no means well defined and the methods of measuring the maximum fluidity length are very much debated. The foundryman is most interested in the property, which, more correctly than maximum fluidity length, ought to be called *mould-filling capacity*.

In this book we have chosen to define maximum fluidity length in the following way:

Maximum fluidity length L_f is the length that a melt manages to flow in a tube or a channel with a given cross-section area before it solidifies.

The maximum fluidity length depends on a great number of material properties of the melt and to a great extent on its temperature. The maximum fluidity length depends mainly on the following properties of the melt:

- temperature;
- solidification mode;
- viscosity;
- composition;
- rate of flow;
- thermal conductivity;
- heat of fusion;
- surface tension;

It is also influenced by the solidification structure of the melt. The influence of these properties is discussed below.

3.7.2 Measurement of Maximum Fluidity Length

A device that is closely related to the definition given above and is used for careful measurement of the maximum fluidity length of a melt, is illustrated in Figure 3.25.

The bent end of a long straight quartz or metal tube is dipped down into the melt. The melt is sucked into the tube by connecting its other end to a vacuum pump. When the melt has solidified in the tube the maximum fluidity length L_f can be read directly.

To get a simple but rough measure of the maximum fluidity length it is common in practice to use a sand mould designed as a flat spiral. In order to control the 'pressure height' h and

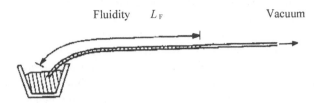

Fluidity L_F Vacuum

Figure 3.25 Equipment for measurement of maximum fluidity length. Reproduced with permission from Elsevier Science.

Pouring cup

h

Figure 3.26 Spiral for simple check of the maximum fluidity length.

consequently also the velocity v of the melt in the gate, a strainer core has been placed at the bottom of the pouring cup.

$$v = \text{const} \sqrt{h} \qquad (3.14)$$

where v = speed of the outflow melt at the entrance of the spiral, and h = height of the upper surface of the melt above the spiral level.

The melt flows into the spiral with the velocity v and stops due to solidification. The maximum fluidity length is the total length of the path from the orifice to the solidification stop point. The reading is facilitated by the nobbies, which are located at the back of the spiral at intervals of 50 mm (Figure 3.26). The uncertainty of the measurements originates mainly from the casting of the metal melt into the spiral.

3.7.3 Temperature Decrease in a Melt Flowing through a Channel

The maximum fluidity length is strongly influenced by the *temperature* of the incoming melt. The maximum fluidity length also depends strongly on the chilling capacity of the mould.

Thermal conduction and temperature decrease of metal melts at cooling and solidification are extensively treated in Chapter 4. Here we will merely give a short description of the temperature of a melt flowing through a casting channel. The temperature of the melt decreases approximately linearly with the distance from the entrance of the channel.

The cross-section of the casting channel is illustrated in the upper part of Figure 3.27. Its temperature is characterized by the temperatures listed below:

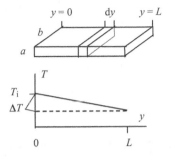

Figure 3.27 Illustration of the temperature decrease in a casting channel.

T_i = temperature of the melt at the entrance gate of the channel at $y = 0$;

T = temperature of the melt at the distance y from the entrance gate;

T_0 = temperature of the mould (not marked in the figure).

Example 3.5
Derive an expression for the temperature decrease in a casting channel according to Figure 3.27 above. The density of the melt ρ_L and its thermal capacitivity c_p^L are known. The heat transfer per unit time across the interface between the melt and the mould is described by a so-called heat transfer coefficient h:

$$\frac{dQ}{dt} = hA\,(T - T_0)$$

where A is the cross-section area.

Solution:
We consider a thin layer of melt at a distance y from the entrance gate and set up a heat balance for the volume element. It is reasonable to neglect the heat transport in the x- and z-directions and consider only the heat that is emitted in the y-direction perpendicular to the surface of the volume element.

The amount of heat emitted by the volume element when its temperature decreases by dT during the time dt is equal to the amount of heat absorbed by the mould. Using the basic formula $cm\,dt = hA(T - T_0)\,dt$ we get the heat balance:

$$c_p^L\,(\rho_L ab\,dy)\,dT = h\,[(2a + 2b)\,dy]\,(T - T_0)\,dt \qquad (1')$$

From Equation (1') we can solve the time derivative of the temperature:

$$\frac{dT}{dt} = \frac{2h(a + b)(T - T_0)}{\rho_L c_p^L ab} \qquad (2')$$

The temperature can be calculated by integration of Equation (2′):

$$\int_{T_i}^{T} \frac{dT}{(T - T_0)} = \frac{2h(a + b)}{\rho_L c_p^L ab} \int_0^t dt \qquad (3')$$

The time it takes for the thin layer to move from $y = 0$ to $y = L$ is:

$$t_L = \frac{L}{v} \qquad (4')$$

where v is the velocity of the melt.

Often the temperature curve can be approximated by a straight line, which has been done in Figure 3.27. In this case it is not necessary to integrate. We can replace dt by t_L and dT by ΔT, and get:

$$\Delta T = \frac{2h\,(a + b)\,(T - T_0)\,L}{\rho_L c_p^L ab\,v} \qquad (5')$$

Answer:
The temperature decrease in the casting channel is approximately determined by the expression:

$$\Delta T = \frac{2h\,(a + b)\,(T - T_0)\,L}{\rho_L c_p^L ab\,v}$$

The designations are given on page 45.

3.7.4 Structures in Metal Melts

The properties of a solid material and the forces that act between the atoms/ions in the crystal lattice are the bases of its macroscopic properties, such as, for example, thermal conductivity, heat of fusion, viscosity and surface tension.

In this section we will discuss the atomic structure of metal melts as a background to the material properties mentioned on page 44, which are important for maximum fluidity length.

The most important and most frequent method for examination of material structure is X-ray diffraction. Within practical metallurgy the Debye–Scherrer powder method is often used.

All crystalline materials, among them solid metals, have X-ray spectra consisting of a number of sharp, narrow, monochromatic lines that are characteristic and specific for the element in question. From wavelengths and intensities of these lines it is possible to calculate:

- the crystal structure of the material;
- the distances to neighbour atoms;
- the number of nearest neighbours, i.e. the coordination number of the element.

The crystal structures of solid metals have been the subject of extensive research. The structures of metal melts have also been examined with an unexpected result.

The common idea of metal melting is that the strong permanent bonds between the positive metal ions in the crystal lattice are completely broken and the ions are able to move freely relative to each other in the melt. Metal melts also have characteristic X-ray diffraction spectra. Instead of sharp lines several wide maxima are obtained from which conclusions can be drawn concerning the structure of the melt. These results modify the conception of the structures of metal melts.

Even in a melt there exists a certain *short distance order*. Each metal atom has a number of 'nearest neighbours', i.e. coordination numbers, in the same way as in a crystal lattice. It has also been possible to calculate the distances to the nearest neighbours in the melt. The metal melt can be described as being easily mobile ions in a plasma of electrons.

Values of nearest-neighbour distances and coordination numbers (number of nearest neighbours in the first 'shell') are given in Table 3.1 for some metals.

It can be seen from Table 3.1 that:

- metals that have densely packed structures (small distances to the nearest neighbour atom) and high coordination numbers change structure comparatively little at fusion;
- metals that have complicated structures and low coordination numbers change structure strongly at fusion. The coordination number increases strongly.

Metals with densely packed structures increase their volume by 3–5 % at fusion. On the other hand, metals with complicated crystal structures may even shrink at fusion. The melt obtains a structure and properties that resemble those of the common, densely packed metals.

3.7.5 Viscosity and Maximum Fluidity Length as Functions of Temperature and Composition

The maximum fluidity length of a metal melt is highly dependent on the viscosity of the melt or how easily it flows. Secondly, the viscosity of the melt is strongly dependent on its temperature and composition. Before discussing these matters we will define the concept of viscosity and derive a few relationships, which will be needed later.

Viscosity

Liquid Flow between two Parallel very Large Plates
The intermediate distance between two large parallel plates is filled with a liquid that flows in plane horizontal layers. The velocity is zero at the lowest plate and increases linearly as seen in Figure 3.28. The liquid particles in each layer move with the same velocity.

TABLE 3.1 Coordination numbers and interatomic distances of metals.

Element	Solid phase		Molten phase	
	Coordination number	Nearest neighbour distance (nm)	Coordination number	Nearest neighbour distance (nm)
Na	8	0.372	9.5	0.370
Mg	12	0.320	10.0	0.335
Al	12	0.286	10.6	0.296
Ge	4	0.245	8.0	0.270
Sn	4	0.303 0.318	8.5	0.327
	2	0.350	8.5	0.270
Pb	12	0.309	8.0	0.340
Bi	3.3	0.353	7.8	0.332
	3.3	0.356	7.8	0.332
Cu	12	0.289	11.5	0.257
Ag	12	0.286	10.0	0.286
Au	12		8.5	0.285

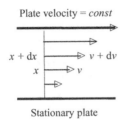

Figure 3.28 Laminar liquid flow between two large parallel plates, one stationary and the other one moving with a constant velocity.

The attractive forces between atoms in layers which move with different velocities cause friction forces between the layers. The friction force F is proportional to the contact area A and the velocity gradient dv/dx.

$$F = \eta A \, \frac{dv}{dx} \qquad (3.15)$$

This is the definition equation of viscosity. η is called the *dynamic viscosity constant*[1] (SI unit $= 1$ poise $= 0.1$ N s/m^2).

The larger the viscosity constant η is, the more slowly flowing is the liquid. The weaker the forces between the atoms in the liquid, the lower is the viscosity. Viscosity can be regarded as internal friction in the liquid.

Liquid Flow in a Tube – Calculation of the Liquid Flux
Not all the liquid that flows through a tube has the same velocity. In the periphery, where the liquid is in contact with the tube wall, the velocity is zero. The velocity has a maximum in the centre. We will determine the particle velocity as a function of the radius, the liquid flow through the tube and derive an expression for the mean flux per unit area.

[1] The kinematic viscosity coefficient is defined as η/ρ.

We will assume that the liquid flows through the tube without turbulence. This means that we have layers that slide in relation to each other just as above. In this case we choose a cylindrical shell as volume element. Within such a shell all particles move with the same velocity. As above, atoms between the shells attract each other and hence friction forces act along the envelope surface. Figure 3.29 shows the stationary flow pattern in a cross-section of the tube.

Figure 3.29 Stationary flow pattern in a cross-section of a tube.

Because the flow is stationary and not free from friction there must be a driving force to maintain it. Between the two ends of the tube there must exist a pressure difference that supplies the work necessary to compensate the friction losses (Figure 3.30). The pressure difference between the

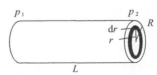

Figure 3.30 Pressures of the fluid at the two ends of a cylinder element.

two ends of the cylinder element is the source of the net force F_p in the direction of the liquid flow:

$$F_p = (p_1 - p_2)\, \pi r^2 \qquad (3.16)$$

where

p_1 = liquid pressure at the left end of the cylinder
p_2 = liquid pressure at the right end of the cylinder
r = radius of the cylinder element.

The retarding friction force F_η, caused by the viscosity, acts along the envelope surface of the cylinder element:

$$F_\eta = -\eta \frac{dv}{dr}\, 2\pi r L \qquad (3.17)$$

The vector sum of these two forces must be zero because the flow is stationary. Their sizes are equal and the directions are opposite and we get the following:

$$(p_1 - p_2)\, \pi r^2 = -\eta\, 2\pi r L \frac{dv}{dr}$$

which can be reduced to:

$$\frac{dv}{dr} = -\frac{\Delta p\, r}{2\eta L} \qquad \text{or} \qquad \int_0^v dv = -\frac{\Delta p}{2\eta L} \int_R^r r\, dr$$

where R is the radius of the tube. We get:

$$v = \frac{\Delta p}{4\eta L}\, (R^2 - r^2) \qquad (3.18)$$

The volume dV, which flows through the cylinder element during the time, dt is:

$$dV = 2\pi\, dr\, v dt$$

Summation of all the cylinder elements gives the total flux, i.e. the liquid volume, that passes each cross-section per unit time:

$$\frac{dV}{dt} = \int_0^R 2\pi\, vr dr = \frac{2\pi\, \Delta p}{4\eta L} \int_0^R r\, (R^2 - r^2) dr$$

which gives the total flux:

$$\frac{dV}{dt} = \frac{\pi R^4 \Delta p}{8\eta L} \qquad \text{Hagen–Poiseuille law} \qquad (3.19)$$

Starting with the Hagen–Poiseuille law we get the average flux per unit area dividing by πR^2. The average flux is:

$$\left(\frac{1}{\pi R^2}\right)\left(\frac{dV}{dt}\right) = \frac{R^2\, \Delta p}{8\eta L} \qquad (3.20)$$

Viscosity of Metal Melts as a Function of Temperature

The viscosity, and thus also the maximum fluidity length, is strongly dependent on the temperature of a metal melt. For pure metals the viscosity constant η_M is a function of the absolute melting point temperature T_M, the atomic weight M and the molar volume V_m (m^3/kmole). If all the quantities are expressed in SI units the expression obtained is:

$$\eta_M = 0.612\, \sqrt{(T_M M)}\, V_m^{-2/3} \qquad (3.21)$$

The agreement between experimental and theoretical values is good as can be seen from Table 3.2.

TABLE 3.2 Comparison between experimental and calculated values of the viscosity coefficient.

Melt	$\eta_{exp}(\times 10^{-3})$	η_{theor} (N s/m^2)
Li	0.60	0.56
Na	0.69	0.62
Sn	2.1	2.1
Cu	4.1	4.2
Ag	3.9	3.9
Fe	5.0	4.9
Ni	4.6	5.0

If η is calculated from viscosity measurements at constant pressure and various values of temperature T, and $\ln \eta$ is plotted against $1/T$ for various metals, a straight line is obtained for each metal. Mathematically the relation between η and T can be written as:

$$\eta = \eta_0 \exp\left(-\frac{Q_\eta}{RT}\right) \qquad (3.22)$$

where Q_η is a constant, characteristic for each metal melt (examples are given in Table 3.3). The higher the temperature

TABLE 3.3 Molar energies Q_η for some common metals.

Metal	Q_η
Hg	5.27×10^7
Na	1.02×10^7
Sn	1.22×10^7
Ag	3.14×10^7
Fe + 2.5 % Cu	7.12×10^7

is, the lower is the viscosity constant. The more easily a melt flows, the longer will be its maximum fluidity length.

Viscosity and Maximum Fluidity Length for Binary Alloys as Functions of their Compositions

The easier a melt flows, the lower is its viscosity and the longer is its maximum fluidity length. Alloys that form intermediate phases (chemical compounds) often show viscosity maxima at compositions that correspond to the constant proportions of the chemical compounds. This can be explained by the fact that the interatomic forces are especially strong in these cases. As an example the system Mg–Sn can be mentioned (Figure 3.31). It has a viscosity maximum at the composition Mg_2Sn in spite of the high temperatures. At the eutectic point (91 at-% Sn) the system has a viscosity minimum.

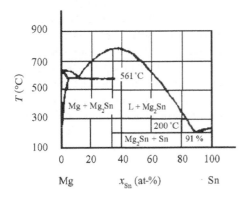

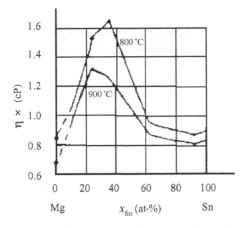

Figure 3.31 Phase diagram and viscosity isotherms for the system Mg–Sn at various temperatures. Viscosity maximum occurs at the composition Mg_2Sn. Viscosity minimum occurs at the eutectic composition 91 at-% Sn. Reproduced with permission from Elsevier Science.

Figure 3.32 shows the maximum fluidity length for Pb–Sn alloys at various temperatures as a function of the Sn content. It can be seen from Figure 3.32(b) that the maximum fluidity length has a sharp maximum at the eutectic composition. Measurements of viscosity and maximum fluidity length

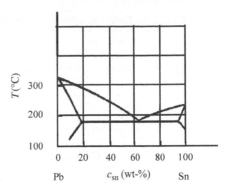

Figure 3.32 (a) Phase diagram for the system Pb–Sn. A comparison between Figures 3.32 (a) and (b) shows that the eutectic composition of the alloy corresponds to the maximum fluidity length. Reproduced with permission from Merton C. Flemings.

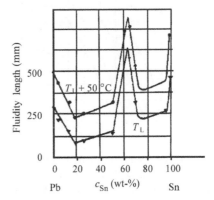

Figure 3.32 (b) Maximum fluidity length isotherms for the system Pb–Sn at liquidus temperature and liquidus temperature +50 °C. Reproduced with permission from Merton C. Flemings.

for all binary alloys always show *low* viscosity and *high maximum fluidity length* for *eutectic* alloys.

Characteristic for all eutectic alloys is that they, like pure metals, have well-defined fusion temperatures that are equal to the solidification temperature.

For other metal mixtures the solid phase and the melt have different compositions. This causes a gradual change of the melting point during the solidification process. A comparison between maximum fluidity length data shows that the *larger* the liquidus–solidus interval is, the *higher* will the viscosity be and the *lower* the maximum fluidity length.

Maximum Fluidity Length of Iron and Steel Alloys as a Function of Carbon Content

Additives of alloying elements initially decrease the maximum fluidity length of a metal melt down to a minimum and then increase it again. Important examples of this are various iron and steel melts with different carbon content. Figure 3.33 shows some typical maximum fluidity length curves for four different iron and steel alloys. Their data are presented in Table 3.4.

TABLE 3.4   Data for some iron and steel alloy melts.

Curve number	Pure iron with additive of		Casting temperature (°C)	Temperature difference above 1299 °C (°C)	Liquidus temperature (°C)	Liquidus temperature difference above 1177 °C (°C)
	% C	% Si				
1	3.6	2.08	1299	0	1177	0
2	3.04	2.10	1342	43	1226	49
3	2.52	2.00	1415	116	1276	99
4	2.13	2.07	1444	145	1312	135

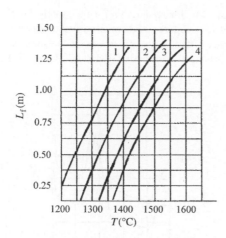

Figure 3.33   Maximum fluidity length as a function of temperature for the four iron and steel alloys in Table 3.4. Reproduced with permission from Addison-Wesley Publishing Co. Inc.

The maximum fluidity lengths of the melts 1–4 have been plotted as a function of the temperature in Figure 3.33. For all of them it has been found that:

- the maximum fluidity length increases strongly and practically linearly with the temperature;
- at a given temperature the cast iron melts 1 and 2 have larger maximum fluidity lengths than the steel melts 3 and 4;
- extrapolation of the curves down to the *T*-axis (maximum fluidity length equal to zero) gives intersection points between the curve and the *T*-axis equal to the liquidus temperature for each curve.

By choosing a suitable temperature of the melt it is possible to obtain a required maximum fluidity length for each alloy.

3.7.6   Models of Maximum Fluidity Length – Critical Length

Models for Maximum Fluidity Length
Experimental results show that there must be pronounced differences in the mechanisms of solidification between on the one hand pure metals and eutectic mixtures with small solidification intervals, and on the other hand alloys with relatively wide solidification intervals (Chapters 4 and 6).

When a melt with a given temperature is poured into a fluidity length channel the walls of the channel will cool it. The heat is conducted away radially and the solidification starts at the walls of the channel close to the gate when the temperature has decreased to the melting point temperature. The solidification process then continues controlled by the heat balance – the solidification heat must be conducted away – while the melt advances. Columnar crystals (Chapter 6) are formed at the channel walls and grow towards the middle. The melt moves forward through a more and more narrow channel in the middle (Figure 3.34). The maximum fluidity length is determined by the final total stop in the channel when the melt solidifies to 100 %.

Model I

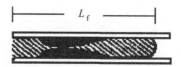

Figure 3.34   Model I – Solidification process for pure metals and eutectic mixtures. Reproduced with permission from Elsevier Science.

The solidification process according to model I (Figure 3.34) is valid for pure metals and eutectic mixtures of metals. If the mean velocity of the melt is designated by v and the solidification time by t_f we have approximately:

$$L_f = vt_f \qquad (3.23)$$

Model II
The solidification process in alloys with wide liquidus–solidus intervals (Chapters 4 and 6) is quite different from that in pure metals. When the temperature of the melt decreases the solidification process starts with dendrite formation all over the melt. Small dendrites whirl round in the melt and make it flow more and more slowly.

The surface tension in the front surface of the melt also contributes in preventing the further advance of the

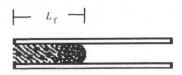

Figure 3.35 Model II – Solidification process for alloys with wide liquidus–solidus intervals. Reproduced with permission from Elsevier Science.

melt (Figure 3.35). It has been found by experiment that when about 50 % of the melt consists of dendrites the melt ceases to move forward. This occurs at half the solidification time.

If the mean velocity of the melt is v we get in this case:

$$L_f = \frac{vt_f}{2} \tag{3.24}$$

Critical Length

The solidification process of a pure metal in a fluidity length channel is illustrated in Figure 3.36. If the length of the channel is less than a certain critical length the melt will dissolve the solid phase and continue to flow freely. The process is described in Figure 3.37.

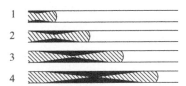

Figure 3.36 Solidification process of a pure metal or a eutectic mixture. Reproduced with permission from Elsevier Science.

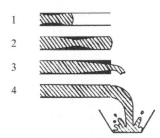

Figure 3.37 Flow of a pure metal when the length of the channel is shorter than the critical length. Reproduced with permission from Elsevier Science.

For alloys with a wide liquidus–solidus interval the critical length is only slightly shorter than the maximum fluidity length at the same temperature. The reason is that the solidification process differs from that of a pure metal.

For channels shorter than the critical length the dendrite precipitation never reaches the 50 % that would cause a

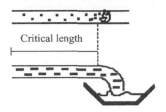

Figure 3.38 Critical fluidity length for an alloy with a wide solidification interval.

total stop in the channel. The melt continues to flow freely. The process is described in Figure 3.38.

3.7.7 Surface Tension and Maximum Fluidity Length

The atoms (ions) in the surface of a liquid are slightly displaced inwards, due to attraction forces from atoms (ions) inside the liquid, until equilibrium is achieved. The surface can be regarded as a film, where the atoms (ions) are packed somewhat more densely than inside the liquid. Addition of energy is required to increase the surface:

Surface tension = surface energy per unit area.

Alternatively surface tension can be regarded as force per unit length:

Surface tension = force per unit length.

If a fictive cut is made in the liquid surface it is necessary to apply a force per unit length, equal to the surface tension σ, to each side of the cut to keep it together (Figure 3.39). Both these models of surface tension are very useful. In the present case with a melt flowing in a tube, either of them can be used.

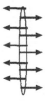

Figure 3.39 Applied forces per unit length necessary to keep the two sides of the surface together after a fictive straight cut. The forces match the surface tension.

We can concentrate on the view that the surface struggles to remain as small as possible and resist an enlargement. Alternatively we can consider the crown of surface tension forces along the periphery of the tube (see figure in Example 3.6). They prevent the advance of the melt. The surface tension represents an obstacle for the metal melt in the fluidity length channel. Its retarding influence is described in Example 3.6 on the next page.

Example 3.6

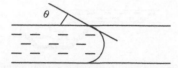

A steel melt flows through a closed circular channel. Calculate the retarding force due to the surface tension as a function of the surface tension σ, the radius r of the channel and the angle θ between the melt and the channel.

Solution:

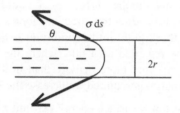

The surface tension forces act along the circular contact line between the free surface of the melt and the channel. They are situated in tangent planes of the surface and have various directions as the figure shows. The forces form a crown on a conical surface, partly seen in the figure. The force on the length element ds is σ ds.

We want the resultant to all these forces. Because they have different directions it is necessary to use vector addition. The simplest way is to add the components of the forces in the direction of the channel. The resultant of the radial forces is zero for symmetry reasons.

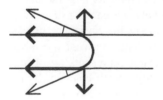

The horizontal component of the force on the length element ds is:

$$dF_\sigma = \sigma \cos\theta \, ds$$

The vector sum of all the surface tension forces can be written as a line integral:

$$F_\sigma = \oint ds = \sigma \, 2\pi \, r \cos\theta$$

Answer:
The retarding force is equal to $2\pi r\sigma \cos\theta$. If the angle θ is small the force will be $2\pi r\sigma$.

3.7.8 Maximum Fluidity Length and Casting

Component casting is a competitive and rapid production method. Examples of products are propeller blades for ships, engine blocks in cars, and numerous smaller or larger articles in general use such as frying pans and sewing machines.

When a casting is to be series produced, the challenge for the material engineer is to design the process in an optimal way. On the basis of the demands of the customers, the range of applications of the product and economical restrictions, he or she has to choose the composition of the material, the temperature of the melt, mould material and a number of other factors. It will always be a matter of compromise between various advantages and disadvantages.

At his/her service he/she has the accumulated knowledge in specialist literature in the form of data concerning heat properties, density, viscosity, surface tension and mechanical and chemical properties of different materials, together with maximum fluidity length curves and phase diagrams.

Choice of Material
The dominant use of cast iron instead of steel is due to the fact that cast iron has a composition close to a eutectic mixture of pure iron and carbon. Consequently it has a much larger maximum fluidity length than does steel at the same temperature (Figure 3.40). Besides, a much lower casting temperature can be used for cast iron than for steel, as can be concluded from Figure 3.41 on page 53 and Figure 3.33 on page 50.

Al–Cu alloys have excellent properties, where mechanical strength is concerned, but have very short maximum fluidity length and are therefore difficult to cast. Al–Si alloys at and close to the eutectic point are easy to cast but have poor mechanical strength. Addition of magnesium or copper increases the strength of the material but decreases its maximum fluidity length.

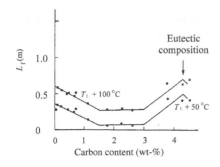

Figure 3.40 Maximum fluidity length as a function of the carbon concentration for some Fe–C alloys, among them those in Figure 3.41. The excess temperatures of the curves are 100 °C and 50 °C, respectively. Reproduced with permission from Elsevier Science.

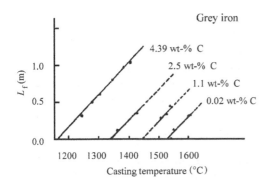

Figure 3.41 Maximum fluidity length as a function of the casting temperature for some different Fe–C alloys. To achieve a certain maximum fluidity length a much higher temperature is required for low carbon alloys (steel) than for high carbon alloys (cast iron). Reproduced with permission from Elsevier Science.

Choice of Casting Temperature

The choice of temperature of the metal melt or rather the temperature above the liquidus temperature is of the utmost importance for the quality of the product. If the temperature is too *low* there is a risk that the mould will not be filled completely and there may be joints and seams in the complete product. If the temperature is too *high*, the walls of the mould will be strongly chemically affected. The reaction products – solids, liquids or gases – may be included in the casting and its surface may be rough and uneven.

It is necessary to specify a temperature interval for the casting temperature to be maintained at during the whole casting process. The narrower this interval is, the better will the casting result be regarding surface finish and homogeneity.

For casting of thin and small products, the casting temperatures are usually somewhere between 175 °C and 275 °C above the liquidus temperature. For bigger and heavier products, such as, for example, machine parts, a lower temperature is better, usually somewhere between 50 °C and 175 °C above the liquidus temperature.

There are possibilities for maintaining a large maximum fluidity length during the casting process other than increasing the temperature of the melt. By reducing the thermal conductivity, for example by the aid of a thin film of soot in the sand mould or a ceramic covering, the melt will cool more slowly. In this way the maximum fluidity length may be increased by a factor of 2 or 3.

SUMMARY

Basic Hydrodynamics
Principle of continuity: $\qquad A_1 v_1 = A_2 v_2$

Bernoulli's Law at Laminar Flow
For laminar flow in metal melts Bernoulli's law is valid if friction losses are neglected:

$$p_1 + \rho g h_1 + \left(\frac{\rho v_1^2}{2}\right) = p_2 + \rho g h_2 + \left(\frac{\rho v_2^2}{2}\right) = \text{const}$$

Special case: The exhaust velocity of the melt is determined by the height of the free surface over the bottom hole of the ladle:

$$v = \sqrt{2gh}$$

Gating Systems in Component Casting

Downhill Casting
Casting Rate
Casting rates can be calculated with the aid of basic hydrodynamic laws. The result depends on the type and design of the gating system.

Penetrable mould	Impenetrable mould
$v_2 = \sqrt{2gh_c}$	$p_2 = p_3 - \rho g h_2$
$v_3 = \sqrt{2gh_{\text{total}}}$	$v_3 = \sqrt{2gh_{\text{total}}}$
$\dfrac{A_3}{A_2} = \dfrac{v_2}{v_3} = \sqrt{\dfrac{h_c}{h_{\text{total}}}}$	

A single gate is enough for small castings. For large castings several gates are required.

Depending of the position of the gates the casting methods are characterized as uphill casting, downhill casting, and casting with lateral sprue. In the latter case special measures have to be taken to achieve an even pressure at all gates.

Casting time
In simple cases the time to empty the ladle and the casting time can be calculated by application of the hydrodynamic

laws in each special case. An empirical relationship between casting time and mass at component or ingot casting is:

$$t = Am_{casting}^n$$

Uphill Casting
Penetrable mould:

$$v = \sqrt{2g(h_{total} - h)}$$

$$t_{fill} = \left(\frac{2A_{mould}}{A_{sprue}\sqrt{2g}}\right)\left(\sqrt{h_{total}} - \sqrt{h_{total} - h_{mould}}\right)$$

Gating System in Ingot Casting
The principles, formulas, position of gating systems and inclusion control are the same for component casting and ingot casting, with reservation for modifications, due to differences in size.

$$\text{Casting time: } t = Am_{casting}^n$$

Gating System in Continuous Casting
A tundish is always used between the ladle and the chill-mould to achieve a constant pressure of the melt at the entrance to the chill-mould. A straight vertical sprue, a so-called casting tube or submerged entry nozzle between the tundish and the chill-mould is used at continuous casting. The casting tube reduces casting defects such as macroslag inclusions.

Inclusion Control in Gating Systems
Three mechanical methods are used: (i) swirl trap; (ii) extended runner, and (iii) ceramic filters.

Viscosity
Definition equation: $\quad F = \eta A \dfrac{dv}{dx}$

Hagen–Poiseuille law:

$$\frac{dQ}{dt} = \frac{\pi R^4 \, \Delta p}{8\eta L} \quad (\text{m}^3/\text{s})$$

Average flux per unit area:

$$\left(\frac{1}{\pi R^2}\right)\left(\frac{dV}{dt}\right) = \frac{R^2 \, \Delta p}{8\eta L} \quad (\text{m/s})$$

Viscosity depends strongly on *temperature*, and low viscosity corresponds to high temperature:

$$\eta = \eta_0 \exp\left(-\frac{Q_\eta}{RT}\right)$$

Viscosity depends strongly on the *composition* of the melt.

Alloys, which form intermediate phases (chemical compounds), have viscosity maxima at compositions which correspond to these. Alloys have viscosity minima at the eutectic point.

Maximum Fluidity Length of a Metal Melt
Maximum fluidity length L_f is the distance that a melt manages to flow in a tube or a band with a given cross-section area before it solidifies.

Influence of External Factors on the Maximum Fluidity Length:
The maximum fluidity length depends mainly on the *viscosity* of the melt and indirectly of its temperature, structure, composition, flow velocity, thermal conductivity, heat of fusion, and surface tension. A large maximum fluidity length corresponds to low viscosity.

Models for Maximum Fluidity Length
See Figures 3.34 and 3.35.

Model I	Model II
$L_f = v\, t_f$	$L_f = \dfrac{v\, t_f}{2}$
Pure metals and eutectic mixtures	Alloys with wide liquidus–solidus intervals.

If the fluidity channel is shorter than a certain critical length the melt will never solidify in the channel. For pure metals and eutectic mixtures the critical length is much shorter than the maximum fluidity length. For alloys with wide liquidus–solidus intervals the critical length is only slightly shorter than the maximum fluidity length.

EXERCISES

3.1 In a foundry you will start to cast small components, which require high accuracy and good surface finish. Suggest a suitable casting method and discuss its advantages and disadvantages.

Hint A1

3.2 At casting of cast iron in a sand mould it is known that the casting time (in seconds) follows the relation:

$$t = 3.4 \, (m_{casting})^{0.42}$$

where $m_{casting}$ is the mass of the casting (in kilograms).

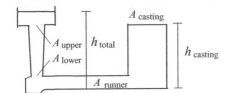

Dimension a sprue (diameters of the upper and lower circular cross-section areas) that will suit a cylindrical casting with a diameter 10 cm and height 23 cm. The total height h_{total} is 28 cm. Uphill casting is used.

Hint A10

3.3 You work in a cast house of a steel company where you intend to start uphill casting of steel. The head of production asks you how long it will take to fill six moulds that are connected to a central gating system (one of them is seen in the figure below). Calculate the filling time.

Hint A35

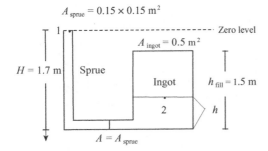

3.4 A steelworks has a continuous casting machine, capable of casting six strands simultaneously. One of them is marked in the figure. The billet castings have the dimensions 140 mm × 140 mm. The machine is equipped with a large tundish, from which the outlet velocity cannot be controlled. The diameter of each of the six exit holes of the tundish is 10 mm. Calculate the casting rate as a function of the height of the melt in the tundish.

Hint A20

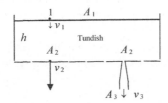

3.5 The illustration shows the principle of a set-up for continuous casting of steel. In a steel plant a machine with two lines is used, for castings with a cross-section of 20 cm × 150 cm. The steel plant has two electric furnaces, each with a melting capacity of 70×10^3 kg/h. The height of the steel bath in the tundish is 50 cm. The density of the steel melt is 7.2×10^3 kg/m^3.

Calculate the casting rate of the machine and the outlet diameter of the tundish.

Hint A53

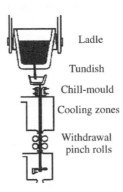

3.6 During ESR remelting of an ingot, a certain amount of sulfur was added to the molten zone in the form of FeS during the remelting process. The figure shows the sulfur distribution after addition of FeS. The decrease in sulfur concentration with time depends on:

(i) slag formation due to the chemical reaction between sulfur and the metal;
(ii) dilution due to drops from the electrode.

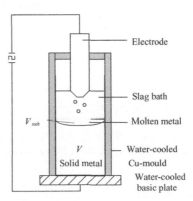

Calculate the distance $y_{1/2}$ where the S concentration amounts to half the initial value after addition of FeS. The slag formation can be neglected.

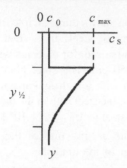

The volume of the molten metal $V_{melt} = 130$ cm^3.
The S concentration of the ingot before addition of FeS $c_0 = 0.030$ wt-%.
The S concentration in the melt immediately after addition of FeS $c_{max} = 0.37$ wt-%.
Diameter of the ingot $D = 100$ mm.

Hint A50

3.7 The maximum fluidity length of metal melts during casting depends on a number of different factors.

(a) Describe how the cooling capacity of the mould, the surface tension, the composition of the alloy and the viscosity of the melt influence the maximum fluidity length.

Hint A212

(b) Are there any other factors that influence the maximum fluidity length? If so, which ones and how?

Hint A300

3.8 The excess temperature, i.e. the temperature above the liquidus temperature T_L, has a strong influence on the maximum fluidity length of a metal melt. The maximum fluidity length can be measured by sucking melt into a quartz tube.

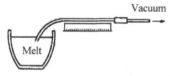

Derive an expression for the excess temperature $T - T_L$ of the melt as a function of time for a steel melt with initial temperature T_i. The melt is sucked into a quartz tube with an inner radius R. All heat to the surroundings is assumed to be emitted by radiation.

Hint A133

3.9 A factor of great importance for the fitness of an intended casting alloy is its maximum fluidity length when it is molten. The concept of maximum fluidity length may not be uniquely defined but the practical founder is mostly interested in information about the form-filling ability of the intended casting metal. The test method, described in the figure below, suits this purpose.

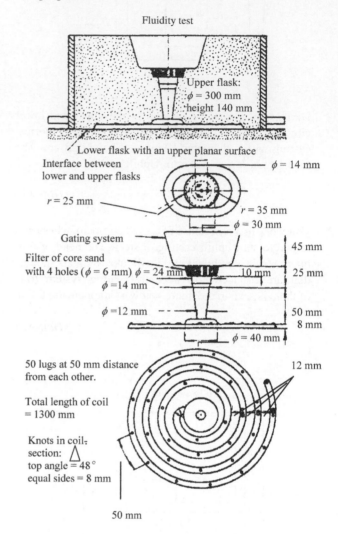

Fluidity test

During the casting process the metal solidifies in the channel shown in the lowest figure. When the inlet of the channel has solidified completely the supply of metal ceases and the spiral length, filled with metal, is a measure of the maximum fluidity length. The cross-section of the spiral is a triangle with two equal sides and a top angle = 48°. The equal sides = 8 8 mm.

Assume that the heat transport occurs entirely by means of heat transfer across the interface between the melt and the coil (see Example 3.5 on page 45). For your calculations you may assume that the velocity of the liquid metal is constant until the inlet has frozen.

(a) Derive the maximum fluidity length of the metal as a function of the initial excess temperature of the melt.

Hint A23

(b) Calculate the maximum fluidity length of aluminium if the initial excess temperature is 30 °C and $h = 300$ W/m^2 K. The temperature of the surroundings is 20 °C. Material data for aluminium are found in standard tables.

Hint A85

3.10 The surface tension between air and metal varies strongly for different metals. For aluminium alloys it is approximately 1.5 J/m^2 and for cast iron it is about 0.50 J/m^2. The high surface tension of aluminium alloys restricts the general possibilities of creating sharp corners.

Regard the metal surface in a corner of a box with a rectangular bottom area as part of a sphere, which touches the perpendicular corner planes at three tangent points. The corner radius R is defined as the radius of curvature of the 'spherical' melt surface in the corner and of the solidified product.

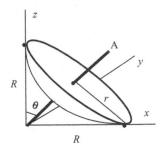

(a) Estimate the corner radius of aluminium as a function of h, the distance from the spherical melt surface to the upper free surface of the melt (not seen in the figure).

Hint A68

(b) Compare the corner radius of aluminium with the corresponding radius of cast iron.

Hint A233

4 Heat Transport during Component Casting

4.1 INTRODUCTION

Casting of metals is closely related to heat release and heat transport during solidification and cooling. The rate of heat removal is very important as it determines the solidification time of the casting and the temperature distribution in the material. These control directly or indirectly the structure of the material (Chapter 6), precipitation of pores and slag inclusions (Chapter 9) and distribution and shape of shrinkage pores (Chapter 10), and hence the qualities and properties of castings.

Materials Processing during Casting H. Fredriksson and U. Åkerlind
Copyright © 2006 John Wiley & Sons, Ltd.

The field of heat transport in casting is very extensive and very important. For these reasons their treatment has been divided into two chapters. The present Chapter 4 contains the general theory of heat transport in casting and applications on component casting. In Chapter 5 heat transport in cast house processes in metal factories is discussed.

Genuine knowledge of the laws of heat transport and their applications within the field of casting is thus of utmost importance for engineers in the process and manufacture industry. This knowledge is the basis for the possibilities of modelling casting methods and designing a casting process for a given purpose. Nowadays there are computer simulation programs as an aid in this work. These modern aids presuppose that the user has a broad knowledge of the laws of heat transport.

Heat transport and solidification processes in various casting processes are often very complicated. The complexity of the problem is illustrated by Figure 4.1.

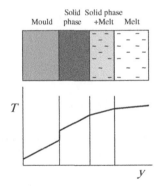

Figure 4.1 Temperature distribution in a cast, solidifying metal melt.

The number of variables is considerable. In order to facilitate calculations of, firstly, the position of the solidification front as a function of time and, secondly, the temperature distribution and the temperature gradient as functions of position, it is necessary to consider the heat transport through four layers: melt, two-phase region of solid + melt, solid, and mould (Figure 4.1).

It is most difficult to treat the heat transport through the two-phase layer as little or nothing is normally known about the state of solidification there. In addition, it is necessary to consider the heat transfer between the mould and the solid casting, and the heat emission at the outer surface of the mould.

The problem is physically and mathematically very complex. In many practical cases it is difficult to find exact solutions. Computer calculations may give acceptable approximate solutions. Luckily it is possible to make simplified but reasonable assumptions in some cases and in this way find relatively good approximate analytical solutions. These simplifications usually imply that the slowest step in a chain of subprocesses, coupled in series, determines the total rate of heat transfer. Efforts are made to identify this step in each case.

In this chapter the fundamental laws of heat transport are extensively discussed and then applied to the various types of component casting that are included in Chapter 1. Some of the simplifications, mentioned above, are treated and approximate solutions are presented.

4.2 BASIC CONCEPTS AND LAWS OF HEAT TRANSPORT

There are three kinds of heat transport: *conduction*, *radiation* and *convection*.

In the case of solidifying metal melts thermal conduction will be the most important way of heat transport. To get the 'tools' required for treatment of heat transport in casting and solidification of metals and alloys, we will give a short survey of the general laws that are valid, and mention when they are applicable.

4.2.1 Thermal Conduction

Basic Law of Thermal Conduction under Stationary Conditions

The amount of heat ΔQ that passes the cross-section area A of a bar during the time Δt, is proportional to the area A and the temperature difference per unit length (Figure 4.2). The basic law of thermal conduction at stationary conditions can therefore be written:

$$\frac{\Delta Q}{\Delta t} = -kA\frac{T_{\mathrm{L}} - T_0}{L} \qquad (4.1)$$

where

$Q =$ amount of heat
$t \ =$ time
$k \ =$ thermal conductivity
$A =$ cross-section area of bar
$L =$ length of bar
$T_{\mathrm{L}} - T_0 =$ temperature difference between the ends of the bar.

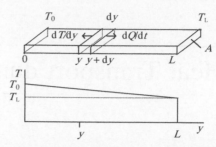

Figure 4.2 Thermal conduction in a metal bar. The temperature is a function of position but not of time (stationary conditions).

Formerly the *thermal conductivity k* used to be called the coefficient of thermal conduction or thermal conduction coefficient. If we introduce the temperature gradient instead of the temperature per unit length the equation takes a more general form:

$$\frac{dQ}{dt} = -kA\frac{dT}{dy} \qquad (4.2)$$

The SI unit of k is J/m s K or W/m K.

Equations (4.1) and (4.2) are valid for a stationary temperature distribution, i.e. T is a function of position but not of time.

Temperature Gradient

The gradient of any *scalar* quantity, here the temperature T, is a *vector*, defined by the following relationship:

$$\mathrm{grad}\ T = \frac{\partial T}{\partial x}\hat{x} + \frac{\partial T}{\partial y}\hat{y} + \frac{\partial T}{\partial z}\hat{z} \qquad (1')$$

where T in the general case is a function of x, y and z.

The gradient is in the direction of increasing temperature and perpendicular to surfaces of equal temperatures. Its magnitude is:

$$|\mathrm{grad}\ T| = \sqrt{\left(\frac{\partial T}{\partial x}\right)^2 + \left(\frac{\partial T}{\partial y}\right)^2 + \left(\frac{\partial T}{\partial z}\right)^2} \qquad (2')$$

For the special case of the temperature gradient being directed in the positive y-direction the temperature gradient will simply be:

$$\mathrm{grad}\ T = \frac{\partial T}{\partial y}\hat{y} \qquad \text{and} \qquad |\mathrm{grad}\ T| = \frac{\partial T}{\partial y} \qquad (3')$$

The coordinate system has been chosen in such a way that the linear temperature decreases when y increases (Figure 4.3). The temperature gradient is equal to the slope of the line. In this case it is negative, i.e. directed towards the *negative* y-axis. The heat always flows from a higher to a lower temperature, namely in the *positive* y-axis direction. For this reason we must have a minus sign in Equations (4.1) and (4.2) above.

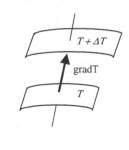

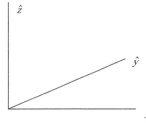

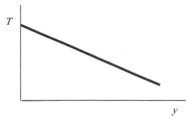

Figure 4.3 Illustration of a temperature gradient in a given coordinate system.

Equation (4.2) can be written in a more general form if we introduce the amount of heat per unit area q instead of Q in Equation (4.2). The heat flux (amount of heat per unit area and unit time) can be written as:

$$\frac{\mathrm{d}q}{\mathrm{d}t} = -k\left(\frac{\mathrm{d}T}{\mathrm{d}y}\right) \quad \text{Fourier's first law} \qquad (4.3)$$

where q = amount of heat per unit area = Q/A.

Table 4.1 gives the k values for some different elements.

TABLE 4.1 Thermal conductivity of some elements and alloys.

Material at 20 °C	k (W/m K)
Cu	390
Cu (s, 1083 °C)	334
Al	237
Al (s, 600 °C)	218
Fe	83
Fe (s, 1535 °C)	31
Fe 0.85%C	45
Fe 18%Cr 8%Ni	~15
Water	0.60
Air (0 °C)	0.024

Heat Transfer across the Interface between Two Materials under Stationary Conditions

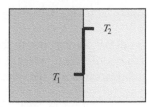

Figure 4.4 Temperature drop across the interface between two materials with different temperatures.

By experience it is known that there is a temperature difference across the interface between two different materials. The larger this temperature difference is the more heat is transferred per unit time from the warmer to the colder side.

$$\frac{\mathrm{d}Q}{\mathrm{d}t} = -h\, A\,(T_2 - T_1) \qquad (4.4)$$

where h = heat transfer constant or heat transfer coefficient or heat transfer number.

The SI unit of h is $\mathrm{J/m^2 s\ K}$ or $\mathrm{W/m^2\ K}$. If we introduce the heat transfer per unit area and replace Q/A with q, we get the equation

$$\frac{\mathrm{d}q}{\mathrm{d}t} = -h(T_2 - T_1) \qquad (4.5)$$

Poor contact between the materials will result in a thin layer of air between them. Air is a poor thermal conductor and the heat transfer will decrease considerably when an air layer appears.

We apply Fourier's law [Equation (4.3)] and the heat transfer Equation (4.5) on the thin layer of air. Identification gives the relation:

$$h = \frac{k}{\delta} \qquad (4.6)$$

where
h = heat transfer coefficient of air
k = thermal conductivity of air
δ = thickness of air layer.

Thermal Conduction through Several Layers and across Several Interfaces, Coupled in Series, under Stationary Conditions

The heat flow through a thermal-conducting layer or across an interface can be compared to an electrical current through a resistor.

$$\frac{\mathrm{d}Q}{\mathrm{d}t} = -\frac{kA}{L}(T_L - T_0) \qquad \text{or} \qquad \frac{\mathrm{d}Q}{\mathrm{d}t} = -hA(T_2 - T_1)$$

is compared to:

$$I = \frac{U}{R}$$

The thermal current dQ/dt corresponds to the electrical current I. The temperature difference, which is the cause of the thermal current, corresponds to the electrical potential difference U.

The 'thermal resistance' L/kA or $1/hA$ corresponds to the electrical resistance R. If several layers are in contact with

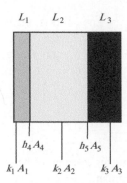

Figure 4.5 Thermal-conducting layers and interfaces, coupled in series. For simplicity the cross-section areas are drawn equal in the figure.

each other (Figure 4.5) they can be considered as coupled in series and the total 'thermal resistance' will be:

$$\frac{L}{kA} = \frac{L_1}{k_1 A_1} + \frac{L_2}{k_2 A_2} + \frac{L_3}{k_3 A_3} + \cdots + \frac{1}{h_4 A_4} + \frac{1}{h_5 A_5} + \cdots \quad (4.7)$$

General Law of Thermal Conduction – Nonstationary Conditions

If the thermal conduction is *transient*, i.e. the temperature is a function of both position and time, the basic Equation (4.3) of thermal conduction is no longer valid and has to be replaced by a more general one.

Consider the volume element in Figure 4.6. The heat amount dQ_y flows into the element across the surface A_y during the time dt. Simultaneously the amount of heat dQ_{y+dy} leaves the element across the surface A_{y+dy}. The difference stays within the volume element and causes its temperature to increase by the amount dT. The law of energy conservation gives:

$$dQ_y - dQ_{y+dy} = -\left[k\rho A_y \left(\frac{\partial T}{\partial y}\right)_y - k\rho A_{y+dy} \left(\frac{\partial T}{\partial y}\right)_{y+dy} \right] dt + c_p \rho A dy \, dT \quad (4.8)$$

We also have $A = A_y = A_{y+dy}$.

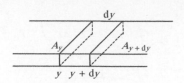

Figure 4.6 Volume element.

After simplification and rearrangement the equation can be written as:

$$\frac{\left(\frac{\partial T}{\partial y}\right)_{y+dy} - \left(\frac{\partial T}{\partial y}\right)_y}{dy} = \left(\frac{\rho c_p}{k}\right)\left(\frac{\partial T}{\partial t}\right) \quad (4.9)$$

At the limit $dy \to 0$ we get:

$$\frac{\partial T}{\partial t} = \alpha \frac{\partial^2 T}{\partial y^2} \qquad \text{Fourier's second law} \quad (4.10)$$

where α is the *thermal diffusitivity* or *coefficient of thermal diffusion*. The derivatives have to be written as partial derivatives as T is a function of both y and t.

Equation (4.10) is the *general equation of thermal conduction* in one dimension. It is valid for both stationary and transient thermal conduction processes. Equation (4.3) is a special case of Equation (4.10) when the partial temperature derivative of time is zero.

Identification of Equations (4.9) and (4.10) gives an expression for the thermal diffusitivity α:

$$\alpha = \frac{k}{\rho c_p} \quad (4.11)$$

where

α = thermal diffusitivity or coefficient of thermal diffusion
k = thermal conductivity
ρ = density
c_p = heat capacity at constant pressure.

The SI unit of α is $m^2 \, s^{-1}$.

4.2.2 Thermal Radiation

All bodies emit electromagnetic radiation or *thermal radiation* to their surroundings. At the same time they absorb such radiation from the surroundings. Figure 4.7 shows the energy radiation per unit time and unit area from a

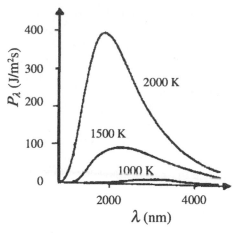

Figure 4.7 The energy emission per unit time and unit area within the wavelength interval λ and $\lambda + d\lambda$ as a function of the wavelength λ at various temperatures.

perfect black body as a function of the wavelength (Planck's radiation law). The higher the temperature is, the more energy is emitted per unit time and unit area, and the more the radiation maximum is moved towards shorter wavelengths.

At $T = 1000\,\mathrm{K}$–$1100\,\mathrm{K}$ ($700\,°\mathrm{C}$–$800\,°\mathrm{C}$) the intensity maximum falls within the visible wavelength region. Our eyes register a dark red light and we say that the body starts to glow. Iron melts often have temperatures of the magnitude $1800\,\mathrm{K}$. The emitted radiation from the melt is perceived by the eye as a bright red light.

The function in Figure 4.7 can be integrated graphically, i.e. the area under the curve is calculated. The result is an expression for the total radiation of energy per unit time and unit area from the body.

It has been calculated theoretically by Boltzmann to give the Stefan–Boltzmann law. During the time interval dt a perfect black body with temperature T and area A emits the amount of energy given by:

$$dQ_{\mathrm{rad}} = \sigma_B A T^4 dt \qquad \text{Boltzmann's law} \qquad (4.12)$$

where σ_B is Boltzmann's constant and T is expressed in kelvin.

All real surfaces radiate less than a perfect black body. This is taken into account by introducing a dimensionless factor ε (<1) called *emissivity* into Equation (4.12). The shinier the surface is, the lower is the value of ε. Table 4.2 gives the ε-values for some different surfaces.

The body absorbs radiation from the surroundings at the temperature T_0. The radiated net energy during the time dt

TABLE 4.2   Emissivity of various surfaces.

Material	$T\ (°\mathrm{C})$	ε
Al film	100	0.09
Oxidized Al	150–500	0.20–0.30
Polished steel	100	0.066
Cast iron	22	0.44
Cast iron	880–990	0.60–0.70
Low carbon steel	230–1065	0.20–0.32
High carbon steel	100	0.074

will then be:

$$dQ_{\mathrm{rad}}^{\mathrm{total}} = \varepsilon \sigma_B A \left(T^4 - T_0^4 \right) dt \qquad (4.13)$$

If T_0 is equal to the room temperature ($\sim 300\,\mathrm{K}$) and T corresponds to the temperature of a metal melt ($\sim 1800\,\mathrm{K}$) the absorbed radiation can be neglected because $T^4 \gg T_0^4$. This is quite clear if, for example, we calculate the difference:

$$\left(T^4 - T_0^4 \right) = \left(1800^4 - 300^4 \right) = 300^4 \left(6^4 - 1 \right)$$
$$= 300^4 (1296 - 1)\,\mathrm{K}^4$$

On the other hand, if the two temperatures do not differ too much alternative approximations may be reasonable and convenient, for example $\left(T^4 - T_0^4 \right) \approx \left(T - T_0 \right) T_0^3$.

It is possible to introduce an effective heat transfer coefficient h_{rad} for radiation, by analogy with the coefficient h for heat transfer. This is illustrated in Example 1.4.1

Example 4.1

An open tundish filled with molten steel has an upper area of $1.0\,\mathrm{m} \times 0.30\,\mathrm{m}$ and a height of $0.60\,\mathrm{m}$. The temperature of the melt is $1500\,°\mathrm{C}$. At this temperature the density of the melt is $7.8 \times 10^3\,\mathrm{kg/m^3}$. Its thermal capacity equals $830\,\mathrm{J/kg\,K}$. The melt stays in the tundish for about 10 minutes.

(a) Calculate the rate of heat loss *by radiation* (part of the total rate of heat loss) from the tundish if the temperature of the surroundings is $20\,°\mathrm{C}$, and the emissivity of molten steel is 0.28. Calculate the temperature decrease of the melt caused by the heat loss due to radiation during its stay in the tundish.

(b) Define the heat transfer coefficient by radiation h_{rad} by analogy with the heat transfer coefficient by conduction h and calculate its value in the present case.

Solution:

(a) The heat loss dQ_{rad} from the upper surface of the tundish during the time dt can be written [Equation (4.13) on page 63]:

$$\frac{dQ_{rad}}{dt} = \varepsilon \sigma_B A \left(T^4 - T_0^4\right) \qquad (1')$$

The values in the text and $\sigma = 5.67 \times 10^{-8}$ W/m^2 K^4 give:

$$\frac{dQ_{rad}}{dt} = 0.28 \times 5.67 \times 10^{-8} \times 1.0 \times 0.30 \left[(1500 + 273)^4 - (20 + 273)^4\right] = 4.7 \times 10^4 \text{ W}$$

The temperature decrease can be calculated by the aid of a material balance.

$$\frac{dQ_{rad}}{dt} = \frac{dT_{rad}}{dt} \rho c_p V \quad \text{which gives}$$

$$\Delta T_{rad} = \frac{dQ_{rad}/dt}{\rho c_p V} \Delta t = \frac{4.7 \times 10^4}{7.8 \times 10^3 \times 830 \times 1.0 \times 0.30 \times 0.60} 10 \times 60 = 24 \text{ K}$$

(b) The definition equation will be: $\frac{dQ_{rad}}{dt} = -h_{rad} A (T_2 - T_1)$
The right hand side of Equation (1') can be split up into factors:

$$\frac{dQ_{rad}}{dt} = \varepsilon \sigma A \left(T^4 - T_0^4\right) = \varepsilon \sigma A \left(T^2 + T_0^2\right)(T + T_0)(T - T_0) \qquad (2')$$

Equations (1') and (2') are identical, which gives:

$$h_{rad} = \varepsilon \sigma_B \left(T^2 + T_0^2\right)(T + T_0)$$
$$h_{rad} = 0.28 \times 5.67 \times 10^{-8} \left(1773^2 + 293^2\right)(1773 + 293)$$
$$= 105 \text{ W/m}^2 \text{ K}$$

Answer:

(a) The rate of heat loss by radiation is 4.7×10^4 W. The temperature loss of the melt, caused by radiation, is 24 K.
(b) $h_{rad} = \varepsilon \sigma_B \left(T^2 + T_0^2\right)(T + T_0) = 1.0 \times 10^2$ W/m^2 K.

4.2.3 Convection

Thermal conduction and thermal radiation result in transport of energy but not of material. Convection means transport of material together with its heat content.

One type of convection present in the casting of a metal melt is caused by cooling the surface of the solidified metal or the mould with water. The heated medium disappears from the neighbourhood of the metal surface and the mould to be continuously replaced by cold air or cold cooling water.

There is a clear distinction between *forced* and *free convection*. *Forced convection* involves external control of the motion of the cooling medium. The cooling of a mould by flowing water is an example of forced convection. The rate of the water flow can be varied arbitrarily.

Free or *natural convection* in a fluid is caused by density differences that are present in the fluid without any influence from outside. One example of natural convection is the cooling of a metal surface in contact with the air without any measures being taken to make the air circulate.

The energy dq per unit area, which is transferred from the surface of the metal and the mould during the time interval dt to a surrounding flowing cooling medium, can be written as:

$$\frac{dq}{dt} = h_{con}(T - T_0) \qquad (4.14)$$

where

h_{con} = heat transfer coefficient of convection
T = temperature of the surface of the metal
T_0 = temperature of cooling medium.

The SI unit of h_{con} is W/m^2 K. The heat transfer coefficient of convection depends on the speed of the flowing cooling medium, the geometry of its channel, and the shape and size of the surface of the casting.

In hydromechanics the heat transfer coefficient is often described by a relationship including Prandtl's and Grashof's numbers for natural convection and by a relationship including Prandtl's and Reynolds's numbers for forced convection. The last two numbers will be defined in

TABLE 4.3 Heat transfer numbers of convection in some specified cases.

Specifications	h_{con} (W/m²K)
Free convection	
Horizontal cylinder with radius 2.5 cm in air	6.5
Horizontal cylinder with radius 1.0 cm in water	890
Forced convection	
Air beam with the velocity 2 m/s over a square plate $(0.2 \text{ m})^2$	12
Flowing water with the 'velocity' 0.5 kg/s in a tube with a diameter of 2.5 cm	3500

connection with the treatment of water cooling in metal moulds in Section 5.4.2 in Chapter 5. Table 4.3 gives some practical examples of the two types of convection. Both natural and forced convection are very important in many casting processes and contribute to the quality of the products.

Equation (4.14) is based on experience but is also in agreement with modern theories of convection. The so-called boundary layer theory of natural convection is described in Section 5.2.1 in Chapter 5 and applied to ingot casting in Section 5.3.2. The corresponding theory of forced convection will not be treated in this book but is applied in Section 5.4.2 and in Section 5.7.6 in connection with heat transport in strip casting.

4.3 THEORY OF HEAT TRANSPORT IN CASTING OF METALS AND ALLOYS

During casting the metal melt has a temperature that generally exceeds its melting point or solidus temperature. After casting, the melt solidifies and cools gradually because the surroundings, mainly the mould material, cool the cast metal. The melt is cooled most rapidly at the surface, which is in contact with the mould. Consequently the solidification starts at this surface and the solidification front, i.e. the interface between solid metal and melt, moves inwards into the melt.

4.3.1 Modes of Solidification

Velocity and Shape of the Solidification Front

The *velocity* of the solidification front depends on the cooling from the surroundings. The stronger the cooling is, the faster the solidification front moves. The temperature gradient is determined by the heat properties of the mould, the cooling capacity of the surroundings and by the thermal conductivity of the cast alloy.

The *shape* of the solidification front is influenced both by the cooling from the surroundings and by the composition and material properties of the cast alloy.

As is seen from Figure 4.1 (page 60) there is a region between the melt and the solid phase that consists of a two-phase layer. The width of this two-phase region is influenced by the temperature gradient in the region and by the width of the solidification interval of the alloy. The latter is determined by the composition of the alloy. The solidification interval of a binary alloy, which solidifies according to the equilibrium laws, is evident from its phase diagram (Figure 4.8).

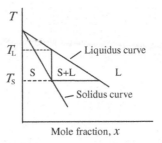

Figure 4.8 Phase diagram of a binary alloy. The solidification interval $= T_L - T_S$.

Pure metals and eutectic alloys have very narrow solidification intervals, while most alloys have wide solidification intervals. The shape of the solidification front is very different in these two cases as is illustrated by Figures 4.9 and 4.10.

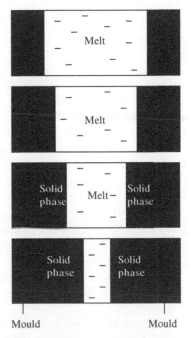

Figure 4.9 Solidification process of a pure metal or an alloy with a narrow solidification interval or a eutectic alloy under the influence of strong cooling. Reproduced with permission from Pergamon Press, Elsevier Science.

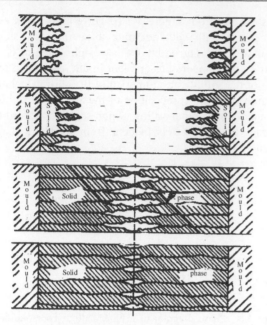

Figure 4.10 Solidification process of an alloy with a wide solidification interval or an alloy under influence of poor cooling. The solid phase areas are marked with dense parallel lines. The light areas represent the melt. Reproduced with permission from Pergamon Press, Elsevier Science.

Solidification Process in Pure Metals and Alloys with Narrow Solidification Intervals in a Strongly Cooled Mould

Figure 4.9 illustrates the solidification process of pure metals and eutectic alloys. The solidification fronts of pure metals and eutectic alloys are planar and well defined. There is no two-phase region present.

Solidification Process in Alloys with Wide Solidification Intervals in a Strongly Cooled Mould

Alloys have solidification intervals of highly variable width, with the result that the solidification process after casting of such an alloy becomes different from that of a pure metal with a well-defined melting point (Figure 4.9).

The solidification starts with a solidification front, corresponding to the liquidus temperature of the alloy (Figure 4.8). The front moves from the surface towards the centre of the melt (Figure 4.10). After some time it is followed by a second solidification front, which indicates that 100 % of the melt has solidified at approximately the solidus temperature. Between these two fronts there is a two-phase region with both solid phase and melt, schematically illustrated in Figure 4.10.

Thus there are three zones in the material, one only with melt, one with a mixture of a solid phase and melt, and one with solid phase only.

The two-phase regions have been marked as lobed areas. The extension of the two-phase area is reduced by factors

that increase the temperature gradient in the solidified material. Such factors, which will be discussed later, are:

- high solidification temperature of the melt;
- poor thermal conductivity of the cast metal;
- high thermal conductivity of the mould.

In those cases when the solidification interval is wide, i.e. when the cooling power of the mould is poor and/or the thermal conductivity of the cast metal is high, the solidification process can be described in the way that is illustrated in Figure 4.11. These conditions are valid for alloys with wide solidification intervals and for alloys that solidify in sand moulds. We will come back to this in Section 4.4.

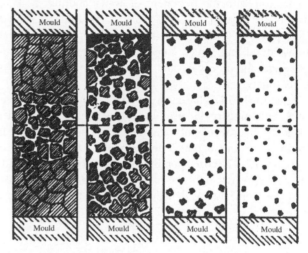

Figure 4.11 Different stages of the solidification process of alloys with wide solidification intervals and/or alloys that solidify in a sand mould. Reproduced with permission from Pergamon Press, Elsevier Science.

The solidification process determines the properties of the cast material and is therefore very important. Below we will treat the solidification process analytically and give examples of calculations of temperature distribution in the material and solidification times in the case of ideal cooling.

4.3.2 Theory of Heat Transport in Casting with Ideal Contact between Metal and Mould

Solution of the Heat Equation

Theoretical calculations of the temperature and temperature changes as a function of position and time in a solidifying alloy melt after casting implies that it is necessary to find a solution of the general law of thermal conduction (Fourier's second law on page 62) in each special case.

$$\frac{\partial T}{\partial t} = \alpha \frac{\partial^2 T}{\partial y^2} \qquad (4.10)$$

where α is the coefficient of thermal diffusion.

The solution of this partial differential equation of second order is the temperature T as a function of position y and time t. The solution contains two arbitrary constants, which are determined by use of given boundary conditions.

The Error Function

If a small amount of heat is distributed in an infinitely large body the solution of Equation (4.10) will be:

$$T = A_0 + \frac{B_0}{\sqrt{t}} \exp\left(-\frac{y^2}{4\alpha t}\right) \qquad (4.15)$$

In many of the approximate solutions it is necessary to integrate the exponential function with respect to y. The integrated function is identical to the so-called *normal distribution function* or *error function*. It will be used in the following theoretical calculations of solidification processes and temperature distributions.

Normal Distribution Function

The normal distribution function appears in many applications and is often associated with problems of a statistical nature, for example, error distribution. This is the reason why it is also called 'error function', shortened 'erf' (Figure 4.12).

$$\text{Definition:} \quad \text{erf}(z) = \frac{2}{\sqrt{\pi}} \int_0^z \exp^{(-y^2)}\,dy \qquad (4.16)$$

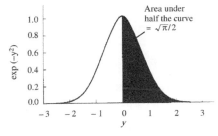

To plot the error function, numerical values are required. They are listed in Table 4.4.

Figure 4.12 Error function.

TABLE 4.4   Error function.

z	erf (z)	z	erf (z)	z	erf (z)	z	erf (z)
0.00	0.0000	0.40	0.4284	0.80	0.7421	1.40	0.9523
0.05	0.0564	0.45	0.4755	0.85	0.7707	1.50	0.9661
0.10	0.1125	0.50	0.5205	0.90	0.7969	1.60	0.9763
0.15	0.1680	0.55	0.5633	0.95	0.8209	1.70	0.9838
0.20	0.2227	0.60	0.6039	1.00	0.8427	1.80	0.9891
0.25	0.2763	0.65	0.6420	1.10	0.8802	1.90	0.9928
0.30	0.3286	0.70	0.6778	1.20	0.9103	2.00	0.9953
0.35	0.3794	0.75	0.7112	1.30	0.9340	∞	1.0000

TABLE 4.5   Some properties of the error function.

1. erf (z) = the area under the curve within the interval $z = 0$ to $z = z$ (part of black area in Figure 4.12).
2. $\dfrac{d\,\text{erf}(z)}{dz} = \dfrac{2}{\sqrt{\pi}} \exp^{(-z^2)}$
3. erf $(0) = 0$ and   erf $(\infty) = 1$
4. erf $(-z) = -$ erf (z) and   erf $(-\infty) = -1$
5. erf$(z) = \frac{2}{\sqrt{\pi}}\left(z - \dfrac{z^3}{3} + \dfrac{z^5}{5}\cdots\cdots\right)$ for small values of z

The error function has the properties listed in Table 4.5.

As a first example we choose the production of a thin metal film by rapid unidirectional cooling.

Example 4.2

In order to achieve very high cooling rates for an alloy melt, small metal droplets are shot towards a copper plate. In this way very good contact between the copper plate and the melt is obtained. The melt is flattened out to a thin film of a thickness of a couple of hundreds of micrometres.

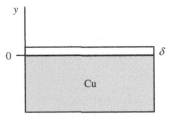

The thickness of the metal layer is exaggerated in the figure compared with the thickness of the copper plate.

Calculate the total solidification time, i.e. the time required for the melt to solidify completely, in terms of the following known data:

Casting temperature of the melt $= T_{\text{cast}}$
Solidus temperature of the melt $= T_S$
Liquidus temperature of the melt $= T_L$
Temperature of the Cu-plate $= T_0$
Thickness of the melt layer $= \delta$
Heat capacitivity of the melt $= c_p$
Heat of fusion of the melt $= -\Delta H$
Thermal diffusivity of the melt $= \alpha$

Solution:

The heat transport into the Cu plate is determined by the general heat equation, Equation (4.10) on page 62:

$$\frac{\partial T}{\partial t} = \alpha\,\frac{\partial^2 T}{\partial y^2} \qquad (1')$$

The solidification front (at the solidus temperature) is a horizontal plane, which moves upwards, starting at the Cu plate. The solidification time is the time when the solidification front reaches the upper surface of the layer.

All thermal energy is transported towards the big Cu plate. Because the layer is thin, the Cu plate receives only a small amount of heat. Under these circumstances the solution of Equation (1') is as in Equation (4.15) on page 67:

$$T = A_0 + \frac{B_0}{\sqrt{t}} \exp\left(-\frac{y^2}{4\alpha t}\right) \qquad (2')$$

If the temperature distribution in the metal melt as a function of time is known we can calculate the time when the solidus temperature is achieved at the upper surface of the layer.

The first thing to do is to determine the constants A_0 and B_0 from the boundary conditions, which are valid in this case.

Boundary Condition 1:

At the time $t = 0$ we have $T = T_0$ for $y = 0$ because the contact between the layer and the Cu-plate is very good. $T = T_0$ is also valid for all values of $y \neq 0$. We insert $T = T_0$ for $y \neq 0$ into Equation (2'). The second term on the right-hand side becomes zero because the exponential term in the numerator approaches zero faster than the square root expression in the denominator and we get $A_0 = T_0$.

Boundary Condition 2:

Two expressions of the total amount of heat, transferred to the Cu plate, are found. They are equal, which gives the second boundary condition.

The whole melt layer cools from the casting temperature T_{cast} to the liquidus temperature T_L, solidifies completely and cools simultaneously from T_L to the solidus temperature T_s. The total amount of heat q per unit area, that is transferred to the Cu plate, consists of cooling heat, when the temperature of the layer decreases from T_{cast} to T_s, and solidification heat $(-\Delta H)$:

$$q = c_p \rho \delta (T_{cast} - T_s) + \rho \delta (-\Delta H) \qquad (3')$$

The amount of heat q per unit area must also be equal to the total 'excess heat' that is stored in the layer. We have to consider the fact that the temperature varies with the distance y from the Cu plate and integrate the excess heat in many infinitesimal layers:

$$q = \int_0^\delta c_p \, \rho \, 1 \, dy (T - T_0) \approx \int_0^\infty c_p \, \rho \left(\frac{B_0}{\sqrt{t}} \exp\left(-\frac{y^2}{4\alpha t}\right) dy \right)$$

$$= c_p \rho B_0 \int_0^\infty \frac{1}{\sqrt{t}} \exp\left(-\frac{y^2}{4\alpha t}\right) dy$$

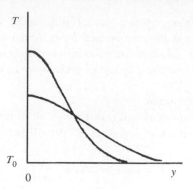

Temperature distribution in the layer at two different times t_1 and t_2. The width of the layer is, for the sake of clarity, greatly enlarged.

Because $A_0 = T_0$ we have $T - T_0 = \frac{B_0}{\sqrt{t}} \exp\left(-\frac{y^2}{4\alpha t}\right)$, which has been inserted into the integral. It does not matter that the upper integral limit has been extended from δ to ∞ because $T = T_0$ within this interval and the contribution to the value of the integral will be zero.

After change of variable $\left(z = \frac{y}{\sqrt{4\alpha t}}\right)$ and $\left(dz = \frac{dy}{\sqrt{4\alpha t}}\right)$ we get:

$$q = \frac{c_p \rho B_0}{\sqrt{t}} \int_0^\infty \exp\left(-\frac{y^2}{4\alpha t}\right) dy = \frac{c_p \rho B_0}{\sqrt{t}} \int_0^\infty \exp^{(-z^2)} \sqrt{4\alpha t} \; dz \qquad (4')$$

$$q = \frac{c_p \rho B_0}{\sqrt{t}} \sqrt{4\alpha t} \int_0^\infty \exp^{(-z^2)} dz = c_p \rho B_0 \sqrt{4\alpha} \left(\frac{\sqrt{\pi}}{2}\right)$$

$$= c_p \rho B_0 \sqrt{\pi \alpha} \qquad (5')$$

The root expression can be moved in front of the integral sign and the value of the integral is obtained from Figure 4.12 on page 67:

If we combine Equations (3') and (5') we get the value of the constant B_0:

$$B_0 = \frac{\left[-\Delta H + c_p (T_{cast} - T_s)\right] \delta}{c_p \sqrt{\pi \alpha}} \qquad (6')$$

Solution of the Heat Equation:

The determined values of B_0 [Equation (6')] and A_0 ($A_0 = T_0$) are inserted into Equation (2') and we get:

$$T = T_0 + \frac{\left[-\Delta H + c_p (T_{cast} - T_s)\right] \delta}{c_p \sqrt{\pi \alpha} \sqrt{t}} \exp\left(-\frac{y^2}{4\alpha t}\right) \qquad (7')$$

The second solidification front of the alloy reaches the upper part of the metal layer at the very end. We get the desired solidification time by inserting $y = \delta$ and $T = T_S$ into Equation (7'). As the layer is very thin we can set the exponent at about zero. Then the exponential factor becomes 1 and we get the answer by solving t.

Answer:
The solidification time of the alloy will be:

$$t = \left(\frac{-\Delta H + c_p(T_{cast} - T_S)}{c_p(T_S - T_0)}\right)^2 \frac{\delta^2}{\pi\alpha}$$

Temperature Distribution of a Metal Melt in a Metal Mould

Production of thin metal films by the method in Example 4.2 is unusual. The normal situation is that both mould and metal melt are extended. The temperature distribution at the beginning of the solidification process is in principle illustrated in Figure 4.13.

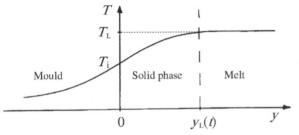

Figure 4.13 Approximate temperature distribution in a solidifying but not superheated metal melt in a metal mould.

In this case the solution of Fourier's second law [Equation (4.10) on page 62] can be found by addition of a great number of subsolutions such as Equation (4.15) on page 67 according to the superposition principle. Each infinitesimal layer dy contributes and we get the following:

$$T - T_L = \int_{-\infty}^{\infty} \frac{B_0}{\sqrt{t}} \exp\left(-\frac{y^2}{4\alpha t}\right) dy \qquad (4.17)$$

This method will be applied to the solidification process of a metal melt in contact with a metal mould, based on the following assumptions:

- The contact between melt and mould is good, which means that there is no resistance against heat transport over the mould/metal interface;
- The metal is not superheated;
- The metal melt has a narrow solidification interval;
- The volume of the mould is very large, approximately infinite.

The schematic temperature profile in Figure 4.13 has been drawn according to these assumptions. The temperatures of metal and mould will adjust to equilibrium at the interface, where they are equal.

The temperature distribution in the metal and mould during solidification will be calculated as a function of y and t. At the time $t = 0$ the metal is poured into the mould. The position of the y-axis is shown in Figure 4.13. y is zero at the interface between metal and mould, positive in the metal and negative in the mould.

Since the metal and the mould are made of different materials with different properties, the differential Equation (4.10) on page 62 must be solved separately for each of them. The symbols for the quantities used are listed below. The differential equations for the mould and the metal will then be:

$$\frac{\partial T_{mould}}{\partial t} = \alpha_{mould}\left(\frac{\partial^2 T_{mould}}{\partial y^2}\right) \qquad (4.18_{mould})$$

$$\frac{\partial T_{metal}}{\partial t} = \alpha_{metal}\left(\frac{\partial^2 T_{metal}}{\partial y^2}\right) \qquad (4.18_{metal})$$

where

T_i = temperature at the interface metal/mould
T_L = temperature of the melt
T_0 = room temperature
$T_{mould}(t, y)$ = temperature of the mould
$T_{metal}(t, y)$ = temperature of the solid metal
α_{mould} = diffusivity of the mould
α_{metal} = diffusivity of the metal
k_{mould} = thermal conductivity of the mould
k_{metal} = thermal conductivity of the metal.

Solution of the General Equation of Thermal Conduction

The solutions to the Equations (4.18) above have the same shape as Equation (4.17). Integration is necessary to add the contributions from all the dy layers [compare Equation (4.17)]. To find the solutions in this case we replace the variable y in Equation (4.17) by $y = \sqrt{4\alpha t}$. With this variable transformation, the factor $\sqrt{t}$ in the denominator in the second term disappears and is thus missing in the solutions of Equations (4.18) [compare Equations (4') and (5') on page 68]. The erf function can be introduced into the solutions, which can be written as:

$$T_{mould} = A_{mould} + B_{mould}\,\text{erf}\left(\frac{y}{\sqrt{4\alpha_{mould}t}}\right) \qquad (4.19_{mould})$$

$$T_{metal} = A_{metal} + B_{metal}\,\text{erf}\left(\frac{y}{\sqrt{4\alpha_{metal}t}}\right) \qquad (4.19_{metal})$$

The four arbitrary constants in the solutions (4.19) will be determined by introducing *boundary conditions*, valid for the present case, as follows.

(1) *Boundary Condition for the Mould*

$$T(t, -\infty) = T_0 \qquad (4.20)$$

(2) *Boundary Condition for the Interface Mould/Metal*

$$T(t, 0) = T_i \qquad (4.21)$$

The relationship $\frac{\partial q}{\partial t} = -k\left(\frac{\partial T}{\partial y}\right)$ [compare Equation (4.3) on page 61] can be used to apply the law of energy conservation at the interface between mould and metal. The heat flux from the solid phase to the mould is equal to the heat flux absorbed by the latter, which gives the next, third, boundary condition [Equation (4.22)].

(3) *Boundary Condition for the Interface Mould/Metal*

$$k_{\text{mould}} \frac{\partial T_{\text{mould}}}{\partial y} = k_{\text{metal}} \frac{\partial T_{\text{metal}}}{\partial y} \qquad (4.22)$$

At the solidification front the temperature is constant, i.e. independent of time, and equal to the liquidus temperature T_L. The solidification front moves, i.e. its position is a function of time. This can be described by the function $y = y_L(t)$ and we get the fourth boundary condition, Equation (4.23).

(4) *Boundary Condition for the Solidification Front*

$$T(t, y_L(t)) = T_L \qquad (4.23)$$

The solidified metal absorbs the heat of solidification, which is generated at the solidification front, because its temperature is lower than that of the melt. The velocity of the solidification front is equal to the derivative of $y_L(t)$ with respect to time, and the heat flux can be written:

$$\frac{dq}{dt} = (-\Delta H)\rho \frac{dy_L(t)}{dt}$$

and we get a fifth boundary condition as Equation (4.24).

(5) *Boundary Conditions for the Metal*

$$k_{\text{metal}} \left(\frac{\partial T_{\text{metal}}}{\partial y}\right)_{y = y_L(t)} = (-\Delta H)\rho \frac{dy_L(t)}{dt} \qquad (4.24)$$

Equations (4.20) to (4.24) are the set of boundary conditions. They will be used to determine the four constants in Equations (4.19). There are four constants and five conditions. Thus the system seems to be overestimated but this is not the case, however, as it is necessary to determine a fifth constant, which will be introduced below.

At the solidification front we have [Equation (4.19$_{\text{metal}}$)]:

$$T_L = A_{\text{metal}} + B_{\text{metal}} \, \text{erf}\left(\frac{y_L}{\sqrt{4\alpha_{\text{metal}} \, t}}\right) \qquad (4.25)$$

Since A_{metal}, B_{metal} and T_L are constants, the erf function must also be a constant, i.e. independent of t. The conclusion is that the variable of the function also must be constant, which gives the condition:

$$y_L(t) = \lambda \sqrt{4\alpha_{\text{metal}} \, t} \qquad (4.26)$$

λ is the fifth constant to be determined from the five boundary conditions. When λ is known the *growth rate* or *solidification rate* can easily be found by derivation of $y_L(t)$ with respect to t:

$$\frac{dy_L(t)}{dt} = \lambda \sqrt{\frac{\alpha_{\text{metal}}}{t}} \qquad (4.27)$$

Determination of the Constants from the Boundary Conditions

The expression for $y_L(t)$ in Equation (4.26) is introduced into Equation (4.25), which gives:

$$T_L = A_{\text{metal}} + B_{\text{metal}} \, \text{erf} \lambda \qquad (4.28)$$

Equation (4.20) is valid for the mould. It can be written as:

$$T_0 = A_{\text{mould}} + B_{\text{mould}} \, \text{erf}(-\infty)$$

or, by use of condition 4 we get:

$$A_{\text{mould}} = T_0 + B_{\text{mould}} \qquad (4.29)$$

At the interface ($y = 0$) the temperature is equal for the metal and the mould [Equation (4.21)] or $T_{\text{metal}} = T_{\text{mould}}$. We use this condition by inserting $y = 0$ into Equations (4.19) and get:

$$A_{\text{mould}} + B_{\text{mould}} \times 0 = A_{\text{metal}} + B_{\text{metal}} \times 0$$

or, by use of Equation (4.29), we get:

$$A_{\text{mould}} = A_{\text{metal}} = T_0 + B_{\text{mould}} \qquad (4.30)$$

The expression for A_{metal} [Equation (4.30)] is introduced into Equation (4.28), which gives:

$$T_L = T_0 + B_{\text{mould}} + B_{\text{metal}} \, \text{erf} \lambda$$

or

$$B_{\text{mould}} = T_L - T_0 - B_{\text{metal}} \, \text{erf} \lambda \qquad (4.31)$$

To be able to use the boundary condition in Equation (4.22) we have to derive the Equations (4.19) with respect to y. We also use condition 2 of the erf function (see page 70):

$$\frac{\partial T_{\text{mould}}}{\partial y} = B_{\text{mould}} \left(\frac{2}{\sqrt{\pi}}\right) \exp\left(-\frac{y^2}{4\alpha_{\text{mould}} t}\right) \frac{1}{\sqrt{4\alpha_{\text{mould}} t}}$$

and

$$\frac{\partial T_{\text{metal}}}{\partial y} = B_{\text{metal}} \left(\frac{2}{\sqrt{\pi}}\right) \exp\left(-\frac{y^2}{4\alpha_{\text{metal}} t}\right) \frac{1}{\sqrt{4\alpha_{\text{metal}} t}}$$

These expressions of the derivatives are introduced into Equation (4.22) and we get, for $y = 0$:

$$k_{\text{mould}} B_{\text{mould}} \frac{1}{\sqrt{4\alpha_{\text{mould}} t}} = k_{\text{metal}} B_{\text{metal}} \frac{1}{\sqrt{4\alpha_{\text{metal}} t}}$$

which can be rewritten as:

$$B_{\text{mould}} = B_{\text{metal}}(k_{\text{metal}} \sqrt{\alpha_{\text{mould}}}/k_{\text{mould}} \sqrt{\alpha_{\text{metal}}}) \quad (4.32)$$

This expression of B_{mould} is introduced into Equation (4.31), which is solved for B_{metal}:

$$B_{\text{metal}} = \frac{T_{\text{L}} - T_0}{(k_{\text{metal}} \sqrt{\alpha_{\text{mould}}}/k_{\text{mould}} \sqrt{\alpha_{\text{metal}}}) + \text{erf}\,\lambda} \quad (4.33)$$

By combining Equations (4.32) and (4.33) we get:

The values of A_{metal} and B_{metal} are introduced into Equation (4.19$_{\text{metal}}$) and we derive the new equation with respect to y. Equation (4.26) is differential with respect to t and the two derivatives obtained are inserted into Equation (4.24). The final result is:

$$\frac{c_{\text{p}}^{\text{metal}} (T_{\text{L}} - T_0)}{-\Delta H} = \sqrt{\pi}\,\lambda \exp^{(\lambda^2)} \left(\sqrt{\frac{k_{\text{metal}} \rho_{\text{metal}} c_{\text{p}}^{\text{metal}}}{k_{\text{mould}} \rho_{\text{mould}} c_{\text{p}}^{\text{mould}}}} + \text{erf}\,\lambda\right)$$

$$(4.36)$$

In Equation (4.36) the relation $\alpha = \frac{k}{\rho c_{\text{p}}}$ [Equation (4.11) on page 62] has been used.

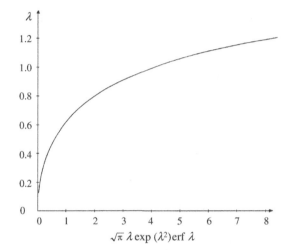

Figure 4.14 The curve in combination with Equation (4.36) can be used to derive a rough value for the constant λ, as illustrated in Example 4.3.

$$B_{\text{mould}} = \frac{T_{\text{L}} - T_0}{(k_{\text{metal}} \sqrt{\alpha_{\text{mould}}}/k_{\text{mould}} \sqrt{\alpha_{\text{metal}}}) + \text{erf}\,\lambda} \times (k_{\text{metal}} \sqrt{\alpha_{\text{mould}}}/k_{\text{mould}} \sqrt{\alpha_{\text{metal}}}) \quad (4.34)$$

Equation (4.30) in combination with Equation (4.34) gives:

$$A_{\text{mould}} = A_{\text{metal}} = T_0 + \frac{T_{\text{L}} - T_0}{(k_{\text{metal}} \sqrt{\alpha_{\text{mould}}}/k_{\text{mould}} \sqrt{\alpha_{\text{metal}}}) + \text{erf}\,\lambda} \times (k_{\text{metal}} \sqrt{\alpha_{\text{mould}}}/k_{\text{mould}} \sqrt{\alpha_{\text{metal}}}) \quad (4.35)$$

Now we have the four constants expressed in known quantities and the unknown constant λ [Equations (4.33), (4.34) and (4.35)]. It can be determined from the last boundary condition, Equation (4.24) in combination with Equation (4.26). The velocity of the solidification front is determined by the temperature gradient in the solid metal.

From Equation (4.36) it is possible to solve the constant λ. It is not possible to find an exact analytical solution to the equation, but it can be solved by iteration and use of a calculator or graphically. The curves in Figures 4.14 and 4.15 may be used to find a reasonable introductory value of λ for iteration. The solution procedure is illustrated in Example 4.3.

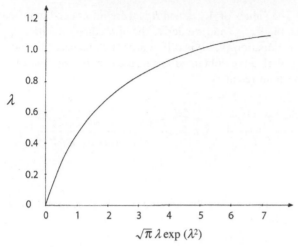

Figure 4.15 The curve can be used with Equation (4.36) to derive a rough value for constant λ as illustrated in Example 4.3.

Example 4.3

The equation below is valid for casting in the case of good contact between mould and metal:

$$\frac{c_p^{metal}(T_L - T_0)}{-\Delta H} = \sqrt{\pi}\,\lambda\,\exp^{(\lambda^2)}\left(\sqrt{\frac{k_{metal}\rho_{metal}c_p^{metal}}{k_{mould}\rho_{mould}c_p^{mould}}} + \mathrm{erf}\,\lambda\right)$$

The constant λ, which can be calculated from this equation, is used to describe the position of the solidification front as a function of time ($y_L(t) = \lambda\sqrt{4\alpha_{metal}\,t}$).

Calculate λ by iteration for the special casting process, which corresponds to the equation:

$$3.20 = \sqrt{\pi}\,\lambda\,\exp^{(\lambda^2)}(0.53 + \mathrm{erf}\,\lambda) \qquad (1')$$

Solution:

We guess an initial value of λ, for example $\lambda = 0.6$, and test it roughly with the aid of Figures 4.14 and 4.15 on pages 71 and 72.

For $\lambda = 0.6$ we read the value $\sqrt{\pi}\,\lambda\,\exp^{(\lambda^2)} \approx 1.6$ in Figure 4.15.

For $\lambda = 0.6$ we read the value $\sqrt{\pi}\,\lambda\,\exp^{(\lambda^2)}\mathrm{erf}(\lambda) \approx 1.0$ in Figure 4.14.

Inserting these values into Equation (1') we get:

$$\sqrt{\pi}\,\lambda\,\exp^{(\lambda^2)}(0.53 + \mathrm{erf}\,\lambda) = 1.6 \times 0.53 + 1.0 = 1.8$$

This value is obviously too low (it should be 3.20) but it is of the right magnitude.

Next we make a table and calculate the values with the aid of a calculator and Table 4.4 on page 67.

Interpolation within the interval $0.75 < \lambda < 0.80$ gives the final value.

λ	erf λ (from Table-4)	$0.53 + \mathrm{erf}\,\lambda$	$\sqrt{\pi}\,\lambda\exp^{(\lambda^2)}$	$\sqrt{\pi}\,\lambda\exp^{(\lambda^2)}(0.53 + \mathrm{erf}\,\lambda)$
0.60	0.6039	1.1339	1.5243	1.73
0.70	0.6778	1.2078	2.0252	2.45
0.75	0.7112	1.2412	2.3331	2.90
0.80	0.7421	1.2721	2.6891	3.42

Answer:

$\lambda = 0.78$

4.3.3 Theory of Heat Transport in Casting with Poor Contact between Metal and Mould. Solidification Rate – Solidification Time

When the contact at the interface between mould and metal is good and the melt is not superheated we get the temperature distribution illustrated in Figure (4.13) on page 69. The temperatures in the mould and the metal are equal at the interface.

With poor contact between the mould and the metal there is a discontinuity of the temperature at the interface. Figure 4.16 shows the temperature in both the metal and the mould some time after the casting. There are two temperatures at the interface, one in the metal, $T_{i\,metal}$, and one in the mould $T_{i\,mould}$. Both temperatures vary with time. T_0 is the temperature of the surroundings of the mould.

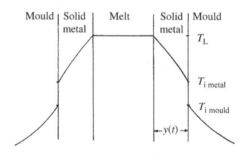

Figure 4.16 Temperature distribution in mould and metal with poor contact at the interface at a time t after casting.

Figure 4.17 shows another example. Effective heat transport in the mould is assumed. Either the thermal conductivity of the mould has to be very good or the outer surface of the mould has to be directly cooled with water or air. In the latter case the heat transport mechanism is convection.

Solidification Rate

The solidification process indicated in Figure 4.17 is very common. The melt solidifies first at the interface between the metal and the mould. Initially the thin metal shell is

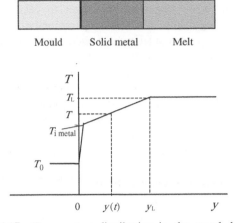

Figure 4.17 Temperature distribution in the metal during the solidification process after casting.

kept in good contact with the mould by the pressure from the melt.

After some time the solidifying metal shell becomes thick enough to resist the pressure from the melt. At the same time it cools and shrinks and suddenly loses contact with the mould wall. At that moment, thermal conduction decreases drastically.

We will analyse this solidification process, i.e. we want the temperature as a function of position and time. We will also derive an expression of the solidification time. As the temperature is a function of two variables we must use partial derivatives.

The amount of heat per unit area that passes a cross-section area perpendicular to the heat flux at the interface between the mould and the solid metal (see Figure 4.17), can be written as:

$$-\frac{\partial q}{\partial t} = \rho\left(-\Delta H\right)\frac{dy_L(t)}{dt} + \rho c_p \int_0^{y_L(t)} -\frac{\partial T}{\partial t}\,dy \quad (4.37)$$

The term on the left-hand side is the heat flux *lost* to the surroundings. For this reason a minus sign has to be added.

The *first* term on the right-hand side describes the amount of solidification heat per unit area and unit time released at position y_L when the solidification front moves with a rate $dy_L(t)/dt$. The heat is transported away by means of thermal conduction through the interface solid/mould.

The *second* term describes the amount of cooling heat in the solid per unit area and unit time, which is transported away by the aid of thermal conduction through the interface solid/mould. Unfortunately $(-\Delta H)$ and c_p are not constant for all metals and all casting processes. This topic is discussed in Section 4.3.4 on page 74. In Equation (4.37)

above and in the following calculations below we have assumed that they are constants.

The total thermal conduction through the interface solid/mould is described by the general law:

$$\frac{\partial q}{\partial t} = -k\left(\frac{\partial T}{\partial y}\right) \quad \text{over the interval } 0 < y < y_L \quad (4.38)$$

This expression is introduced into Equation (4.37) to give:

$$k\frac{\partial T}{\partial y} = \rho(-\Delta H)\frac{dy_L(t)}{dt} + \rho c_p \int_0^{y_L(t)} -\frac{\partial T}{\partial t}\,dy \quad (4.39)$$

The differential dT is a function of both dy and dt.

$$dT = \frac{\partial T}{\partial y}\,dy + \frac{\partial T}{\partial t}\,dt \quad (4.40)$$

If $\partial T/\partial t$ is small, the second term in Equations (4.37), (4.39) and (4.40) can be neglected and the rest of Equation (4.39) can be integrated. The partial derivative $\partial T/\partial y$ is replaced by dT/dy. In order to find a simple approximate solution for Equation (4.39) we initially assume that $\partial q/\partial t$ is constant and independent of y. As a consequence of this assumption and from Equation (4.37) we conclude that $dy_L(t)/dt$ is also approximately independent of y. If these assumptions are valid we get:

$$\int_{T_{i\,metal}}^{T_L} k\,dT = \int_0^{y_L(t)} \rho(-\Delta H)\frac{dy_L(t)}{dt}\,dy \quad (4.41)$$

Provided that the heat of solidification $(-\Delta H)$ is constant we get:

$$k(T_L - T_{i\,metal}) = \rho(-\Delta H)\frac{dy_L(t)}{dt}y_L(t) \quad (4.42)$$

In Equation (4.42), the temperature T_L is constant. To be able to calculate the solidification rate, i.e. the velocity of the solidification front, $dy_L(t)/dt$, we have to know the amount of heat transported across the interface per unit time and unit area, or the heat flux dq/dt. This heat flux is described by Equation (4.5) on page 61, which can be applied here:

$$\frac{dq}{dt} = h(T_{i\,metal} - T_0) \quad (4.43)$$

where h is the coefficient of heat transfer. It depends on many quantities and is hard to calculate theoretically. It is normally determined by experimental methods.

We have two expressions for the heat flux given in Equations (4.38) and (4.43). If we replace the derivative in Equation (4.38) by a linear function we get:

$$h(T_{i\,metal} - T_0) = k\frac{T_L - T_{i\,metal}}{y_L(t)} \qquad (4.44)$$

$T_{i\,metal}$ can be solved from Equation (4.44):

$$T_{i\,metal} = \left[\frac{T_L - T_0}{1 + \left(\dfrac{h}{k}\right)y_L(t)}\right] + T_0 \qquad (4.45)$$

By introducing this expression for $T_{i\,metal}$ into Equation (4.42) and combining it with Equation (4.44), we get the velocity of the solidification front:

$$\frac{dy_L}{dt} = \left[\frac{T_L - T_0}{\rho\,(-\Delta H)}\right]\left[\frac{h}{1 + \left(\dfrac{h}{k}\right)y_L}\right] \qquad (4.46)$$

This is the desired expression for the solidification rate, provided that $(-\Delta H)$ is constant.

Solidification Time

To find the solidification time we transform Equation (4.46) by separating the variables and integrating:

$$\int_0^t dt = \int_0^{y_L} \frac{\rho\,(-\Delta H)}{h\,(T_L - T_0)}\left(1 + \left(\frac{h}{k}\right)y_L\right)dy_L \qquad (4.47)$$

which gives the time it takes to achieve a shell of thickness y_L:

$$t = \frac{\rho\,(-\Delta H)}{(T_L - T_0)}\left(\frac{y_L}{h}\right)\left(1 + \left(\frac{h}{2k}\right)y_L\right) \qquad (4.48)$$

Equation (4.48) can be used to calculate the total solidification time when the dimensions of the metal melt are known and $(-\Delta H)$ is constant.

Theory versus Practice

The theory of heat transport at casting is treated in Sections 4.3.1 to 4.3.3. The temperature distribution in metal and mould, the position of the solidification front, the solidification rate and the solidification time are quantities that can be calculated as functions of time, heat of fusion of the metal, temperature of the melt and material constants.

The aim of such calculations is to design optimal casting processes in order to get casting products of high quality. Reliable calculations for design of casting processes require two fulfilled conditions:

- reliable models for different types of casting processes;
- reliable values of the material constants.

Cooling is an important parameter in casting design. Strong cooling results in a high temperature gradient, which promotes the heat transport, and vice versa. As has been mentioned several times above, it is important to remember and bear in mind that $(-\Delta H)$ and c_p are not constant for all metals and alloys. This topic is discussed in Section 4.3.4.

It is well known that accurate heat measurements are difficult, mainly due to heat losses to the surroundings. At the end of the Twentieth century an enormous development in instrumentation enabled new and very accurate methods of thermal measurements to be made. The common name of these methods is *thermal analysis*.

The problems with heat loss to the surroundings are almost eliminated thanks to the new technique in *microscale*. The instruments measure for example the *difference* in temperature between the sample and a reference of the same 'thermal mass', which is exposed to the same heating process in a furnace. The heat losses to the surroundings will be about the same in both cases.

4.3.4 Heat of Fusion and Thermal Capacity as Functions of the Solidification Rate

Solidification and cooling processes have a strong influence of the structure of cast metals. Such processes can be studied experimentally by temperature registration of the casting as a function of time. The result is a so-called cooling curve of the type illustrated in Figure 4.18. Three parameters determine the shape of the curve: the heat capacities of the superheated melt and of the solid metal, and the heat of fusion of the metal. The values of these parameters can be determined from the cooling curve if the heat flow is known (page 156 in Chapter 6).

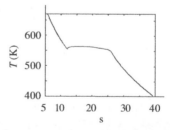

Figure 4.18 c_p^L and c_p^s can be derived from the slopes of the straight parts of the curve. $-\Delta H$ is derived from the horizontal part of the curve.

These matters will be further discussed in Chapter 6. Here we will concentrate on and explain the experimental fact that the heat of fusion and the thermal capacity, which in most cases are material constants at a given temperature, vary in some cases with the solidification rate, i.e. the growth rate of the solidification front.

The heat flow from the casting during the cooling processes before (c_p^L) and after (c_p^s) the solidification process can be written as:

$$\frac{dQ}{dt} = -V\rho c_{\mathrm{p}}\left(\frac{dT}{dt}\right) \qquad (4.49)$$

where V is the volume of the casting. It can be seen from Figure 4.18 that the cooling rates are approximately parallel, which means that the heat flow dQ/dt from the casting is approximately constant even during the solidification period, when the temperature is constant. This gives the following relationship between the cooling rate dT/dt and the solidification rate dV/dt:

$$\underset{\substack{\text{Flow}\\\text{rate}}}{\frac{dQ}{dt}} = \underset{\substack{\text{Flow rate}\\\text{during cooling}}}{-V\rho c_{\mathrm{p}}\frac{dT}{dt}} = \underset{\substack{\text{Flow rate}\\\text{during solidification}}}{\rho(-\Delta H)\frac{dV}{dt}} \qquad (4.50)$$

In case of one-dimensional solidification, the solidification front is planar. This is a common case that has been treated extensively in Sections 4.3.1 to 4.3.3. With one-dimensional solidification, $dV = A\,dy_{\mathrm{L}}$ where A is the cross-section area of the casting and y_{L} the position of the solidification front. In this case, Equation (4.50) can be written as:

$$\frac{dQ}{dt} = -V\rho c_{\mathrm{p}}\frac{dT}{dt} = \rho(-\Delta H)A\frac{dy_{\mathrm{L}}}{dt}$$

or

$$-\frac{dT}{dt} = \frac{-\Delta H}{Vc_{\mathrm{p}}}A\frac{dy_{\mathrm{L}}}{dt} \qquad (4.51)$$

The first factor on the right-hand side is approximately constant. Thus we have for one-dimensional solidification:

The cooling rate and the growth rate dy_{L}/dt are proportional.

If we want to study a material quantity, for example the heat of fusion or the thermal capacity as a function of the solidification rate, conclusions can be drawn even if we plot the quantity as a function of the cooling rate instead of the solidification rate. This fact will be used below and in Chapters 6 and 7.

Influence of Solidification Rate on Heat of Fusion

Lattice Defects

As an explanation of the experimentally observed variation of the latent heat of solidification as a function of the cooling rate, it has been proposed that:

The variation of the heat of solidification/heat of fusion is associated with the presence of lattice defects in the metal or alloy.

Lattice defects or irregularities in the crystal lattice are present in all crystals at all temperatures above $T = 0$ K. The number of lattice defects increases strongly with the temperature. Here we will restrict the discussion to vacancies, which are the dominating defects in metals and alloys. A *vacancy* is a missing atom in the crystal lattice. Vacancies can move within the crystal.

Influence of Vacancies on Heat of Fusion

At a given temperature there are equilibrium concentrations of vacancies of the metal or alloy in the solid and liquid states. The vacancy concentration is higher in the liquid than in the solid phase. At solidification of a melt there is no time for achieving equilibrium and an excess of vacancies is assumed to be trapped at the interface between the melt and the solid phase.

At a low solidification rate (soft cooling = low cooling rate) the solidification front moves slowly and comparatively few vacancies get trapped at the interface per unit time. At a high solidification rate (strong cooling = high cooling rate) the reverse is true and a large number of vacancies get trapped per unit time.

Formation of vacancies and other lattice defects requires energy. The energy required to form 1 kmol of vacancies will here be designated by $-\Delta H_{\mathrm{m}}^{\mathrm{vac}}$. It should be emphasized that $-\Delta H_{\mathrm{m}}^{\mathrm{vac}}$ depends on the material and can thus be regarded as a *material constant*.

At the solidification front, solidification energy per unit mass $(-\Delta H^{\mathrm{e}})$ is released. Some of this energy is used *internally* to form vacancies in the solid phase. The rest is transported away and can be observed and measured externally as $(-\Delta H^{\mathrm{eff}})$. If we consider the solidification of a unit mass = 1 kmol of melt instead of 1 kg we get the molar quantities:

$$-\Delta H_{\mathrm{m}}^{\mathrm{e}} = (x_{\mathrm{vac}} - x_{\mathrm{vac}}^{\mathrm{e}})(-\Delta H_{\mathrm{m}}^{\mathrm{vac}}) + (-\Delta H_{\mathrm{m}}^{\mathrm{eff}})$$

or

$$-\Delta H_{\mathrm{m}}^{\mathrm{eff}} = (-\Delta H_{\mathrm{m}}^{\mathrm{e}}) - (x_{\mathrm{vac}} - x_{\mathrm{vac}}^{\mathrm{e}})(-\Delta H_{\mathrm{m}}^{\mathrm{vac}}) \qquad (4.52)$$

where

$-\Delta H_{\mathrm{m}}^{\mathrm{e}}$ = molar (kmol) heat of solidification at equilibrium concentration of vacancies. This is normally the tabulated value

x_{vac} = mole fraction of vacancies in the crystal

$x_{\mathrm{vac}}^{\mathrm{e}}$ = equilibrium concentration of vacancies

$-\Delta H_{\mathrm{m}}^{\mathrm{vac}}$ = heat of solidification of 1 kmol vacancies in the crystal

$-\Delta H_{\mathrm{m}}^{\mathrm{eff}}$ = effective molar (kmol) heat of solidification, measured at a given cooling rate

$-\Delta H_{\mathrm{m}}^{\mathrm{e}} =$ is a material constant, characteristic for the metal or alloy.

As $x_{\mathrm{vac}} > x_{\mathrm{vac}}^{\mathrm{e}}$, it can be concluded from Equation (4.52) that the effective heat of solidification is *smaller* than the corresponding equilibrium value. As mentioned above, $-\Delta H_{\mathrm{m}}^{\mathrm{vac}}$ is also a material constant.

Influence of Vacancies on Thermal Capacity

The excess vacancies trapped at the solidification front annihilate or *condense* gradually, for example, at dislocations or grain boundaries within the material. Thus, heat due to condensation of trapped vacancies is released *during* and also *after* the solidification process. This heat release contributes to the effective thermal capacity. On the other hand, vacancies are formed when the temperature increases, which requires energy. The vacancy concentration increases strongly with temperature.

Contribution to C_{p} from Change of the Vacancy Concentration with Temperature

Consider 1 kmole of a metal with the vacancy concentration x_{vac} at temperature T. When the temperature is raised by an amount $\mathrm{d}T$ the vacancy concentration increases by $\mathrm{d}x_{\mathrm{vac}}$. More energy is required than normal heating of the metal to the higher thermal energy. The additional energy consists of formation energy of vacancies. Thus the effective molar thermal capacity, i.e. the energy per kmole and unit degree, can be written as:

$$C_{\mathrm{p}}^{\mathrm{eff}} = C_{\mathrm{p}}^{\mathrm{e}} + \frac{\mathrm{d}x_{\mathrm{vac}}}{\mathrm{d}T}\left(-\Delta H_{\mathrm{m}}^{\mathrm{vac}}\right) \quad (4.53)$$

where

$C_{\mathrm{p}}^{\mathrm{eff}} =$ effective molar (kmol) thermal capacity of the metal at temperature T and vacancy concentration x_{vac}

$C_{\mathrm{p}}^{\mathrm{e}} =$ molar (kmol) thermal capacity of the metal at temperature T and vacancy concentration $x_{\mathrm{vac}}^{\mathrm{e}}$.

During the cooling of a casting, the temperature change is negative. A certain fraction x_{vac} of vacancies has been caught by the solid phase. This value is larger than the equilibrium value $x_{\mathrm{vac}}^{\mathrm{e}}$.

During the subsequent cooling of the already formed solid, the system tries to achieve equilibrium, which results in a decrease in the vacancy concentration with decreasing temperature. This is achieved by condensation of supersaturated vacancies and condensation energy is released. This contribution to the thermal capacity corresponds to the last term in Equation (4.53).

Contribution to C_{p} from Condensation of Vacancies

The condensation of vacancies starts immediately in the thin solid shell as soon as the solidification front has been formed. The system tries to achieve equilibrium. The condensation of excess vacancies goes on during the solidification time, when the temperature is constant, and continues during and even after the cooling period.

If the vacancy concentration is known as a function of time instead of temperature during the cooling time, an alternative expression can be used for the effective thermal capacity. The contribution to the thermal capacity is a function of the condensation rate of vacancies $\mathrm{d}x_{\mathrm{vac}}/\mathrm{d}t$ and the cooling rate $\mathrm{d}T/\mathrm{d}t$. The vacancy derivative with respect to temperature, $\mathrm{d}x_{\mathrm{vac}}/\mathrm{d}T$, in Equation (4.53) can be replaced by:

$$\frac{\mathrm{d}x_{\mathrm{vac}}}{\mathrm{d}T} = \frac{\mathrm{d}x_{\mathrm{vac}}}{\mathrm{d}t}\frac{1}{(\mathrm{d}T/\mathrm{d}t)} \quad (4.54)$$

where $\mathrm{d}T/\mathrm{d}t$ is the cooling rate. If we consider 1 kmole of the material the effective molar thermal capacity can either be described by Equation (4.53) or by:

$$C_{\mathrm{p}}^{\mathrm{eff}} = C_{\mathrm{p}}^{\mathrm{e}} + \frac{\mathrm{d}x_{\mathrm{vac}}}{\mathrm{d}t}\left(-\Delta H_{\mathrm{m}}^{\mathrm{vac}}\right)\frac{1}{(\mathrm{d}T/\mathrm{d}t)} \quad (4.55)$$

The first and the third factors in the last term in Equation (4.55) are both negative on cooling. Thus the last term gives a positive contribution to the effective molar thermal capacity during the cooling time. It becomes zero when the vacancy condensation is over.

Modification of Analytical Models of Solidification during Casting

In many solidification processes the effects discussed above are negligible. For example, in pure aluminium an increase of the vacancy concentration by a factor of 5, compared with its equilibrium value, would be required in order to reduce the latent heat of solidification by only 3%.

In other cases, for example the aluminium-base alloy silumin (Example 6.2 on pages 149–150 in Chapter 6), the influence of the vacancies and other lattice defects formed during solidification processes cannot be neglected.

Figure 4.19 shows results from thermal analysis experiments. The observed heats of solidification per unit mass are plotted as functions of the cooling rate for silumin. It can be seen from the figure that the latent heat decreases considerably with increasing cooling rate. In such cases the changes in heat of solidification and thermal capacity must be considered and the analytical models have to be revised. As a single example, Equation (4.37) on page 73 will be

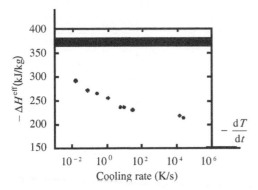

Figure 4.19 Effective heat of solidification of silumin as a function of the cooling rate. Equilibrium values [$(-\Delta H^{\text{eff}})$ at the cooling rate = 0], which have been derived from many independent measurements, fall within the dark region.

mentioned. This equation:

$$\frac{dq}{dt} = \rho(-\Delta H) \cdot \frac{dy_L(t)}{dt} + \rho c_p \int_0^{y_L(t)} -\frac{\partial T}{\partial t}\,dy \qquad (4.37)$$

has to be replaced by:

$$\frac{dq}{dt_y} = \rho\left[(-\Delta H^e) - (x_{\text{vac}} - x_{\text{vac}}^e)(-\Delta H^{\text{vac}})\right]\frac{dy_L(t)}{dt} + \rho \int_0^{y_L(t)} -\left(c_p^e + \frac{dx_{\text{vac}}}{dT}(-\Delta H^{\text{vac}})\right)\frac{\partial T}{\partial t}\,dy \qquad (4.56)$$

Analogous modifications have to be made in *all other* equations, where the heat of fusion and the thermal capacity are involved, for example in Equations (4.46) and (4.48) on page 74. However, this is only necessary in some cases.

In the rest of this chapter we will assume that the heat of fusion and the thermal capacity are constants. In Chapter 5, when we discuss rapid solidification processes, it will be necessary to consider the effect of the vacancies.

4.4 HEAT TRANSPORT IN COMPONENT CASTING

4.4.1 Temperature Distribution in Sand Mould Casting

In the preceding section we have treated the temperature distribution in mould and metal with ideal cooling, but in reality the contact between melt and mould is very seldom ideal. During casting in sand moulds the solution, in Section 3.4.2, of the general law of thermal conduction, which presumes good thermal contact between mould and melt, can be applied even if this condition is not fulfilled. The reason is the poor thermal conductivity of sand. Below we will discuss how the temperature T_i at the interface between mould and metal varies as a function of the properties of the mould material.

Calculation of the Temperature T_i between Sand Mould and Metal

The contact between metal and sand mould is not ideal. Owing to the poor thermal conductivity of the sand mould (of the magnitude 1 W/mK) it is reasonable, however, to assume that the temperature distribution is about the same as if the cooling were ideal.

In Section 4.3 we have calculated the temperature distribution in mould and metal during casting with ideal thermal contact between mould and melt. The result of these calculations is given by Equations (4.19) on page 69 in combination with Equations (4.33) to (4.35) on page 71. Equations (4.33), (4.34), and (4.35) contain a common constant λ,

which is determined from Equation (4.36) on page 71:

$$T_{\text{mould}} = A_{\text{mould}} + B_{\text{mould}}\,\text{erf}\left(\frac{y}{\sqrt{4\alpha_{\text{mould}}\,t}}\right) \qquad (4.19_{\text{mould}})$$

$$T_{\text{metal}} = A_{\text{metal}} + B_{\text{metal}}\,\text{erf}\left(\frac{y}{\sqrt{4\alpha_{\text{metal}}\,t}}\right) \qquad (4.19_{\text{metal}})$$

where A mould, B mould, A metal, and B metal are all constants.

$$B_{\text{metal}} = \frac{T_L - T_0}{\left(\dfrac{k_{\text{metal}}\sqrt{\alpha_{\text{mould}}}}{k_{\text{mould}}\sqrt{\alpha_{\text{metal}}}}\right) + \text{erf}\,\lambda} \qquad (4.33)$$

$$B_{\text{mould}} = \frac{T_L - T_0}{\left(\dfrac{k_{\text{metal}}\sqrt{\alpha_{\text{mould}}}}{k_{\text{mould}}\sqrt{\alpha_{\text{metal}}}}\right) + \text{erf}\,\lambda}\,\frac{k_{\text{metal}}\sqrt{\alpha_{\text{mould}}}}{k_{\text{mould}}\sqrt{\alpha_{\text{metal}}}} \qquad (4.34)$$

$$A_{\text{mould}} = A_{\text{metal}} = T_0 + \frac{T_L - T_0}{\dfrac{k_{\text{metal}}\sqrt{\alpha_{\text{mould}}}}{k_{\text{mould}}\sqrt{\alpha_{\text{metal}}}} + \text{erf}\,\lambda}\,\frac{k_{\text{metal}}\sqrt{\alpha_{\text{mould}}}}{k_{\text{mould}}\sqrt{\alpha_{\text{metal}}}} \qquad (4.35)$$

$$\frac{c_p^{\text{metal}}(T_L - T_0)}{-\Delta H} = \sqrt{\pi}\,\lambda\exp^{(\lambda^2)}\left(\sqrt{\frac{k_{\text{metal}}\rho_{\text{metal}}c_p^{\text{metal}}}{k_{\text{mould}}\rho_{\text{mould}}c_p^{\text{mould}}}} + \text{erf}\,\lambda\right) \qquad (4.36)$$

These equations will be applied to the present case. At the interface between the sand mould and metal we have:

$$T_i = A_{mould} = A_{metal} \qquad (4.57)$$

where T_i is the temperature at the interface. These relationships are found by replacing y by zero in Equations (4.19_{mould}) and (4.19_{metal}). Equation (4.35) can be transformed into:

$$A_{mould} = A_{metal} = T_i$$
$$= T_0 + \frac{T_L - T_0}{1 + \dfrac{\mathrm{erf}\,\lambda}{\left(k_{metal}\sqrt{\alpha_{mould}}/k_{mould}\sqrt{\alpha_{metal}}\right)}} \qquad (4.58)$$

Sand has poor thermal conductivity compared with metals, which means that $k_{metal} \gg k_{mould}$.

Because $\alpha = k/\rho c_p$ [Equation (4.11) on page 62] we get:

$$\frac{k_{metal}\sqrt{\alpha_{mould}}}{k_{mould}\sqrt{\alpha_{metal}}}$$
$$= \frac{k_{metal}\sqrt{\dfrac{k_{mould}}{\rho_{mould}c_p^{mould}}}}{k_{mould}\sqrt{\dfrac{k_{metal}}{\rho_{metal}c_p^{metal}}}} = \frac{\sqrt{k_{metal}\rho_{metal}c_p^{metal}}}{\sqrt{k_{mould}\rho_{mould}c_p^{mould}}} \gg 1 \qquad (4.59)$$

The value of erf λ in Equation (4.58) lies between 0 and 1. The second term in the denominator in Equation (4.58) will thus be very small and the denominator approximately equal to 1. In this case we get:

$$T_i = T_{i\,metal} = T_{i\,mould} \approx T_L \qquad (4.60)$$

During casting in a sand mould the temperature at the interface between the metal and the mould is approximately equal to the liquidus temperature of the melt.

4.4.2 Solidification Rate and Solidification Time in Sand Mould Casting – Chvorinov's Rule

Dry Sand Mould
The result in Equation (4.60) above that $T_i = T_{i\,metal} = T_{i\,mould} \approx T_L$ is *very* important and is frequently used for casting in sand moulds. The solidification process occurs mainly in the way illustrated by Figure 4.10 on page 66. The temperature distribution in the sand mould and in the metal is illustrated in Figure 4.20.

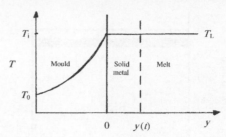

Figure 4.20 Sketch of the temperature profile during casting in a sand mould.

Because sand has a poor thermal conductivity, the heat transport through the sand mould is a 'bottle neck'. The solidification process is completely controlled by thermal conduction through the sand mould.

We start with the following assumptions:

- The thermal conductivity of the metal is very large compared to that of the sand mould
- During casting the temperature of the mould wall immediately becomes equal to the temperature T_L of the melt and maintains this temperature throughout the whole solidification process
- At large distances from the interface the temperature of the mould is equal to the room temperature T_0.

The temperature distribution in the metal is simple in this case and given by:

$$T_{metal} = T_L \qquad (4.61)$$

It is not necessary to solve any equation in this case in order to find the temperature in the metal unless the thickness is very large and/or the mould is comparatively small. Instead we can restrict the calculations to the thermal conduction in the mould.

The solution of the general law of thermal conduction:

$$\frac{\partial T_{mould}}{\partial t} = \alpha_{mould}\frac{\partial^2 T_{mould}}{\partial y^2} \qquad (4.62)$$

can, by use of customary designations, be written as:

$$T_{mould} = A_{mould} + B_{mould}\,\mathrm{erf}\left(\frac{y}{\sqrt{4\alpha_{mould}t}}\right) \qquad (4.63)$$

In order to determine the constants A_{mould} and B_{mould} we will use known boundary conditions, which have to be fulfilled.

Boundary Condition 1:
At the interface $\quad y = 0 \quad T(0,t) = T_i$
Boundary Condition 2:
In the mould $\quad y = -\infty \quad T(-\infty, t) = T_0$
These pairs of values are inserted into Equation (4.63):

$$T_i = A_{\text{mould}} + B_{\text{mould}} \,\text{erf}(0) = A_{\text{mould}} \qquad (4.64)$$

$$T_0 = A_{\text{mould}} + B_{\text{mould}} \,\text{erf}(-\infty) = A_{\text{mould}} - B_{\text{mould}} \qquad (4.65)$$

We can solve A_{mould} and B_{mould} from this equation system:

$$\begin{aligned} A_{\text{mould}} &= T_i \\ B_{\text{mould}} &= T_i - T_0 \end{aligned} \qquad (4.66)$$

If we insert these values into Equation (4.63) we get the solution to the general law of thermal conduction, i.e. the temperature distribution in the mould as a function of position and time:

$$T_{\text{mould}}(y,t) = T_i + (T_i - T_0)\,\text{erf}\left(\frac{y}{\sqrt{4\alpha_{\text{mould}}\,t}}\right) \qquad (4.67)$$

The amount of heat per unit area, which is transported into the mould per unit time in the negative direction through the interface between the mould and the metal, is given by Equation (4.3) on page 61:

$$\frac{\partial q(0,t)}{\partial t} = -k_{\text{mould}}\frac{\partial T_{\text{mould}}(0,t)}{\partial y} \qquad (4.68)$$

Equation (4.67) is derived with respect to y and the derivative is inserted into Equation (4.68):

$$\frac{\partial q}{\partial t} = -k_{\text{mould}}(T_i - T_0)\frac{2}{\sqrt{\pi}}\exp\left(-\frac{y^2}{4\alpha_{\text{mould}}\,t}\right)\frac{1}{\sqrt{4\alpha_{\text{mould}}\,t}}$$

If we insert $y = 0$ the exponential factor becomes equal to 1 and we get:

$$\frac{\partial q}{\partial t} = -k_{\text{mould}}(T_i - T_0)\frac{1}{\sqrt{\pi\alpha_{\text{mould}}t}} \qquad (4.69)$$

The relation $\alpha_{\text{mould}} = k_{\text{mould}}/\rho_{\text{mould}}\,c_p^{\text{mould}}$ is inserted into Equation (4.69) and we get:

$$\frac{\partial q}{\partial t} = -\sqrt{\frac{k_{\text{mould}}\,\rho_{\text{mould}}\,c_p^{\text{mould}}}{\pi t}}(T_i - T_0) \qquad (4.70)$$

The amount of heat that passes the interface per unit area and unit time in the negative direction into the mould consists only of solidification heat because the temperature

(T_L) is constant in the solid phase and the melt. The solidification heat per unit area and unit time will be:

$$\frac{\partial q}{\partial t} = -\rho_{\text{metal}}(-\Delta H)\frac{dy_L}{dt} \qquad (4.71)$$

where dy_L/dt is the solidification rate, i.e. the velocity of the solidification front in the melt. We can get the thickness of the solidifying layer as a function of time by inserting the expression for the partial time derivative of q [Equation (4.70)] into Equation (4.71), solve the solidification rate and integrate:

$$y_L(t) = \int_0^t \frac{T_i - T_0}{\rho_{\text{metal}}(-\Delta H)}\sqrt{\frac{k_{\text{mould}}\rho_{\text{mould}}\,c_p^{\text{mould}}}{\pi}}\frac{dt}{\sqrt{t}}$$

or

$$y_L(t) = \frac{2}{\sqrt{\pi}}\frac{T_i - T_0}{\rho_{\text{metal}}(-\Delta H)}\sqrt{k_{\text{mould}}\rho_{\text{mould}}c_p^{\text{mould}}}\,\sqrt{t} \quad (4.72)$$

It is important to notice that the *second* factor contains data for the *metal* and the *third* factor data for the *mould*.

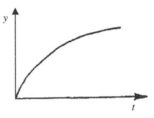

Figure 4.21 The thickness y of a solidifying shell as a function of time during casting in a sand mould.

It can be seen from Equation (4.72) that *the thickness of the solidifying layer is a parabolic function of time*. The solidification rate is rapid at the beginning of the solidification process but decreases successively when the solidified layer gets thicker.

The geometrical shape of the mould wall also influences its capacity to absorb heat. Heat is transferred more rapidly in a *concave* mould area than in a planar one because heat is spread into a *larger* volume than in the planar case. The opposite is valid for a convex surface. Heat is transferred more slowly into a *convex* mould/casting area than in a planar one because heat is spread into a *smaller* volume than in the planar case. The differences are rather small for simpler moulds, though. Two examples will be discussed below.

Thermal Conduction at Sharp Corners

Cast components often have very irregular shapes and more or less sharp corners. Figure 4.22 (a) shows the isotherms in a mould in the neighbourhood of and at a convex corner, seen from the melt (outer corner). The heat from the walls of the

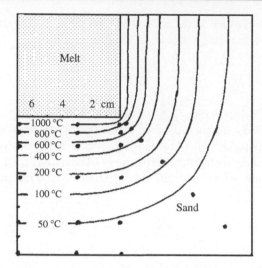

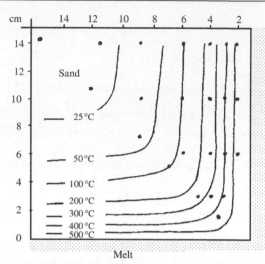

Figure 4.22 (a)   Isotherms in the sand mould outside an outer corner of a casting 15 minutes after the casting process. The cast metal is cast iron. The temperature of the inner surface is 1083 °C. The points in the figure represent thermoelements. The length scale in the figure corresponds to the distances from the corner.

Figure 4.22 (b)   Isotherms in the sand mould inside an inner corner of a casting 15 minutes after the casting process. The cast metal is an aluminium-base alloy. The temperature of the inner surface is 548 °C. The points in the figure represent thermoelements. The length scale in the figure corresponds to the distances from the corner.

casting can only be transported in the direction of the temperature gradient, i.e. perpendicular to the isotherms. The heat transport per unit area and unit time is proportional to the temperature gradient.

independent of its geometrical shape and its position on the surface of the casting. If we make this assumption we can replace $y_L(t)$ in Equation (4.72) with V_{metal}/A to give:

$$\frac{V_{metal}}{A} = \frac{2}{\sqrt{\pi}} \frac{T_i - T_0}{\rho_{metal}} \sqrt{k_{mould}\rho_{mould}c_p^{mould}} \sqrt{t_{total}} \tag{4.73}$$

The closer the isotherms are situated, the larger is the temperature gradient. It can be seen from Figure 4.22 (a) that the thermal conduction is larger at an outer corner in the direction of the diagonal than at a planar surface.

Figure 4.22 (b) shows the isotherms in a mould in the neighbourhood of and at a concave corner, seen from the melt (inner corner). The larger the distances between the isotherms, the smaller the temperature gradient and the smaller is the heat flux from the casting. Figure 4.22 (b) shows that the distances between the isotherms in the direction of the diagonal are larger than those perpendicular to a planar surface. Thermal conduction is thus smaller at an inner corner in the direction of the diagonal than that at a planar surface.

Chvorinov's Rule

A useful approximation is to assume that every unit area of the mould wall has a constant ability to absorb heat

where
V_{metal} = total volume of the solidified casting
A = total area of the interface between the mould and the metal
t_{total} = total solidification time.

Equation (4.73) can be written more simply as:

$$t_{total} = C \left(\frac{V_{metal}}{A}\right)^2 \qquad \text{Chvorinov's rule} \tag{4.74a}$$

The total solidification time of a casting is proportional to the square of the volume of the casting and inversely proportional to the square of the contact area between sand mould and the casting.

The constant C is obtained in terms of material constants by identification of Equations (4.73) and (4.74a). The result is:

$$C = \frac{\pi}{4} \frac{\rho_{\text{metal}}^2 (-\Delta H)^2}{(T_i - T_0)^2 \, k_{\text{mould}} \rho_{\text{mould}} c_p^{\text{mould}}} \tag{4.74b}$$

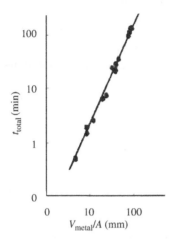

Figure 4.23 Chvorinov's experimental results of the solidification time of castings as a function of their volume/area ratio. The scale is logarithmic on both axes and the slope of the line is 2. Reproduced with permission from Pergamon Press, Elsevier Science.

Equation (4.74a) is very well known and extraordinarily useful for relative comparisons of the solidification times of castings made of the same material.

Chvorinov has verified his rule by experiments on sand mould castings of varying shapes and sizes, from 10-mm castings to 65-ton ingots (Figure 4.23). Later it is shown that a similar rule is valid for other castings as well, such as some (but not all) ingot castings. This subject will be further discussed on page 96 in Chapter 5.

Example 4.4
Thin castings are produced in a foundry. The thickness of the cast iron is just enough to prevent white solidification. A new mould production method is introduced, which gives denser packing of the sand in the mould. This increases the risk of white solidifying, which gives poor quality of the casting. The density of the mould material increases by 20 % and the thermal conductivity of the sand increases by 10 %.

The casting was earlier done in such a way that the temperature of the melt in the mould at the end of the casting process was exactly equal to the solidus temperature. By how many degrees must the casting temperature be increased if you tolerate the same maximum risk of white solidification as before? This condition means that the solidification time should be the same in the two cases.

$$-\Delta H = 170 \text{ kJ/kg} \qquad \text{and} \qquad c_p^{\text{metal}} = 0.42 \text{ kJ/kg K}.$$

Solution:
We assume that the casting temperature has to be increased by the amount ΔT. The thermal capacitivity c_p^{mould} will not change because of the new casting method. We apply Chvorinov's rule to calculate the solidification time at casting according to the old method (index 1) and the new method (index 2).

$$t_1 = \left(\frac{\sqrt{\pi}}{2} \frac{\rho_{\text{metal}} (-\Delta H)}{T_i - T_0} \frac{1}{\sqrt{k_{\text{mould1}} \rho_{\text{mould1}} c_p^{\text{mould1}}}} \right)^2 \left(\frac{V_{\text{metal}}}{A} \right)^2 \tag{1'}$$

$$t_2 = \left(\frac{\sqrt{\pi}}{2} \left(\frac{\rho_{\text{metal}} \left(-\Delta H + c_p^{\text{metal}} \Delta T \right)}{T_i - T_0} \right) \frac{1}{\sqrt{k_{\text{mould2}} \rho_{\text{mould2}} c_p^{\text{mould2}}}} \right)^2 \left(\frac{V_{\text{metal}}}{A} \right)^2 \tag{2'}$$

The solidification time must be the same in both cases, i.e. $t_1 = t_2$, to prevent an increased risk of white solidification. This gives the equality:

$$\frac{-\Delta H}{\sqrt{k_{\text{mould1}} \rho_{\text{mould1}}}} = \frac{-\Delta H + c_p^{\text{metal}} \Delta T}{\sqrt{k_{\text{mould2}} \rho_{\text{mould2}}}} \tag{3'}$$

or

$$\frac{-\Delta H + c_p^{\text{metal}} \Delta T}{-\Delta H} = \frac{\sqrt{k_{\text{mould2}} \, \rho_{\text{mould2}}}}{\sqrt{k_{\text{mould1}} \, \rho_{\text{mould1}}}} = \sqrt{1.10 \times 1.20} \tag{4'}$$

The values given in the text are inserted and we get $\Delta T = 60 \text{ °C}$.

Answer:
The casting temperature increase should be 60 °C.

Relation between Solidification Time and the Ratio V/A for Spherical and Cylindrical Moulds

For spherical and cylindrical moulds it is possible to derive a more exact expression than Chvorinov's rule, which gives a relationship between the casting time and the ratio *V/A*. In these cases the partial differential equation for thermal conduction can be written as:

$$\frac{\partial T}{\partial t} = \alpha_{mould} \left(\frac{\partial^2 T}{\partial r^2} + \frac{n}{r} \frac{\partial T}{\partial r} \right) \qquad (4.75)$$

where

r = the radius of the casting
$n = 1$ for a cylinder
$n = 2$ for a sphere.

If a similar derivation is performed like that which led to Equation (4.73), the result will be:

$$\frac{V_{metal}}{A} = \left(\frac{T_i - T_0}{\rho_{metal}(-\Delta H)} \right) \left(\frac{2}{\sqrt{\pi}} \sqrt{k_{mould}\rho_{mould}c_p^{mould}} \sqrt{t_{total}} + \frac{nk_{mould}t_{total}}{2r} \right) \qquad (4.76)$$

A comparison between Equations (4.76) and (4.74) shows that the more k_{mould} decreases and r increases, the better will Chvorinov's simple approximation be valid.

There is better agreement for a cylinder than for a sphere. For a given *V/A* ratio a sphere solidifies more rapidly than a cylinder, which in turn solidifies more rapidly than a plate.

Example 4.5

Determine the solidification time for a steel cylinder with a diameter of 15 cm that is cast in a sand mould. The height of the cylinder is much larger than its diameter. The sand mould and the steel have the following material constants:

The thermal conductivity of the sand	$k_{mould} = 0.63$ J/m K s
The density of the sand	$\rho_{mould} = 1.61 \times 10^3$ kg/m^3
The thermal capacity of the sand	$c_p^{mould} = 1.05 \times 10^3$ J/kg K
Solidification temperature of the steel	$T_L = T_i = 14:90$ °C
Temperature of the surroundings	$T_0 = 23$ °C
Solidification heat of the steel	$-\Delta H = 272$ kJ/kg.

Solution:
We apply Equation (4.76):

$$\frac{V_{metal}}{A} = \frac{T_i - T_0}{\rho_{metal}(-\Delta H)} \left(\frac{2}{\sqrt{\pi}} \left(\sqrt{k_{mould}\rho_{mould}c_p^{mould}} \right) \sqrt{t_{total}} + \frac{nk_{mould}t_{total}}{2r} \right) \qquad (1')$$

We introduce the given values and the height L of the cylinder:

$$\frac{T_i - T_0}{\rho_{metal}(-\Delta H)} = \frac{1490 - 23}{(7.8 \times 10^3) \times (272 \times 10^3)} = 0.691 \times 10^{-6} \text{ K m}^3/\text{J}$$

$$\frac{2}{\sqrt{\pi}} \sqrt{k_{mould}\rho_{mould}c_p^{mould}} = \frac{2}{\sqrt{\pi}} \sqrt{0.63 \times (1.61 \times 10^3) \times (1.05 \times 10^3)} = 1.16 \times 10^3 \text{ J/m}^2 \text{ K s}^{0.5}$$

$$\frac{V_{metal}}{A} = \frac{\pi r^2 L}{2\pi r L} = \frac{r}{2} = \frac{7.5}{2} = 3.75 \times 10^{-2} \text{ m}$$

and

$$\frac{nk_{mould}}{2r} = \frac{1 \times 0.63}{2 \times 0.075} = 4.2 \text{ J/m}^2\text{s K}$$

A dimension check shows that the dimensions agree. We introduce the calculated and known values into Equation (4.76):

$$3.75 \times 10^{-2} = 0.691 \times 10^{-6} \left(1.16 \times 10^3 \sqrt{t_{\text{total}}} + 4.2\, t_{\text{total}}\right) \tag{1'}$$

This is a second-order equation of $\sqrt{t_{\text{total}}}$ which can be written as:

$$t_{\text{total}} + 276 \sqrt{t_{\text{total}}} - 12921 = 0 \tag{2'}$$

This equation has the roots:

$$\sqrt{t_{\text{total}}} = -138 \pm \sqrt{138^2 + 12921} \tag{3'}$$

The solidification time is associated with the positive root:

$$\sqrt{t_{\text{total}}} = -138 + 179 = 41; \quad t_{\text{total}} = 1681\,\text{s} = 28\,\text{min}$$

Answer:
The steel cylinder solidifies completely after approximately 28 minutes.

Temperature Distribution in a Moist Sand Mould – Velocity of the Evaporation Front/Waterfront

Sand moulds often contain water, which is evaporated when the sand mould is heated by the melt. This evaporation contributes to the cooling due to the high evaporation heat of water and influences the solidification process for this reason.

The water vapour will condense when it reaches the level in the mould where the temperature is below $100\,°\text{C}$. A waterfront, which moves inside the mould, is formed. Figure 4.24 describes graphically the temperature distribution in the mould in this case. The evaporation front and the waterfront are identical. Mathematically the temperature distribution can, even in this case, be described by a solution to the law of thermal conduction of the type (for

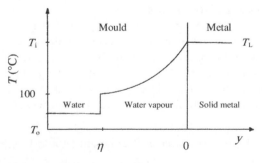

Figure 4.24 Temperature distribution in a moist sand mould. η = the coordinate of the waterfront.

simplicity the index mould has been dropped):

$$T = A + B \operatorname{erf}\left(\frac{y}{\sqrt{4\alpha t}}\right) \tag{4.77}$$

Boundary conditions:
(1) At $y = 0$ the temperature at the interface between mould and metal will be:

$$T = T_i \tag{4.78}$$

(2) At the waterfront $y = \eta$ (η is always negative, see Figure 4.24) the temperature will be:

$$T = 100\,°\text{C} \tag{4.79}$$

The first boundary condition gives the constant $A = T_i$. The second boundary condition permits calculation of the constant B:

$$100 = A + B \operatorname{erf}\left(\frac{\eta}{\sqrt{4\alpha t}}\right) \tag{4.80}$$

For the same reasons as before (page 70) the ratio $\eta / \sqrt{4\alpha t}$ must be independent of time:

$$\eta = -\lambda \sqrt{4\alpha t} \tag{4.81}$$

where λ is a positive constant, which is determined [compare Equation (4.36) on page 71] by:

$$\sqrt{\pi}\, \lambda \exp^{(\lambda^2)} \operatorname{erf}\lambda = \frac{c_p \,(T_i - 100)}{(-\Delta H^{\text{vapour}})\,(\text{fraction of moisture})} \tag{4.82}$$

where the fraction of moisture is the weight per cent water based on the weight of the mould. The values of the constants become:

$$A = T_i \quad \text{and} \quad B = \frac{T_i - 100}{\operatorname{erf}\lambda}$$

λ can be determined by iteration or graphically with the aid of Figures 4.14 and 4.15 on pages 71–72.

λ determines the velocity of the waterfront in the mould. We get an expression of this velocity by deriving Equation

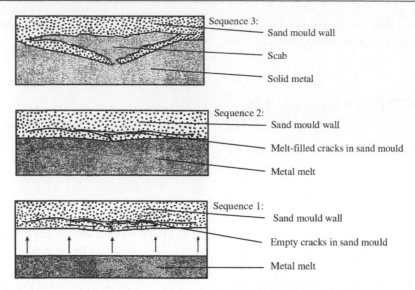

Figure 4.25 (a)   The melt creeps into, enlarges and fills the cracks in the sand mould. A so-called *scab* is formed. Reproduced with permission from the Butterworth Group, Elsevier Science.

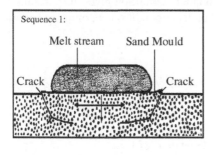

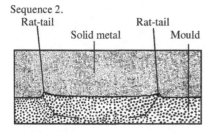

Figure 4.25 (b)   The melt cannot creep into very narrow cracks. Instead so-called *rat tails* are formed. Reproduced with permission from the Butterworth Group, Elsevier Science.

(4.81) with respect to *t* [compare Equation (4.27) on page 70]:

$$\frac{d\eta}{dt} = -\lambda \sqrt{\frac{\alpha}{t}} \qquad (4.83)$$

where $d\eta/dt$ is the velocity of the water front. $d\eta/dt$ is negative for all values of *t*. The position of the waterfront is defined by Equation (4.81). It moves in the direction of the negative *y*-direction with a velocity defined by Equation (4.83). Equations (4.81) and (4.83) are analogous with Equations (4.26) and (4.27) on page 70.

As mentioned on page 83 a waterfront, which moves forward, is formed in moist sand moulds. The amount of water at the waterfront increases successively, which has a negative effect on the mechanical properties of the mould. Figure 4.25 gives two examples of casting faults, which appear at the formation of a waterfront.

Figure 4.25 (a) shows a case where the upper sand mould wall is heated by the melt during the mould-filling process and a waterfront is formed in the wall. The moist sand in the mould has very bad mechanical properties in the layer that contains the waterfront. When the sand in this layer expands during heating thermal stresses force the sand layer to split up (sequence 1). Later the cracks in the sand get filled with melt (sequence 2) and get enlarged. The appearance of the rough surface after solidification (sequence 3) associates to scab.

Figure 4.25 (b) shows a similar case. Here the sand mould gets heated inhomogeneously when a melt stream flows along the lower surface of the mould during the filling process. In this case the poor mechanical properties of the waterfront layer result in narrow cracks (sequence 1). A mirror pattern of the thin cracks is formed in the casting. The pattern is reminiscent of rat tails (sequence 2).

4.4.3   Heat Transport in Permanent Mould Casting

The most important permanent mould casting processes are squeeze casting, pressure die casting and gravity die casting, where the metal is poured into the mould. All these methods are described in Chapter 1. Permanent mould casting is used for a variety of cast components, the dominating cast metal being aluminium.

A sand mould can only be used once, while permanent moulds, made of metal, can be used thousands or millions of times depending on the type of cast metal. Such moulds have 30–50 times better thermal conductivity than do sand moulds and allow high cooling rates and rapid cooling. They are therefore most useful for thin component casting.

Heat Transfer at the Interface Mould/Metal

Heat transfer in permanent mould casting mainly depends on the interface between the mould and the cast metal. The heat flux from the hot casting to the cooler mould varies with time and temperature. It depends on:

- the air gap between mould and casting;
- surface roughness and interfacial air or gas films;
- mould coating.

The *air gap* is caused by solidification shrinkage and thermal contraction during cooling of the casting. It is discussed in Section 4.3 and will be further discussed in connection with continuous casting and in Chapter 10, where we discuss thermal contraction.

The *surface roughness* of both the mould and the solidifying metal causes very uneven contact between the two surfaces. The voids are filled with air or some other gas in special cases. Both these effects contribute to the thermal resistance of the interface.

The mould is often covered with a thin film of some ceramic powder, held together by a binder, usually water glass (silicate of sodium). The coating in an aqueous dispersion is normally sprayed on the interior surface of the hot mould ($\sim$200 °C). When the dispersion hits the mould surface the water evaporates immediately, leaving big voids inside the coating (Figure 4.26).

The coating layer has low thermal conductivity, mainly due to being about 50% voids, and forms a thin insulating layer. It provides several functions. It provides thermal insulation, which helps to fill the mould completely before the solidification starts, prevents the casting from too early

solidification, and reduces the formation of cold shots. The coating absorbs inclusions and controls the heat transfer from casting to mould. It also protects the mould surface from wear and thermal fatigue, which prolongs the lifetime of the mould.

4.4.4 Water Cooling during Die Casting

Several measures have to be taken to design a die casting machine in an optimal way. Vents and draw pockets are necessary to prevent air inclusions at the filling of the mould, for example.

The irregular shapes of many castings require special precautions to maintain the temperature at an even and optimal level in all parts of the casting. Some sections, for example gating points where the speed of the injected metal raises the temperature, must be cooled with water to maintain the correct temperature.

The water runs in special water tubes, which are drilled into the die block. In some cases they are drilled through the whole block. In other cases the cooling is selective. The water may enter through one short pipe, circulate around the regions to be cooled and leave through another short pipe or nipple.

The design of proper water cooling is extensively discussed in Section 5.4 in Chapter 5. This is common for all casting methods with water cooling.

4.4.5 Nussel's Number – Temperature Profile at Low Values of Nussel's Number

On page 73 we found that the heat transfer across the interface metal/mould drastically decreases when the solid metal loses contact with the mould wall. The solidification process was analysed, which among other things resulted in an expression for the temperature of the metal at the interface as function of h, k and the distance y_L of the solidification front from the interface. The expression is Equation (4.45) on page 74:

$$T_{i\,\text{metal}} = \frac{T_L - T_0}{1 + \dfrac{h}{k}\, y_L(t)} + T_0 \qquad (4.45)$$

We will analyse this relationship more closely here. If the heat transfer at the interface between metal and mould is very slow (h very small) and/or the thermal conductivity of the solid cast metal is large (k very large), then the second term in the denominator will be small.

At complete solidification, y_L has reached its maximum value, which we will call s. At constant values of the temperatures of the melt and the surroundings in Equation (4.45), the temperature gradient in the solidifying shell is obviously determined by the so-called *Nussel's number*,

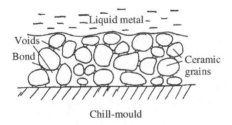

Figure 4.26 Ceramic mould coating. The thickness of the layer is normally of the magnitude 1–0.1 mm.

defined by the relationship:

$$Nu = \frac{hs}{k} \quad (4.84)$$

where

$h =$ heat transfer number for the interface between the mould and the metal

$s =$ value of y_L at complete solidification

$k =$ thermal conductivity of the metal.

Nussel's number is frequently used as a criterion on the choice of temperature distribution model. If $Nu \ll 1$ the simple temperature distribution, illustrated in Figure 4.27, is valid and can safely be used.

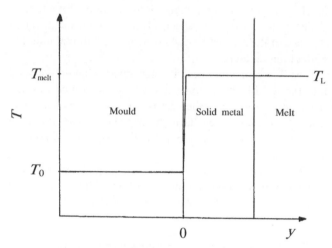

Figure 4.27 Temperature distribution in mould and metal for low values of Nussel's number.

If $Nu \ll 1$ Equation (4.48) on page 74 can be simplified to:

$$t = \frac{\rho\,(-\Delta H)}{T_L - T_0}\,\frac{y_L}{h} \quad (4.85)$$

The total solidification time is obtained if y_L in Equation (4.85) is replaced by the thickness of the casting for unidirectional cooling. Equation (4.85) can also be applied to *bilateral* cooling of castings. In this case s means *half* the thickness of the casting.

The solidification process described above is valid for thin component casting in permanent moulds, i.e. when $Nu \ll 1$.

Example 4.6

You will produce a thin component for the car industry by pressure die casting. To make production cheap, the article should be cast as rapidly as possible. You can either cast the article in a magnesium alloy or in an aluminium alloy. Which one should you use? The heat transfer con-

stant can be regarded as equal in both cases. Material constants for the two metals are found in standard reference tables.

Solution:

In order to apply Equation (4.85) and calculate the solidification time in each case, we have to use the material constants for Mg and Al, taken from standard reference tables:

$$-\Delta H_{Al} = 354 \times 10^3 \text{ J/kg} \qquad -\Delta H_{Mg} = 208 \times 10^3 \text{ J/kg}$$

$$\rho_{Al} = 2.69 \times 10^3 \text{ kg/m}^3 \qquad \rho_{Mg} = 1.74 \times 10^3 \text{ kg/m}^3$$

The temperature of the melt is approximately equal to the melting point temperature of the metal:

$$T_M^{Al} \approx 658\,°C \qquad T_M^{Mg} \approx 651\,°C.$$

y_L is the same in both cases. The values above and $T_0 = 20\,°C$ are inserted into Equation (4.85), which gives:

$$\frac{t_{Al}}{t_{Mg}} = \frac{\left(\dfrac{\rho_{metal}(-\Delta H)}{T_M - T_0}\right)_{Al}}{\left(\dfrac{\rho_{metal}(-\Delta H)}{T_M - T_0}\right)_{Mg}}$$

$$= \frac{(2.69 \times 10^3)(354 \times 10^3)}{658 - 20} \times \frac{651 - 20}{(1.74 \times 10^3) \cdot (208 \times 10^3)} = 2.6$$

Answer:

The Mg alloy should be chosen. It has the shortest solidification time.

SUMMARY

■ Thermal energy is transported by means of conduction, radiation and convection.

■ *Thermal Conduction*

Thermal conduction in a layer – stationary conditions

$$\frac{dq}{dt} = -k\,\frac{dT}{dx} \qquad \text{Fourier's first law}$$

Heat transfer at the interface between two different materials – stationary conditions

$$\frac{dQ}{dt} = -hA\left(T_2 - T_1\right) \quad or \quad \frac{dq}{dt} = -h\left(T_2 - T_1\right)$$

h = heat transfer coefficient,

Thermal conduction through several layers and interfaces in series – stationary conditions

$$\frac{L}{kA} = \frac{L_1}{k_1 A_1} + \frac{L_2}{k_2 A_2} + \frac{L_3}{k_3 A_3} + \cdots \frac{1}{h_4 A_4} + \frac{1}{h_5 A_5} \cdots\cdots$$

Relationship between h and k at an air gap:

$$h = \frac{k}{\delta}$$

■ General Law of Thermal Conduction – Nonstationary Conditions

$$\frac{\partial T}{\partial t} = \alpha \frac{\partial^2 T}{\partial y^2} \qquad \text{Fourier's second law}$$

The constant $\alpha = \frac{k}{\rho c_p}$ is called *thermal diffusivity*

■ Thermal Radiation

$$dW_{total} = \varepsilon \sigma A \left(T^4 - T_0^4\right) dt$$
ε = *emissivity*, a dimensionless factor < 1.

■ Convection
Water cooling in an example of *forced* convection. The motion of the flowing water is controlled.

Natural or *free* convection occurs under the influence of gravitational forces without external influences.

$$\frac{dq}{dt} = h_{con}\left(T - T_0\right)$$

■ Heat Transport in Casting with Ideal Contact between Metal and Mould
The temperature distribution is calculated by solving the general law of thermal conduction and determining the constants with the aid of boundary conditions such as given temperatures of the melt, of the interface mould/metal and of the surroundings.

Five constants appear in the solution. One of them is λ. It is determined by means of iteration from the equation

$$\frac{c_p^{\text{metal}}\left(T_L - T_0\right)}{-\Delta H} = \sqrt{\pi}\,\lambda\,\exp^{(\lambda^2)}\left(\sqrt{\frac{k_{\text{metal}}\rho_{\text{metal}}c_p^{\text{metal}}}{k_{\text{mould}}\rho_{\text{mould}}c_p^{\text{mould}}}} + \text{erf}\,\lambda\right)$$

When λ is known some important quantities can be calculated:

Position of the solidification front: $\quad y_L(t) = \lambda\sqrt{4\alpha_{\text{metal}}\,t}$

Solidification rate: $\qquad\qquad \dfrac{dy(t)}{dt} = \lambda\sqrt{\dfrac{\alpha_{\text{metal}}}{t}}$

■ Heat Transport at Casting with Poor Contact between Solid Metal and Mould
At poor contact between mould and metal there is a discontinuity of the temperature at the interface. This is a very common case. When the solidifying shell solidifies and cools, it shrinks and loses contact with the mould wall. The heat contact between mould and metal becomes suddenly poor.

Solidification Processes with a Temperature Decrease Across the Interface
When the temperature of the melt, the conductivity, the heat transfer number of the air gap and the temperature of the surroundings are known, it is possible to calculate the following:

Heat flux at the interface:

$$\frac{dq}{dt} = h\left(T_{i\,\text{metal}} - T_0\right)$$

Temperature of the metal at the metal/mould interface:

$$T_{i\,\text{metal}} = \frac{T_L - T_0}{1 + \dfrac{h}{k}\,y_L(t)} + T_0$$

Solidification rate:

$$\frac{dy_L}{dt} = \frac{T_L - T_0}{\rho(-\Delta H)} \cdot \frac{h}{\left(1 + \dfrac{h}{k}\right)y_L}$$

Relationship between the solidification time and the thickness of the solidified layer:

$$t = \frac{\rho\left(-\Delta H\right)}{T_L - T_0}\frac{y_L}{h}\left(1 + \frac{h}{2k}\,y_L\right)$$

At unilateral cooling $y_L = s$ and the time t is equal to *the total solidification time*.

At symmetrical bilateral cooling $y_L = s/2$, which gives a shorter total solidification time.

Influence of Vacancies and Other Lattice Defects on Heat of Fusion and Thermal Capacity

(i) *Heat of Fusion*

It has been found that the heat of fusion $(-\Delta H)$ is not always constant but depends on the solidification rate or the cooling rate during the solidification process. For some alloys the observed heat of fusion $(-\Delta H^{eff})$ decreases with increasing growth rate:

$$-\Delta H_m^{eff} = (-\Delta H_m^e) - \left(x_{vac} - x_{vac}^e\right)(-\Delta H_m^{vac})$$

(ii) *Thermal Capacity*

A similar influence of vacancies and other lattice defects on the thermal capacity has been observed:

$$C_p^{eff} = C_p^e + \frac{dx_{vac}}{dT}\left(-\Delta H_m^{vac}\right)$$

$$= C_p^e + \frac{dx_{vac}}{dt}\left(-\Delta H_m^{vac}\right) 1/(dT/dt)$$

■ *Heat Transport in Component Casting – Casting in Sand Moulds*

Temperature Distribution in a Dry Sand Mould

The contact between the sand mould and the metal is poor. However, due to the poor thermal conductivity of the sand mould, there is no discontinuity in temperature at the metal/mould interface. In this respect it appears as if the contact were ideal.

The solution of the general law of thermal conduction is thus analogous to the one in the case of ideal cooling. λ is determined by means of iteration from the equation:

$$\frac{c_p^{metal}(T_L - T_0)}{-\Delta H} = \sqrt{\pi}\,\lambda\exp^{(\lambda^2)}\left(\sqrt{\frac{k_{metal}\rho_{metal}c_p^{metal}}{k_{mould}\rho_{mould}c_p^{mould}}} + \mathrm{erf}\,\lambda\right)$$

Calculations show that $T_i = T_{i\,metal} = T_{i\,mould} \approx T_L$

When casting in a sand mould the temperature at the interface between the metal and the mould is approximately equal to the liquidus temperature of the melt. In a moist sand mould a waterfront or evaporation front is formed.

Thickness of Solidified Shell

Thickness of the solidified layer is a parabolic function of time:

$$y_L(t) = \frac{2}{\sqrt{\pi}}\frac{T_L - T_0}{\rho_{metal}(-\Delta H)}\sqrt{k_{mould}\rho_{mould}c_p^{mould}}\,\sqrt{t}$$

Chvorinov's Rule:

At casting in a sand mould the solidification time is proportional to the square of the ratio volume/area of the casting (Chvorinov's rule):

$$t_{total} = C\left(\frac{V_{metal}}{A}\right)^2$$

where

$$C = \frac{\pi}{4}\frac{\rho_{metal}^2(-\Delta H)^2}{(T_i - T_0)^2\,k_{mould}\rho_{mould}c_p^{mould}}$$

■ *Heat Transport in Casting Using Permanent Moulds*

Permanent moulds are made of metal. At solidification of the casting, an air gap appears between the mould and the casting, which results in poor thermal conduction.

Heat Transport Across an Interface with an Air Gap

Definition of Nussel's number: $Nu = hs/k$

Temperature of the metal/mould interface if $Nu \ll 1$:

$$T_{i\,metal} = \frac{T_L - T_0}{1 + \frac{h}{k}y_L(t)} + T_0$$

Relationship between the solidification time and the thickness of the solidified layer if $Nu \ll 1$:

$$t = \frac{\rho(-\Delta H)}{T_L - T_0}\frac{y_L}{h}$$

With unilateral cooling $y_L = s$ and the time t is equal to *the total solidification time*.

With symmetrical bilateral cooling $y_L = s/2$, which gives a shorter total solidification time.

Calculation of Nussel's number is very useful in choosing a model for calculations on a given casting process.

If Nu ≪ 1 the simplified equations are valid. If the Nussel criterion is not fulfilled then the normal equations for heat transport in casting with poor contact between metal and mould are valid.

EXERCISES

4.1 Turbine blades for jet engines are made by precision casting of highly resistant Fe- and Ni-base alloys. The performance and lifetime of the blades can be increased by aligning the solidification at casting and the following solidification process in such a way that the crystals are directed along the lengthwise direction of the blades. One way to do this is illustrated in the figure.

Calculate the solidification time of a blade with a length of 10 cm. For the calculation the material constants of pure iron can be used. The melt is not superheated. The contact between the Cu-plate and the casting is assumed to be ideal.

Hint A2

Material constants and other data are found in the table below.

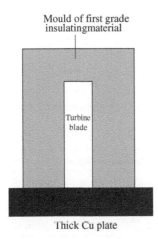

Mould of first grade insulatingmaterial

Turbine blade

Thick Cu plate

Quantity	Fe (metal)	Cu (mould)
k	32 W/m K (700 °C)	350 W/m K (200 °C)
ρ_s	7.88×10^3 kg/m^3 (25 °C)	8.94×10^3 kg/m^3 (25 °C)
c_p^s	830 J/kg K ($\sim$1100 °C)	397 J/kg K ($\sim$400 °C)
$-\Delta H$	272 kJ/kg	
T	$T_L = 1808$ K (no excess temperature)	$T_0 = 373$ K

4.2 Stainless steel tube castings are often cast in Cu chill-moulds by centrifugal casting. A well-balanced quantity of metal, suitable for casting, is supplied through a channel in the inner part of the chill-mould. The centrifugal force presses the melt towards the chill-mould during the whole casting process. Solidification of the stainless steel melt occurs from the chill-mould surface and inwards towards the centre. The melt is not superheated.

Calculate an approximate value of the solidification time of a tube casting with a thickness of 10 cm. Material constants are found in the table below.

Hint A30

Quantity	Fe (stainless steel)	Cu (chill-mould)
k	30 W/m K (1325 °C)	398 W/m K (25 °C)
ρ_s	7.50×10^3 kg/m^3 (25 °C)	8.94×10^3 kg/m^3 (25 °C)
c_p^s	650 J/kg K ($\sim$500 °C)	384 J/kg K ($\sim$25 °C)
$-\Delta H$	300 kJ/kg °C	
T	$T_L = 1598$ K (1325 °C) (no excess temperature)	$T_0 = 298$ K (25 °C)

4.3 (a) Calculate the solidification time when you cast pure aluminium in a sand mould. The size of the casting is $900 \times 100 \times 900$ mm. The melt is not superheated. Material constants for the sand mould are given in the table below.

Hint A137

Material constants
Aluminium:
T_L = 660 °C
ρ_{Al} = 2.7×10^3 kg/m^3
$-\Delta H$ = 398 kJ/kg
Sand mould:
k_{mould} = 0.63 W/m K
ρ_{mould} = 1.61×10^3 kg/m^3
c_p^{mould} = 1.05×10^3 J/kg K

Use the material constants for steel given in Exercise 4.1.

(b) Calculate the solidification time for a steel casting, with the same dimensions as in (a), which is cast in a sand mould. Discuss and compare the results in (a) and (b) in terms of driving force of heat transport in the two cases. The temperature of the surroundings is 25 °C.

Hint A327

4.4 An aluminium cube with the side 25 cm is cast in a sand mould. Calculate approximately the cooling curve (temperatures and time intervals down to 50 °C below the liquidus temperature) in the centre of the cube, where a thermoelement has been located. The melt has an excess temperature of 50 °C. The temperature of the surroundings is 25 °C.

Hint A295

Material constants
Aluminium:
$c_p^L = 1.18$ kJ/kg K at 660 °C
$c_p^S = 1.25$ kJ/kg K at 660 °C
Other material constants for aluminium and constants for the sand mould are given in Exercise 4.3.

4.5 A straight cylinder with a diameter of 30 cm and a height of 60 cm is to be produced in cast iron. The cylinder will be cast in a sand mould. Calculate the solidification time if the casting temperature is 1160 °C, which corresponds to the liquidus temperature. Material constants are listed in the table below.

Hint A62

Material constants	
ρ_{Fe}	$= 7.2 \times 10^3$ kg/m^3
$-\Delta H_{Fe}$	$= 162$ kJ/kg
c_p^{Fe}	$= 420$ J/kg K
ρ_{sand}	$= 1.5 \times 10^3$ kg/m^3
c_p^{sand}	$= 1.05$ kJ/kg K
k_{sand}	$= 0.63$ J/m s K

4.6 In order to increase the production capacity at casting of thin wall Al castings, a foundry has decided to change from sand mould casting to metal mould casting of a product with a thickness of 5.0 mm. The heat transfer coefficient between metal and mould at the mould casting is 900 W/m^2 K. Material data are listed in the table. The room temperature is 20 °C.

Compare the solidification time of the product when cast in a sand mould and in a metal mould.

Hint A40

Material constants	
T_M^{Al}	$= 660$ °C
ρ_{Al}	$= 2.7 \times 10^3$ kg/m^3
$-\Delta H_{Al}$	$= 398$ kJ/kg
k_{Al}	$= 0.23 \times 10^3$ W/m K
c_p^{Al}	$= 1.25$ kJ/kg K
ρ_{sand}	$= 1.6 \times 10^3$ kg/m^3
c_p^{sand}	$= 1.05$ kJ/kg K
k_{sand}	$= 0.63$ J/ms K

4.7 You must make wedge-shaped details of an Al–Si alloy. To be able to design a cyclic casting process you must know the solidification time as a function of the distance from the top of the wedge. Find this function and calculate the total solidification time.

Assume that the heat transport in the length direction of the wedge (perpendicular to the plane seen in the figure) and through the bottom areas can be neglected. The width OB of the wedge plane is 10 cm and the top angle is 10°. The heat transfer coefficient between air and alloy is assumed to be 2.0×10^3 W/m^2 K. The room temperature is 20 °C.

Hint A101

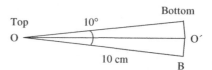

Material constants	
ρ_s	$= 2.6 \times 10^3$ kg/m^3
$-\Delta H$	$= 373$ kJ/kg
T_L	$= 853$ K
k	$= 1.84 \times 10^4$ W/m K

4.8 Aluminium cylinders with various diameters are to be cast. The heat transfer coefficient between mould and casting is 1.68 kW/m^2 K. The excess temperature of the melt can be neglected in the calculations. The room temperature is 25 °C. Material constants for aluminium are given in the table.

Material data for aluminium

T_L	$= 660\ °C$
k	$= 220\ W/m\ K\ at\ T_L$
$-\Delta H$	$= 390\ kJ/kg$
ρ	$= 2.7 \times 10^3\ kg/m^3$

(a) Derive an equation that describes the solidification time as a function of the radius of the cylinder and illustrate the function in a diagram.

Hint A135

(b) Plot the solidification rate as a function of the distance from the outer surface and inwards for a cylinder with a diameter of 20 cm.

Hint A76

4.9 It can be seen from Figure 4.1 on page 60 that the heat transport during the solidification process of a casting can be described as a number of steps, coupled in series. The step or steps that correspond to the largest heat transfer resistance will determine the whole temperature distribution.

The step that normally offers the largest heat transfer resistance is the air gap between the mould and the casting. Depending on the circumstances, the temperature distribution can either be described by Figure 4.17 on page 73 or by Figure 4.27 on page 86.

The heat transfer coefficient h varies from 2×10^2 up to 2×10^3 W/m² K in casting processes of technical interest. The thermal conductivity varies strongly, depending on the choice of alloy.

(a) Discuss the conditions for the temperature distributions in Figure 4.17 and Figure 4.27.

Hint A49

(b) Calculate the surface temperature $T_{i\ metal}$ of steel and copper castings as a function of the thickness y_L of the solidified shell. Use two values of the heat transfer coefficient, $2 \times 10^2\ W/m^2\ K$ and $2 \times 10^3\ W/m^2\ K$ respectively, for the respective metal. Show the results in two diagrams, one for steel and one for copper. The temperature of the surroundings is 20 °C.

Hint A268

Material constants

Steel:
$T_L = 1530\ °C$
$k\ = 30\ W/m\ K$
Copper:
$T_L = 1083\ °C$
$k\ = 398\ W/m\ K$

4.10 In Chapter 4 the solidification rates of various types of casting processes are discussed. Using measurements of solidification rates of a steel melt, cast in a cast iron mould, the thickness of the solidified layer as a function of time, illustrated in the figure, has been found experimentally.

(a) Explain the shape of the curve.

Hint A100

(b) Try to estimate the heat transfer coefficient from the experimental data.

Hint A48

The room temperature is 25 °C.

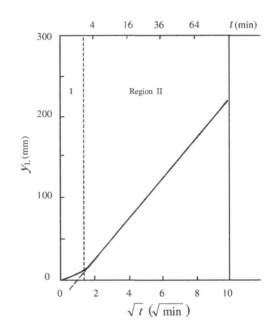

Material constants

ρ_{steel}	$= 7.9 \times 10^3\ kg/m^3$
k_{steel}	$= 30\ W/m\ K$
$-\Delta H_{steel}$	$= 270\ kJ/kg$
T_L	$\approx 1500\ °C$

4.11 The figure shows the result of temperature measurements during the solidification process at various positions and various times in a square Al ingot with the dimensions 18 cm × 18 cm. The temperature is plotted as a function of position in the ingot at various times (0.25−3.5 min) after casting. A manifold of curves with the time as parameter is obtained.

(a) Use the figure to estimate the heat transfer coefficient between the mould and the metal, as a

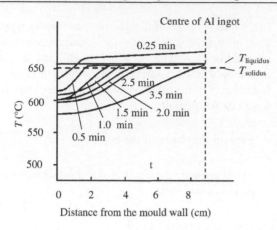

Material data for aluminium

T_L	$= 660\ °C$
k	$= 220\ \text{W/m K at } T_L$
$-\Delta H$	$= 390\ \text{kJ/kg}$
ρ	$= 2.7 \times 10^3\ \text{kg/m}^3$

function of the distance of the solidification front from the interface mould/metal. Discuss the result.

Hint A331

(b) Use the experimental values to sketch y_L as a function of time t and explain the discontinuity of the curve. Explain why the solidification rate increases at the end of the solidification.

Hint A120

For the calculations the data for Al in the table can be used. The temperature of the surroundings is 20 °C.

5 Heat Transport in Cast House Processes

5.1 INTRODUCTION

The major part of Chapter 4 is devoted to the basic theory of heat transport during casting, the rest concerns applications to component casting. The theory of heat transport of cast house processes, which we need in Chapter 5, is the same as that given in Chapter 4, with a few additions, consisting mainly of an extended theory of natural convection.

The solidification process for ingots and other castings is completely controlled by the heat transport through the materials and thus by the basic laws of heat transport. The major part of this chapter consists of applications of the heat transport laws on cast house processes. In ingot casting the natural convection in the melt is strong and very important. For continuous casting water cooling is most essential. Several near net shape casting methods are also discussed. The chapter ends with a short discussion of spray casting methods.

5.2 NATURAL CONVECTION IN METAL MELTS

In connection with casting in permanent moulds we have briefly discussed water cooling, which is an example of convection. The heat flow is absorbed by running water and transported away. Such convection is forced because the water flow is controlled from outside.

Transport of energy by simultaneous motion of matter without external influence is called *natural* or *free* convection. Natural convection arises in liquids as a consequence of density differences within the liquid, caused by temperature or concentration variations. In this section we will study natural convection in metal melts. This process is of importance for crystal formation during the solidification process. It will be applied in Chapter 6.

The density of the metal melt decreases at increasing temperature due to the volume dilation of the melt. When an ingot or a casting solidifies, natural convection arises when heat is transported away through the mould. The solidification front is cooled and the melt becomes cooler there. The density of the melt close to the solidification front increases and the melt will therefore move downwards as a consequence of natural convection. It is replaced by hotter melt, which moves upwards.

In the melt a ring of cyclic movements arises that leads to constant motion of the melt, which transports heat from the interior of the melt to the solidification front. The process is illustrated in Figure 5.1.

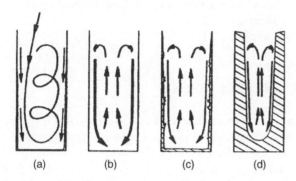

Figure 5.1   Casting of an ingot. (a) Convection during the casting operation; (b)–(d) natural convection in the ingot during the solidification process.

When an ingot solidifies, the metal melt is coolest close to the solidification front, which is connected via the solid phase to the mould wall. A metal flow directed downwards along the solidification front arises, which continues along the bottom and then upwards at the centre and then radially outwards along the upper surface. The greater the height is, the stronger will the flow be, thus making the flow in the melt more or less turbulent.

The flow is also affected by the conditions at the upper surface. A strong cooling of the upper surface leads to an increase of the convection flow. If the upper surface is heated the consequence will be that the natural convection decreases.

The temperature distribution in the melt controls the velocity of the flowing melt and thus the heat transport. To get an idea of the extension of heat transport within the melt we have to study the temperature distribution in the melt and the flow rate emanating from the temperature differences.

5.2.1   Theory of Natural Convection in Metal Melts

The American scientists Eckert and Drake developed a theory for natural convection which, according to the scientific literature, agrees reasonably well with experimental investigations.

Natural convection occurs in most cases under the influence of gravitation forces. Eckert and Drake discussed several simple cases of natural convection under the influence of gravitation, such as natural convection at a planar vertical surface, between two planar parallel walls and between two concentric cylinders of different temperatures.

For application of natural convection in a solidifying ingot the simplest mathematical model of a planar vertical surface (Figure 5.2) is the one that fits best. According to the theory the vertical wall can be hotter or cooler than the neighbouring medium, which can be a gas or a liquid. We will restrict the discussion below to liquids and will apply Eckert's and Drake's mathematical model to metal melts.

If the vertical wall, i.e. the solidification front, is colder than the metal melt heat is transferred from the melt to the wall. The melt close to the wall starts to flow downwards due to temperature and volume decrease and associated density increase. Eckert and Drake assumed that the flow occurs within a thin boundary layer of variable thickness and that its thickness is zero at the upper surface of the melt and increases downwards along the vertical wall.

The driving force of natural convection is the density difference caused by temperature differences in the melt. Eckert and Drake made the assumption that the temperature within the thin boundary layer increases from T_{solid} close to the vertical solid phase where $y = 0$, to the temperature T_{melt}, of the melt beyond the thin boundary layer. The temperature T_{solid} of the solid, close to the thin boundary layer, is assumed to be constant along the whole vertical surface (Figure 5.2).

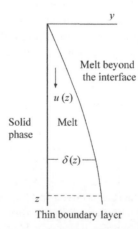

Figure 5.2   Thickness of the thin boundary layer in natural convection, close to a vertical wall, as a function of the distance z from the upper surface of the melt.

The approximate temperature distribution within the thin boundary layer is described by the function:

$$T = T_{melt} - (T_{melt} - T_{solid})\left(1 - \frac{y}{\delta}\right)^2 \qquad (5.1)$$

where

    y = distance from the solid within the thin boundary layer

    T = temperature within the thin boundary layer at distance y from the solid

    T_{melt} = temperature of the melt close the thin boundary layer

    T_{solid} = temperature of the solid phase close to the thin boundary layer

    δ = thickness of the thin boundary layer.

It can be seen from Equation (5.1) that the function fulfils the boundary conditions:

$$T = T_{solid} \quad \text{for } y = 0 \quad \text{and} \quad T = T_{melt} \quad \text{for } y = \delta.$$

The function (5.1) is illustrated in Figure 5.3. It can be seen from the figure that the temperature at constant z rises continuously from the vertical solid phase towards the melt.

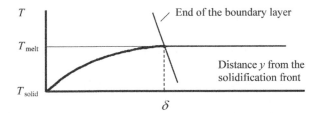

Figure 5.3 Temperature of the thin boundary layer as a function of the distance y from the vertical solid/liquid interface during natural convection close to the solid phase.

On the basis of the temperature distribution within the boundary layer, given in Equation (5.1), Eckert and Drake were able to derive the following expression for the flow rate u within the boundary layer at a point (y, z):

$$u = u_0(z)\frac{y}{\delta}\left(1 - \frac{y}{\delta}\right)^2 \qquad (5.2)$$

where $u_0(z)$ is a function of the distance z below the upper surface of the melt. The function (5.2) is illustrated in Figures 5.4 (a) and (b). The direction of u is the same as in Figure 5.2.

The flow rate shows a maximum within the interval δ. By a conventional maximum/minimum examination ($du/dy = 0$ at a given value of z) the coordinates of the maximum value of the flow rate can be determined. The result of the maximum/minimum calculations is:

$$y = \frac{\delta(z)}{3} \qquad (5.3)$$

$$u_{max} = \frac{4}{27} \times u_0(z) \qquad (5.4)$$

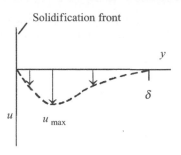

Figure 5.4 (a) The velocity flow vector as a function of the distance y from the solid phase for a given value of z.

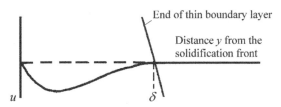

Figure 5.4 (b) The flow rate u inside the thin boundary layer during natural convection as a function of the distance y from the vertical solid phase for a given value of z.

The maximum thickness of the boundary layer is assumed to be the same for the temperature and the flow rate Equations (5.1) and (5.2). This is not in complete agreement with reality, but it is a most acceptable approximation that simplifies the further calculations considerably and gives good agreement between theory and experiments.

Both the flow rate u and the distance δ are functions of the vertical coordinate z (Figure 5.2 on page 94). The temperature Equation (5.1) and the flow rate Equation (5.2) are combined with one equation of momentum (the total momentum is constant in the absence of outer forces) and one equation of heat transport (the law of energy conservation). After elimination of quantities of less interest, two differential equations remain, which can be used to solve u_{max} and δ. The equations are given in the box on page 96 together with the outlines of the solutions. The box does not contain the complete solution in detail but illustrates the principles.

The expression of u_{max} is seldom used and will not be given here but the maximum boundary layer thickness is a concept that is most useful:

$$\delta(z) = 3.93\left(\frac{v_{kin}}{\alpha}\right)^{-\frac{1}{2}}\left(\frac{20}{21} + \frac{v_{kin}}{\alpha}\right)^{\frac{1}{4}}\left[\frac{g\beta(T_{melt} - T_{solid})}{v_{kin}{}^2}\right]^{-\frac{1}{4}}z^{\frac{1}{4}}$$

$$(5.5)$$

α, β and v_{kin} ($= \eta/\rho$) are material constants. For each alloy melt, the expression (5.5) for the maximum boundary layer thickness can be written:

$$\delta(z) = B\left[\frac{g}{z}(T_{melt} - T_{solid})\right]^{-\frac{1}{4}} \quad (5.6)$$

where B is a summarized material constant. Examples of B values of some common metal melts are given in Table 5.1.

TABLE 5.1 *B* values for some different metals.

Metal	$B\,(\mathrm{m}^{3/4}\,\mathrm{K}^{1/4})$
Fe	5.2×10^{-2}
Al	11.9×10^{-2}
Cu	13.8×10^{-2}

Determination of u_{max} and δ

The momentum and energy equations can, after transformations, be written as:

$$\frac{1}{105}\frac{d}{dy}(u_{max}^2 \delta) = \frac{1}{3}g\beta\delta(T_{melt} - T_{solid}) - v_{kin}\left(\frac{u_{max}}{\delta}\right) \quad (1')$$

$$\frac{1}{30}\frac{d}{dy}(u_{max}\delta) = \frac{2\alpha}{\delta} \quad (2')$$

where

$\alpha =$ the constant in the general law of thermal conduction

$\beta =$ volume dilatation coefficient of the melt

$v_{kin} =$ kinematic coefficient of viscosity of the melt η/ρ.

In order to solve u_{max} and δ the following solutions are suggested:

$$u_{max} = C_1 z^p \quad \text{and} \quad \delta = C_2 z^q \quad (3')\text{ and }(4')$$

The expressions (3') and (4') are inserted into Equations (1') and (2') and the constants C_1, C_2, p and q are determined.

The values of p and q are found to be: $p = \frac{1}{2}$ and $q = \frac{1}{4}$

As is seen from Equation (5.5) the constants C_1 and C_2 are relatively complicated functions of the material constants α, β, and v_{kin}.

5.3 HEAT TRANSPORT IN INGOT CASTING

5.3.1 Experimental Examination of Heat Transport

Ingot casting has been described briefly in Chapter 2. When an ingot solidifies and cools it shrinks and an air gap is formed between it and the mould wall. This air gap results in a sudden and drastically reduced rate of heat transfer. Heat transport under these circumstances has been treated theoretically in Section 4.3.3 in Chapter 4.

The heat transport process at casting in sand moulds can be used as a basis for understanding the solidification process in ingots. On page 81 in Chapter 4 it is mentioned that Chvorinov's rule has been tested for ingots and found to be valid in some but not all cases. The explanation is the following.

The mould is normally made of cast iron. If copper is cast in such a mould thermal conductivity is much *better* in the copper metal than in cast iron. Therefore, in this case, Chvorinov's rule is valid, as the situation is analogous to the circumstances when a metal solidifies in a sand mould.

However, Chvorinov's rule is *not* valid when steel is cast in such a mould. The thermal conductivity is *lower* for steel than for cast iron and heat transport through the solidified layer will thus be slow. This layer will be the step that controls the solidification rate. The result is a large temperature gradient across the solidified steel layer, which promotes the solidification of the ingot more than Chvorinov's rule predicts for sand mould castings. In addition, there will also be a temperature gradient in the cast iron mould. The solidification rate will anyway be larger than that described by Chvorinov's rule, as the thermal conductivity of steel is greater than that of sand.

The solidification and cooling process of an ingot influences the quality of the steel, i.e. its structure, distribution of alloying elements and slag inclusions, pore inclusions, and tendency to crack formation. These matters will be extensively discussed in the following chapters.

In spite of considerable practical difficulties, due to the high temperature of the metal melt, a few experimental investigations have been performed in steelworks into the solidification and cooling of steel ingots. These experiments are a complement to, or rather a check on, the conventional theoretical calculations with computers. In particular, two direct methods have been used: (i) temperature measurements with thermocouples, and (ii) use of radioactive tracer elements.

Experimental Observations

Figure 5.5 illustrates an experiment performed by Jonsson, who used a combined method. Three thermocouples, shielded by high-temperature-resistant tubes, were placed halfway from the bottom in a 2-ton ingot at the centre

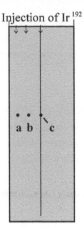

Figure 5.5 Two-ton ingot with thermocouples at 71 mm, 154 mm, and 225 mm (centre) from the mould wall. The total solidification time for the ingot is equal to the time it takes for the solidification front to reach the central thermocouple.

and at two given distances from the mould wall. The temperatures at the three positions were measured as a function of time and cooling curves were drawn. In addition, the position of the solidification front y_L was plotted as a function of the square root of the solidification time. The diagrams of such measurements on the 2-ton ingot in Figure 5.5 are shown in Figures 6 (a) and (b).

The motion of the melt in the 2-ton ingot as a function of time was also registered by the aid of three injections of a radioactive tracer element (Ir^{192}) just below the upper surface of the melt at the end of the teeming operation. The path of the melt was followed. The motion stopped at the position where the melt solidified. After cooling the ingot was cut into two parts and the radioactivity was measured as a function of position. The plot of the three y_L positions is shown in Figure 5.6 (b).

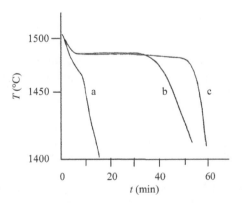

Figure 5.6 (a) Cooling curves for the three thermocouple points in the 2-ton ingot described in Figure 5.5. Reproduced with permission from Kjell-Olof Jonsson.

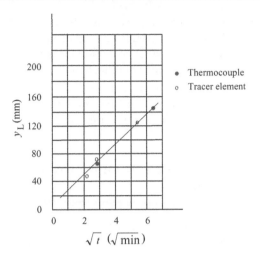

Figure 5.6 (b) Position of the solidification front y_L as a function of the square root of time for the three thermocouple points in the 2-ton ingot described in Figure 5.5. The slope of the straight line is $2.58 \times 10^{-2} \text{m}/\sqrt{\text{min}} \approx 1 \text{ inch}/\sqrt{\text{min}}$. Reproduced with permission from Kjell-Olof Jonsson.

Discussion and Interpretation of Results
The radioactive tracer experiment showed that:

- The internal heat transport mechanism is natural convection.
- There is no or a very small temperature gradient in the liquid during the solidification process except at the very beginning when the temperature decreases very rapidly due to natural convection. There is no heat transport by the aid of thermal conduction.
- The temperature decreases very quickly when the solidification front reaches the point of measurement. This indicates a large temperature gradient in the solidified shell.

> **Rule of Thumb for Ingots**
> $y_L = \sqrt{t}$
> inch   min
> or
> $y_L = 2.5\sqrt{t}$
> cm   min

- The thermocouple measurements and the radioactive tracer measurements agree very well for two of the thermocouple points. The equation of the straight line in Figure 5.6(b) corresponds to the equation:

$$y_L = C\sqrt{t} + D$$

where D is small, which agrees well with Equation (4.72) on page 79 in Chapter 4.

- These facts can be interpreted as a linear growth law at the beginning and a parabolic growth law later. When the parabolic law is valid, heat transport through the solidified shell controls the solidification process.

The slope of the straight line $y = \text{const}\sqrt{t}$ for ingots is often of the magnitude 1 inch $/\sqrt{\min}$ [compare Figure 5.6 (b)]. This gives the simple 'rule of thumb', given in the box above.

5.3.2 Heat Transport by Natural Convection in Solidifying Ingots

In Section 5.2.1 we introduced a model for natural or free convection and derived an expression for the thickness of the boundary layer. These results will now be used to examine heat transport by convection in an ingot and describe how it influences the temperatures of the solid/liquid interface and the melt, none of which is constant during the solidification process.

The model below, based on convection in the melt, will be used in Chapter 6. The disappearance of superheat in ingot and other casting processes before solidification is very important for formation of different types of crystal structures in the material.

Step 1: Calculation of the Heat Flux dq/dt

We use the basic equation for heat transport $\frac{dq}{dt} = -k\frac{dT}{dy}$ and introduce the derivative of Equation (5.1) on page 94:

$$T = T_{\text{melt}} - (T_{\text{melt}} - T_{\text{solid}})\left(1 - \frac{y}{\delta}\right)^2 \quad (5.1)$$

with respect to time. Equation (5.1) is valid within the boundary layer where natural convection occurs. After this operation we get the heat flux at the distance y from the solidification front:

$$\frac{dq}{dt} = -k\left[-(T_{\text{melt}} - T_{\text{solid}}) \times 2\left(1 - \frac{y}{\delta}\right)\frac{-1}{\delta}\right]$$
$$= -2k\left(1 - \frac{y}{\delta}\right)\frac{T_{\text{melt}} - T_{\text{solid}}}{\delta} \quad (5.7)$$

The *effective thermal conductivity* k_y depends on the position y within the thin boundary layer:

$$k_y = 2k\left(1 - \frac{y}{\delta}\right) \quad (5.8)$$

If we use an average value over the whole boundary layer, i.e. in the interval $y = 0$ to $y = \delta$, we get $k_{\text{av}} = k$.

In this case we are interested in the conditions valid at the solid/liquid interface, and have to use the k-value for $y = 0$:

$$k_{y=0} = 2k \quad (5.9)$$

The expression for the heat flux at the wall will then be:

$$\frac{dq}{dt} = -2k\frac{(T_{\text{melt}} - T_{\text{solid}})}{\delta} \quad (5.10)$$

Step 2: Calculation of the Heat Flow dQ/dt

The next step is to calculate the total heat flow, which is transported from the interior of the ingot through the total surface of the solidification front.

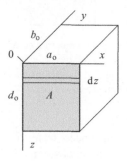

Figure 5.7 Dimensions of the ingot.

Figure 5.7 gives:

$$\frac{dQ}{dt} = \iint_A \frac{dq}{dt}\,dA = \int_0^{d_0} \frac{dq}{dt} a_0 dz \quad (5.11)$$

where $a_0 = $ width of melt in the x-direction, and $d_0 = $ height of melt.

The expression of dq/dt [Equation (5.10)] is inserted into Equation (5.11) to give:

$$\frac{dQ}{dt} = \int_0^{d_0} -2k\frac{(T_{\text{melt}} - T_{\text{solid}})}{\delta} a_0\,dz \quad (5.12)$$

where δ is a function of z. An expression of δ is obtained from Equation (5.6) on page 96 and introduced into equation (5.12) to give:

$$\frac{dQ}{dt} = \int_0^d -2k(T_{\text{melt}} - T_{\text{solid}})\frac{a_0\,g^{\frac{1}{4}}z^{-\frac{1}{4}}}{B(T_{\text{melt}} - T_{\text{solid}})^{-\frac{1}{4}}}dz \quad (5.13)$$

The value of the integral is:

$$\frac{dQ}{dt} = -\frac{8k\,g^{\frac{1}{4}}}{3B}a_0\,d_0^{\frac{3}{4}}(T_{\text{melt}} - T_{\text{solid}})^{\frac{5}{4}} \quad (5.14)$$

The heat flow is a function of the time t because both T_{melt} and T_{solid} depend on t. Thus we have to determine both the temperature of the solidification front and the temperature of the melt as functions of time in order to be able to use Equation (5.14). T_{melt} and T_{solid} are shown in Figure 5.8 (a).

Step 3: Calculation of T_{solid}

A necessary condition for solidification of a metal melt is *undercooling*. The driving force of solidification is normally related to a growth temperature T_{solid} at the solidification front, which is lower than the liquidus temperature T_L. The higher the growth rate is, the lower the growth temperature will be. The faster the solidification rate is, the larger is the undercooling. The simplest kinetic relation, which describes the growth rate as a function of the growth temperature, is:

$$\frac{dy_L}{dt} = \mu(T_L - T_{\text{solid}})^n \qquad (5.15)$$

where μ is a growth constant, which has a characteristic value for each alloy. n is a dimensionless constant, the value of which is often equal to 2.

According to Equation (4.72) on page 79 in Chapter 4, verified by Figure 5.6 (b) on page 97, the relationship between the thickness of the solidifying shell and the time can be written as:

$$y_L = C\sqrt{t} \qquad (5.16)$$

where C is a growth constant. By differentiating Equation (5.16) with respect to time we get an expression for the solidification rate or growth rate:

$$\frac{dy_L}{dt} = \frac{C}{2\sqrt{t}} \qquad (5.17)$$

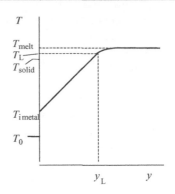

Figure 5.8 (a)   Temperature profile of the ingot.

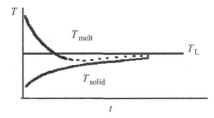

Figure 5.8 (b)   The temperatures T_{solid} and T_{melt}, as functions of the time. T_{solid} is the temperature at the solid/liquid interface. T_{melt} is the temperature in the interior of the melt.

Step 4: Calculation of T_{melt}

The temperature T_{melt} at the centre of the ingot can be determined by the aid of the total heat flow from the interior of the ingot. Using the basic equation for temperature increase in a material at heat supply ($dQ = cmdT$) we get:

$$\frac{dQ}{dt} = c_p\, \rho\, a_0\, b_0\, d_0\, \frac{dT_{\text{melt}}}{dt} \qquad (5.19)$$

where b is the extension of the melt in the y-direction and $a_0\, b_0\, d_0$ is the volume of the melt (Figure 5.7).

We combine Equations (5.14), (5.16), (5.18) and (5.19) to give:

$$\frac{dT_{\text{melt}}}{dt} = \frac{-8k\, g^{\frac{1}{4}}}{3B(b_0 - 2C\sqrt{t})(d_0 - C\sqrt{t})^{\frac{1}{4}}\rho c_p} \left[(T_{\text{melt}} - T_L) + \left(\frac{C}{2\mu\sqrt{t}}\right)^{\frac{1}{2}} \right]^{\frac{5}{4}} \qquad (5.20)$$

This expression of the solidification rate is introduced into Equation (5.15) with $n = 2$. We solve T_{solid} and get:

$$T_{\text{solid}} = T_L - \left(\frac{C}{2\mu\sqrt{t}}\right)^{\frac{1}{2}} = T_L - \frac{C}{\sqrt{2\mu y_L}} \qquad (5.18)$$

The lower curve in Figure 5.8 (b) corresponds to the temperature of the solid metal close to the melt as a function of time. As expected it approaches the liquidus temperature at large values of t.

where

$b_0 =$ the width of the mould in y-direction

$d_0 =$ the height of the mould

$C =$ a growth constant.

In Equation (5.20), care has been taken with regard to the shell growth at the faces and the bottom of the mould. The shell growth has been calculated with the aid of Equation (5.16):

$$b = b_0 - 2C\sqrt{t} \quad \text{and} \quad d = d_0 - C\sqrt{t}$$

Equation (5.20) is a differential equation, which has no simple analytical solution. T_{melt} is most conveniently solved numerically, which gives T_{melt} as a function of time. Such a solution is shown in Figure 5.8 (b) above.

The figure shows that the temperature of the melt drops rapidly at the beginning (compare Figure 5.6 (a) on page 97). It falls below the liquidus temperature because the interface temperature is lower than the liquidus temperature in accordance with Equation (5.15).

The temperature of the melt passes a minimum and rises to the dotted line. The reason for depicting part of the temperature curve as a dotted line is that it is no longer described by the integrated Equation (5.20). The reason for this is that new crystals are formed in the melt and the heat of solidification released from them results in a temperature increase. This phenomenon will be further discussed in Chapter 6.

By combining the solutions of steps 2, 3 and 4 we have solved the problem of calculating the total heat flow from the interior of the ingot, caused by natural convection, as a function of time.

Example 5.1

When ingots solidify, the change from so-called columnar crystals to so-called equiaxial crystals is determined by the rate at which the excess temperature ahead of the solidification front disappears. The decrease in the temperature of the melt is promoted by natural convection. The lengths of the columnar crystals can be determined by the aid of the time required for the excess temperature in the melt to disappear.

Calculate this time for a steel ingot with a height of 1.5 m and a cross-section area of 1.0 m × 0.3 m. The melt of the ingot has an excess temperature of 20 °C. Assume that the solidification front has an undercooling of 3 °C and that this undercooling is constant during the whole solidification process.

In order to get an analytical solution you may assume that the undercooling is constant, which is not quite correct according to Equation (5.15) on page 99. To simplify the calculations you may also disregard the decrease in volume due to the motion of the solidification front toward the centre of the ingot.

Solution:

Provided that no external agitation is applied, the convection pattern will, after a short time, be determined by the natural convection. It causes a flow directed downwards along the contact area solid metal/melt at the solidification front within its boundary layer. The maximum thickness δ is a function of the distance z from the upper surface. We

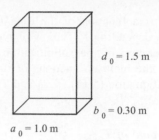

$d_0 = 1.5$ m

$b_0 = 0.30$ m

$a_0 = 1.0$ m

apply Equation (5.5) on page 95 or Equation (5.6) on page 96.

$$\delta(z) = 3.93 \times \left(\frac{v_{kin}}{\alpha}\right)^{-\frac{1}{2}} \left(\frac{20}{21} + \frac{v_{kin}}{\alpha}\right)^{\frac{1}{4}} \left[\frac{g\beta(T_{melt} - T_{solid})}{v_{kin}^2}\right]^{-\frac{1}{4}} z^{\frac{1}{4}} \tag{1'}$$

or [equation (5.6) on page 96]

$$\delta(z) = B\left[\frac{g}{z}(T_{melt} - T_{solid})\right]^{-\frac{1}{4}} \tag{2'}$$

where

$B =$ a material constant, specific for the steel alloy (Table 5.1 on page 96)

$T_{melt} =$ temperature of the melt far from the solidification front

$T_{solid} =$ temperature of the solid metal at the solidification front.

The heat flux through the solidification front from the melt can be written using Equations (5.10) on page 98 and Equation (2'):

$$\frac{dq}{dt} = -\frac{2k(T_{melt} - T_{solid})}{\delta} = -\frac{2kg^{\frac{1}{4}}(T_{melt} - T_{solid})^{\frac{3}{4}}}{B} z^{-\frac{1}{4}} \tag{3'}$$

If we integrate Equation (3') above with respect to z, i.e. over the whole height d_0, and simultaneously multiply with the width a_0 of the melt, we get the total heat flow through the solidification front [Equation (5.14) on page 98]:

$$\frac{dQ}{dt} = -\left(\frac{8kg^{\frac{1}{4}}}{3B}\right) a_0 d_0^{\frac{3}{4}}(T_{melt} - T_{solid})^{\frac{3}{4}} \tag{4'}$$

The cooling rate of the melt dT_{melt}/dt can be calculated from the total heat flow [Equation (5.19) on page 99]:

$$\frac{dQ}{dt} = c_p \rho a_0 b_0 d_0 \left(\frac{dT_{melt}}{dt}\right) \tag{5'}$$

The heat flow through the solidification front is equal to the total heat flow from the ingot. Thus the right-hand side expressions of Equations (5.4′) and (5.5′) are equal. As was suggested in the text we disregard the volume change which means that a_0, b_0 and d_0 are considered to be constant. We also disregard the variation of T_{solid} with time and get:

$$\frac{dT_{\text{melt}}}{dt} = -\frac{8kg^{\frac{1}{4}}}{3Bb_0d_0^{\frac{1}{4}}\rho c_p}\left(T_{\text{melt}} - T_{\text{solid}}\right)^{\frac{5}{4}} \qquad (6')$$

By use of reasonable values for the constants, taken from general reference tables, and $B = 5.2 \times 10^{-2}$ m$^{3/4}$ K$^{1/4}$ from Table 5.1 on page 96 we get:

$$\text{const} = \frac{8kg^{\frac{1}{4}}}{3Bb_0d_0^{\frac{1}{4}}\rho c_p} \qquad (7')$$

$$\text{const} = \frac{8 \times 71 \times 9.81^{\frac{1}{4}}}{3 \times 5.2 \times 10^{-2} \times 0.30 \times (1.5)^{\frac{1}{4}} \times 7 \times 10^3 \times 750}$$

$$= 3.69 \times 10^{-3}$$

We integrate Equation (6′). The two pairs of values determine the integration limits:

At $t = t$ we have $T_{\text{solid}} = T_{\text{L}} -$ undercooling

$$= 1453\,\text{K} - 3\,\text{K} = 1450\,\text{K}$$

where T_{L} is the liquidus temperature of steel.

At $t = 0$ we have $T_{\text{melt}} = T_{\text{L}} +$ excess temperature

$$= 1453\,\text{K} + 20\,\text{K} = 1473\,\text{K}$$

Thus we get:

$$\int_{T_{\text{melt}}}^{T_{\text{L}}} \frac{dT_{\text{melt}}}{(T_{\text{melt}} - 1450)^{\frac{5}{4}}} = -\int_{1473}^{1453} \frac{dT_{\text{melt}}}{(T_{\text{melt}} - 1450)^{\frac{5}{4}}} = \text{const}\int_0^t dt$$

which gives:

$$\left[4(T_{\text{melt}} - 1450)^{-\frac{1}{4}}\right]_{1453}^{1473} = 3.69 \times 10^{-3}t$$

or

$$t = \frac{1.213}{3.69 \times 10^{-3}} = 329\,\text{s} = \frac{329\,\text{s}}{60} = 5.48\,\text{min}$$

Answer:
The excess temperature disappears after a time interval of the magnitude 5–6 minutes.

5.4 WATER COOLING

5.4.1 Dimensioning of Water Cooling

In Chapter 4 we treated the heat transport between solid metal and mould under various conditions, casting with perfect cooling of alloys with both narrow and wide solidification intervals, casting in moulds with ideal or poor thermal conductivity, and casting with poor contact between metal and mould.

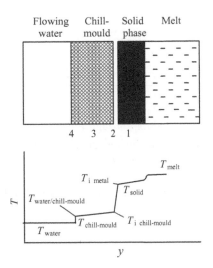

Figure 5.9 Temperature distribution during heat transport during casting. The air gap between the solid phase and the chill-mould is strongly exaggerated in the figure.

The heat flow between cast metal and mould is only part of the total heat transport. As can be seen from Figure 5.9 it consists of several steps.

The various steps are:

(1) Thermal conduction through the solidifying metal shell.
(2) Heat transfer from the solidifying shell to the inside of the chill-mould via the narrow air gap, which arises from the solidification and cooling shrinkage of the metal shell.

Calculation of the Total Heat Transfer Coefficient

Consider the four steps, coupled in a series and described in Figure 5.9 on page 101. The same amount of heat per unit time passes each of them at stationary conditions. With short and general designation we obtain:

$$\frac{dQ}{dt} = k_{metal}A_1\left(\frac{T_{initial} - T_1}{l_{metal}}\right) = k_{air}A_2\left(\frac{T_1 - T_2}{\delta}\right) = k_{mould}A_3\left(\frac{T_2 - T_3}{l_{mould}}\right) = h_{H_2O}A_4(T_3 - T_{final}) = const \quad (1')$$

The effective or total heat transfer number is defined with the aid of the equation:

$$\frac{dQ}{dt} = h_{total}A_{total}(T_{initial} - T_{final}) = const \quad (2')$$

The temperature difference in Equation (2') can be written as:

$$T_{initial} - T_{final} = (T_{initial} - T_1) + (T_1 - T_2) + (T_2 - T_3) + (T_3 - T_{final}) \quad (3')$$

Equations (1') and (2') inserted into Equation (3') give:

$$\frac{const}{h_{av}A_{total}} = \frac{const \times l_{metal}}{k_{metal}A_1} + \frac{const \times \delta}{k_{air}A_2} + \frac{const \times l_{mould}}{k_{mould}A_3} + \frac{const}{h_w A_4} \quad (4')$$

We assume that $A_{total} = A_1 = A_2 = A_3 = A_4 = A$
and dividing Equation (4') by the common factors, we get:

$$\frac{1}{h_{total}} = \frac{l_{metal}}{k_{metal}} + \frac{\delta}{k_{air}} + \frac{l_{mould}}{k_{mould}} + \frac{1}{h_w} \quad (5')$$

Equation (5') can easily be generalized.

---

(3) Thermal conduction through the chill-mould.

(4) Heat transfer from the outside of the chill-mould to stagnant or flowing water or air.

The total heat transfer coefficient is derived in box above:

$$\frac{1}{h_{total}} = \frac{l_{metal}}{k_{metal}} + \frac{\delta}{k_{air}} + \frac{l_{mould}}{k_{mould}} + \frac{1}{h_w} \quad (5.21)$$

where
$h =$ heat transfer coefficient (W/m^2 K)
$k =$ thermal conductivity (W/m K)
$\delta =$ width of air gap.

If the simplifications, discussed in Section 4.3.3 in Chapter 4, are valid the temperature of the metal at the interface metal/chill-mould is determined by Equation (4.45) on page 74 in Chapter 4.

A conclusion from Equation (5.21) is that the heat transfer is controlled by the slowest step, that is, by the term that has the smallest denominator or the poorest heat transfer. We have analysed steps (1) to (3) in Chapter 4. Below we will treat the fourth step, the heat transfer from the mould to the flowing water.

The function of the cooling water is to transfer the heat absorbed by the inner mould wall from the solid metal without exposing the mould to too high temperatures. The heat transfer between the outer mould wall and the flowing water is analysed below. On the basis of the acquired results we will make practical conclusions on the claims to the mould construction and to the dimensioning of the cooling system of the mould.

5.4.2 Water Cooling at Low Heat Flow and/or Strong Cooling

Initially we treat the case when the heat flow is low or the cooling system so strongly dimensioned that the temperature of the outer chill-mould wall is lower than the boiling point of water.

As a concrete example we will deal with a construction where the water flows vertically in a long concentric space (Figure 5.10). There are no special surge diverters between the inner and outer mould walls.

Reynold's number is defined by the relationship:

$$Re = \frac{MD_E}{\eta} \quad (5.22)$$

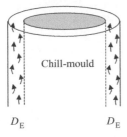

Figure 5.10 Water cooling of a cylindrical chill-mould.

where

$M =$ mass velocity per unit area of the water (kg/m² s)

$D_E =$ thickness of the water column

$\eta =$ viscosity coefficient of water (kg m/s).

A rough estimate shows that for all normally cooled moulds of the actual type Re > 10 000, which is the criterion for fully developed turbulence. In that case the following expression for the heat transfer coefficient at the water/chill-mould interface is valid, according to hydrodynamics:

$$h_w = \frac{k}{D_E} \times 0.023 \times \frac{u_w D_E^{0.8}}{v_{kin}} Pr^{0.33} \qquad (5.23)$$

where

$h_w =$ heat transfer coefficient at the water/chill-mould interface

$k =$ thermal conductivity of water

$v_{kin} =$ kinetic viscosity coefficient of water (η/ρ);

$u_w =$ linear velocity of water

$Pr =$ Prandls' number $= c_p \eta / k$.

TABLE 5.2 Physical data for air-saturated water.

Quantity (unit)	At 10 °C	At 40 °C
k (W/m K)	0.587	0.633
v (m²/s)	1.31×10^{-6}	6.86×10^{-7}
Pr	9.41	4.52

Using of the values in Table 5.2, Equation (5.23) can be written as Equation (5.24) for a water temperature of 10 °C.

$$h_w = 81.3 \times u_w^{0.8} D_E^{-0.2} \quad \text{W/m}^2\text{ K} \qquad (5.24)$$

At a water temperature of 40 °C Equation (5.23) will be:

$$h_w = 126 \times u_w^{0.8} D_E^{-0.2} \quad \text{W/m}^2\text{ K} \qquad (5.25)$$

On the basis of Equations (5.24) and (5.25), the heat transfer number has been calculated as a function of the water velocity u_w for various values of the width of the water slit D_E at 10 °C and 40 °C. The result is shown in Figure 5.11.

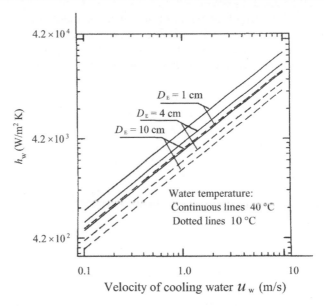

Figure 5.11 The heat transfer coefficient for mould wall/cooling water as a function of the linear velocity of water at different water temperatures and widths of the water slit. Reproduced from P. O. Mellbery's thesis.

On the basis of Figures 5.11 and 5.12 some general observations and comments can be made.

- The heat transfer coefficient depends primarily on the linear velocity of the water. An increase in the water velocity by a factor of 10 gives an increase in the heat transfer number by a factor of about 7.
- Bad cooling, which is expected as a consequence of increased temperature of the cooling water, is counteracted by the fact that the heat transfer coefficient of the forced convection increases with increasing temperature: $h_w = k/\delta_w$, where δ_w is the thickness of the temperature boundary layer (Figure 5.12). k increases with the water temperature. An increase of the water temperature from 10 °C to 40 °C results in an increase in the heat transfer coefficient by approximately 50 %.

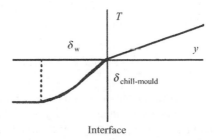

Figure 5.12 The temperature boundary layer is defined as the distance from the interface to the position where the temperature has achieved a stationary value. An analogous definition is valid for other quantities, for example velocities (compare natural convection, page 95).

- An increase of the width of the water slit causes a minor decrease in the efficiency of the cooling when the linear velocity of the water is kept constant. The water flow has to be increased to keep the linear velocity constant when the width of the water slit is increased.

Once more it has to be mentioned that the conclusions above are valid if there is a fully developed turbulence [see Equation (5.25)] and the temperature at the mould wall is lower than the boiling point of water.

It should also be noted that the calculations are approximate, primarily because that the quality of the mould surface affects the heat transfer, and this has been neglected. For a more precise analysis, the water temperatures in Figure 5.11 should be interpreted as mean temperatures within the interface layer mould/cooling water rather than the mean temperature of the water.

After this general analysis we will try to make more quantitative conclusions concerning the requirements for dimensioning of water cooling of chill-moulds. The water cooling must mainly fulfil three demands:

(1) The volume flow, normally measured in m³/min, must be large enough to absorb a certain amount of heat without too large a temperature increase.

(2) The temperature of the departing cooling water must be lower than the boiling point of water at the prevailing pressure.

(3) The dimensioning of the cooling system must be such that the temperature at each point on the mould surface does not become too high with regard to the mechanical strength of the mould material.

As we will see later, the last demand is usually the determining factor for the dimensioning of the cooling. Heat transfer when the water boils locally within the temperature interface layer will be treated later in this section. We will now analyse the conditions necessary for the requirement that the water temperature *must not* exceed the boiling point at any point.

Assume that the maximum heat flux dq/dt in the mould is 1.05×10^6 J/m² s. We will use the relationship

$$\frac{dq}{dt} = -h_w \Delta T \qquad (5.26)$$

where ΔT is the difference between the temperatures of the outer mould wall and the cooling water. By use of Figure 5.11 we can calculate the temperature of the mould wall at the cooling water temperatures 10 °C and 40 °C as a function of the water velocity. The result is shown in Figure 5.13 for a width of the water slit of 5 mm, i.e. $D_E = 10$ mm. We ignore the influence of the pressure and the dissolved air on the boiling point of water and assume that it is 100 °C.

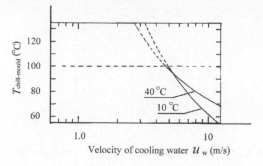

Figure 5.13 The surface temperature of the outer mould wall as a function of the linear velocity of the water at different water temperatures when $D_E = 1.0 \times 10^{-2}$ m and the heat flux $= 1.05 \times 10^6$ J/m² s and no surface boiling occurs. Reproduced from P. O. Mellbery's thesis.

Figure 5.13 shows that the water must have a linear velocity of at least 5 m/s to keep the mould surface temperature below the boiling point. This value is practically independent of the mean temperature of the water within the interval 10 °C to 40 °C.

5.4.3 Heat Transfer in Surface Boiling

Several scientists have investigated heat transfer from a solid body to a liquid that has a temperature equal or nearly equal to its boiling point. They found that the heat transfer depends strongly on the surface temperature of the solid body.

If the surface temperature of the solid body exceeds the boiling point of the liquid by only a few degrees, vapour bubbles are nucleated on the solid surface. They grow and rise successively to the liquid surface. Within this temperature interval, the heat transfer coefficient increases strongly with increasing surface temperature of the solid body and the boiling becomes more and more violent at the same time. At a sufficiently high overheating of the solid surface, a continuous vapour film is formed, which separates the solid surface from the liquid. Heat transfer decreases drastically in this case.

In order to apply this process to mould cooling, we will consider the heat transfer from a solid surface, with a temperature higher than the boiling point of water, to flowing water at a temperature far below its boiling point. Such a process is usually called 'local boiling in an undercooled liquid' or simply 'surface boiling'. In such cases the bubbles are able to grow and 'survive' only in the part of the temperature boundary layer (Figure 5.12, page 103) close to the solid surface, which has a temperature that exceeds the boiling point at the actual pressure. As soon as a bubble moves into the colder water it condenses and disappears. Because of this and due to the change of the structure of the interface layer, which is caused by the bubbles, an enormous increase of the thermal conduction is initiated, compared to the case when no surface boiling is present.

No quantitative estimations of the heat transfer at surface boiling have been found in the scientific literature. With the aid of given data for surface boiling at forced convection in tubes at a nonreported water velocity, the mould temperature can be estimated as 10–20 °C above the boiling point of water for a heat flux of 1.05×10^6 J/s m^2. These values agree very well with the measured temperatures on the outside of the mould.

This additional information has been introduced into Figure 5.13 and the result is presented in Figure 5.14. Corresponding values at a water pressure of 2 atm are also included. At this pressure the boiling point of water is about 120 °C. In the figure we have assumed that the heat transfer depends slightly on the velocity of the cooling water and that the curves successively coincide with the curves for nonboiling. The exact relationship between temperature and water velocity at surface boiling cannot be estimated in the latter case.

The following question is of great interest concerning mould cooling: Is the maximum heat flow through the chill-mould wall enough to give such a superheating of the chill-mould surface that a stable vapour layer can form at low water velocity? The answer is: Experience shows that a superheating of about 60 °C is required for the formation of a stable vapour layer at stagnant water-cooling and a water temperature of 100 °C.

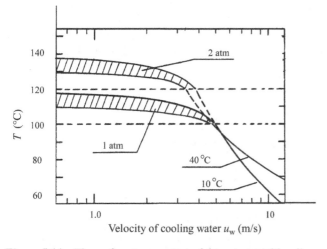

Figure 5.14 The surface temperature of the outer mould wall as a function of the linear velocity of the water. The dashed areas are valid for surface boiling at 1 atm and 2 atm water pressure, respectively. $D_E = 10$ mm. The heat flux $= 1.0 \times 10^6$ J/s m^2. Reproduced from P. O. Mellbery's thesis.

If the cooling water has a temperature that is essentially lower than the boiling point of water, a much higher level of superheating is required in order to form a water vapour layer. The answer also depends on the linear velocity of the cooling water. If such a vapour layer formation were to be formed, the heat transfer would decrease drastically

and the temperature of the mould would increase. It would certainly cause a breakthrough, i.e. the solidified shell would be very thin and burst and the melt would rush out. In the worst case the mould would melt.

5.5 HEAT TRANSPORT DURING CONTINUOUS CASTING OF STEEL

Continuous casting is based on casting of a metal in a vertical chill-mould. The metal flows from the ladle via the tundish down into the vertical, water-chilled, copper mould. During the passage into the chill-mould the melt starts to solidify and a solid shell is formed. This shell is drawn continuously out of the chill-mould into the chill-zone where complete solidification occurs. The velocity of the shell is called casting velocity or *casting rate*.

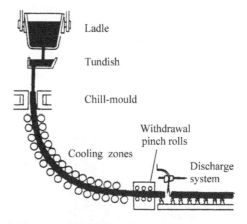

Figure 5.15 Modern machine for continuous casting with a curved path below the chill-mould. Reproduced with permission from Pergamon Press, Elsevier Science.

A necessary condition for continuous casting is that the shell has such mechanical properties that it is rigid outside the chill-mould. Water cooling is therefore very important at continuous casting (Figure 5.15). In order to design it properly the general principles, given in Section 5.4 above, are applied. Continuous casting is a typical example of heat transport with poor contact between chill-mould and metal (Section 4.3.3 in Chapter 4). In this section heat transport at continuous casting will be presented in more detail.

To obtain maximum yield the highest possible production velocity is required. This demands careful control of the cooling and casting conditions.

5.5.1 Construction of the Chill-Mould

The chill-mould and the process in the chill-mould are very important for the final result of the casting. During the short time the melt stays in the chill-mould it has to solidify rapidly on the surface to get such a rigidity that it can be drawn out of the chill-mould for further solidification inside. The motion and wear and tear of the chill-mould influences the risk of crack formation in the casting. This problem is treated in Chapter 10.

Because copper has a very good thermal conductivity nearly all chill-moulds are made of pure copper or of precipitation-hardened copper alloyed with chromium. The mould surfaces are normally coated electrolytically with a thin layer of nickel in order to increase their wear resistances. There are two types, tube chill-moulds and block or plate chill-moulds.

Tube chill-moulds (Figure 5.16) are used in most cases for small square sections, for example $100 \, cm^2$ up to $200 \, cm^2$.

Block chill-moulds (Figure 5.17) are used for large square sections or for rectangular cross sections. The upper part of the chill-mould is frequently covered with some protecting material, for example burnt brick or plating, which protects the chill-mould from damage.

Chill-moulds are often made slightly conical to compensate for the solidification and cooling shrinkage of the casting strand. This aspect is further discussed in Section 5.5.2 and in Chapter 10.

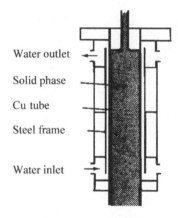

Figure 5.16 Tube chill-mould. Reproduced with permission from Pergamon Press, Elsevier Science.

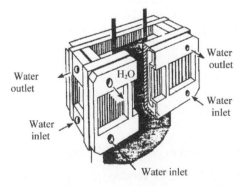

Figure 5.17 Block chill-mould or plate chill-mould. Reproduced with permission from Pergamon Press, Elsevier Science.

5.5.2 Solidification Process in the Chill-Mould

The chill-mould has two purposes: it should (i) *define the shape and cross section of the casting*, and (ii) *remove heat and facilitate formation of a solid shell*. When the melt has passed the chill-mould, the shell must have such a thickness that it can resist the ferrostatic pressure from the melt in the interior. The solidification process is illustrated in Figures 5.18 and 5.19.

Imagine that we cut a thin slice of the casting, perpendicular to the direction of the casting rate, and follow it on its way through the chill-mould. At the time $t = 0$ all metal is liquid. The solidification starts close to the chill-mould at the upper part of the vertical surface of the melt. At first the thickness of the solidifying shell continues to grow rapidly. Due to solidification and cooling shrinkage, the shell contracts and loses contact with the chill-mould wall. The ferrostatic pressure counteracts this process and the shell is deformed permanently. As the shell thickness grows continuously its power to resist the ferrostatic pressure increases with time.

When the shell has become thick enough it loses contact with the chill-mould wall. This happens initially at the corners where the cooling is strongest. Finally the contact between shell and chill-mould is lost and an air gap is formed. This decreases heat transport suddenly and strongly. The solidification rate decreases because heat can no longer be removed at the same rate as before.

The growth rate of the shell at the solidification front, the interface between the solid metal and the melt, are illustrated in Figure 5.18. It shows the shell thickness of our mobile slice at various positions and times. It can be seen

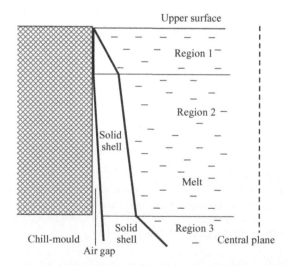

Figure 5.18 Shell growth in the melt close to the chill-mould wall in continuous casting. Close to the chill-mould the air gap is seen. The solidification front in region 1 was formed while the metal was still in touch with the chill-mould. The shell growth is small due to poor heat transport through the air gap in region 2.

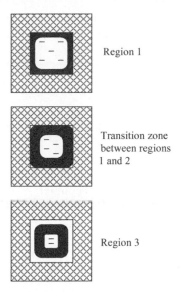

Region 1

Transition zone
between regions
1 and 2

Region 3

Figure 5.19 Formation of an air gap in the chill-mould in continuous casting. The mould is seen from above.

that the solidification process can be separated into three steps or three regions (compare Example 5.2 below):

(1) region 1 with close contact between shell and chill-mould;
(2) region 2 with an air gap between shell and chill-mould;
(3) region 3 outside the chill-mould when the shell is strongly cooled by water.

The difference in temperature distribution in the zones 1 and 2 is illustrated in Figure 5.20 (a).

The heat transport occurs in four steps:

(A) thermal conduction through the solidified shell;
(B) heat transfer across the air gap between the shell and the chill-mould;
(C) thermal conduction through the chill-mould;
(D) heat transfer between the chill-mould and the cooling water.

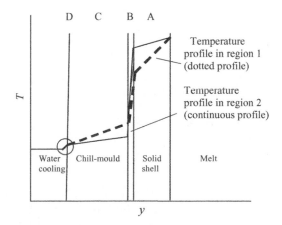

Figure 5.20 (a) Temperature distribution in the chill-mould, solid metal and melt during continuous casting.

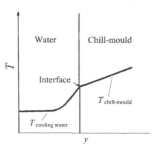

Figure 5.20 (b) Enlargement of the part of Figure 5.20 (a) that is enclosed in a circle.

Example 5.2

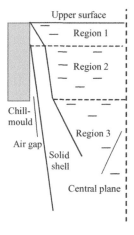

Reproduced with permission from the Institute of Materials, The Metal Society.

Make an approximate calculation of the shape of the solidification front at continuous casting of low-carbon steel. The casting rate is 60 cm/min, the length of the chill-mould is 90 cm and the dimensions of the strand are 100 cm × 10 cm. Assume for the calculation that the casting temperature is very close to the melting point of the low-carbon steel.

Perform the calculation in three steps, i.e. determine times and distances for the three regions:

(1) the shape of the solidification front in region 1 when the steel strand stays in touch with the chill-mould wall;
(2) the shape of the solidification front in region 2 when the strand has lost contact with the chill-mould wall;
(3) the appearance of the solidification front in region 3 all the way to complete solidification.

The thickness of the strand shell is 5.0 mm when it loses contact with the chill-mould wall. It is reasonable to assume that the temperature of the chill-mould is 100 °C.

The heat transfer coefficients for the three regions are:

$$h_1 = 0.168 \times 10^4 \, \text{W/m}^2 \, \text{K}$$
$$h_2 = 0.0042 \times 10^4 \, \text{W/m}^2 \, \text{K}$$
$$h_3 = 0.042 \times 10^4 \, \text{W/m}^2 \, \text{K}$$

Material constants of steel	
ρ	7.80×10^3 kg/m^3
$-\Delta H$	280×10^3 J/kg
T_L	1520 °C
k	0.50×10^2 W/m K

Solution:
We will apply Equation (4.48) from page 74 in Chapter 4 to each of the zones. It is important to consider that the heat transfer coefficient has different values in the three regions. In the above table we find the data for steel.

Region 1
The time required for the shell to grow from 0 to $d_1 = 5$ mm is:

$$t_1 = \frac{\rho(-\Delta H)}{T_L - T_0} \frac{y_L}{h_1}\left(1 + \frac{h_1}{2k}y_L\right) \qquad (1')$$

$$t_1 = \frac{(7.8 \times 10^3)(280 \times 10^3)}{1520 - 100} \frac{(5 \times 10^{-3})}{0.168 \times 10^4}\left(1 + \frac{0.168 \times 10^4}{2 \times (0.50 \times 10^2)} \times 5 \times 10^{-3}\right) = 5.0\,\text{s}$$

Upper surface

The height l_1 from the upper surface of the melt before the shell loses contact with the mould wall is obtained by multiplying the casting rate and the time t_1:

$$l_1 = \frac{0.60\,\text{m}}{60\,\text{s}} \times 5.0\,\text{s} = 0.01\text{m/s} \times 5.0\,\text{s} = 0.050\,\text{m}$$

Region 2
The time required for the shell to pass the chill-mould after it has lost contact with the chill-mould wall is found by

dividing the length of the chill-mould by the casting rate and subtracting the time for passage of zone 1 (see figure below):

$$t_2 = \frac{0.90\,\text{m}}{0.01\,\text{m/s}} - 5\,\text{s} = 85\,\text{s}$$

During time t_2 the shell thickness has grown from d_1 to $d_1 + d_2$. We introduce the value $t_2 = 85$ s and $h_2 = 0.0042\,\text{W/m}^2\text{K}$ into Equation (1') and get $y_2 = d_2$ when we solve the second-order equation in y_L. The result is $d_2 = 2 \times 10^{-7}$ m, which is negligible. The total thickness of the shell when it has passed the chill-mould will be:

$$d = d_1 + d_2 = 5 \times 10^{-3}\,\text{m} + 2 \times 10^{-7}\,\text{m} = 5 \times 10^{-3}\,\text{m}$$

Region 3
Within zone 3 the shell thickness grows from 5 mm to 5 cm. The time for 45 mm growth can be calculated by inserting

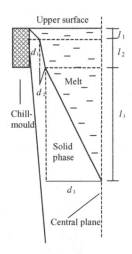

Upper surface

(This figure is not to scale)

$y_L = 45 \times 10^{-3}$ m into Equation (1') and calculating the value of t_3:

$$t_3 = \frac{(7.80 \times 10^3)(280 \times 10^3)}{1520 - 100} \frac{0.045}{(0.042 \times 10^4)}\left(1 + \frac{0.042 \times 10^4}{2(0.50 \times 10^2)} \times 0.045\right) = 165\,\text{s}$$

which gives $t_{\text{total}} = t_1 + t_2 + t_3 = 5 + 85 + 165 = 255\,\text{s}$

The total solidification length, called the metallurgical depth (see pages 113–114), is equal to

$$l_{\text{total}} = v_{\text{cast}}\, t_{\text{total}} = \frac{0.60}{60}\,\text{m/s} \times 255\,\text{s} = 2.55\,\text{m}$$

Answer:
See the above figure:

$$d_1 = 5\,\text{mm} \qquad l_1 = 0.05\,\text{m} \qquad t_1 = 5\,\text{s}$$

$$d_2 = 0\,\text{mm} \qquad l_2 = 0.85\,\text{m} \qquad t_2 = 85\,\text{s}$$

$$d_3 = 45\,\text{mm} \qquad l_3 = 1.65\,\text{m} \qquad t_3 = 165\,\text{s}$$

The total solidification time is 255 s. The casting will be completely solid at about 2.5 m below the upper surface.

To prevent the shell sticking to the chill-mould wall, a lubricating substance is added. Lubrication is discussed on pages 17 and 23 in Chapter 2. In addition, the chill-mould is forced to oscillate during the casting process. This oscillation decreases the friction between the mould and the solidified shell.

Air Gap Formation in Casting in Chill-Moulds

Example 5.2 illustrates the influence of the air gap in continuous casting. The heat transfer resistance of the air gap is very high. According to Equation (4.6) on page 61 in Chapter 4 the heat transfer coefficient of the air gap can be written as:

$$h = \frac{k_{\text{air}}}{\delta} \tag{5.27}$$

where h = heat transfer coefficient; k_{air} = thermal conductivity of air, and δ = width of the air gap.

With the aid of Equations (4.45), (4.46) and (4.48) on page 74 in Chapter 4, the shell thickness and outer temperature of the shell can be calculated if we assume that the simplifications, discussed in Section 4.3.3 of Chapter 4, are valid.

A wide air gap gives a high outer temperature and hence poor shell thickness growth in the chill-mould. In order to reduce this disadvantage, the solidification and cooling shrinkage is taken into consideration by making the chill-mould slightly *conical* (a cone upside down). This compensates for the cooling shrinkage of the metal during passage through the chill-mould. In this way a constant air gap can be achieved instead of an increasing one.

Figure 5.21 shows the shell thickness s and the surface temperature T_s of the metal as functions of the width of

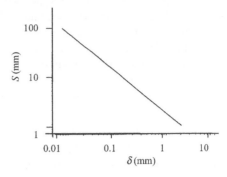

Figure 5.21 (a)   Shell thickness S as a function of the width of the air gap at the exit of the chill-mould in continuous casting.

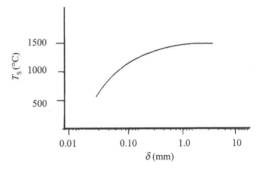

Figure 5.21 (b)   Surface temperature of a metal as a function of the width of the air gap at the exit of the chill-mould in continuous casting.

the air gap, provided that it is the same in the whole chill-mould. This is achieved by making the chill-mould slightly conical.

Chill-moulds get worn when they are used, which results in an increase of the width of the air gap (Figure 5.22). The consequence is that the outer temperature of the shell increases by several hundred degrees and the shell growth decreases compared with that in a new chill-mould.

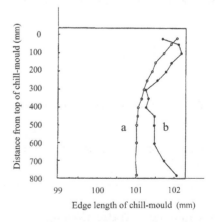

Figure 5.22   Wear and tear of a billet chill-mould. (a) Profile of a new chill-mould. (b) Profile of a chill-mould that has been used 350 times. Reproduced with permission from the Scandanvian Journal of Metallurgy, Blackwell.

An uneven air gap between the shell and the chill-mould results in uneven shell growth and an uneven surface temperature of the shell. This causes thermal stress in the shell, which in serious cases may result in rupture of the shell, resulting in a hole through which the melt may rush out.

Example 5.3

Stainless steel has been cast in a slab machine. The temperature of the melt was 1460 °C, the surface temperature of the chill-mould was 100 °C and a casting rate of 1.2 m/min was used. A shell burst and was examined. Its thickness was measured as 19 mm at a distance of 65 cm below the upper surface of the melt.

Calculate the heat transfer coefficient h of the air gap between the shell and the chill-mould during the casting process.

Solution:

The relationship between the shell thickness y_L and the time required to achieve this is given by Equation (4.48) on page 74 in Chapter 4:

$$t = \left(\frac{\rho(-\Delta H)}{T_L - T_0}\right) \frac{y_L}{h} \left[1 + \left(\frac{h}{2k}\right) y_L\right] \quad (1')$$

The time t can be calculated from the information about the casting rate ($1.2/60 = 0.020$ m/s) and the distance below the upper surface of the melt:

$$t = \frac{\text{distance from the upper level}}{\text{casting rate}} = \frac{0.65}{0.020} = 32.5 \text{ s} \quad (2')$$

The material constants for stainless steel are:

$$\rho = 7.8 \times 10^3 \text{ kg/m}^3 \quad -\Delta H = 276 \times 10^3 \text{ J/kg} \quad k = 46 \text{ W/mK}.$$

These values are inserted into Equation (4.48):

$$32.5 = \frac{(7.8 \times 10^3)(276 \times 10^3)}{(1460 - 100)} \times \frac{0.019}{h}\left(1 + \frac{h}{2 \times 46} \times 0.019\right) \quad (3')$$

h can be solved from this equation of the first degree.

Answer:

The desired heat transfer coefficient h is approximately 1.1×10^3 W/m^2 K.

5.5.3 Dimensioning of Water Cooling of the Chill-Mould

In order to dimension the water cooling in continuous casting properly we have to discuss each step of the total heat transport (Figure 5.23) and compare each with the others [Equation (5.21) on page 102]:

$$\frac{1}{h_{\text{total}}} = \frac{l_{\text{metal}}}{k_{\text{metal}}} + \frac{\delta}{k_{\text{air}}} + \frac{l_{\text{mould}}}{k_{\text{mould}}} + \frac{1}{h_w} \quad (5.21)$$

In continuous casting chill-moulds made of copper are used that have a very good thermal conductivity (large k_{mould}). As a concrete example we assume that the cast metal is steel. The thermal conductivity of the solidified shell is

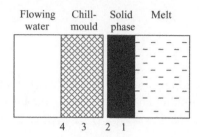

Figure 5.23 Heat transport from melt to flowing water through a water-cooled chill-mould.

also good, both for steel and other metal alloys (k_{metal} is comparatively large). On the other hand, the thermal conductivity of air is very poor compared with those of metals.

Next we compare the first three terms on the right-hand side in Equation (5.21). We assume reasonable values for the thickness of the chill-mould wall, the air gap and the solidified shell, and use known values of the thermal conductivity coefficients of the alloy, the air and copper. Then we realize that two terms, which represent the chill-mould and the metal shell, are relatively small compared with the term that represents the air gap.

In order to dimension the water cooling in an optimal way the total shell thickness s_{steel} after growth in the chill-mould is calculated first of all. The width of the air gap is a very important parameter in the calculations. A heat balance is then used to calculate the amount of water that has to pass the chill-mould per unit time.

$$\frac{dq}{dt} = c_p^w \rho_w D_E \, a \, u_w \, \Delta T_w = \rho \, v_{\text{cast}} \, s_{\text{steel}} \, a \, (-\Delta H) \quad (5.28)$$

Heat flux removed by the aid of the cooling water.	Heat flux emitted from the casting strand.

where

D_E = width of water gap

a = width of a mould side

ρ_w = density of water

ρ = density of steel

c_p^w = thermal capacity of water

ΔT_w = temperature increase of cooling water after passage of the chill-mould

u_w = velocity of cooling water

v_{cast} = casting rate or the velocity of the casting strand

s_{steel} = thickness of steel shell at the exit from the chill-mould

$-\Delta H$ = heat of fusion per kg steel.

Material constants and v_{cast} are known. The shell thickness is normally of the magnitude 1–1.5 cm, which influences the design of the mould. Three unknowns remain: D_E, u_w and ΔT_w.

The removed heat flux can alternatively be written as:

$$\frac{dq}{dt} = h_{steel}(T_{steel\ surface} - T_{mould}) = h_w(T_{mould}^w - T_w) \quad (5.29)$$

The maximum value of $T_{steel\ surface}$ is equal to the melting point of steel. T_{mould}^w must not exceed 100 °C at any point, in accordance with earlier discussions (Section 5.4.3 on page 104). Maximum heat transfer occurs close to the meniscus of the melt, where h_{steel} is of the magnitude 4×10^3 W/m² K. At this level there is a maximum risk of getting too high a water temperature. Therefore, the water flow has to be designed in such a way that the water temperature at the mould wall cannot exceed 100 °C at this level. The water velocity is calculated as follows:

(1) Reasonable temperature values and $h_{steel} = 4 \times 10^3$ W/m² K are inserted into Equation (5.29) and a minimum value of h_w is calculated.

(2) The calculated value of h_w, a reasonable value of D_E and material constants are inserted into Equation (5.23) (page 103):

$$h_w = \frac{k}{D_E} \times 0.023 \times \frac{u_w D_E^{0.8}}{v_{kin}} Pr^{0.33} \quad (5.23)$$

and a minimum value of u_w is calculated.

(3) The calculated value of u_w and the chosen value of D_E are inserted into Equation (5.28), together with material constants, and the value of ΔT_w is obtained. The temperature increase ΔT_w of the cooling water must not be unreasonably large (≤ 10 °C). If it is too large, a new value of D_E is chosen and the calculations are repeated until a satisfactory value of ΔT_w is reached.

5.5.4 Secondary Cooling

The Function of Secondary Cooling

The casting rate in continuous casting is chosen in such a way that the solidified shell, formed in the chill-mould,

is thick enough to withstand the pressure from the melt. Secondary cooling is required. It has three purposes:

(1) control of the casting rate in such a way that the core has solidified before the casting leaves the pinch rolls;

(2) control of the surface temperature in order to avoid serious cracks;

(3) cooling of the machine foundation.

After passage of the chill-mould, the casting runs in the so-called casting bow, where it will solidify completely. This consists of:

(a) a frame, which stabilizes the construction;

(b) rolls, which steer the casting into its proper position;

(c) casting bow nozzles that spray water on the surface of the casting.

Three different types of nozzles (Figure 5.24) are used: (i) full cone nozzles; (ii) flat nozzles, and (iii) air atomizing nozzles.

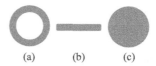

(a)   (b)   (c)

Figure 5.24 The cross-section of the water beam in (a) a full cone nozzle, (b) a flat nozzle, and (c) an air atomizing nozzle.

Full cone nozzles produce ring-shaped sprays, and flat nozzles give rectangular spray shapes. None of these types of nozzle distributes the water uniformly over the surface of the casting, with the result that the temperature on the casting surface will vary [see Figure 5.25]. Full cone nozzles and flat nozzles require a certain minimum water pressure in order to work. If the pressure is too low, only water drops will come from the nozzle and the cooling is inhibited. The air atomizing nozzle works in such a way that air is forced into the water, which then becomes

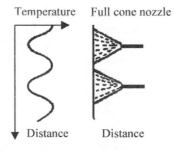

Temperature   Full cone nozzle

Distance   Distance

Figure 5.25 (a) Variation of the surface temperature of the strand while a full cone nozzle is used in the cooling chamber.

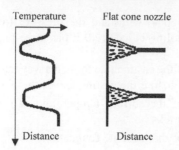

Figure 5.25 (b) Variation of the surface temperature of the strand while a flat cone nozzle is used in the cooling chamber.

atomized and gives many small droplets. The amount of water can be reduced in this case.

Dimensioning of Secondary Cooling in Slabs
During continuous casting of steel and other metals their thermal conductivities will influence the solidification process. The surface temperature of the shell is very sensitive to the cooling conditions, which have to be varied during the cooling time. For this reason the secondary cooling is divided into several different zones with cooling capacities that successively decrease with the distance from the chill-mould (see Example 5.4).

In order to achieve well-dimensioned secondary cooling, one has to avoid large temperature variations and keep the surface temperature as constant as possible. At dimensioning of the secondary cooling each zone is considered separately. Within each zone conditions are created that are as uniform as possible.

It is reasonable to assume that the heat transfer coefficient is constant within each separate zone and is a function of (i) the water flow through the nozzles in the zone, (ii) the length of the zone, and (iii) the dimensions of the casting.

Water-cooling has to be dimensioned in such a way that the core has solidified completely before the casting leaves the withdrawal pinch rolls. Water-cooling is generally the only parameter that can be adjusted during the casting process. The surface temperature and the solidification rate of the casting are calculated from Equations (4.45), (4.46) and (4.48) in Chapter 4 using known material constants and the heat transfer coefficient h_w between the surface of the casting and the flowing water.

The heat transfer coefficient h_w depends on the water flow and the temperature of the water. In the scientific literature several empirical relations are given. Here we will use the one in Equation (5.30).

$$h_w = \frac{1.57 \times w^{0.55}[1 - (0.0075 \times T_w)]}{\alpha} \times 10^3 \quad (5.30)$$

where

h_w = heat transfer number (W/m^2 K)

w = water flux (litres/m^2 s)

T_w = temperature of the cooling water

α = machine-dependent parameter (magnitude approximately 4).

The machine-dependent parameter is of course different for different casting machines. For the sake of simplicity it has been given a constant value for all machines here.

Example 5.4
In order to avoid cracks in the strand at continuous casting it is vital to keep the surface temperature of the casting constant during secondary cooling. Find the heat transfer coefficient h_w as a function of casting rate v_{cast} and the distance z from the chill-mould when the surface temperature condition is fulfilled. All material constants are known and the casting rate is assumed to be constant.

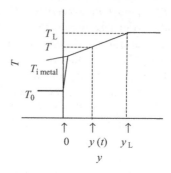

Solution:
Using the basic laws of thermal conduction and heat transfer we get the developed heat of fusion, which is independent of the casting per unit area and unit time:

$$\frac{dq}{dt} = -h_w(T_{i\,metal} - T_o) = -k\frac{(T_L - T_{i\,metal})}{y} = -(-\Delta H)\rho\frac{dy}{dt} \quad (1')$$

The temperature $T_{i\,metal}$ is constant according to the text. We can integrate the last two terms in Equation (1') and solve y from the new equation:

$$y = \sqrt{\frac{2k(T_L - T_{i\,metal})t}{\rho(-\Delta H)}} \quad (2')$$

We solve Equation (1') for h_w by use of the first equality:

$$h_w = \frac{k(T_L - T_{i\,metal})}{y(T_{i\,metal} - T_0)} \quad (3')$$

and introduce the value of y [Equation (2′)] into Equation (3′):

$$h_w = \frac{k(T_L - T_{i\,metal})}{T_{i\,metal} - T_0}\sqrt{\frac{\rho(-\Delta H)}{2k(T_L - T_{i\,metal})t}}$$

$$= \frac{\sqrt{k(T_L - T_{i\,metal})\rho(-\Delta H)}}{\sqrt{2t}(T_{i\,metal} - T_0)} \qquad (4')$$

or

$$h_w = \frac{\sqrt{k(T_L - T_{i\,metal})\rho(-\Delta H)}}{\sqrt{2}(T_{i\,metal} - T_0)}\left(\frac{1}{\sqrt{t}}\right) = \frac{const}{\sqrt{t}} \qquad (5')$$

As the casting rate is constant the time t can be replaced by the distance z in Equation (5′) if we use the relationship $t = z/v_{cast}$. The result of this operation is:

$$h_w = \frac{\sqrt{k(T_L - T_{i\,metal})\rho(-\Delta H)v_{cast}}}{\sqrt{2}(T_{i\,metal} - T_0)}\left(\frac{1}{\sqrt{z}}\right) = \frac{const}{\sqrt{z}} \qquad (6')$$

Answer:
The heat transfer coefficient has to be inversely proportional to the square root of the distance from the chill-mould in order to keep the surface temperature of the casting constant. The desired relationship is given in Equation (6′).

Equation (2′) in Example 5.4 (Equation 5.31):

$$y = \sqrt{\frac{2k(T_L - T_{i\,metal})t}{\rho(-\Delta H)}} \qquad (5.31)$$

can be used to calculate the thickness of the solidified shell as a function of time. The function is illustrated in Figure 5.26 for low carbon steel at three different surface temperatures of the strand. For the calculations of the function we have assumed that the surface temperatures are kept constant during secondary cooling and that the heat transfer coefficients decrease according to Equation (6′) in Example 5.4. The parabolic function is analogous to that in Figure 4.21 on page 79 in Chapter 4.

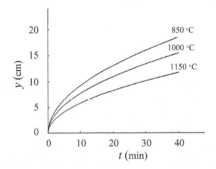

Figure 5.26 Shell thickness as a function of time after start of casting for three alternative surface temperatures of the strand.

The temperature of the cooling water affects its cooling power strongly. Some of the water that hits the hot strand surface is evaporated in the cooling chamber. The higher the water temperature is, the less efficient is the cooling power of the water (Figure 5.27). The efficiency is even more reduced if the water contains impurities and oil.

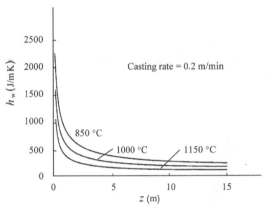

Figure 5.27 Ideal heat transfer coefficient at the surface of the strand as a function of the distance from the chill-mould for three different surface temperatures of the strand.

In order to avoid cracks in the strand during continuous casting it is desirable to *keep the surface temperature constant* during the whole of the secondary cooling process. This is approximately achieved in practice by use of several different heat transfer coefficients, one for each zone, calculated from Equations (5.32) and (5.30) (see also Example 5.4 on page 112).

$$h_w = \left(\sqrt{\frac{k(T_L - T_{i\,metal})\rho(-\Delta H)v_{cast}}{\sqrt{2}\cdot(T_{i\,metal} - T_0)}}\right)\frac{1}{\sqrt{z}} = \frac{const}{\sqrt{z}} \qquad (5.32)$$

The water cooling is then designed according to the calculations.

Metallurgical Length
Water-cooling has to be adjusted in such a way that the casting has solidified completely before it leaves the withdrawal pinch rolls. Using Equation (5.31) the solidification time of the strand of a slab can be calculated. The strand is water cooled from all sides. The solidification is complete when the two solidification fronts meet at the centre of the strand, i.e. when y equals *half* the thickness s (Figure 5.28). The value $y = s/2$ is inserted

into Equation (5.31), which gives the solidification time t_{sol} as a function of the thickness s:

$$t_{sol} = \frac{s^2 \rho (-\Delta H)}{8k(T_L - T_{i\,metal})} \qquad (5.33)$$

When the time for complete solidification is known it is easy to calculate the *metallurgical depth* or the *metallurgical length*, which is defined as follows:

The metallurgical length L is the distance from the top of the chill-mould to the point in the casting where the core has just solidified (Figure 5.28).

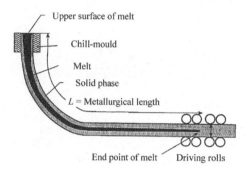

Figure 5.28   A necessary condition is that the strand has solidified completely before it leaves the lower pinch rolls.

Provided that the casting rate is constant the metallurgical length can be calculated as a function of material constants, the casting rate and the surface temperature $T_{i\,metal}$.

$$L = v_{cast}t_{sol} = \frac{v_{cast}\ s^2\ \rho(-\Delta H)}{8k(T_L - T_{i\,metal})} \qquad (5.34)$$

Figure 5.29 shows the metallurgical length of a slab as a function of the casting rate for the three different surface temperatures used in Figure 5.26.

Relation between Shell Thickness and Time for Continuous Castings with a Square Cross Section

The calculations behind Figures 5.26 and 5.29 are valid only for slabs. For square castings the area of the solidification front *decreases* with increasing distance from the top of the chill-mould and distance from the surface. The result is that the solidification rate initially decreases and then increases.

A very large fraction of continuous castings have square cross sections. For this reason it is vital to derive a simplified relationship between shell thickness and time for this case. This has been done in Example 5.5, which shows in detail how such calculations can be performed.

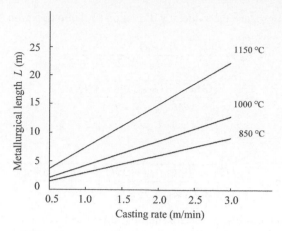

Figure 5.29   The metallurgical length as a function of the casting rate for three different surface temperatures of the strand.

Example 5.5

In a continuous casting process, a square casting with the dimensions $a \times a$ is cooled after passage of the chill-mould in a cooling chamber.

(a) Determine the surface temperature $T_{i\,metal}$ as a function of the shell thickness y of the solidified layer. The temperature of the melt T_L and the water temperature T_0 are known.

(b) Derive the relationship between solidification rate and shell thickness.

(c) Derive the relationship between shell thickness and time.

(d) Calculate the total solidification time.

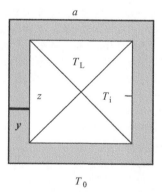

Solution:

The basis for the solution is the square cross section of the casting and the coordinate system given in the figure.

At the time $t = 0$ the solidification front is in the position $y = 0$. At the time t it has moved towards the centre and has the position $y = y(t)$. The area of the solidification front decreases more and more the closer to the centre it comes. This means that the solidification rate increases with time. We will try to estimate the size of this increase.

The square can be divided into four triangles and we assume that these solidify separately without mutual heat exchange. We also assume that the heat flux through the shell dq/dt is constant at every level of the shell. The deeper we come into the casting the smaller will the square and its side z be.

We can conclude from the figure that the side z of the small square, at the distance y from the large square with the side a, is:

$$z = a - 2y \qquad (1')$$

At the solidification front the solidification heat dQ is developed during the time dt and we get:

$$\frac{dQ}{dt} = -A\rho(-\Delta H)\frac{dy}{dt} = -(a-2y)b\rho(-\Delta H)\frac{dy}{dt} \qquad (2')$$

where b is the height of the casting.

The heat flow through the solid shell can be written as:

$$\frac{dQ}{dt} = -kA\frac{dT}{dy} = -k(a-2y)b\frac{dT}{dy} \qquad (3')$$

The heat transfer between chill-mould and cooling water can be written as:

$$\frac{dQ}{dt} = -h_w A(T_{i\,metal} - T_0) = -h_w ab(T_{i\,metal} - T_0) \qquad (4')$$

We have four equations. The unknown quantities are $T_{i\,metal}$, z, y and its time derivative.

(a): The first step is to determine the temperature T_i as a function of y. The left-hand side of Equation $(3')$ is equal to the left-hand side of Equation $(2')$. The same must be true for the right-hand sides of the equations:

$$(a-2y)bk\frac{dT}{dy} = (a-2y)b\rho(-\Delta H)\frac{dy}{dt}$$

which can be simplified to:

$$k\frac{dT}{dy} = \rho(-\Delta H)\frac{dy}{dt} \qquad (5')$$

If we assume that dy/dt is relatively independent of time, Equation $(5')$ can be integrated from $y = 0$ to an arbitrary y value. The corresponding values of T are the desired temperature $T_{i\,metal}$ and T_L, the temperature of the melt:

$$k\int_{T_{i\,metal}}^{T_L} dT = \rho(-\Delta H)\frac{dy}{dt}\int_0^{y(t)} dy \qquad (6')$$

which can be written as:

$$T_L - T_{i\,metal} = \rho(-\Delta H)\frac{dy}{dt}\frac{y}{k} \qquad (7')$$

Equation $(3')$ can be transformed into:

$$\frac{dQ}{dt}\frac{dy}{a-2y} = -bk\,dT \qquad (8')$$

Equation $(8')$ is integrated and simplified:

$$\frac{dQ}{dt}\int_0^y \frac{dy}{a-2y} = -bk\int_{T_{i\,metal}}^{T_L} dT$$

$$\frac{dQ}{dt}\left(\frac{-1}{2}\right)\ln\frac{a-2y}{a} = -bk(T_L - T_{i\,metal})$$

or

$$\frac{dQ}{dt} = \frac{-2bk(T_L - T_{i\,metal})}{\ln\left(\dfrac{a}{a-2y}\right)} \qquad (9')$$

The expression $(9')$ is inserted into Equation $(4')$:

$$\frac{-2bk(T_L - T_{i\,metal})}{\ln\left(\dfrac{a}{a-2y}\right)} = -h_w ab(T_{i\,metal} - T_0) \qquad (10')$$

We solve $T_{i\,metal}$ from Equation $(10')$, which gives:

$$T_{i\,metal} = \frac{2kT_L + ah_w T_0\ln\left(\dfrac{a}{a-2y}\right)}{2k + ah_w\ln\left(\dfrac{a}{a-2y}\right)} \qquad (11')$$

which is the required relationship. It shows that the growth rate initially decreases like a one-dimensional casting but increases at the end of the solidification when y approaches the value $a/2$.

(b): The next step is to derive an expression for dy/dt as a function of y. We realize that the right-hand sides of Equations $(2')$ and $(4')$ must be equal:

$$(a-2y)b\rho(-\Delta H)\frac{dy}{dt} = h_w ab(T_{i\,metal} - T_0) \qquad (12')$$

We divide Equation $(12')$ with the factor b, introduce the value of $T_{i\,metal}$ (Equation $(11')$ above) and solve the solidification rate dy/dt. After reduction we get:

$$\frac{dy}{dt} = \frac{ah_w}{(a-2y)\rho(-\Delta H)}\cdot\frac{2kT_L + ah_w T_0\ln\left(\dfrac{a}{a-2y}\right) - T_0\left[2k + ah_w\ln\left(\dfrac{a}{a-2y}\right)\right]}{2k + ah_w\ln\left(\dfrac{a}{a-2y}\right)}$$

which can be reduced to:

$$\frac{dy}{dt} = \frac{T_L - T_0}{\rho(-\Delta H)} \times \frac{a}{(a-2y)} \times \frac{h_w}{1 + \left(\frac{ah_w}{2k}\right) \times \ln\left(\frac{a}{a-2y}\right)} \quad (13')$$

This is the required expression.

(c): A relationship between y and t is wanted. It is obtained by separation of the variables in Equation $(13')$ and integrating:

$$\int_0^y (a-2y)\left[2k + ah_w \ln\left(\frac{a}{a-2y}\right)\right] dy$$

$$= \frac{2kah_w}{\rho(-\Delta H)}(T_L - T_0)\int_0^t dt$$

which can be transformed into:

$$t = \frac{\rho(-\Delta H)}{2k(T_L - T_0)}\int_0^y (a-2y)\left(\frac{2k}{ah_w} + \ln a - \ln(a-2y)\right) dy \quad (14')$$

and divided into two integrals:

$$t = \frac{\rho(-\Delta H)}{2k(T_L - T_0)}(I_1 + I_2) \quad (15')$$

where

$$I_1 = \int_0^y (a-2y)\left(\frac{2k}{h_w} + \ln a\right) dy = \left(\frac{2k}{h_w} + \ln a\right)\int_0^y (a-2y) dy$$

After integration we get:

$$I_1 = \left(\frac{2k}{ah_w} + \ln a\right)\frac{(a-2y)^2 - a^2}{2(-2)} = \left(\frac{2k}{ah_w} + \ln a\right)y(a-y)$$

The second integral, which contains a logarithm function, is solved by partial integration:

$$I_2 = \int_0^y -(a-2y)\ln(a-2y)\,dy$$

$$= -\left[\frac{(a-2y)^2}{2(-2)}\ln(a-2y)\right]_0^y - \int_0^y \frac{(a-2y)^2}{4}\left(\frac{-2}{a-2y}\right) dy$$

or

$$I_2 = \left(\frac{(a-2y)^2}{4}\ln(a-2y) - \frac{a^2}{4}\ln a\right) - \left[\frac{(a-2y)^2}{(-2)2(-2)}\right]_0^y$$

Further calculations give:

$$I_2 = \frac{(a-2y)^2}{4}\ln(a-2y) - \frac{a^2}{4}\ln a + \frac{y(a-y)}{2}$$

If we introduce the solutions of the integrals into the expression of t [Equation $(15')$] we finally get the desired relationship between time and shell thickness:

$$t = \frac{\rho(-\Delta H)}{2k(T_L - T_0)}\left[\left(\frac{2k}{ah_w} + \ln a\right)y(a-y) + \left(\frac{(a-2y)^2}{4}\ln(a-2y)\right) - \left(\frac{a^2}{4}\ln a + \frac{y(a-y)}{2}\right)\right] \quad (16')$$

(d): Inserting $y = a/2$ into Equation $(16')$ gives the desired time.

Answer:
The desired relationships are as follows:

(a) Equation $(11')$ above;
(b) Equation $(13')$ above;
(c) Equation $(16')$ above.
(d) The total solidification time is $t = \frac{\rho(-\Delta H)}{(T_L-T_0)}\frac{a^2}{8k}\left(\frac{2k}{ah_w} + \frac{1}{2}\right)$.

Comparing Equation $(13')$ and Equation (4.46) on page 74 in Chapter 4 shows that the solidification rate in this three-dimensional case *increases* at the end of the solidification process, while it *decreases* at the end in the one-dimensional case.

Temperature Field at Secondary Cooling

Example 5.5 above is approximately valid for a casting with a square cross section. The real conditions are often so complicated that it is necessary to use a computer for careful calculation of the temperature fields at the surface of the casting at various times and under various cooling conditions. Several types of result can be derived from these calculations, for example:

- metallurgical length;
- isotherms in a cross section of the casting;
- isotherms in a length section of the casting;
- temperature in an arbitrary point during the whole casting process;
- shell thickness.

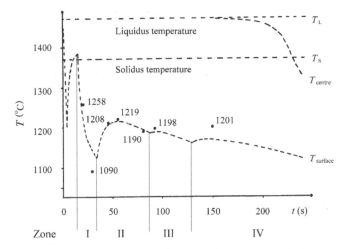

Figure 5.30 The temperature at the centre and at the middle of a face of a square casting as a function of time in continuous casting. The casting rate was 2.8 m/min. Reproduced with permission from the Scandanvian Journal of Metallurgy, Blackwell.

A concrete example is given in Figure 5.30. The calculations were made for casting of billets of the size 100 mm × 100 mm with the following material data:

liquidus temperature = 1470 °C
solidus temperature = 1370 °C
carbon concentration = 0.57 wt-%.

The *upper curve* shows the temperature at the centre as a function of time. It can be seen that the temperature deviated from the liquidus temperature after about 150 s. The whole casting had solidified, i.e. reached the solidus temperature, after about 230 s.

The metallurgical length was calculated to be:

$$\left[\frac{2.8\,\text{m}}{60\,\text{s}} \times 230\,\text{s} \approx 11\text{m} \right].$$

The *lower curve* describes the surface temperature at the middle of a face. The secondary cooling is in this case divided into four zones, marked by the designations I–IV in Figure 5.30. The amount of cooling water per unit time decreased with the distance from the chill-mould. The larger the distance was, the smaller was the water flow. During the casting process the surface temperature of the casting was measured at the entrance to each zone. It increased initially because the cooling water was reduced from one zone to the next one. The result of such measurements has been introduced into Figure 5.30 as points with temperature figures.

The zone before zone I is identical to the chill-mould.

In zone I, closest to the chill-mould, the entrance temperature was measured as 1258 °C. The surface temperature decreased strongly because the water cooling was strong. The heat transfer coefficient was large.

In zone II the entrance temperature was measured to 1090 °C. In zone II, three temperatures were measured, 1208 °C, 1219 °C and 1190 °C. A strong reheating of the surface at the entrance of zone II was obtained. A somewhat lower value of the heat transfer number than in zone I was obtained.

In zone III a temperature of 1198 °C was recorded. A tiny reheating of the surface at the entrance of zone III indicated that the water cooling was somewhat weaker than in zone II but the surface temperature continued to decrease. The heat transfer coefficient was somewhat lower than in zone II.

In zone IV the heat capacity has decreased strongly. In spite of this the temperature decreases slightly.

After passing through the last zone the casting was cooled with air only. The surface temperature continued to decrease in spite of very weak cooling and remaining melt in the interior of the casting. The solidification rate was so low that the weak cooling of the surface more than balanced the developed solidification heat.

The curves in Figure 5.30 are based on computer calculations. They show good agreement with the measured temperature values, which are marked with dots in Figure 5.30.

5.6 HEAT TRANSPORT IN THE ESR PROCESS

On pages 27–28 in Chapter 2 the electro-slag refining (ESR) process for refining ingots is treated. Figure 5.31 illustrates the principle of ESR but it is very schematic and simplified as compared with reality. Here a more realistic version of the ESR process, based on heat flow during solidification, will be briefly discussed.

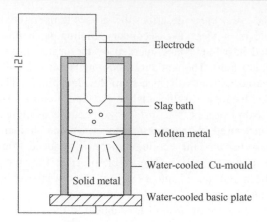

Figure 5.31 The principle of the ESR process.

5.6.1 Temperature Distribution in the Metal Bath

We will concentrate on the temperature distribution in the upper part of the refined ingot and examine the shape of the solidification front. This will give an indication of the temperature distribution in the upper part of the ingot.

Figure 5.32 illustrates the shape of the refined melt, the solidification front and some isotherms in the solid phase. The solidification front is identical to the T_L-isotherm.

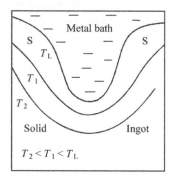

Figure 5.32 The shape of the melt and some isotherms in the melt at remelting with the aid of ESR. Reproduced from P. O. Mellbery's thesis.

Experimental evidence shows that:

- the shape of the metal bath primarily depends on the amount and composition of the slag and to a minor extent on the remelting rate;
- a high remelting rate always gives a deep metal bath but a deep bath can also be attained at a low remelting rate if the slag amount is sufficiently small.

A deep metal bath suggests a high temperature and a large vertical temperature gradient (dense isotherms) in the centre of the melt. As Figure 5.32 shows there is also a comparatively large temperature gradient at the centre of the remelted ingot at temperatures below the liquidus temperature.

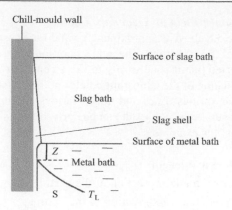

Figure 5.33 Shapes of the solidification front and the metal bath at the wall of the chill-mould. Reproduced from P. O. Mellbery's thesis.

Figure 5.33 illustrates the appearance of the slag bath and the metal bath close to the chill-mould wall. It can be seen from this figure that:

- The metal bath achieves a certain height Z above the solid ingot close to the chill-mould wall and stays in direct contact with the solid slag shell.
- This height Z is important. At remelting conditions with a small height or none at all the surface of the ingot becomes very rough.

5.6.2 Temperature Distribution on the Inside and Outside of the Chill-Mould Wall

Figure 5.34 shows the result of a series of temperature measurements on the *inside* and *outside* of the chill-mould wall, for ESR of an ingot with a diameter of 100 mm. The vertical distances between the measurement points were 10 mm. The figure shows the measured temperatures within the different regions in contact with the chill-mould.

The radial heat flux can be written as:

$$\frac{dq}{dt} = -k\frac{(T_{inner} - T_{outer})}{\Delta d} \quad (5.35)$$

where Δd is the constant thickness of the chill-mould wall. The radial heat flux between the inner and outer chill-mould walls is proportional to the temperature difference between them. The heat flux within the different zones, i.e. the temperature difference in the diagram, can be interpreted as follows.

The source of the heat flux in region I above the slag bath is heat radiation. The heat flux increases gradually at lower heights. At the slag bath surface it increases rapidly. In region II the heat flux from the slag bath through the chill-mould is high and practically constant. The resistance in the slag bath is high and electrical heat (RI^2) is developed.

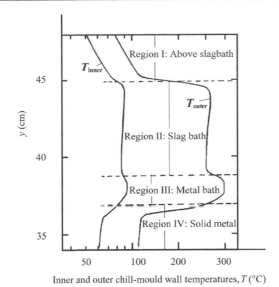

Figure 5.34 Temperature profile of the inside and outside of the chill-mould wall in ESR of steel. Reproduced from P. O. Mellbery's thesis.

The average heat flow in region III exceeds the heat flow in region II slightly. The reason is that the solid slag layer, close to the mould, is thinner where it is in contact with the melt bath than where it is in contact with the slag bath. This is the consequence of a higher heat transfer from the metal melt than from the slag. The metal melt has higher thermal conductivity and lower viscosity than the slag bath.

At the change from melt in region III to solid metal in region IV the heat flow is drastically decreased within a narrow height interval. The reason is that an air gap is formed between the solidified metal and the mould due to cooling shrinkage.

The calculation of the shape of the solidification front and of the solidification rate is very complicated because heat is transferred from the slag bath to the metal bath by the aid of strong convection in both these melts. These calculations are beyond the scope of this book.

5.7 HEAT TRANSPORT IN NEAR NET SHAPE CASTING

In the 1970s and 1980s a large number of direct casting methods were developed and refined to facilitate continuous production of fibres, strips and wires. These methods were based on the fact that the melt solidifies rapidly at very high cooling rates by bringing a small amount of melt into intimate contact with a material with very high conductivity. In this way most metal alloys obtain a radically changed structure and improved material properties. Good examples are modern amorphous materials, which are discussed in Chapter 6.

At sufficiently high cooling rates, normally of the magnitude 10^5 K/s for metal alloys, crystallization is suppressed

and a noncrystalline phase, i.e. an *amorphous* metal alloy is formed. Each material has its own critical cooling rate, which must be exceeded in order to allow an amorphous structure to be formed. An amorphous alloy can be regarded as a viscous liquid and has no regular crystalline structure.

Amorphous metal alloys have excellent properties. They have, for example, high corrosion resistance, very good soft-magnetic properties and particularly good mechanical properties, such as high ductility and mechanical strength, close to the theoretical value for the material in question.

Below we discuss heat transport during strip casting both without (Sections 5.7.2 and 5.7.3) and with (Sections 5.7.4 to 5.7.6) respect to convection in the melt during the solidification process.

5.7.1 Heat Transport in Strip Casting

With the aid of computer calculations it is possible to analyse heat transport through the solidifying strip and the solidification process during strip casting in detail. It is important to know and be able to control the solidification process because the properties of the material depend strongly on the solidification rate. The more rapidly the strip solidifies, the finer will the structure of the material be and the better its properties.

If the strip is thin and the heat transport rapid, the strip solidifies very quickly and without fluctuations in the alloy composition. This ideal process is called *rapid solidification*. With the aid of new rapid solidification methods, strips of a few millimetres in thickness can be cast, ready for direct use or cold rolling. There is great interest in casting thin plates in this way in order to gain considerable production advantages and save energy.

The solidification rate in the melt is controlled by various factors depending on Nussel's number (pages 85–86 in Chapter 4). Heat transport is entirely controlled by the heat flux through the surface of the strip if Nu ≪ 1. Otherwise the convection in the melt also influences the heat transport through the solidifying strip and its solidification rate. In Section 5.7.4 we will return to convection during rapid solidification.

Below, a simple but illustrative model of the heat transport through a solidifying strip, with no regard to the convection in the melt, will be described. We will also discuss production rates of strip casting machines.

5.7.2 Thermal Conduction in a Solidifying Strip – Solidification Time and Solidification Length

In Chapter 1 different types of strip casting processes are described. In order to analyse their solidification processes we will start with a simple one, the model of a single strip casting machine as illustrated in Figure 5.35. Here we will assume that Nu ≪ 1. In section 5.7.3 we will

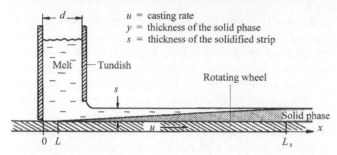

Figure 5.35 Simple model of a strip casting machine. Tundish with a gap in contact with a rotating wheel. Reproduced by permission from N. Jacobsson.

discuss other types of machines with no restriction on Nussel's number.

Solidification Time of a Strip – No Convection and Nu ≪ 1

In the following derivation we will assume that the thermal conductivity of the melt is very large and that the temperature is the same everywhere in the melt because all possible temperature differences will even out instantly. The temperature of the melt is equal to the liquidus temperature throughout (Figure 5.36). We will further assume here that the solidification front has a temperature close to the liquidus temperature (Nu ≪ 1). In Chapter 6 we discuss a case where the front temperature varies with the growth rate.

The heat flux from the solidified strip at the interface between melt and strip can be described with the aid of the relationship given in Section 4.4.5. If Nussel's number ≪ 1, which is normally the case for thin strip castings, we can apply Equation (4.85) on page 86 in Chapter 4 and set $T_{i\,metal}$ equal to T_L. In this case the solidification time will be:

$$t = \frac{\rho_{metal}(-\Delta H)}{(T_L - T_0)} \frac{y}{h} \tag{5.36}$$

where

y = thickness of the solidified layer

h = heat transfer coefficient between melt and solid phase

T_L = temperature of the melt (= liquidus temperature)

T_i = temperature of the interface melt/solid strip

T_0 = surface temperature of the rotating wheel

t = time

ρ_{metal} = density of the solid metal.

$-\Delta H$ = heat of fusion of the metal (J/kg)

When the thickness y is equal to the strip thickness s, the time t is equal to the solidification time t_{total}.

$$t_{total} = \frac{\rho_{metal}(-\Delta H)}{(T_L - T_0)} \frac{s}{h} \tag{5.37}$$

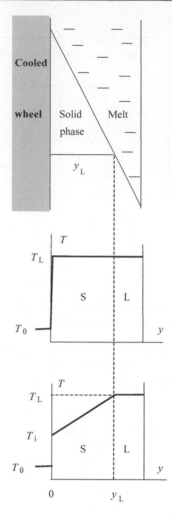

Figure 5.36 Top: the solidifying strip; Centre: temperature distribution in the strip in the case of Nu ≪ 1; bottom: temperature distribution in the strip in the general case.

Equation (5.37) is valid for low values of Nussel's number, up to Nu < 0.2. If the strip is cooled from both sides the thickness s is equal to *half* the strip thickness.

Solidification Length of a Strip – No Convection and Nu ≪ 1

If we assume that y increases linearly, it is easy to calculate the distance L_s where the strip has solidified completely. The strip moves with velocity u. Because $x = ut$ we get:

$$L_s = u \frac{\rho_{metal}(-\Delta H)}{(T_L - T_0)} \frac{s}{h} \tag{5.38}$$

The function L_s or the *solidification length* of the strip at a wheel temperature of 200 °C and varying strip thickness as a function of the heat transfer coefficient h is given in Figure 5.37 for steel and in Figure 5.38 for aluminium.

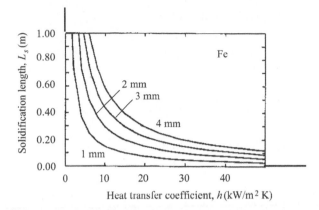

Figure 5.37 The solidification length as a function of the heat transfer coefficient for steel strips of various thicknesses. The casting rate is 1 m/s. Reproduced with permission from N. Jacobsson.

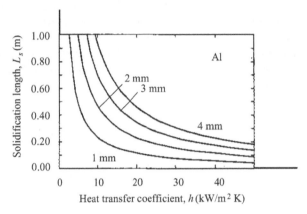

Figure 5.38 The solidification length as a function of the heat transfer coefficient for aluminium strips of various thicknesses. The casting rate is 1 m/s. Reproduced with permission from N. Jacobsson.

The temperature of the melt was in both cases equal to the liquidus temperature of the relevant metal.

If the strip thickness is small and the heat transfer coefficient is low this simple model gives approximately the same results as do careful computer calculations where the thermal conduction in the strip is considered as well. Equation (5.38) can be used for values of Nussel's number up to 0.2, i.e. Nu < 0.2.

In most cases the temperature of the melt T_{melt} exceeds the liquidus temperature. However, experimental investigations show that this temperature increase affects the solidification length very little. The reason is that the heat of solidification contribution is much larger than the thermal capacity contribution, which can be neglected.

Figures 5.39 and 5.40 illustrate this fact. In both cases the excess temperature was 100 °C. A comparison between Figures 5.39 and 5.40 and the Figures 5.37 and 5.38 shows that the excess temperature in this case has no influence outside the tundish.

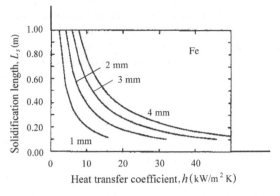

Figure 5.39 The solidification length as a function of the heat transfer coefficient for steel strips of various thicknesses. The excess temperature of the melt is 100 °C. Reproduced with permission from N. Jacobsson.

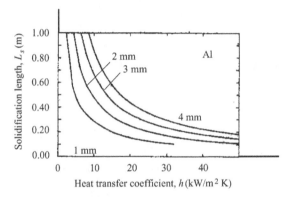

Figure 5.40 The solidification length as a function of the heat transfer coefficient for aluminium strips of various thicknesses. The excess temperature of the melt is 100 °C. Reproduced with permission from N. Jacobsson.

5.7.3 Casting Rate and Production Capacity as Functions of Strip Thickness for Single- and Double-Roller Machines with no Consideration Given to Convection

Casting Rate

In order to judge the factors that influence production capacity we will discuss a strip casting machine of the single-roller type (Figure 2.15 on page 24).

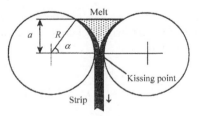

Figure 5.41 Strip casting machine of double-roller type. Reproduced with permission from the Scandanvian Journal of Metallurgy, Blackwell.

The calculations can also be applied to the double-roller process (Figure 2.16 on page 24 and Figure 5.41).

The solidification length L_s is the distance that the cast strip moves before it solidifies completely. Figure 5.42 illustrates the following simple geometrical relationship:

$$\alpha = \frac{L_s}{R} \qquad (5.39)$$

where the contact angle α is expressed in radians. According to Figure 5.42 we also have:

$$\sin \alpha = \frac{a}{R} \qquad (5.40)$$

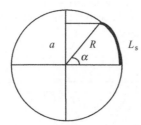

Figure 5.42　Definition of a, R, α and L_s.

The angle α is solved using Equation (5.40) and introduced into the left-hand side of equation (5.39) *expressed in degrees*. Then the right-hand side must also be expressed in degrees:

$$\left(\arcsin \frac{a}{R}\right)^{\circ} = \frac{L_s}{R}\frac{180}{\pi} \qquad (5.41)$$

and we get:

$$L_s = \frac{\pi R}{180}\left(\arcsin \frac{a}{R}\right)^{\circ} \qquad (5.42)$$

In a *double-roller machine* the cast strip solidifies both from the upper and lower sides. The solidification time is thus the time it takes for the strip to solidify to half the thickness. We use Equation (4.48) on page 74 and introduce $y = s/2$ and get the solidification time for a double-roller machine:

$$t_{\text{double}} = \frac{\rho_{\text{metal}}(-\Delta H)}{(T_{\text{metal}} - T_0)}\frac{s}{2h}\left[1 + \frac{h}{2k}\frac{s}{2}\right] \qquad (5.43)$$

For a *single-roller machine* the thickness y of the solidified layer is replaced by the strip thickness s. The solidification-time of the strip will be:

$$t_{\text{single}} = \frac{\rho_{\text{metal}}(-\Delta H)}{(T_{\text{metal}} - T_0)}\frac{s}{h}\left[1 + \frac{h}{2k}s\right] \qquad (5.44)$$

where, here and above:

　　L_s = solidification length
　　α = contact angle

　　　R = radius of the roller
　　　a = height of the melt
　$-\Delta H$ = heat of fusion (J/kg)
　ρ_{metal} = density of the solid metal
　　　h = heat transfer coefficient
　　　k = thermal conductivity of the solid metal
　　　s = strip thickness
　　T_L = liquidus temperature
　　T_0 = temperature of the roller.

We earlier derived an expression for the solidification time for a single-roller machine [equation (5.37) on page 120]. The two expressions in Equations (5.37) and (5.44) are different. Equation (5.44) is generally valid, i.e. for the case when Nu > 0.2, while Equation (5.37) is a special case, valid only if Nu < 0.2. Equation (5.44) becomes equal to Equation (5.37) for small values of Nussel's number.

The casting rate u, i.e. the velocity of the cast strip in the double-roller machine can be calculated when the solidification time and the solidification length are known. Using Equations (5.42) and (5.43) we get:

$$u_{\text{double}} = \frac{L_s}{t_{\text{double}}} = \frac{2h(T_L - T_0)}{\rho_{\text{metal}}(-\Delta H)s}\frac{1}{1 + \frac{hs}{4k}}\frac{\pi R}{180}\left(\arcsin \frac{a}{R}\right)^{\circ}$$

$$(5.45)$$

or

$$u_{\text{double}} = \frac{2h(T_L - T_0)}{\rho_{\text{metal}}(-\Delta H)s}\frac{1}{1 + \frac{hs}{4k}}\frac{\pi R}{180}\left(\arcsin \frac{a}{R}\right)^{\circ} \qquad (5.46)$$

The corresponding equation for a single-roller machine is obtained by replacing $s/2$ by s:

$$u_{\text{single}} = \frac{h(T_L - T_0)}{\rho_{\text{metal}}(-\Delta H)s}\frac{1}{1 + \frac{hs}{2k}}\frac{\pi R}{180}\left(\arcsin \frac{a}{R}\right)^{\circ} \qquad (5.47)$$

If the casting rate is increased, the strip thickness s decreases in accordance with Equation (5.34) on page 114 because the solidification length is constant. Several alternative strip casting processes have been developed. For direct casting of aluminium foil the Hunter method is often used.

Example 5.6

In the production of aluminium strips, according to the Hunter method, molten Al is pressed vertically upwards

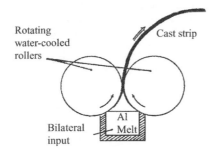

Rotating water-cooled rollers

Cast strip

Bilateral input

Al Melt

Strip casting machine according to Hunter.

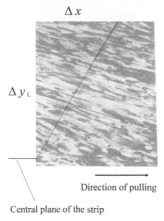

Δx

Δy_L

Direction of pulling

Central plane of the strip

Structure of a cast aluminium strip. A short piece of *half* the horizontal strip is shown in the figure.

between two water-cooled rollers that are located at a distance adjusted to the thickness of the strip. The Al metal solidifies on each of the cooled rollers and the two shells are pressed together at the contact area and cooled simultaneously. In this way a dense structure for the complete strip is obtained. The strip is moved vertically upwards and then removed horizontally for immediate rolling.

The right-hand Figure above shows the structure of such an aluminium strip. Its thickness is 6.0 mm. The casting rate was 1.0 m/min.

(a) Calculate the direction of the solidification front and illustrate it in the figure.

(b) Estimate the heat transfer coefficient between the strip and the cooled rollers.

Solution:

(a): Equation (4.46) on page 74 is applied for the case that Nu ≪ 1. In this case we get:

$$\frac{dy_L}{dt} = \frac{h(T_L - T_0)}{\rho_{metal}(-\Delta H)} \qquad (1')$$

The solidification rate and the casting rate are related by the expression:

$$\frac{dy_L}{dt} = \frac{dy_L}{dx}\frac{dx}{dt} \qquad (2')$$

where dy_L/dx is the slope of the solidification front. The solidification front moves in the direction of the 'structure' and is perpendicular to the bright structures in the right-hand figure. A line, perpendicular to the structure and corresponding to a solidification front, has already been drawn in the Figure above. The slope of the line is deter

mined by measurement in the figure.

$$\frac{dy_L}{dx} \approx 2 \quad \text{The casting rate} = \frac{dx}{dt} = 1.0\,\text{m/min} = \frac{1}{60}\,\text{m/s}.$$

(b): The values in (a) are inserted into Equation (2′) and Equations (1′) and (2′) are combined to give:

$$\frac{dy_L}{dt} = \frac{dy_L}{dx}\frac{dx}{dt} \approx 2 \times \frac{1}{60} = h\frac{T_L - T_0}{\rho_{metal}(-\Delta H)}$$

$$= h\left[\frac{658 - 20}{(2.69 \times 10^3)(322 \times 10^3)}\right]$$

which gives $h \approx 4.5 \times 10^4\,\text{W/m}^2\,\text{K}$.

Answer:

(a) The appearance of the solidification front is seen in the right-hand figure above.

(b) $4 \times 10^4\,\text{W/m}^2\,\text{K} < h < 5 \times 10^4\,\text{W/m}^2\,\text{K}$.

Another very common strip casting method is the Hazelett process. This is used for casting of Zn, Al, Cu and their alloys. The machine can cast bars, billets and slabs of numerous and varying cross sections. It is particularly convenient for casting of large quantities of thin, wide slabs and strips.

Example 5.7

The Figure in this exercise shows a strip casting machine for the Hazelett method. Melt is cast between two endless steel belts and is allowed to solidify between them. The cast strip must solidify completely before it leaves the belt loops. The distance between the driving wheels is 1.0 m and the heat transfer number between steel belt and casting is 900 W/m^2 K.

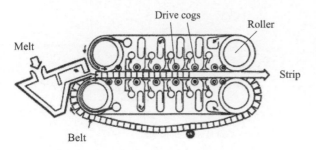

Reproduced with permission from Pergamon Press, Elsevier Science

Calculate the casting rate (m/min) as a function of the maximum strip thickness for a copper alloy, which can be cast in the machine. Material constants for Cu are found in standard references.

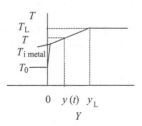

Solution:
The condition that the strip must be completely solid before it leaves the driving belts can be expressed as the casting rate times the solidification time $\leq$ the distance between the driving wheels:

$$ut_s \leq l_{\text{belt}} \qquad (1')$$

To get the casting rate u as a function of the strip thickness s, we must calculate the total solidification time of the strip.
Using Equation (4.48) on page 74:

$$t = \frac{\rho(-\Delta H)}{T_L - T_0} \frac{y_L}{h} \left(1 + \frac{h}{2k} y_L\right)$$

we calculate the total solidification time:

$$t = t_s \quad \text{for} \quad y_L = \frac{s}{2} \quad \Rightarrow \quad t_s = \frac{\rho(-\Delta H)}{T_L - T_0} \frac{s}{2h} \left(1 + \frac{h s}{4k}\right) \qquad (2')$$

T_L	1083 °C (Cu)
T_0	100 °C
ρ_{Cu}	8.94×10^3 kg/m^3
$-\Delta H$	206 kJ/kg
h	900 W/m^2 K
k_{Cu}	398 W/m K
L_{belt}	1.0 m
s	thickness of strip
y_L	$s/2$
u	casting rate

The expression for t_s is introduced into Equation (1'), which gives:

$$u = \frac{2\, l_{\text{belt}}\, h\, (T_L - T_0)}{s\, \rho_{\text{metal}}(-\Delta H)\left(1 + \dfrac{hs}{4k}\right)} \qquad (3')$$

Inserting the material constants from the table above and other known quantities we get:

$$u = \left[\frac{2 \times 1.0 \times 900 \times (1083 - 100)}{s(8.94 \times 10^3) \times 206 \times \left(1 + \dfrac{900s}{4 \times 398}\right)}\right] \times 60 \text{ m/min}$$

Answer:
The relationship between u and s is $u = \frac{57.7}{s(1+0.57s)}$ m/min (s is expressed in metres).

Production Capacity
A quantity of economical interest is the *production capacity P per hour for strips with the width* 1 metre. It is measured in units of kg/hour and can be calculated from the casting rate, and the thickness and density of the strip:

$$P = \rho\, u\, s \times 60 \times 60 \qquad (5.48)$$

It can be seen from Equations (5.46) and (5.47) on page 122 that the production capacity depends strongly on the radius of the roller and the contact angle α. The maximum value of the angle in practice is 90° for both single and double machines. This value, together with the strip thickness, determines the maximum casting rate in accordance with Equations (5.45) and (5.47). Equation (5.48) is valid for both single- and double-roller machines. An increased casting rate corresponds to a decreased strip thickness because the two solidification fronts must meet at the 'kissing point' (Figure 5.41 on page 121). Otherwise the strip will not solidify before it leaves the rolls.

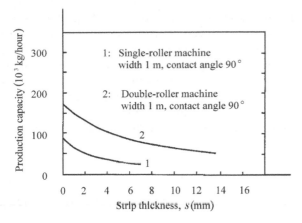

Figure 5.43 The production capacity as a function of strip thickness for (1) a single-roller process, and (2) a double-roller process. Reproduced with permission from the Scandanvian Journal of Metals, Blackwell.

In Figure 5.43 the production capacity is illustrated as a function of the strip thickness for two different strip casting machines. It can be seen from the figure that a *double-roller machine has a higher production capacity than a single-roller machine*, and that *the production capacity increases when the strip thickness decreases*. The conclusion is that the production capacity increases with decreasing strip thickness and it is *more* profitable to cast *thin* strips than *thick* ones. The number of production steps after casting can be reduced if an adequate final strip dimension is cast directly. Besides, there is a gain in quality with casting thin strips as the crystal structure of the material will become finer and better the faster the strip solidifies.

5.7.4 Convection and Solidification in Rapid Strip Solidification Processes

Influence of Convection on Strip Thickness in Strip Casting

In Section 5.7.3 we derived the relationships between the casting rate u and the strip thickness s and expressions for the solidification time and the solidification length in strip casting. The assumptions for these calculations were that the temperature was the same over the whole strip and that the heat transport was controlled entirely by the heat flow through the surface of the strip (page 120). These conditions are supposed to be valid if Nussel's number Nu < 0.2. In this analysis we assumed that the strip thickness s was determined by the slit width of the machine (Figure 5.35 on page 120).

In many rapid solidification processes the casting rate is so high that we have to consider the convection in the melt during the casting process. We will analyse the influence of convection on the solidification process below.

Figure 5.44 shows the principle of a strip casting machine. The melt in the tundish with a baffle is in contact with the rotating wheel. When the wheel rotates it catches melt metal,

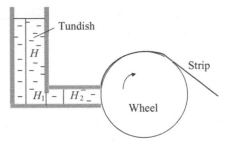

Figure 5.44 Principle of a strip casting machine.

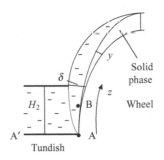

Figure 5.45 Partial enlargement of Figure 5.44.

which accompanies the wheel and then solidifies rapidly on the wheel and forms a thin metal strip. Figure 5.45 is a partial enlargement of Figure 5.44, where the solidification process is illustrated in more detail.

As the wheel rotates, a thin layer δ of the melt is drawn up towards the wheel and follows it. The thickness of this layer is given by the boundary layer theory for laminar convection (Section 5.2.1 on pages 94–96). The effective cooling results in a very high cooling rate, and the strip solidifies rapidly. The thickness s of the solidified strip is obviously equal to the thickness of the thin layer. It is vital in this case to design the process in such a way that δ_{slit} becomes equal to the exit. Otherwise the process will be uncontrolled and the desired thickness is not achieved.

5.7.5 Strip Thickness as a Function of Tundish Slit Width and Periphery Velocity of the Wheel with Consideration to Convection

We will apply the principle of continuity of incompressible liquids to the melt that passes the tundish slit, and the melt that is carried away with the wheel out of the tundish. With the aid of Figures 5.44 and 5.45 we get:

$$A_1 u_1 = A u_{\text{wheel}} \qquad (5.49)$$

where

$A_1 =$ cross-section area of the slit (height $H_1 \times$ width l)

$A =$ cross-section area of the strip $= \delta \times$ width l

$u_1 =$ exit velocity of the melt from the tundish

$u_{\text{wheel}} =$ periphery velocity of the wheel.

Because the width of the tundish and the wheel are equal, Equation (5.49) can be written as:

$$H_1 u_1 = \delta u_{wheel} \qquad (5.50)$$

where H_1 is defined in Figure 5.44. The exit velocity can be calculated using hydrodynamics [Equation (3.3) on page 30 in Chapter 3] where the height in this case is equal to the difference in level between the free liquid surfaces $(H - H_2)$:

$$u_1 = \sqrt{2gh} = \sqrt{2g(H - H_2)} \qquad (5.51)$$

H_1 and H_2 are marked in Figures 5.44 and 5.45.

To be able to calculate *the thickness δ of the layer* carried away from the melt we make the assumption that it *is determined by the convection in the melt*. The convection pattern will be similar to that seen when a melt or liquid phase passes a stationary cold wall. A velocity boundary layer is developed where the velocity of the melt is zero at the surface of the wall by analogy with the one described on page 94 and in Figures and 5.46. In the present case it is the wall (the wheel) that moves and the liquid phase that is stationary (Figure 5.47). The same hydrodynamic laws are valid in both cases.

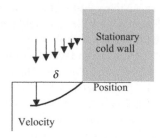

Figure 5.46 Flow pattern due to convection in the thin boundary layer of a metal melt close to a vertical cold stationary wall.

In Section 5.2.1 the theory of the velocity boundary layer close to the interface, between the melt and the wall, was analysed. The driving force of the convection flow was the density difference, caused by the temperature difference between the interior of the melt and the melt at the cooled wall (Figure 5.46).

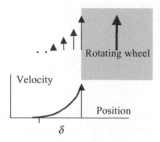

Figure 5.47 Flow pattern due to forced convection in a stationary metal melt in the thin layer close to a rotating wheel.

In the present case the convection in the melt in the tundish close to the wheel is *forced* as the velocity boundary layer is developed due to the rotation of the wheel. Anyway it is reasonable to assume that the withdrawn layer of melt or the thickness of the strip is equal to the thickness δ of the velocity boundary layer (Figure 5.47). The theory of convection flow gives the following tentative solutions for the differential equations, see the boxed material on page 96 ($p = \frac{1}{2}$ and $q = \frac{1}{4}$):

$$u_{max} = C_1 z^{\frac{1}{2}} \qquad (5.52)$$

$$\delta = C_2 z^{\frac{1}{4}} \qquad (5.53)$$

If we form the product $u_{max} \times \delta^2$ we get:

$$u_{max}\delta^2 = C_1 z^{\frac{1}{2}} \left(C_2 z^{\frac{1}{4}} \right)^2 \qquad (5.54)$$

which can be written as follows:

$$\delta = C_3 \left(\frac{z}{u_{max}} \right)^{\frac{1}{2}} \qquad (5.55)$$

The basic formula (5.51) is applied in the present case. The distance z corresponds to the height H_2, i.e. the distance from the lower part of the tundish $A'A$ (Figures 5.48 and 5.45) where the layer leaves the tundish. According to the theory of convection the velocity u is the relative velocity between the wall and the liquid close to the wall. In the present case the melt has no vertical velocity component relative to the tundish at the entrance slit and the vertical relative velocity u between the melt and the wheel corresponds to the periphery velocity u_{wheel}. As the periphery velocity is constant, u_{wheel} corresponds to both u and u_{max}. It can be shown that $C_3 \approx 2(v_{kin})^{\frac{1}{2}}$. Thus Equation (5.55) can be written as:

$$\delta = 2 v_{kin}^{\frac{1}{2}} \left(\frac{H_2}{u_{wheel}} \right)^{\frac{1}{2}} \qquad (5.56)$$

where

$\delta =$ thickness of the velocity boundary layer and the thickness of the strip

$v_{kin} =$ kinematic viscosity coefficient (η/ρ) of the melt

$H_2 =$ height of the entrance slit of the melt

$u_{wheel} =$ periphery velocity of the wheel.

Equation (5.56) expresses the strip thickness as a function of the slit height in the tundish and the periphery velocity of the wheel. It is evident from this relationship that the velocity of the wheel determines the thickness of the strip. Equation (5.56) differs strongly from Equation (5.6) on page 96 because the convection is forced in this case.

Equation (5.56) is combined with Equations (5.50) and (5.51). The expression of u_1 in Equation (5.51) is introduced into equation (5.50), which gives:

$$H_1 \sqrt{2g(H - H_2)} = \delta\, u_{\text{wheel}} \qquad (5.57)$$

H_1 and H_2 are given by the design of the machine but the height H of the melt in the tundish can be varied. We have to calculate H to be able to design the casting process properly. The expression of δ in Equation (5.57) is inserted into Equation (5.56), which gives the relationship:

$$H_1 \sqrt{2g(H - H_2)} = 2 v_{\text{kin}}^{1/2} \left(\frac{H_2}{u_{\text{wheel}}} \right)^{\frac{1}{2}} u_{\text{wheel}} \qquad (5.58)$$

We can solve H from Equation (5.58):

$$H = H_2 \left(1 + \frac{2 v_{\text{kin}}\, u_{\text{wheel}}}{g H_1^2} \right) \qquad (5.59)$$

The height H of the melt in the tundish is a function of the periphery velocity of the wheel u_{wheel}. The height of the melt must be adapted to the peripheral velocity of the wheel in such a way that Equation (5.59) is fulfilled during the casting.

5.7.6 Heat Transport through the Strip During Casting with Respect to Convection

Convection occurs only in liquids. In the case of a solidifying strip convection occurs in the melt close to a cold wall. Before the sucked melt close to the rotating wheel has started to solidify, the cold wall is the rotating wheel. During solidification of the strip, the cold wall is the solid layer close to the melt. When the strip loses contact with the tundish, the boundary conditions and temperature distribution will change again. We will treat the first alternative below and the other ones in Chapter 6.

Temperature Distribution before Solidification
The temperature distribution in the strip in Figure 5.45 before the solidification process has started is illustrated in Figure 5.48. At point A in Figure 5.45 the temperature is the same as in the melt in the tundish. The temperature of the melt that has been sucked to the wheel decreases successively down to the liquidus temperature at point B where the solidification process starts. According to the theory given above, the strip thickness is determined by the thickness of the velocity boundary layer when the melt leaves the tundish, i.e. when the distance z is equal to the slit height H_2.

In Section 5.2 we discussed the temperature boundary layer during natural convection and pointed out that the

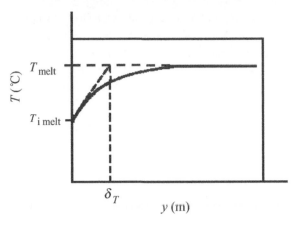

Figure 5.48 Temperature distribution in the temperature boundary layer before the start of the solidification process as a function of the distance from the wheel. The y-axis is perpendicular to the wheel and directed towards the melt in the figure. The thickness of the temperature boundary layer is constructed in the way shown in the figure.

temperature boundary layer is not identical with the velocity boundary layer. Below we will discuss the temperature distribution within the temperature boundary layer in the melt before it leaves the tundish. In this case the differences between the temperature boundary layer during natural convection and forced convection are large. Heat is transported from the melt beyond the boundary layer, through the boundary layer and then transferred to the wheel.

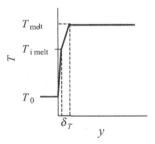

Figure 5.49 Schematic temperature distribution of the melt in the tundish, close to the wheel before the solidification of the strip has started.

The heat transport through the temperature boundary layer, i.e. through the thin layer of sucked melt on the wheel, occurs by thermal conduction [Figures (5.48) and (5.49)]. The heat flux can approximately be written as:

$$\frac{dq}{dt} = k \left(\frac{T_{\text{melt}} - T_{\text{i melt}}}{\delta_T} \right) \qquad (5.60)$$

where

$T_{\text{melt}} = $ temperature of the melt in the tundish beyond the temperature boundary

$T_{\text{i melt}}$ = temperature of the melt close to the wheel

δ_{T} = thickness of temperature boundary layer

k = thermal conductivity of the melt.

The heat flux at the interface between the melt and the wheel can be written as:

$$\frac{\mathrm{d}q}{\mathrm{d}t} = h\,(T_{\text{i melt}} - T_0) \qquad (5.61)$$

where

h = heat transfer coefficient at the interface melt/ wheel

$T_{\text{i melt}}$ = temperature of the melt close to the wheel

T_0 = temperature of the wheel.

According to the theory of forced convection, the thickness of the temperature boundary layer can in this case be written as:

$$\delta_{\text{T}} = 3.09 (v_{\text{kin}})^{\frac{1}{3}} \left(\frac{k c_{\text{p}} v_{\text{kin}}}{\rho} \right)^{\frac{1}{2}} \left(\frac{z}{u_{\text{wheel}}} \right)^{\frac{1}{2}} \qquad (5.62)$$

where

c_{p} = heat capacity of the melt

ρ = density of the melt

v_{kin} = kinematic viscosity coefficient (η/ρ) of the melt

z = distance from the bottom of the tundish.

Equation (5.62) can be summarized as:

$$\delta_{\text{T}} = \text{const}\sqrt{t} \qquad (5.63)$$

where t is equal to z/u_{wheel}. The time $t = 0$ refers to the start at A in Figure 5.45. Equation (5.63) is valid as long as the solidification process has not started, i.e. to point B.

By combining Equations (5.59), (5.60) and (5.61) we can calculate $T_{\text{i melt}}$ as a function of time and the temperature of the melt. We get:

$$T_{\text{i melt}} = \left(\frac{k\,T_{\text{melt}}}{\text{const}\sqrt{t}} + h T_0 \right) \bigg/ \left(\frac{k}{\text{const}\sqrt{t}} + h \right) \qquad (5.64)$$

Obviously $T_{\text{i melt}}$ varies with time. The time t_{cool} required to cool the melt from temperature T_{melt} to the liquidus temperature T_{L} can be calculated with the aid of a heat balance. t_{cool} is a function of the temperature T_{melt} of the melt, its conductivity k, and its heat transfer coefficient h.

The calculations are shown in the box on this page.

Calculation of the Cooling Time

The boundary layer has approximately the shape of a prism.

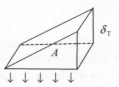

Direction of heat flow

The area A is perpendicular to the heat flow.

$V = (A\,\delta_{\text{T}}/2)$ where

V = volume of the boundary layer

A = area of boundary layer cross section.

$$\frac{\mathrm{d}Q}{\mathrm{d}t} = -\frac{A\delta_{\text{T}}}{2}\rho c_{\text{p}} \frac{\mathrm{d}T_{\text{i melt}}}{\mathrm{d}t} = Ah(T_{\text{i melt}} - T_0) \qquad (1')$$

Cooling heat from   Heat transported across the inter—

the boundary layer   face metal/mould to the mould

Equation (1') can be combined with Equation (5.63) and integrated:

$$\frac{-\rho c_{\text{p}}}{2h} \int_{T_{\text{melt}}}^{T_{\text{L}}} \frac{\mathrm{d}T_{\text{i metal}}}{T_{\text{i metal}} - T_0} = \int_0^{t_{\text{cool}}} \frac{\mathrm{d}t}{\delta_{\text{T}}} = \int_0^{t_{\text{cool}}} \frac{\mathrm{d}t}{\text{const}\sqrt{t}} \qquad (2')$$

which gives, after reduction, Equation (3') below:

$$\sqrt{t_{\text{cool}}} = \frac{\rho c_{\text{p}}\text{const}}{4} \left(\ln \frac{T_{\text{melt}} - T_0}{T_{\text{L}} - T_0} \right) \frac{1}{h} \qquad (3')$$

Figure 5.50 illustrates Equation (3') for an iron-base alloy with the liquidus temperature 1150 °C and various temperatures the melt. The *lower* the temperature of the melt is and the *larger* the thermal conductivity is, the *shorter* the time it takes to come down to the liquidus temperature and the start of the solidification process.

Temperature Distribution in the Strip during Solidification

The temperature distribution in the strip during solidification is closely related to, and will be discussed in connection with, the structure of the solid phase. For this reason we will return to the topic in Section 6.10 in Chapter 6.

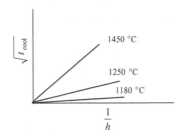

Figure 5.50 The square root of the time required to cool the melt close to the wheel to the liquidus temperature 1150 °C of an iron-base alloy, as a function of $1/h$ where h is the heat transfer coefficient [Equation (3')]. The curve has been plotted for three different alternative temperatures of the melt in the tundish.

5.8 HEAT TRANSPORT IN SPRAY CASTING

The spray casting method is described in Section 2.5.4. Superheated metal melt is atomized into droplets in narrow nozzles by the aid of a driving gas, normally nitrogen. The droplets solidify rapidly and are then compressed to very compact castings at high temperature and pressure.

The heat transport from the droplets to the surrounding gas during the short solidification process is of essential interest for the design of the spray casting process. Many theoretical models for predicting the droplet size have been suggested, but the large number of simplifying assumptions limits their application considerably for real processes.

5.8.1 Droplet Size

A large number of empirical models for the droplet size have also been published. Lubanska suggested the most well-known empirical model, which he based on experiments on disintegration of molten tin, iron, and low-melting alloys using a gas-spray ring. The ring consisted of discrete gas nozzles, symmetrically distributed around the axis of a vertical stream of a metal melt.

Lubanska found that the gas-spray ring atomizer produced droplets in a wide range of different sizes. He described the droplet size distribution by using two statistical quantities, namely average diameter of the droplets and the average deviation from this value. The sizes of the solidified droplets depend on a large number of variables. Lubanska's model involves the diameter of a droplet with average mass, the velocities of the droplets and the gas, viscosity coefficients, and quantities related to the heat transfer. It can be written as:

$$\frac{D_{ave}}{D_0} = C\left[\frac{v_{melt}}{v_{gas}We}\left(1 + \frac{J_{melt}}{J_{gas}}\right)\right]^{0.5} \quad (5.65)$$

where

D_{ave} = diameter of a droplet with average mass
D_0 = metal melt stream diameter
v_{melt} = kinematic viscosity coefficient of the melt
$\quad (\eta_{melt}/\rho_{melt})$
v_{gas} = kinematic viscosity coefficient of the gas
$\quad (\eta_{gas}/\rho_{gas})$
We = Weber's number (see below)
J_{melt} = mass flux of the melt
J_{gas} = total mass flux of the gas through the nozzles.

Weber's number is one of the many 'numbers' in hydromechanics. It is defined by the relationship:

$$We = \frac{v_{gas}^2\,\rho_{melt}\,D_0}{\sigma_{melt}} \quad (5.66)$$

where ρ_{melt} = density of the melt, and σ_{melt} = surface tension of the melt.

Lubanska claimed that the factor $(1 + J_{melt}/J_{gas})$ is independent of the design of the atomizer and generally applicable to various melts and atomizer types. This statement is supported by Figure 5.51, which reports a great number of experiments with various metal melts and atomizers. The straight line indicates that Equation (5.65) is valid, i.e. C is a constant.

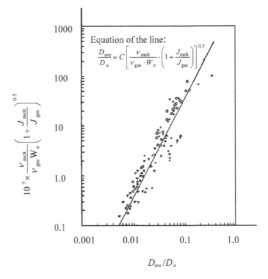

Figure 5.51 Atomization data for various experiments with iron, tin, and other alloys (published by different authors, but not reported here in detail). Reproduced with permission from John Wiley & Sons, Ltd.

5.8.2 Heat Transport

The heat transport from the liquid droplets to the surrounding gas is due to *radiation* and *convection*. Heat is formed when the droplets solidify and cool, and the heat is transferred across the droplet surface/gas interface.

To describe the heat flux we introduce the total heat transfer number:

$$h = h_{con} + h_{rad} \qquad (5.67)$$

Heat Transfer Coefficient of Radiation

The radiation heat flux (Section 4.2.2 on page 62) from the droplet to the surrounding gas can be written in two ways:

$$\left(\frac{dq}{dt}\right)_{rad} = \varepsilon \sigma_B (T_{melt}^4 - T_{gas}^4) = h_{rad}(T_{melt} - T_{gas}) \quad (5.68)$$

where $\sigma_B^{(1)}$ is the constant in Stefan–Boltzmann's radiation law. The left-hand side of Equation (5.68) can be written as:

$$\left(\frac{dq}{dt}\right)_{rad} = \sigma_B \varepsilon (T_{melt} - T_{gas})(T_{melt} + T_{gas})(T_{melt}^2 + T_{gas}^2) \qquad (5.69)$$

The identity of Equations (5.68) and (5.69) gives:

$$h_{rad} = \sigma_B \varepsilon (T_{melt} + T_{gas})(T_{melt}^2 + T_{gas}^2) \qquad (5.70)$$

The radiation heat transfer number obviously depends strongly on the temperatures involved.

Heat Transfer Coefficient of Convection

The heat transfer coefficient of convection h_{con} was introduced in Section 4.2.3 on pages 64–65. It depends on a number of quantities, which will be discussed briefly below.

To find a useful expression for the heat transfer coefficient of convection we need to define a couple of dimensionless numbers, often used in hydromechanics. *Nussel's number* is introduced in Section 4.4.5 on pages 85–86. Here it can be written as:

$$Nu = \frac{h_{con}D}{k_{gas}} \qquad (5.71)$$

where D = diameter of the droplet, and k_{gas} = thermal conductivity of the gas.

Reynold's number is defined by:

$$Re = \frac{(V_{gas} - V_{melt})D}{\nu_{gas}} \qquad (5.72)$$

where

$(V_{gas} - V_{melt})$ = relative velocity of the gas and the droplet,

ν_{gas} = kinematic viscosity of the gas (η_{gas}/ρ_{gas}).

Prandl's number is defined by:

$$Pr = \frac{c_p^{gas}\eta_{gas}}{k_{gas}} \qquad (5.73)$$

where

c_p^{gas} = thermal capacitivity of the atomizing gas

η_{gas} = dynamic viscosity coefficient of the gas.

The heat transfer coefficient of convection across a droplet/gas interface is obtained by means of the so-called Ranz–Marshall correlation:

$$Nu = 2 + \left(0.6 \times Re^{1/2}Pr^{1/3}\right) \qquad (5.74)$$

The heat transfer coefficient of convection is obtained from Equations (5.71) and (5.74):

$$h_{con} = \frac{k_{gas}}{D}\left[2 + \left(0.6 \times Re^{1/2}Pr^{1/3}\right)\right] \qquad (5.75)$$

or

$$h_{con} = \frac{k_{gas}}{D}\left[2 + 0.6\left(\frac{(V_{gas} - V_{melt})D}{\nu_{gas}}\right)^{1/2}\left(\frac{c_p^{gas}\eta_{gas}}{k_{gas}}\right)^{1/3}\right] \qquad (5.76)$$

Total Heat Transfer Coefficient

If we combine Equations (5.65), (5.68), and (5.74) we get the total heat transfer coefficient:

$$h = \sigma_B \varepsilon (T_{melt} + T_{gas})\left(T_{melt}^2 + T_{gas}^2\right) + \frac{k_{gas}}{D}\left[2 + 0.6\left(\frac{(V_{gas} - V_{melt})d}{\nu_{gas}}\right)^{1/2}\left(\frac{c_p^{gas}\eta_{gas}}{k_{gas}}\right)^{1/3}\right] \qquad (5.77)$$

[1]The constant is here denoted σ_B in order to distinguish it clearly from surface tension σ.

5.8.3 Production of Spray Cast Materials

Figure 5.52 shows a spray casting machine. Atomization of a liquid melt and rapid solidification of small droplets results in a very fine structure of the droplets. The droplets are caught on a substrate during the solidification process.

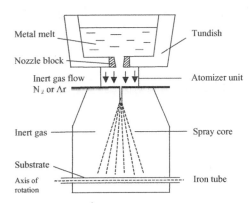

Figure 5.52 Production of spray-cast material on a substrate in the shape of a rotating tube.

The powders produced in this way must be sintered at high temperature and high pressure to form a material of high mechanical strength and excellent mechanical properties. As the production rate is very low the material is very expensive. Products made of spray cast materials are often used for producing complex shapes but also for simple shapes such as plates and tubes for special purposes. The sintering process is integrated with the solidification process. The final product is built up layer by layer of semisolidified droplets that have been exposed to high temperature and high pressure.

The production of spray cast materials can be compared with conventional powder process materials. The number of process steps is smaller in a conventional powder process and the production rate is much higher than in a spray cast process at the expense of the quality of the product. The structures of spray cast materials are briefly discussed in Chapter 6.

SUMMARY

■ Natural Convection

Natural Convection in Metal Melts
The temperature differences within the melt cause natural convection in ingots. Within a thin layer close to the solidification front melt moves downwards. Temperature boundary layer: $T = T_{\text{melt}} - (T_{\text{melt}} - T_{\text{i metal}})\left(1 - \frac{y}{\delta}\right)^2$
Velocity boundary layer: $u = u_0(z)\frac{y}{\delta}\left(1 - \frac{y}{\delta}\right)^2$. Maximum

of the temperature boundary layer:

$$\delta(z) = 3.93\left(\frac{v_{\text{kin}}}{\alpha}\right)^{-\frac{1}{2}}\left(\frac{20}{21} + \frac{v_{\text{kin}}}{\alpha}\right)^{\frac{1}{4}}\left[\frac{g\beta(T_{\text{melt}} - T_{\text{i metal}})}{v_{\text{kin}}^2}\right]^{-\frac{1}{4}}z^{\frac{1}{4}}$$

or

$$\delta(z) = B\left[\frac{g}{z}(T_{\text{melt}} - T_{\text{i metal}})\right]^{-\frac{1}{4}}$$

Heat Transport in Ingot Casting
Experiments have shown that the heat transport mechanisms in ingot casting are internal convection and radiation. Relationship between the thickness of the ingot shell and time: $y_{\text{L}} = C\sqrt{t}$. Rate of solidification:

$$u = \frac{dy_{\text{L}}}{dt} = \frac{C}{2\sqrt{t}}$$

Solidification front temperature of the ingot:

$$T_{\text{i melt}} = T_{\text{L}} - \frac{C}{\sqrt{2\mu y_{\text{L}}}}$$

Cooling rate of the melt at the centre of the ingot:

$$\frac{dT_{\text{melt}}}{dt}$$
$$= \frac{-8kg^{\frac{1}{4}}}{3B(b_0 - 2C\sqrt{t})(d_0 - C\sqrt{t})^{\frac{1}{4}}\rho c_{\text{p}}}\left[T_{\text{melt}} - T_{\text{L}} + \left(\frac{C}{2\mu\sqrt{t}}\right)^{\frac{1}{2}}\right]^{\frac{5}{4}}$$

T_{melt} is calculated numerically.

■ Water Cooling

Dimensioning of Water Cooling during Casting
The heat transport is controlled by the layer that has the poorest heat transport

$$\text{Heat transport:} \quad \frac{1}{h_{\text{total}}} = \frac{l_{\text{metal}}}{k_{\text{metal}}} + \frac{\delta}{k_{\text{air}}} + \frac{l_{\text{mould}}}{k_{\text{mould}}} + \frac{1}{h_{\text{w}}}$$

The heat transfer number for chill-mould wall/cooling water depends primarily on the velocity of the cooling water but also on its temperature. Increased temperature does not necessarily mean worse cooling because the heat transfer coefficient increases.

The flow of cooling water should be dimensioned in such a way that:

(1) the temperature increase becomes reasonable and well below the boiling point;

(2) the temperature of the mould surface must not, at any point, become too high, bearing the strength of the material in mind.

Water Cooling during Continuous Casting

For continuous casting, chill-moulds of copper are used (large k). The flux that passes the chill-mould

$$\frac{dq}{dt} = c_p^w \, \rho_w \, D_E \, a \, u_w \, \Delta T_w = \rho v_{cast} s_{steel} a (-\Delta H)$$

or

$$h_{steel}(T_{steel\,surface} - T_{mould}) = h_w \left(T_{mould}^w - T_w\right)$$

Combining these Equations with an empirical equation for h_w, the lowest possible velocity for the cooling water and the temperature increase after passage of the chill-mould can be calculated.

Secondary Cooling in Continuous Casting

Secondary cooling has three functions:

(1) control of the casting rate in such a way that the core has solidified before the casting leaves the withdrawal pinch rolls;
(2) control of the surface temperature in order to avoid unnecessary cracks;
(3) cooling of the machine foundation.

The heat transfer number h can be determined using empirical relationships as functions of water flow and water temperature. h has to be inversely proportional to the square root of the distance from the chill-mould to keep the surface temperature of the casting constant.

Metallurgical Length

The metallurgical length $L =$ the distance from the top of the chill-mould to the point in the casting where the core just has solidified.

An expression for the metallurgical length is derived in the text.

■ Heat Transport in Near Net Shape Casting

Heat Transport in Strip Casting with no Consideration to Convection

If Nussel's number $Nu \ll hs/k$ the heat transport is controlled entirely by the heat flux through the surface of the strip:

$$t_{single} = \frac{\rho_{metal}(-\Delta H)}{(T_L - T_{su})}\frac{s}{h} \qquad L_{single} = \frac{\rho_{metal}(-\Delta H)}{(T_L - T_{su})}\frac{s}{h}u$$

| Solidification time of the strip | Solidification length of the strip |

Otherwise:

$$t_{single} = \frac{\rho_{metal}(-\Delta H)}{(T_{metal} - T_0)}\frac{s}{h}\left[1 + \left(\frac{h}{2k} \times s\right)\right] \qquad L_s = \frac{\pi R}{180}\left(\arcsin\frac{a}{R}\right)^\circ$$

$$t_{double} = \frac{\rho_{metal}(-\Delta H)}{(T_{metal} - T_0)} \times \frac{s}{2} \, h \times \left[1 + \left(\frac{h}{2k} \times \frac{s}{2}\right)\right]$$

where

$$L_s = \frac{\pi R}{180}\left(\arcsin\frac{a}{R}\right)^\circ$$

Casting rates as a function of strip thickness:

$$u_{single} = \frac{L_s}{t_{single}}^\circ$$

$$u_{double} = \frac{L_s}{t_{double}}^\circ$$

Production Capacity in Strip Casting

The production capacity per hour strip width 1 m: $P = \rho u \, s \times 60 \times 60$.

A double-roller machine has a higher production capacity than an single-roller machine. The production capacity increases when the strip thickness decreases, hence it is more profitable to cast thin strips than thick ones.

Heat Transport in Strip Casting with Consideration to Convection

At high casting rates the convection during the casting process must be taken into consideration.

The strip thickness is equal to the thickness of the boundary layer.

Time to cool the melt from T_{melt} to T_{cool}:

$$\sqrt{t_{cool}} = \text{const}\left(\ln\frac{T_{melt} - T_0}{T_L - T_0}\right)\frac{1}{h}$$

Velocity Boundary Layer
Strip thickness is given by:

$$s = \delta = 2v_{kin}^{\frac{1}{2}}\left(\frac{H_2}{u_{wheel}}\right)^{\frac{1}{2}}$$

where choice of H is given by:

$$H = H_2\left(1 + \frac{2v_{kin}u_{wheel}}{gH_1^2}\right)$$

Temperature Boundary Layer

Thickness of temperature boundary layer is given by $\delta = \text{const}\sqrt{t}$, where

$$t = (Z/U_{\text{wheel}})^{\frac{1}{2}}$$

Hence:

$$\delta_T = 3.09(\nu_{\text{kin}})^{\frac{1}{3}}\left(\frac{kc_p\nu_{\text{kin}}}{\rho}\right)^{\frac{1}{2}}\left(\frac{z}{u_{\text{wheel}}}\right)^{\frac{1}{2}}$$

Temperature Distribution in the Melt Close to the Wheel

Heat flux:

$$\frac{dq}{dt} = k\left(\frac{T_{\text{melt}} - T_{\text{i melt}}}{\delta_T}\right) \quad \text{and} \quad \frac{dq}{dt} = h(T_{\text{i melt}} - T_0)$$

Temperature of the melt close to the wheel:

$$T_{\text{i melt}} = \frac{\dfrac{kT_{\text{melt}}}{\text{const}\sqrt{t}} + hT_0}{\dfrac{k}{\text{const}\sqrt{t}} + h}$$

Time required to cool the melt close to the wheel from T_{melt} to T_L is:

$$\sqrt{t_{\text{cool}}} = \text{const}\left(\ln\frac{T_{\text{melt}} - T_0}{T_L - T_0}\right)\frac{1}{h}$$

The lower the temperature of the melt is and the larger h is, the shorter is the time required to come down to the liquidus temperature and the start of the solidification process.

■ Heat Transport in Spray Casting

Superheated metal melt is atomized into droplets in narrow nozzles by the aid of a driving gas, normally nitrogen. The droplets solidify rapidly and are then compressed to very compact castings at high temperature and pressure. Many theoretical models have been suggested for predicting the droplet size, but the large number of simplifying assumptions limits their application to real processes considerably.

Heat Transport

Heat transport from the liquid droplets to the surrounding gas is due to *radiation* and *convection*. Heat is formed when the droplets solidify and cool and the heat is transferred across the droplet surface/gas interface $h = h_{\text{con}} + h_{\text{rad}}$

Production of Spray Cast Materials

Atomization of a liquid melt and rapid solidification of small droplets results in a very fine structure of the droplets. The droplets are caught on a substrate during the solidification process.

The sintering process is integrated with the solidification process. The final product is built up, layer by layer, of semisolidified droplets that are exposed to high temperature and high pressure.

EXERCISES

5.1 A 10-ton steel ingot with a cross-section area of 400 mm × 1000 mm is cast at an excess temperature of 100 °C. The superheat disappears in 10 minutes.

If the upper surface of the ingot is unshielded, primarily part of the excess heat and later part of the solidification heat is lost through the upper surface by radiation. The radiation from the rest of the ingot can be neglected as the cast iron mould walls are very thick and the temperature of its outer surface is much lower than the temperature of the upper steel surface.

(a) Calculate the fraction of the total excess heat that disappears due to radiation. What mechanism is responsible for the rest of the heat transport?

Hint A3

(b) Calculate the amount of heat lost by radiation from the upper steel surface during the solidification process of the ingot.

Hint A73

(c) The unshielded upper surface of the ingot solidifies during the solidification. Calculate the maximum thickness of the solidified layer if the solidification heat at the upper surface is assumed to be lost by radiation.

Hint A87

(d) The upper surfaces of ingots are often insulated as it is desirable that the central part of an ingot solidifies before the upper surface in order to avoid a so-called pipe (Chapter 10). What is the effect of the insulation of the upper surface on the solidification process?

Hint A38

The radiation constant of an uninsulated upper steel melt surface is 5.67×10^{-8} W/K^4 m^2. The factor ε in the radiation law is 0.2. The temperature of the surroundings is 20°C. Material constants of the steel are $\rho = 7.88 \times 10^3 \text{ kg/m}^3$; $-\Delta H = 272 \times 10^3 \text{ J/kg}$; $c_p^s = 420 \text{ J/kg K}$, and $T_L = 1450 °C$.

5.2 The sketch below illustrates the so-called Watts' continuous casting process for steel slabs. A critical part of the process is to keep a sufficiently large part of the central material at the inlet molten during the whole casting process. This can be controlled using an excess temperature of the melt. The excess temperature is required to keep the channel continuously open.

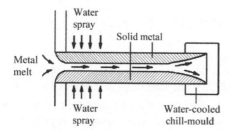

(a) Derive a relationship that shows how the shell thickness varies with excess temperature of the melt under stationary conditions.

Hint A55

(b) What is the minimum excess temperature of the melt that has to be used if you want to cast slabs of thickness 20 cm in the machine?

Hint A129

Assume that the heat transfer coefficient between the melt and the solidified material is constant. The width of the casting is much larger than its thickness.

The liquidus temperature of the steel is 1450°C. The heat transfer coefficient between the slab surface and the cooling water is 1.0 kW/m² K and between the steel melt and solid phase it is 0.80 kW/m² K. The thermal conductivity of steel is 30 W/m K.

5.3 The left-hand figure below shows an ingot that solidifies in the direction from left to right. The heat flow is horizontal and unidirectional. The figure illustrates the experimental and calculated interface profiles during the casting process.

Explain the shape of the solidification front.

Hint A12

5.4 The right-hand Figure below shows the heat flux as a function of position in the chill-mould in continuous casting for a steel alloy with 0.7 wt-% carbon. The chill-mould is made of copper. The dotted vertical line is the level of the upper surface of the melt.

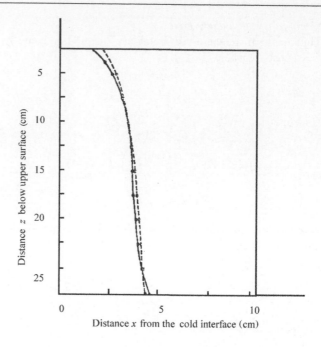

The heat flux is small at the beginning but increases strongly up to a maximum (region 1). After the maximum the heat flux decreases continuously (region 2).

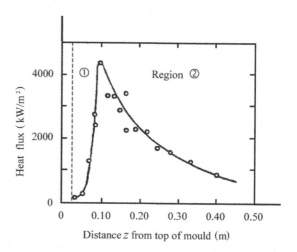

(a) Explain the shape of the curve in region 1.

Hint A105

(b) Describe the situation in the chill-mould at the maximum of the curve.

Hint A278

(c) Explain the shape of the curve in region 2.

Hint A235

5.5 The figure below shows a sketch of a continuous casting machine, especially designed for steel. The cross-section area of the casting is 1500 mm × 290 mm. mm. The machine contains three secondary cooling zones.

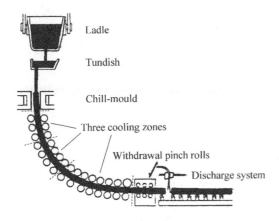

Ladle

Tundish

Chill-mould

Three cooling zones

Withdrawal pinch rolls

Discharge system

Cooling zone	Heat transfer numbers ($W/m^2 K$)	Zone length (m)
Chill-mould	1000	1.0
1	440	4.0
2	300	5.0
3	200	10.0

Material constants	
ρ	7.88×10^3 kg/m^3
$-\Delta H$	272 kJ/kg
k	30 W/mK
T_L	1470 °C

Calculate

(a) the total solidification time;

Hint A104

(b) the maximum casting rate

Hint A147

for continuous casting of steel using the machine.

Information about the cooling zones and material constants are given in the tables. The solution can be simplified by calculation of a 'weighted' average value of the heat transfer coefficient for the whole machine.

5.6 During casting in a certain machine for continuous casting the following data are valid:

Dimension of casting $a \times a$: 100 mm×100 mm
Casting rate: 3.0 m/min
Temperature of cooling water: 40 °C.

The data for the cooling zones are given in the table. Calculate the heat transfer coefficient between the metal surface and the cooling water for each zone.

Hint A15

Cooling zone	Length (mm)	Water flow (litre/min)
Spray zone	200	80
Zone 1	1280	175
Zone 2	1850	150
Zone 3	1900	175

5.7 The figure illustrates a continuous casting method for aluminum strips. The process can briefly be described in the following way.

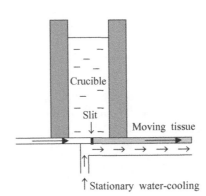

Crucible

Slit

Moving tissue

Stationary water-cooling

Material constants for aluminium	
T_L	660 °C
k	220 W/m K at T_L
$-\Delta H$	390 kJ/kg
ρ_s	2.7×10^3 kg/m^3

An Al melt with a minor excess temperature is kept in a crucible. Through a rectangular slit at the bottom of the crucible melt is transferred to a moving glass fibre gauze in contact with a plate. Immediately after leaving the crucible the Al melt is cooled from below by the plate and starts to solidify. Thermal radiation through the upper surface of the Al melt contributes to the heat transport.

The equipment was once used to cast strips of 6.0 mm thickness. Calculate the distance Y_L from the upper side of the strip to the plane where the two solidification fronts meet.

Hint A60

The heat transfer coefficient between the strip and the plate is 1.0 kW/m² K. The temperature of the cooling water and the air is 20 °C. Material constants are given in the table.

5.8 The figure illustrates a continuous-strip casting process for steel. The melt runs down from a nozzle to a roller and a strip is formed on the surface of the roller when the melt solidifies. The strip remains in contact with the roller for a certain distance before it loses contact.

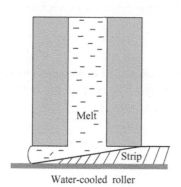

Water-cooled roller

Such a strip casting machine is used to cast strips of a final thickness of 100 μm. The melt has practically no excess temperature. The heat losses by radiation can be neglected. The solidification temperature is 1400 °C. The heat transfer coefficient between the strip and the roller is 2.0 kW/m² K. Material constants for steel are $\rho = 7.8 \times 10^3$ kg/m³ and $c_p = 650$ J/kg K.

Derive a relationship for the cooling rate of the strip at its solidification temperature and calculate the value of the cooling rate.

Hint A110

5.9 Aluminium foils are often cast according to the so-called Hunter method, where the melt is forced in between two rotating rollers (upper figure) and brought to solidify between the rollers during a simultaneous movement upwards.

The lower figure (turned 90° relative to the upper figure) illustrates the solidification process of the melt between the rollers. The figure shows that the solidification fronts will meet where the roller slit is thinnest. The consequence of this is that the width of the

Material constants for aluminium	
T_L	660 °C
k	220 W/m K at T_L
$-\Delta H$	398 kJ/kg
ρ_s	2.7×10^3 kg/m³

roller slit will correspond to the maximum thickness of the casting.

The rollers are very large, i.e. no attention need be paid to the curvature of the rollers in the calculations. The heat transfer number between the casting and the roller is 3.0 kW/m² K. The distance between the exit of the nozzle and the point where the solidification fronts meet is 50 mm.

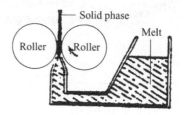

Hunter engineering

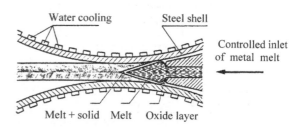

Calculate the maximum casting rate (m/min) as a function of the thickness of the casting and illustrate the function graphically.

Hint A150

5.10 Copper castings for wire production are often cast in a process developed by Properzi. The figure on the next page illustrates the process.

Square castings are cast on a roller with tracks. The roller track is covered with a cooled steel belt. The casting stays in contact with the roller along more than half the periphery. The melt is not superheated.

(a) Calculate the maximum casting rate in a machine with a roller diameter of 2.00 m to make sure that castings with the dimension 60 mm × 60 mm will be able to solidify completely before they leave the roller.

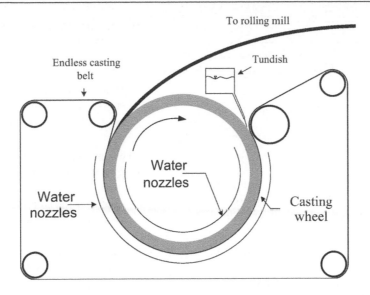

The asymmetry in the heat flow due to different heat transfer numbers can be neglected. The heat transfer number at the roller/casting interface is $1.0 \, \mathrm{kW/m^2 \, K}$. At the strip/casting interface $h = 0.70 \, \mathrm{kW/m^2 \, K}$. It is reasonable to use a weighted average value of h to simplify the calculations. Material constants are found in the table.

Hint A131

Material constants for copper	
ρ_s	$8.94 \times 10^3 \, \mathrm{kg/m^3}$
k	$398 \, \mathrm{W/mK}$
$-\Delta H$	$206 \, \mathrm{kJ/kg}$
T_L	$1083 \, ^\circ\mathrm{C}$

(b) The same question but with the difference that attention must be paid to the asymmetry in heat flow.

Hint A74

5.11 The figure in Example 5.7 on page 124 shows a sketch of a continuous casting machine, according to Haze-lett, for casting of copper slabs.

Assume that the heat transfer number between slab and belt is $400 \, \mathrm{J/m^2 \, s \, K}$. The maximum slab thickness that the machine can cast is $20 \, \mathrm{cm}$. The melt is not

superheated. Material constants for copper are given in Exercise 5.10.

Calculate the maximum casting rate (m/min) that can be used as a function of the slab thickness and illustrate it in a diagram.

Hint A125

5.12 The figure below illustrates the so-called Michelin process that is used for casting of thin steel wires (radius 100–$500 \, \mu\mathrm{m}$). After leaving the nozzle, the wire solidifies and cools in the air. The melt has no excess temperature.

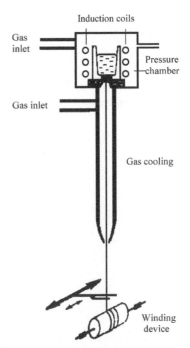

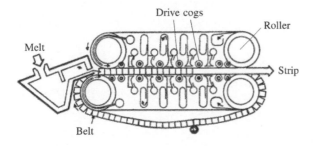

Reproduced with permission from the American Institute of Chemical Engineers.

Material constants for steel	
ρ	$7.8 \times 10^3 \, \text{kg/m}^3$
$-\Delta H$	$276 \, \text{kJ/kg}$
T_L	$1753 \, \text{K}$
c_p	$450 \, \text{J/kg K}$
σ_B	$5.67 \times 10^{-8} \, \text{J/m}^2 \, \text{s K}^4$
ε	1

(a) All solidification and cooling heat is assumed to be transported by the aid of heat radiation. Try to judge whether this is a reasonable assumption or not.

Hint A21

(b) Calculate the metallurgical length of the wire, i.e. the distance required for total solidification, if the wire moves at a rate of 8.0 m/s.

Hint A130

(c) Calculate the cooling rate of a wire with a diameter of 100 μm.

Hint A307

(d) What parameters are likely to be dominant for the success of the method, i.e. if the produced wire will be of uniform thickness?

Hint A94

The casting rate depends on the solidification time of the wire that is a function of its radius.

(e) Calculate the solidification time for a pure iron wire as a function of the radius if we assume that the cooling occurs entirely with the aid of radiation and that the excess temperature of the melt is small. Illustrate the function graphically.

Hint A240

The material constants and other data in the table may be used. The temperature of the surroundings is 300 K.

6   Structure and Structure Formation in Cast Materials

6.1   INTRODUCTION

6.1.1   Structures of Cast Materials

Metals are crystalline materials, i.e. they have regular structures. The crystals have various shapes and sizes and in addition the crystal structure varies from one metallic material to another.

The crystal structure determines most of the material properties of the metal and it is therefore of greatest importance to study metal structures in detail and try to identify the relationship between structure and material properties. The formation process determines the structure of a metal.

If metallic materials with certain specific properties need to be produced, it is necessary to understand the relationship between the method of formation and different process parameters in order to find an efficient way of controlling the structure and the properties.

Metal and alloy components are often produced by casting. The material properties of ingots and other castings are to a great extent determined by the method of solidification and by the choice of parameters, i.e. casting method, casting temperature and rate of solidification or the cooling process.

The cooling rate determines the coarseness of the metal structure. By combining thermal conduction equations with relationships that describe coarseness of structure, it is possible to analyse the formation of the structure mathematically for different casting processes. The analysis results in predictions of the material properties of, for example, an ingot or of a continuously cast product. Conversely the analysis can be used to control the casting process and for reducing the casting defects to a minimum.

In this chapter we will discuss nucleation, appearance and formation of casting structures, and the relationship between the structure, formation in various casting processes and the corresponding solidification processes. We will also discuss the possibility of controlling casting processes in order to give a material the desired properties.

6.2 STRUCTURE FORMATION IN CAST MATERIALS

6.2.1 Nucleation

A necessary condition for formation of crystals in a melt is that it is *undercooled*, i.e. it has a temperature that is lower than the melting point T_M or the liquidus temperature T_L.

Forces of attraction are acting between the atoms or ions in a *melt*. They are weak enough to permit the atoms to move freely relative to each other. The binding energy is low. In a *solid crystalline phase* the forces between the atoms or ions are stronger than those in a liquid phase. The atoms or ions are arranged in a crystal lattice and can only make small deviations from their equilibrium positions in the form of vibrations. The binding energy of the lattice atoms is higher than that between liquid atoms.

All solidification starts by formation of so-called *nuclei* at various positions in the melt and the crystals grow from these nuclei. There is a distinction between homogeneous and heterogeneous nucleation. *Heterogeneous* nucleation implies that foreign particles in the melt or at the surface act as nuclei for subsequent growth of the solid phase. The process occurs spontaneously but may, if special measures are taken, be controlled and contribute to the required structure of the solidified material.

Homogeneous Nucleation in Pure Metal Melts
The kinetic motion in a melt results in incessant collisions between its atoms. The particles, so-called *embryos* that consist of several atoms ordered in a crystalline way, arise spontaneously and at random. Between the embryos and the melt there is a continuous exchange of atoms. In many cases the embryos are dissolved and disappear.

If the kinetic motion is not too violent – the temperature is lower than the melting point – the attraction between the

atoms may sometimes be so strong that some embryos 'survive' in the melt. An embryo that achieves a certain critical minimum size continues to grow as a crystal.

Embryos that are bigger than the critical size required for continuous growth are called *nuclei*. Embryos that successively increase their sizes form nuclei by random additions of atoms from the melt. As can be concluded by the statements above, an embryo has a higher free energy than the melt. This excess free energy $-\Delta G_i$ can be written:

$$-\Delta G_i = \left[\frac{V(-\Delta G_m)}{V_m} \right] + \sigma A \qquad (6.1)$$

where

$-\Delta G_i$ = free energy required to form an embryo with the volume V and the area A
$-\Delta G_m$ = change in free energy per kmole at transformation of melt into solid phase
V_m = molar volume = M/ρ (m^3/kmole)
σ = surface energy per unit area of the embryo.

For a spherical nucleus we have:

$$V = \frac{4\pi r^3}{3} = \frac{4\pi r^2 \times r}{3} = \frac{A \times r}{3}$$

The surface energy is caused by the surface tension forces, which work in the interface between the embryo and the melt. These try to keep the embryo together and minimize its area.

By common maximum–minimum calculations on the function $(-\Delta G_i)$ (the curve in Figure 6.1) we obtain both the critical size r^* of a nucleus, capable of growing, and an expression for the energy $(-\Delta G^*)$ that is required to form such a nucleus. The calculations are not performed here but the final result is:

$$-\Delta G^* = \frac{16\pi}{3} \times \frac{\sigma^3 V_m^2}{(-\Delta G_m)^2} \qquad (6.2)$$

where $-\Delta G^*$ = activation energy for formation of a nucleus of the critical size r^*.

An embryo has to grow at least to the critical size r^* to form a nucleus capable of further growth.

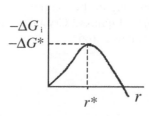

Figure 6.1 The free energy required to form an embryo as a function of its radius.

If we can find an expression for $(-\Delta G_m)$ in terms of heat of fusion and temperature or in terms of composition differences we may understand the formation of new crystals. For *pure* metals there is a simple relationship between the molar free energy of the melt and the molar heat of fusion, which involves the temperature of the melt and the melting point temperature:

$$-\Delta G_m = \frac{(T_M - T)}{T_M}\left(-\Delta H_m^{fusion}\right) \qquad (6.3)$$

where

$-\Delta G_m$ — change in free energy per kmole at transformation of melt into solid phase
$-\Delta H_m^{fusion}$ = molar heat of fusion of the metal
T_M = melting point temperature of the pure metal
T = temperature of the melt
$T_M - T$ = undercooling.

It is important to emphasize that Equation (6.3) is *not valid for alloys* as they have a solidification interval instead of a well-defined melting point. The phase diagrams of alloys have to be involved. The simple phase diagram in Figure 6.2 describes the relationship between undercooling and supersaturation. Supersaturation is an alternative variable for describing $(-\Delta G_m)$ for alloys instead of undercooling. This topic will be discussed in the next section.

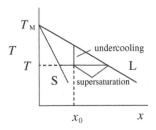

Figure 6.2 Phase diagram of a binary alloy.

Equations (6.2) and (6.3) do not give any information concerning *when* or *at what temperature* the nucleation occurs in a pure metal melt. However, by use of Boltzmann's statistical mechanics in combination with laws that describe the capacity of the nuclei to grow and not shrink and become embryos again, it is possible to derive:

- the number of nuclei per unit volume;
- the number of nuclei formed per unit volume and unit time;
- the average activation energy, required to form a nucleus of the critical size.

By means of such calculations it can be seen that a large number of nuclei per unit volume are formed at a certain

critical temperature T^*. The average activation energy for forming a nucleus at this temperature is calculated as:

$$-\Delta G^* = 60\,k_B T^* \qquad (6.4)$$

where k_B is Boltzmann's constant. The critical temperature T^*, when a large number of nuclei are formed, is found by combining Equations (6.2) and (6.4) with Equation (6.3), applied at the critical temperature (T is replaced by T^*). Elimination of the molar free transformation energy $(-\Delta G_m)$ gives:

$$60\,kT^* = \frac{16\pi}{3} \times \frac{\sigma^3 V_m^2}{\left(\dfrac{(T_M - T^*)}{T_M}\left(-\Delta H_m^{fusion}\right)\right)^2} \qquad (6.5)$$

The solution of this equation gives the critical temperature T^*.

Below the critical temperature very few nuclei per unit volume and unit time are formed. *Above* and *at* this temperature a great number of nuclei per unit time are formed. The critical temperature is also called the *nucleation temperature*. The undercooling $(T_M - T)$ is the driving force in solidification. The lower the temperature T of a pure metal melt is, the faster will the nucleation process be.

Heterogeneous Nucleation in Metal Melts in the Presence of Small Amounts of Foreign Elements

If the nucleation temperature T^* is calculated using Equation (6.5), it is found that the value is *much* lower than the melting point temperature. A very large undercooling is required in order to produce nucleation in an absolutely pure metal melt. Such large undercoolings are quite unrealistic in normal casting and solidification processes. In reality nucleation occurs at considerably higher temperatures, i.e. at much lower undercoolings than the calculations indicate. The explanation of this is that even 'pure' metals contain small amounts of foreign elements that influence the nucleation temperature strongly.

Nucleation occurs on foreign particles or crystals, so-called *heterogeneities*, which are precipitated in the melt. When crystals are formed on these small heterogeneities, some of their surface energy supplies the required formation energy of the new crystal. The last term on the right-hand side in Equation (6.1) will then decrease. Using analogous calculations for pure metals it is found that the critical temperature T^* in the presence of heterogeneities becomes much higher than for pure metals.

Often these small foreign particles nucleate homogeneously. The metal melt, including the foreign element, which is difficult to dissolve, constitute a two-phase system with a phase diagram such as the one in Figure 6.3.

At the nucleation temperature T^* small crystals precipitate with a composition x^s that is approximately equal to 100 % of element B. These small crystals grow somewhat during continued cooling.

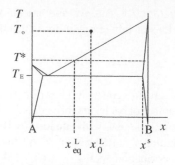

Figure 6.3 Phase diagram of a metal with a low concentration of a foreign element B that is difficult to dissolve.

For elements with very low solubility in the melt, the relationship between the change in molar free energy of the melt at nucleation (solidification) is:

$$-\Delta G_{\mathrm{m}} = RT^* \ln \left(\frac{x_0^{\mathrm{L}}}{x_{\mathrm{eq}}^{\mathrm{L}}} \right) \qquad (6.6)$$

where $x_{\mathrm{eq}}^{\mathrm{L}}$ = equilibrium concentration (mole fraction) of the foreign element in the melt at temperature T^*, and x_0^{L} = original concentration (mole fraction) of the foreign element in the melt. In this case the critical temperature T^* is found by combining Equations (6.2), (6.4) and (6.6). Elimination of $(-\Delta G_{\mathrm{m}})$ gives:

$$60 \, k_B T^* = \frac{16\pi}{3} \times \frac{\sigma^3 V_{\mathrm{m}}^2}{\left(RT^* \ln \dfrac{x_0^{\mathrm{L}}}{x_{\mathrm{eq}}^{\mathrm{L}}} \right)^2} \qquad (6.7)$$

The solution of this equation gives the critical temperature T^*. It is found to be much higher than that of pure metal melts and corresponds to low undercooling. It is low enough to permit easy formation of new crystals in normal casting and solidification processes.

6.2.2 Inoculation

Calculations using Equation (6.7) show that the required concentration of foreign elements in the melt for formation of heterogeneities is very low. This fact is used in so-called *inoculation* of metals to start the solidification process. Small amounts of elements are added to the melt and small crystals are formed by homogenous nucleation. These crystals are the heterogeneities on which new crystals nucleate. The mechanism is called *heterogeneous nucleation*.

The properties of cast materials are often improved by increasing the number of formed crystals. Thus it is of great interest to increase the number of crystals during the solidification in technical processes. Inoculation is often used in casting of aluminium to get a great number of small crystals, which prevent formation of bad texture

(bad plastic deformation in certain crystallographic directions) in plastic forming after casting. Inoculation is also used in casting of cast iron to reduce the risk of white solidification (page 151). In addition, in many casting processes an increase in the number of crystals occurs through so-called *crystal multiplication*, when a single crystal splits up into two or more new crystals. This phenomenon is treated in Section 6.3.3 on page 144.

At nucleation on heterogeneities in the melt it is difficult to define a specific nucleation temperature. Instead it has to be based on experimental observations. In many cases a relationship between the number of nucleated small crystals per unit volume N and the undercooling $\Delta T = T_{\mathrm{L}} - T$, which is the difference between the liquidus temperature T_{L} and the temperature T of the melt, has been found to be:

$$N = A(T_{\mathrm{L}} - T)^B = A(\Delta T)^B \qquad (6.8)$$

A and B are constants that are determined experimentally.

Example 6.1

During inoculation of metals an alloying element is added to the melt. An example of this is addition of FeTi to a steel melt with about 2.3 at-% C. When FeTi is added TiC is precipitated.

As FeTi is added, Ti atoms will diffuse into the melt at the same time as C-atoms diffuse inwards towards the FeTi grains (the underline denotes atom dissolved in the melt). The processes lead to the concentration profile given in the figure below. The solubility product of the C and Ti concentrations is constant at equilibrium. This product is illustrated in the figure by a straight line.

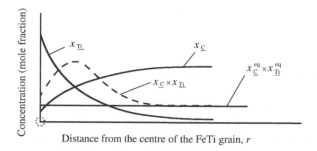

Distance from the centre of the FeTi grain, r

The solubility product of TiC at the given temperature 1500 °C is $x_{\mathrm{Ti}}^{\mathrm{eq}} x_{\mathrm{C}}^{\mathrm{eq}} = 5.2 \times 10^{-6}$ (mole fraction)^2. Other data have to be taken from standard references.

(a) Calculate the concentration of FeTi (at-%) that has to be reached to give the desired inoculation effect.

(b) The product of the Ti and C concentrations is illustrated in the figure by the dotted curve. It varies strongly with the distance from the grain. Why is it not constant and equal to the product $x_{\mathrm{Ti}}^{\mathrm{eq}} x_{\mathrm{C}}^{\mathrm{eq}}$?

Solution:

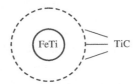

The precipitation is illustrated schematically in the figure. Each grain of FeTi gives many dispersed TiC particles that serve as heterogeneities in the melt. Thus FeTi serves as an effective inoculation agent. Besides, TiC lowers the solubility of $\underline{C}$ in the melt.

Data

$$R = 8.31 \times 10^3 \, \text{J/kmole K}$$
$$T = 1500 + 273 = 1773 \, \text{K}$$
$$V_m = 8 \times 10^{-3} \, \text{m}^3/\text{kmole}$$
$$\sigma = 0.5 \, \text{J/m}^2$$
$$k_B = 1.38 \times 10^{-23} \, \text{J/K}$$
$$x_{\underline{C}}^o = 2.3 \, \text{at-\%} \ (\text{at-\%} = \text{mole fraction} \times 100)$$
$$x_{\underline{Ti}}^{eq} x_{\underline{C}}^{eq} = 5.2 \times 10^{-6} \ (\text{mole fraction})^2$$

Equation (6.4) on page 141 is applied in combination with Equation (6.6) on page 142. If we assume that $T^* = T$ we get:

$$-\Delta G^* = 60 \, k_B T^* = \frac{16\pi}{3} \times \frac{\sigma^3 V_m^2}{(-\Delta G_m)^2} \qquad (1')$$

where

$$-\Delta G_m = RT^* \ln \frac{x_{Ti} x_{C}}{x_{Ti}^{eq} x_{C}^{eq}} \qquad (2')$$

as both C and Ti are alloying elements. We insert the expression $(2')$ into Equation $(1')$ and solve $x_{\underline{Ti}}$:

$$x_{\underline{Ti}} = \left(\frac{x_{Ti}^{eq} x_{C}^{eq}}{x_{\underline{C}}} \right) \exp\left(\frac{1}{RT^*} \sqrt{\frac{16\pi\sigma^3 V_m^2}{3 \times 60 \, k_B T^*}} \right) = \left(\frac{x_{Ti}^{eq} x_{C}^{eq}}{x_{\underline{C}}} \right) \exp(0.648)$$

$$(3')$$

Inserting the given data gives $x_{\underline{Ti}} = 4.36 \times 10^{-4}$ mole fraction.

Answer:

(a) An amount of FeTi must be added such that the initial $\underline{Ti}$ concentration in the melt will be 0.044 at-%.

(b) A few crystals of TiC are primarily nucleated at the surface of the added FeTi particles and stay there. $\underline{C}$ and $\underline{Ti}$ atoms diffuse inwards and, respectively, outwards. The difference between the dotted curve and the equilibrium line represents the supersaturation of TiC. It is seen from the figure in the text that no supersaturated zone is initially present close to the FeTi particles because the supersaturation there is too low for nucleation of TiC.

The diffusion of the atoms results in an increase of the concentration of $\underline{C}$ and a decrease of $\underline{Ti}$ around the FeTi particle with time. The product $x_{\underline{C}} x_{\underline{Ti}}$ increases and exceeds the solubility product. Thus a driving force is developed and a spherical shell of small TiC crystals around the FeTi particle is formed.

The answer of question (b) is that the diffusion shows that the system is not at equilibrium. Equilibrium is achieved when all the FeTi particles are dissolved.

6.3 DENDRITE STRUCTURE AND DENDRITE GROWTH

Dendrite Structure

Because structure is extremely important for the material properties of metals, structure studies have been the object of scientists' interest for a very long time. The most apparent features of the structure of ingots are the sizes and shapes of the crystals. Already by the Eighteenth century the French scientist Grignon accomplished a detailed study. He found that needle-shaped crystals with branchings were formed when cast iron melts solidify. A great number of modern investigations have been done to investigate the influence of various factors on the casting structure.

Today the needle-shaped crystals that were detected in the Eighteenth century are called *dendrites*. Most technically interesting alloys solidify by a primary precipitation of dendrites.

Figure 6.4 illustrates a dendrite. Such a crystal aggregate is formed because its growth in certain specific crystal directions is favoured. From a crystal nucleus a dendrite tip grows, which forms the *main branch* or *primary arm*. Immediately behind this tip, *lateral arms* or *secondary dendrite arms* are formed. For metals with cubic structures they are normally

Figure 6.4 Diagram of a dendrite crystal aggregate with primary and secondary dendrite arms. From the secondary arms tertiary dendrite arms grow, which is apparent from the figure. Reproduced with permission from the Scandinavian Journal of Metallurgy, Blackwell Publishing.

situated in two mutually perpendicular planes. The lateral arms are perpendicular to each other and the main branch. The lateral arms are successively formed behind the growing tip and reach different lengths for this reason. The oldest ones are longest.

6.3.1 Relationship between Dendrite Arm Distance and Growth Rate

It has been found that the distance between the secondary dendrite arms is not constant but increases with the distance from the cooled surface at directed solidification and at growth towards the centre in an ingot. It has been shown both experimentally and theoretically that the following relationship between the distance λ_{den} for secondary or primary dendrite arms and the growth rate v_{growth} is valid:

$$v_{growth}\lambda_{den}^2 = \text{const} \qquad (6.9)$$

The dendrite arm distance also depends on a number of other factors, among them the composition of the alloy and its phase diagram. The influence of these factors affects the value of the constant in Equation (6.9).

In most experimental investigations, authors have chosen to report relationships between the dendrite arm distance and some other experimental variable, for example, the total solidification time θ. In this case the relationship will be:

$$\lambda_{den} = K\theta^n \qquad (6.10)$$

where K and n are two constants. The constant n has a value between 1/3 and 1/2 for different types of steel alloys. Figure 6.5 shows the result of an investigation of a low-carbon alloy where the primary and secondary dendrite arm distances have been measured as functions of the cooling rate of the melt by microscope studies.

When a network of dendrite arms has been formed there is remaining melt left between the arms. This melt solidifies

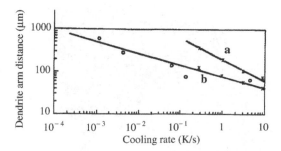

Figure 6.5 The primary and secondary dendrite arm distances as functions of the cooling rate for a steel alloy containing 25 % Ni. (a) Primary dendrite arms; (b) secondary dendrite arms. The growth and the cooling rates are proportional. Reproduced with permission from Merton C. Flemings.

during the continued temperature decrease by precipitation of solid phase on the dendrite arms. The solid phase and the melt of the alloy have different compositions. During the solidification process the melt will successively be enriched by the alloying element and the last solidified melt will therefore have a higher concentration of the alloying element than the one that solidified during an earlier stage. This phenomenon is called *microsegregation* and will be treated in Chapter 7.

6.3.2 Relationship between Growth Rate and Undercooling

In Section 6.2.1 we point out that a condition for formation of nuclei is that the melt has a temperature that is lower than its melting point, i.e. the melt is *undercooled*. Another necessary condition for growth of a dendrite tip is that the melt is undercooled. In addition, the solidification heat must be transported away from the solidification front. If this last condition is not fulfilled the solidification process will stop. The heat flux from the solidification front determines the growth rate. The higher the undercooling is, the greater will the growth rate be. In most cases the following simple relationship between the growth rate of the dendrite tip and the undercooling of the melt is valid:

$$v_{growth} = \mu(T_L - T)^n \qquad (6.11)$$

where $T_L =$ liquidus temperature of the melt, and $T =$ temperature of the melt at the solidification front. μ and n are constants. The value of n is usually between 1 and 2.

6.3.3 Crystal Multiplication

In Section 6.2.2 we treated inoculation as a method of facilitating the formation of crystals. A second method is so-called *crystal multiplication*. This means that parts of the dendrite skeleton are carried into the melt and serve as nuclei for new crystals. Different mechanisms for spontaneous crystal multiplication have been suggested, one of which is that fragments are torn off purely mechanically from the growing dendrite tips, for example, by influence of the natural convection in the melt. Examples of such a process are the broom structure of cementite in white cast iron (page 151), the corresponding phenomenon for silicon in silumin (pages 147–148), and feather crystal growth in aluminium.

This mechanical mechanism of crystal multiplication can be used deliberately in many ways. It has been found that many materials can be forced to fine-grain solidification by ultrasonic treatment. In order to get the full effect, it is necessary to combine the ultrasonic treatment with violent convection to make sure that the torn fragments are carried out into the melt effectively. This method has been successful for both aluminium alloys and stainless steel.

A rotating magnetic field has also proved to be an effective way to produce a fine-grain structure in magnetic metal melts. The electromagnetic field is applied in a circle around the ingot. By choosing a convenient frequency for the alternating current in the magnetic coils it is possible to force the melt to rotate with the magnetic field. To increase the fine-grain structure the direction of the magnetic field is changed regularly. With the forced rotation of the melt, shear forces arise on the dendrite arms that are great enough to break them and then carry them into the melt. This method has been used successfully for both static and continuous casting.

The second mechanism for crystal multiplication is based on the principle of *melting off dendrite arms*. Papapetrou introduced this method in the 1930s. He found that certain dendrite arms are melted off under the influence of surface tension. The reason for this is that surface tension causes an excess pressure of the melt inside the interface that is inversely proportional to the radius of curvature. This overpressure causes a melting point decrease that is proportional to the pressure. The excess pressure, caused by the surface tension of a small droplet, is larger than the pressure caused by a big droplet (Figure 6.6).

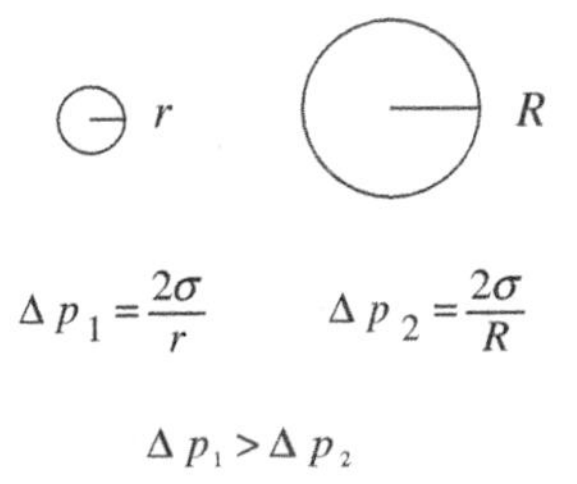

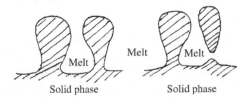

$$\Delta p_1 = \frac{2\sigma}{r} \qquad \Delta p_2 = \frac{2\sigma}{R}$$

$$\Delta p_1 > \Delta p_2$$

Figure 6.6 Pore pressure related to pore radius.

The secondary dendrite arms are always thinnest at the root, where the radius of curvature is smallest, and the excess pressure at its maximum. Thus the dendrite arms preferably melt off at the roots (Figure 6.7). If this happens the released dendrite fragments may be distributed into the melt by convection and serve as nuclei for new crystals.

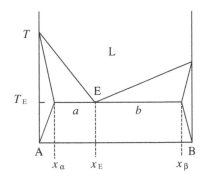

Figure 6.7 Melting off of dendrite arms.

In practice, the melting off method can be realized in a very direct way by increasing the temperature during the solidification. This can be done by the convection in the melt, which may direct a flow of hot material to a growing crystal. The simplest way to cause a desired temperature

increase is to add a small amount of hot melt during the solidification process.

6.4 EUTECTIC STRUCTURE AND EUTECTIC GROWTH

In Section 6.3 we discussed dendrite solidification. In the last part of the Twentieth century, eutectic alloys were getting used as composite materials. Two of the most frequently used cast alloys, silumin and cast iron, are eutectic. It is therefore important to understand the eutectic solidification process and its influence on the structure of this type of alloy too, so that it is possible to give them good material properties.

A eutectic alloy is a binary alloy with a composition that corresponds to the eutectic point in the phase diagram of the binary system.

Eutectic Reactions – Eutectic Alloys
Figure 6.8 shows a simplified version of the phase diagram of a binary alloy.

Figure 6.8 Schematic phase diagram of a binary alloy.

If the temperature decreases in a melt having a eutectic composition, no solid phase is precipitated until the temperature has reached the eutectic temperature T_E. The alloy solidifies by precipitation of two solid phases with the compositions x_α and x_β at the constant proportions:

$$\frac{N_A}{N_B} = \frac{b/(a+b)}{a/(a+b)} = \frac{b}{a} \qquad (6.12)$$

These constant proportions correspond to a constant chemical composition of the solid and the solidification process is called a *eutectic reaction* for this reason. During the whole solidification process the temperature remains constant and close to T_E. When all melt has solidified the temperature continues to sink.

No microsegregation is present in eutectic alloys. Figure 6.9 shows a typical temperature–time curve at eutectic solidification.

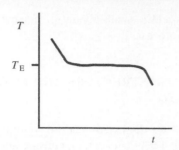

Figure 6.9 Temperature–time curve of a binary eutectic alloy during cooling and solidification.

Figure 6.10 (a) Graphite crystal, which grows radially from a centre, forms a cell.

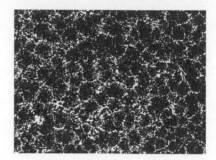

Figure 6.10 (b) Eutectic macrostructure of a large number of cells of the type shown in Figure 6.10 (a). Reproduced with permission from Pergamon Press, Elsevier Science.

Normal Eutectic Structure

The structures of eutectic alloys show many different morphologies. The main types are described in Table 6.1. Several of the morphologies may occur within one and the same alloy, depending on the growth mechanism of the individual phases. It is primarily the conditions at the interface between the phases that determine the structure. The growth of the phases is thus the basis of classification.

TABLE 6.1 Eutectic structures.

Designation	Description
Lamella eutectic structure	The two solid phases are tied in separate planar layers.
Rod eutectic structure	One of the phases is precipitated as rods and is surrounded by the other phase.
Spiral eutectic structure	One of the phases is precipitated as spirals and is surrounded by the other phase.
Flake-like eutectic structure	One of the phases is precipitated as plates, separated from each other and surrounded by the other phase.
Nodular eutectic structure	One of the phases is precipitated as spherical particles and is surrounded by the other phase.

For a *degenerated* eutectic reaction one of the two phases grows more rapidly than the other phase. For a *normal* eutectic reaction both phases grow in close cooperation with each other and with the *same* growth rate. This is the definition of normal eutectic structure. We will restrict further discussion to this case.

The microstructure of a eutectic alloy has many different appearances. Figure 6.10 and Figure 6.11 give some examples. The topic will be discussed in more detail later. During several casting processes the growth occurs radially from centres when the two phases are formed (Figure 6.10 (a)). These centres are called *eutectic cells* or just *cells*. They grow radially until they meet and fill the whole molten volume.

Lamellar Eutectic Structure

A lamellar structure is formed when both precipitated phases grow side by side as in Figure 6.12 (a). During the growth of an α-lamella, the B atoms will continuously concentrate in the melt in front of the solidification front of the α lamella. In the same way the A atoms concentrate in front of the solidification front of the β lamella. This leads to diffusion of A and B atoms as shown in Figure 6.12 (b).

It can be shown that that the three-phase equilibrium along the lines where the α-phase, β-phase and melt meet results in curved surfaces instead of planes. The formation of a lamellar eutectic structure is associated with the surface tension conditions of the α- and β-phases.

6.4.1 Relationship between Lamella Distance and Growth Rate

The distance between two neighbouring lamellas is called λ_{eut}. The relationship between the lamella distance λ_{eut} and the growth rate v_{growth} of the eutectic colony (pile of lamellas) is in most cases described by the equation:

$$v_{growth}\lambda_{eut}^2 = const \qquad (6.13)$$

by analogy with the relationship that is valid for dendritic growth. The structure will be coarse (large λ_{eut}) when the growth is slow, and fine (small λ_{eut}) at rapid growth. The finer the structure, the better will the mechanical properties of the material be.

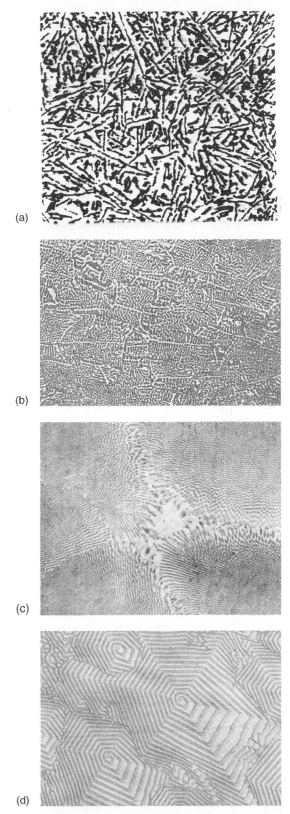

(a)

(b)

(c)

(d)

Figure 6.11 (a) Disk-shaped Si plates in a matrix of Al in a cutectic Al–Si alloy. (b) 'Chinese script' structure in a eutectic Bi–Sn alloy. (c) Mixed lamella and rod structure in a eutectic Pb–Cd alloy with 0.1 % Sn. (d) Spiral structure in a eutectic Zn–Mg alloy.

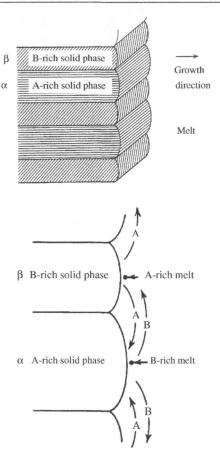

Figure 6.12 (a) and (b) Growth mechanism for a plate-like (lamellar) eutectic alloy. Reproduced with permission from Robert E. Krieger Publishing Co., John Wiley & Sons, Inc.

6.4.2   Relationship between Growth Rate and Undercooling

For eutectic growth there is a relationship between the growth rate v_{growth} and the undercooling corresponding to that which is valid for dendritic growth:

$$v_{growth} = \mu(T_E - T)^n \qquad (6.14)$$

where T_E = eutectic temperature of the melt, and T = temperature of the melt at the solidification front. μ and n are constants.

As concrete examples of eutectic alloys we will discuss silumin and cast iron. Both have great technical importance and are used to a great extent by the casting industry.

6.4.3   Eutectic Structures of Silumin and Cast Iron

Eutectic Structures of Silumin

Aluminium–silicon alloys are widely used for producing various commercial products in foundries, for example engine blocks. In particular the eutectic Al–Si alloy *silumin* is of great technical importance.

Figure 6.13 shows the phase diagram for the Al–Si system. It contains a eutectic point at 12.6 wt-% Si. The eutectic temperature is 577 °C. The Al-phase dissolves a maximum 1.65 wt-% Si while the solubility of aluminium in silicon is very low and can be neglected.

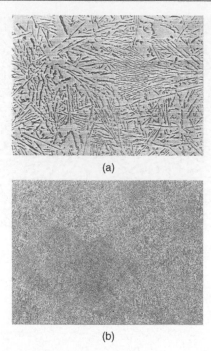

(a)

(b)

Figure 6.14 (a) Structure of unmodified eutectic silumin. (b) Structure of modified eutectic silumin.

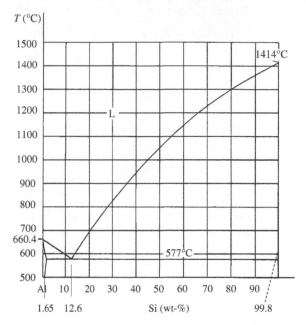

Figure 6.13 Phase diagram for the system Al–Si.

Figure 6.14 a shows typical microstructure of a eutectic Al–Si alloy that has been formed at a relatively low cooling rate. The structure consists of relatively coarse plates of Si imbedded in a matrix of Al phase.

It can be seen in Figure 6.14 (a) that these plates often have a broom-like shape. The reason for this is that the disc- and flake-shaped Si crystals easily break at the solidification front, which results in crystal multiplication. The broken crystal fragments often turn somewhat before they grow. The result is that the silicon plates have diverging directions and a fan-shaped structure is formed.

The American metallurgist Pacz discovered in 1920 that addition of small amounts of sodium (a couple of hundredths of a per cent) to a silumin melt before solidification changed the flaky, plate-like and branched microstructure of normal silumin into a much finer, more regular and fibrous microstructure [Figure 6.14 (b)]. The product is called *sodium-modified silumin*. If the solidification is rapid (quench modification), the structure of the alloy will be even finer, i.e. the lamella distances become smaller than those without inoculation. Modified silumin has better mechanical properties, ductility and hardness than does plain silumin.

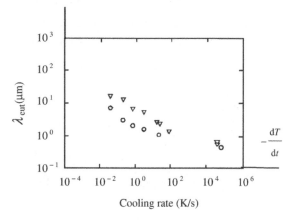

Figure 6.15 The distances between the Si plates as a function of the cooling rate for ∇ unmodified silumin, and ○ silumin, modified with 250 ppm Sr.

One complication with sodium modification of silumin is that the added sodium fades away rapidly, due to volatilization and oxidation, and it is difficult to control the sodium content in the melt. For this reason strontium is used as an alternative modifier with almost the same improvement of the mechanical properties of the alloy as can be achieved by use of sodium.

The distance between the Si plates depends on the cooling rate and the extent of modification. The higher the cooling rate is, the finer will the structure be, i.e. the smaller will the distances between the Si plates be (Figure 6.15).

The figure also shows that the more modified the Al–Si alloy is, the finer will the structure be at lower cooling rates.

Silumin is one of the alloys where the heat of fusion varies considerably as a function of the growth rate due to lattice defects formed at the solidification front. This topic was treated in Chapter 4 on pages 76–77. The growth rate depends on the cooling rate, and the heat of fusion decreases with increasing cooling rate.

Example 6.2

A eutectic unmodified Al–Si alloy is cast in a water-cooled copper mould into a plate with a thickness of 10 mm. The melt is superheated by 100 K. Calculate the solidification time and check if the solidification time will be the same if a modified Al–Si alloy replaces the unmodified alloy.

The heat transfer coefficient h between the melt and the mould is 200 W/m^2 K. The thermal capacitivity c_p of both alloys is 1.16 kJ/kg K. Other data are taken from standard reference tables and the phase diagram of the Al–Si system.

Solution:

The total solidification heat is transported away through the metal/mould interface. The energy law gives:

$$th(T_E - T_{mould}) = \rho(-\Delta H)_{eff}\, s \qquad (1')$$

where s is the thickness of the cast plate. To calculate the solidification time we will need the value of the heat of fusion, which varies considerably with the cooling rate. To find the solidification time we must first calculate the cooling rate and then read the heat of fusion from Figure 4.19 on page 77 in Chapter 4.

Calculation of the Cooling Rate

The eutectic temperature (577 °C) is found in the phase diagram on page 148. The silumin melt cools from 677 °C to 577 °C before solidification. The cooling rate of the melt at the eutectic temperature can be calculated from Equation (2'):

$$h(T_E - T_{mould}) \quad = \quad \rho c_p \left(\frac{-dT}{dt}\right) s \qquad (2')$$

heat flux through the interface metal/mould   cooling heat per unit time and unit area

where s is the thickness of the cast plate.

$$\frac{-dT}{dt} = \frac{h(T_E - T_{mould})}{\rho c_p\, s}$$

$$= \frac{200(577 - 373)}{(2.6 \times 10^3)(0.88 \times 10^3) \times 0.010} = 1.57\,\text{K/s}$$

It is reasonable to assume that the cooling rate is the same for both alloys.

Calculation of the Heat of Solidification

The effective heat of solidification per mass unit of silumin at the cooling rate 1.57 K/s is read from the figure below. The scale is logarithmic ($\log 1.57 = 0.20$). The heat of solidification is found to be:

$$-\Delta H \approx 250\,\text{kJ/kg for the unmodified alloy,}$$

$$-\Delta H \approx 220\,\text{kJ/kg for the modified alloy.}$$

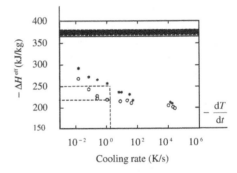

Calculation of the Solidification Time

The solidification time is calculated from Equation (1') which gives for the unmodified alloy:

$$t = \frac{\rho(-\Delta H)_{eff}\, s}{h(T_E - T_{mould})}$$

$$= \frac{(2.6 \times 10^3)(250 \times 10^3) \times 0.010}{200(577 - 373)} = 159\,\text{s} = 2.66\,\text{min}$$

and for the modified alloy:

$$t = \frac{\rho(-\Delta H)_{eff}\, s}{h(T_E - T_{mould})}$$

$$= \frac{(2.6 \times 10^3)(220 \times 10^3) \times 0.010}{200(577 - 373)} = 140\,\text{s} = 2.33\,\text{min}$$

Answer:

The solidification times for the two alloys differ. It is 2.7 min for the unmodified Al–Si alloy and 2.3 min for the modified alloy.

Eutectic Structures of Cast Iron

Some Definitions

Fe–C alloys with <2 wt-% C are called *steels*. Due to additives such as Cr, Ni, Mn, Mo, V, Si and S, there is an extensive variety of steel alloys with various properties. Examples are the stainless steel alloys, which form an important group of materials. Fe–C alloys with carbon content >2 wt-% C are called *cast iron*.

The cast iron alloys that are used in industry, normally contain 2.5–4.3 wt-% C. In most cases cast iron has a carbon concentration close to the eutectic composition (point E in Figure 6.16). The alloys also contain various concentrations of Si and Mn. Just like the alloyed steels there are special cast iron qualities that contain particular additives with the purpose of giving the original cast iron alloy one or several specific properties.

Cast iron occurs with many different structures, due to differences in composition and various solidification and cooling processes. The materials consist in most cases of a mixture of different phases. The dominating phases are *austenite* (γ-Fe), *graphite*, and *cementite*, which has the composition Fe_3C.

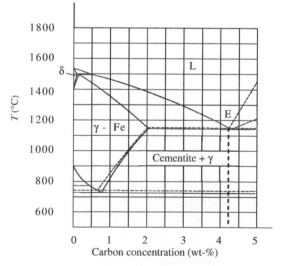

Figure 6.16   Part of the phase diagram for the system Fe–C.

In cast iron the carbon occurs either as free graphite or chemically bound as *cementite*. It is possible to control the solidification processes or composition of cast iron melts in such a way that the carbon either occurs as free carbon with graphite structure (grey iron solidification) or bound as Fe_3C (white iron solidification). Grey and white iron are discussed in detail below.

Grey Cast Iron

Grey cast iron generally shows a variety of different morphologies. Attempts have been made to classify them as different types, called A-, B-, C- and D-graphite. The composition of the melt and the cooling rate – or rather the undercooling of the melt – controls the formation of these different structures.

A-graphite is coarse so-called *flake graphite*, which often has a disc-like appearance. It consists of spherically grown crystal aggregates of graphite and austenite. Figure 6.17 shows an aggregate of austenite and graphite growing from a nucleus in the centre. The growth of the cell occurs mainly radially but with a rather uneven solidification front. The solidification has been interrupted by quenching and the remaining liquid has solidified as white iron (see next section). The cell structure can also be seen in Figures 6.10 (a) and (b) on page 146.

The graphite leads the growth and the austenite is mainly formed behind the tips of the graphite flakes. This growth process can hardly be characterized as a normal eutectic reaction. One can rather say that the graphite is precipitated primarily and that the austenite is formed as a secondary phase.

Figure 6.17   Structure of grey cast iron. The background is solidified white iron.

The graphite flakes do not reach their final thickness primarily, but grow successively by diffusion of carbon atoms from the melt across the austenite layer (Figure 6.18).

B-graphite or *nodular graphite* is also called *spheroidal graphite* (SG), which characterizes its structure. The graphite nodules are formed in the melt and are surrounded

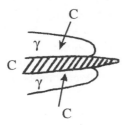

Figure 6.18 Sketch of the growth mechanism in flake graphite. In the centre there is primarily precipitated graphite. It thickens by diffusion of C atoms through the austenite layer.

by an austenite shell. The growth of the graphite mainly occurs by diffusion of carbon through this shell.

Spheroidal graphite in cast iron is obtained by adding *cerium* to the cast iron melt as was first described by Morrogh and Williams in 1948. Later a corresponding magnesium process was suggested. This technique improves the quality of cast iron greatly. Nodular structure in cast iron results in products with measurable degrees of ductility, which has extended its usefulness enormously.

C-graphite or *vermicular graphite* (VG) has a cord-like structure and can be regarded as an intermediate form between flake graphite and spheroidal graphite. The grains are characterized by low *l/g* ratio (*l* = length and *g* = thickness) in comparison with flaky graphite, which has an *l/g* ratio > 50. Vermicular graphite has generally an *l/g* ratio between 2 and 10. The *l/g* ratio of spherical graphite is 1.

D-graphite, which has a very fine graphite structure compared with flake graphite, is also called *undercooled graphite*. In D-graphite the graphite occurs in the shape of rods (Figure 6.19). During the formation of undercooled graphite the solidification front is relatively even. The formation process can be characterized as a normal eutectic reaction with good cooperation between the two phases of graphite and austenite. The formation of undercooled graphite is favoured by high cooling rate but is also promoted by various impurities. Low concentrations of sulfur and oxygen favour the formation of undercooled graphite.

Figure 6.19 Structure of a eutectic colony in undercooled graphite. Reproduced from the Georgi Publishing Company.

White Cast Iron

White cast iron consists of a mixture of cementite and austenite. The eutectic structure can appear in several different morphologies (shapes), mainly depending on whether the cast iron is eutectic or not. An example of the structure of a supereutectic white cast iron is shown in Figure 6.20.

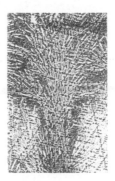

Figure 6.20 Structure of white cast iron. The broom structure is caused by crystal multiplication of fragile cementite discs. Reproduced with permission from Mats Hillert.

White cast iron is hard and fragile. It is difficult to work, for instance, to turn in a lathe and to countersink. For this reason it is in most cases vital to avoid formation of white iron, the exception being applications where materials with high wear resistance are wanted.

Grey cast iron is easier to machine and has much better mechanical properties than white iron. It is therefore desirable that cast iron solidifies as grey cast iron. Measures to control the solidification processes of cast iron melts so that the result is grey and not white iron will be discussed below.

Solidification Control of Cast Iron

Three parameters control the formation of white and grey iron of a cast iron melt: (i) composition; (ii) cooling rate, and (iii) inoculation.

Composition There are two types of composition effect: bulk element effects and tracer element effects. *Bulk element effects* imply that alloying elements with concentrations >0.5 wt-% affect the thermodynamic stability of the phases.

The stable eutectic equilibrium corresponds to point E in Figures 6.16 and 6.21, where the horizontal line AE corresponds to the eutectic temperature of the (γ + C) eutectic. The eutectic point E′ corresponds to the eutectic (γ + Fe$_3$C). It can be seen from Figure 6.21 that the latter has a eutectic temperature that is normally 6 °C *lower* than the eutectic temperature of the stable (grey) eutectic phase. E′ corresponds to the metastable (white) eutectic phase, which is achieved by *undercooling*. The larger the undercooling is the easier will white iron be formed.

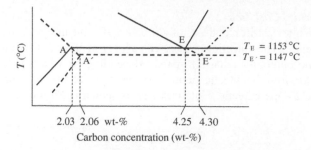

Figure 6.21 Enlargement of the central part of the phase diagram of the system Fe–C in Figure 6.16 (not to scale). Metastable levels have *higher* energy than do stable levels. In a phase diagram the metastable temperature level is consequently *lower* than the stable one.

Alloying elements change the eutectic temperature in different ways. The effects of some alloying elements on the eutectic temperatures E and E′ are shown in Figure 6.22.

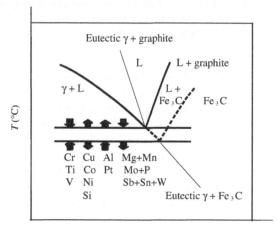

Figure 6.22 The effects of some alloying elements on the stable and metastable Fe–C eutectic temperatures. Reproduced with permission from Plenum Press, Springer.

The most evident effects are found for Cr, Ti and V, which promote formation of white iron. Cu, Co, Si and Ni have a preventive effect. The alloying elements also affect the carbon concentration and carbon distribution during the eutectic reaction. This effect is not shown in the figure.

Tracer element effects means that alloying elements in very small concentrations strongly affect the growth kinetics of graphite in spheroidal cast iron and eutectic cell growth in grey iron. Examples of elements that promote formation of grey iron in this way are S, As, Se, Sb, Te, Pb and Bi. The effects of the individual elements are additive.

Cooling Rate   A low cooling rate leads to low growth rate, which corresponds to a low undercooling in accordance with Equation (6.14) on page 147. This favours stable eutectic

solidification, i.e. formation of grey iron. On the other hand, a high cooling rate leads to a high growth rate and consequently to high undercooling, which promotes the metastable eutectic reaction. The more rapid the cooling rate is the greater will the undercooling be and the more easily will the lower metastable eutectic temperature be reached, which increases the risk of white solidification.

By analogy with dendritic growth, the eutectic lamella distance is a function of the growth rate and also of the cooling rate. The result, shown in Figure 6.5 on page 144, is valid for eutectic solidification as well. The lower the cooling rate is the coarser will the graphite eutectic structure be.

Inoculation   The main purpose of inoculation is to reduce the risk of white solidification. To reduce this risk and promote grey iron formation at solidification of cast iron, it is preferable to increase the number of nuclei in the melt. This is done by *inoculation*, i.e. addition of special alloying elements, immediately before casting. By increasing the number of cells the cooling area increases. At a constant heat flux both the growth rate and the undercooling decrease, which results in a reduced risk of white solidification. This topic is discussed in Exercise 6.5.

By mixing inoculation additives (normally 0.10–0.40 wt-%) into a cast iron melt, many heterogeneities are created, which gives many more growing eutectic cells than in a noninoculated melt. The area of the solidification front becomes larger and the heat of solidification will be released more rapidly. This will reduce the growth rate and the risk of white structure formation decreases. Simultaneously the graphite structure will be coarser, due to the lower undercooling.

Inoculation is particularly important for superheated melts. The majority of the particles that form the heterogeneities for nucleating crystals are dissolved and disappear from the melt at high temperatures. No new heterogeneities are formed during the cooling. Only a few heterogeneities grow to cells during the solidification process with the result that the risk of white structure formation increases. Thus superheating of a melt increases the risk of formation of a metastable solid phase and white solidification. To guarantee the presence of additional heterogeneities that will grow to crystals in the melt, it is important to inoculate a superheated melt *before* the solidification process starts.

There is a relatively large number of possible additives with different compositions. Most of them are based on silicon. Pure graphite is also used as an inoculation agent. In most inoculation additives varying amounts of Ca and Al and small amounts of Ba, Sr, Zr or Ce are included. Some of them have a nucleating effect while others may react with their surroundings and form products that are good nuclei.

The inoculation technique has greatly contributed to improving the reputation of cast iron as a reliable and useful material for engineering construction.

Example 6.3

For a certain quality of cast iron, it has been found that a cast iron melt solidifies as white cast iron at a cooling rate above 0.60 K/s, close to the eutectic temperature. Calculate the smallest thickness of a casting of this quality that can be cast and cooled from two sides, if white solidification is to be avoided. Assume that the melt has an excess temperature of 60 °C when it fills the cavity of the sand mould. Material constants are given in the table. The temperature of the surroundings is 25 °C.

Materials constants

Cast iron:

	T_E	1153 °C
	ρ_{metal}	$7.2 \times 10^3 \, \text{kg/m}^3$
	c_p^{metal}	420 J/kg K

Sand mould:

	k_{mould}	0.63 J/m K s
	ρ_{mould}	$1.61 \times 10^3 \, \text{kg/m}^3$
	c_p^{mould}	$1.05 \times 10^3 \, \text{J/kg K}$

Solution:

The melt cools from $T = T_E + \Delta T$ at $t = 0$ down to $T = T_E$ during the time t. By analogy with Equation (4.70) on page 79 in Chapter 4, the heat flow from the melt towards the sand mould can be written:

$$\frac{\partial Q}{\partial t} = A \sqrt{\frac{k_{mould} \, \rho_{mould} \, c_p^{mould}}{\pi t}} (T - T_0) \qquad (1')$$

The heat flow removed from the cooling melt is:

$$\frac{\partial Q}{\partial t} = -c_p^{metal} A y \rho_{metal} \left(\frac{dT}{dt}\right) \qquad (2')$$

Equations (1') and (2') describe the same heat flow. After division by the area A we get:

$$\sqrt{\frac{k_{mould} \rho_{mould} c_p^{mould}}{\pi t}} (T_i - T_0) = -\rho_{metal} c_p^{metal} y \left(\frac{dT}{dt}\right) \qquad (3')$$

After separation of the variables we integrate Equation (3'):

$$\int_0^t \frac{dt}{\sqrt{t}} = -\frac{\sqrt{\pi} \, \rho_{metal} \, c_p^{metal} y}{\sqrt{k_{mould} \, \rho_{mould} \, c_p^{mould}}} \int_{T_E + \Delta T}^{T_E} \frac{dT}{(T - T_0)} \qquad (4')$$

The factor $(T - T_0)$ is fairly constant during the cooling time and can be replaced by the average value $(T_E + \Delta T/2 - T_0)$ and be moved outside the integral sign. Integration gives:

$$2\sqrt{t} = \left(\frac{\sqrt{\pi} \, \rho_{metal} \, c_p^{metal} y}{(T_E + \Delta T/2 - T_0) \sqrt{k_{mould} \, \rho_{mould} \, c_p^{mould}}}\right) \Delta T \qquad (5')$$

We solve $1/\sqrt{t}$ from Equation (5') and insert it into Equation (3'). If we solve y in the new equation we get the minimum thickness of the casting:

$$y = \sqrt{\frac{2k_{mould} \, \rho_{mould} \, c_p^{mould}}{\pi \Delta T \left(-\frac{dT}{dt}\right)} \left[\frac{T_E + (\Delta T/2 - T_0)}{\rho_{metal} \, c_p^{metal}}\right]} \qquad (6')$$

Inserting the given numerical values, we get:

$$y = \sqrt{\frac{(2 \times 0.63)(1.61 \times 10^3)(1.05 \times 10^3)}{\pi \times 60 \times 0.60} \left[\frac{1153 + 30 - 25}{(7.2 \times 10^3)420}\right]}$$

$$= 0.052 \, \text{m}$$

Answer:

The thickness of the casting must be at least 11 cm in order to avoid white solidification.

6.4.4 Eutectic Growth of Cast Iron in a Sand Mould

Below we will discuss the crystal growth during casting in a sand mould. As an example we choose to study the growth of a eutectic cast iron alloy. There are two reasons for this choice: First, the theoretical analysis is comparatively simple, and second, cast iron is a very common metal in industrial production.

When cast iron that has been cast in a sand mould solidifies, the solidification occurs by precipitation of graphite flakes as described by Figures 6.23 (a)–(c) where Figure 6.23 (c) is an illustration of the macroetched structure shown in Figure 6.10 (b) on page 146.

The coarseness of the structure is determined by the growth rate v_{growth} of the solidifying shell [Equation (6.13)].

$$v_{growth} \lambda_{eut}^2 = \text{const} \qquad (6.13)$$

where λ_{eut} is the lamella distance.

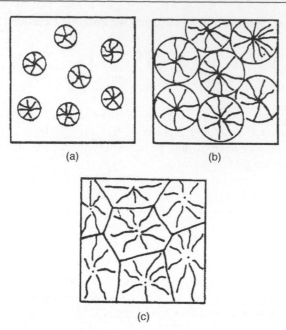

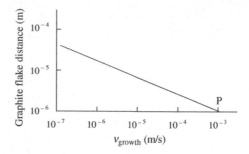

(a)   (b)

(c)

Figure 6.23 Eutectic growth of grey iron. Reproduced with permission from North-Holland, Elsevier Science.

It is impossible to register the growth rate of each cell but it is possible to calculate the average growth rate. The properties of cast iron depend on the internal structure, the coarseness of each colony. Hence, we want to estimate the average distance between the graphite flakes within the colonies and, in addition, find out how this distance varies from the centre to the periphery of the colonies. It is not possible to do this experimentally because a casting is too big. Instead, theoretical calculations are performed based on some reasonable assumptions. We will illustrate these calculations by a concrete example (Example 6.4).

Example 6.4
Cast iron is cast in a sand mould. The casting has the shape of a plate with the thickness 10 cm. When the molten cast iron starts to solidify, 1.0×10^9 growing eutectic colonies per m^3 are immediately formed without undercooling. No further colonies are formed during the continued solidification process. From other observed solidification experiments it is known that in similar cases the distance between the graphite flakes varies as a function of the growth rate in the way illustrated in the figure above. Theoretical analysis shows that the amount of heat transported per unit area and unit time can be described by the relationship:

$$\frac{dq}{dt} = \frac{0.75 \times 10^6}{\sqrt{t}} \ \text{W/m}^2 \qquad (1')$$

For N colonies/m^3 that are nucleated simultaneously and grow like spheres, the fraction of solid phase is described by the so-called Johnson–Mehl equation:

$$f = 1 - \exp\left(-N\frac{4\pi R_c^3}{3}\right) \qquad (2')$$

where R_c is the average radius of a colony. The Johnson–Mehl equation is derived from statistical laws and takes the collisions between cells [Figures 6.23 (b) and 6.23 (c)] into consideration. Material constants are given below.

The straight line intersects the v_{growth} axis in the point P. The coordinates of P are (10^{-3} m/s and 10^{-6} m). The scales on both axes are logarithmic (base 10).

Materials constants	
Cast iron:	
ρ	7.0×10^3 kg/m^3
$-\Delta H$	1.55×10^5 J/kg

(a) Calculate the total solidification time of the casting.
(b) Calculate the volume fraction of solid phase f^s as a function of the time t after the beginning of the casting process.
(c) Determine the growth rate dR_c/dt of the colonies as a function of the time t, provided that they are spherical.
(d) Calculate the graphite flake distances during the solidification process as a function of time and illustrate the function graphically.

Solution:
(a) We integrate the Equation (1′) and get:

$$q = 1.5 \times 10^6 \sqrt{t} \qquad (\text{J/m}^2) \qquad (3')$$

The solidification is bilateral, which gives $y = 0.102/$ m.

$$q = y\rho(-\Delta H) = \frac{0.10}{2}(7.0 \times 10^3)(155 \times 10^3)$$

$$= 1.5 \times 10^6 \sqrt{t} \qquad (4')$$

or $t = 1.3 \times 10^3 \quad s \approx 21.8$ min

(b) The volume fraction of solidified solid phase at the time t can be calculated by use of two equal expressions for the heat flow per unit area and unit time:

$$\frac{dq}{dt} = y\rho(-\Delta H)\frac{df}{dt} \qquad (5')$$

or

$$\frac{df}{dt} = \frac{\dfrac{dq}{dt}}{y\rho(-\Delta H)} = \frac{(0.75 \times 10^6)/\sqrt{t}}{0.05(7.0 \times 10^3)(1.55 \times 10^5)}$$

$$= \frac{1.4 \times 10^{-2}}{\sqrt{t}} \qquad (6')$$

Integration gives:

$$f = (1.4 \times 10^{-2})2\sqrt{t} = (2.8 \times 10^{-2})\sqrt{t} \qquad (7')$$

(c) When several colonies grow simultaneously they will sooner or later collide with each other during growth. We have to consider this when we will calculate the volume fraction of the solid phase. This is done by means of the Johnson–Mehl equation above [Equation (2')].

We solve R_c from Equation (2'):

$$\ln(1 - f) = \left(-\frac{4}{3}\right)\pi N R_c^3$$

$$R_c^3 = \frac{-\ln(1 - f)}{\left(\dfrac{4}{3}\right)\pi N} \Rightarrow R_c = \left(\frac{-3\ln(1 - f)}{4\pi N}\right)^{\frac{1}{3}} \qquad (8')$$

Equation (8') is derived with respect to the time t:

$$\frac{dR_c}{dt} = \left(\frac{-3}{4\pi N}\right)^{\frac{1}{3}}\frac{1}{3}[\ln(1 - f)]^{-\frac{2}{3}}\left(\frac{1}{1 - f}\right)\left(-\frac{df}{dt}\right) \qquad (9')$$

If we introduce the expressions for f and df/dt, which we derived in (b) above [Equations (6') and (7')] we get:

(d) dR_c/dt is equal to the growth rate v_{growth} of the colonies. By the aid of the line in the diagram we can determine the constant in Equation (6.13). We see that the point P is situated on the line. P has the coordinates $v_{growth} = 10^{-3}$ m/s and $\lambda_{eut} = 10^{-6}$ m. If we insert these values into Equation (6.13) we get:

$$\text{const} = v_{growth}\,\lambda_{eut}^2 = 10^{-3}(10^{-6})^2 = 10^{-15} \qquad (11)$$

When the constant is known it is easy to calculate λ_{eut} for all values of $v_{growth} = dR_c/dt$. We choose a number of values of t between 0 and 1300 s and calculate the corresponding values of $v_{growth} = dR_c/dt$ using (c) above and finally λ_{eut} for each of them:

$$\lambda_{eut} = \sqrt{\frac{10^{-15}}{v_{growth}}} \qquad (12)$$

These calculations have been performed in the table below.

t	$\dfrac{df}{dt}$		$v_{growth} = \dfrac{dR_c}{dt}$	$\lambda_{eut} = \sqrt{\dfrac{10^{-15}}{v_{growth}}}$
(s)	($\times 10^{-4}$)	f	($\times 10^{-7}$)	($\times 10^{-5}$)
10	44.27	0.0885	389.6	0.51
100	14.00	0.2800	12.42	2.84
200	9.90	0.3960	4.445	4.74
300	8.08	0.4850	2.457	6.38
400	7.00	0.5600	1.627	7.84
500	6.26	0.6261	1.193	9.16
600	5.72	0.6859	0.936	10.34
700	5.29	0.7408	0.772	11.38
800	4.95	0.7920	0.665	12.26
900	4.67	0.8400	0.599	12.92
1000	4.43	0.8854	0.568	13.27
1100	4.22	0.9287	0.585	13.03
1200	4.04	0.9699	0.755	11.51

Answer:

(a) The total solidification time of the casting is about 22 min

(b) $f = (2.8 \times 10^{-2})\sqrt{t}$

$$\frac{dR_c}{dt} = \left(\frac{-3}{4\pi N}\right)^{\frac{1}{3}}\frac{1}{3}[\ln(1 - (2.8 \times 10^{-2})\sqrt{t})]^{-\frac{2}{3}}\left(\frac{1}{1 - (2.8 \times 10^{-2})\sqrt{t}}\right)\left(\frac{1.4 \times 10^{-2}}{\sqrt{t}}\right) \qquad (10')$$

(c) $\dfrac{\mathrm{d}R_c}{\mathrm{d}t} = \left(\dfrac{-3}{4\pi N}\right)^{\frac{1}{3}}\left(\dfrac{1}{3}\right)[\ln(1-(2.8\times10^{-2})\sqrt{t})]^{-\frac{2}{3}}$

$\left(\dfrac{1}{1-(2.8\times10^{-2})\sqrt{t}}\right)\left(\dfrac{1.4\times10^{-2}}{\sqrt{t}}\right)$

(d) The graphite flake distance in the cells in a sand mould cast iron plate, as a function of time during the solidification process, is seen in the figure. The values in the above table on page 155 have been used to draw the curve.

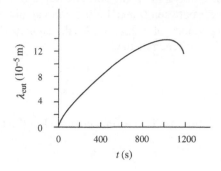

From Example 6.4 we can conclude that the lamella distance, i.e. the distance between the graphite flakes, in cast iron initially increases to a maximum during the solidification process and then decreases again. The explanation of this behaviour is given below.

The growth rate is rapid initially because the growing colonies originally have small areas. When the area of each growing cell increases the growth rate decreases, as the heat extraction is practically constant. The growth rate increases again at the end of the solidification process when the colonies collide with each other and the area of the solidification front that emits solidification heat decreases. This is

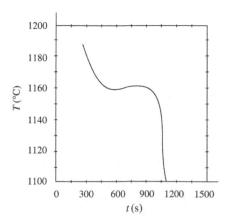

Figure 6.24 Temperature–time curve for casting solidification.

illustrated in Figure 6.23 (a) and (b). This can also be concluded from the cooling curve of the process in Figure 6.24, which will be discussed below.

6.5 COOLING CURVES AND STRUCTURE

In Section 4.3.4 in chapter 4, the heats of fusion and the thermal capacities of metals and alloys were discussed as functions of the solidification rate, which is closely related to the cooling rate. Here we will analyse somewhat more closely the information that can be obtained from cooling curves. With the aid of thermal analysis accurate cooling curves can be registered.

6.5.1 Basic Information from Cooling Curves

A cooling curve of a casting, such as that in Figure 4.18 on page 74, can roughly be divided into three regions. Region I extends from the initial temperature of the superheated melt to the temperature at which the solidification starts. Region II corresponds to the solidification time of the casting. This is characterized by an almost constant temperature. When the solidification process is finished the temperature starts to decrease again, which characterizes region III.

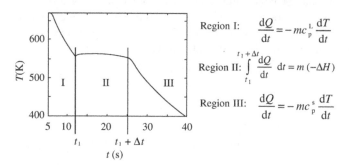

Figure 6.25 By simultaneous registration of temperature and emitted heat per unit time as functions of time, c_p^L, c_p^s and the heat of fusion $(-\Delta H)$ (J/kg) can be derived by the aid of the equations given in the figure. Reproduced with permission from John Wiley & Sons, Inc.

The heat capacities of the melt and the solid phase and the heat of fusion can be derived as indicated in Figure 6.25 by using the following information for the specimen of a metal or alloy:

- mass of the specimen;
- heat flow as a function of time;
- temperature as a function of time.

6.5.2 Information about Structure from Cooling Curves

In addition to the derivations of material constants described above, cooling curves also give qualitative information about the structure of metals and alloys. The temperature during the solidification of a metal or alloy is only roughly constant; rather it varies slightly with time during the solidification process. The variation in the temperature reflects the formation process of crystals and their structure in the melt. This will be discussed below for cast iron as an example and is also discussed in connection with crystal formation in ingot melts on page 171.

6.5.3 Cooling Curves and Structures of Cast Iron

When cast iron alloys of different compositions solidify, the process starts with the precipitation of austenite crystals (Figure 6.16 on page 150). These are nucleated at random in the melt in the mould, whirl around due to convection, grow, collide with each other, and a network of dendrites is formed. The solidification of austenite can easily be observed from the first deviation from a constant cooling rate on the solidification curve. Figure 6.26 gives an example.

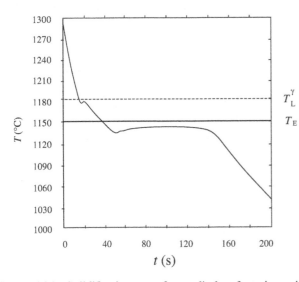

Figure 6.26 Solidification curve for a cylinder of cast iron with a diameter of 20 mm and a composition of 3.8 % C. The solidification starts with primary precipitation of austenite, followed by a eutectic reaction. Reproduced with permission from John Wiley & Sons, Inc. and the Metallurgical Society.

When the temperature of the melt after solidification continues to decrease, the eutectic point of the Fe–C system is reached and *eutectic reactions* occur (page 145). The eutectic reactions in cast iron are very complex and depend on several factors. Models for eutectic reactions have been developed over the last part of the Twentieth century using various alternative assumptions about the heat transfer and eutectic growth. The models, derived with the aid of computer programs for calculation of solidification curves for alternative eutectic reactions in the Fe–C system, are illustrated below. It has been found that the appearances of the solidification curves differ, due to the structure of the solid that is formed as a consequence of the eutectic reaction.

As mentioned on page 150, there are five different types of substructure for graphite in cast iron: flake graphite, under-cooled graphite, nodular graphite (spherical cells), vermicular graphite (thread-like structure) and so-called white structure, cementite. The properties of cast iron are strongly influenced both by the coarseness and the amounts of these structures included in the ingot.

The calculations in the above five cases have resulted in four different solidification curves, each characteristic of the corresponding structure. Undercooled graphite and flake graphite have identical solidification curves. In all the curves there is a constant level, preceded by a temperature minimum. The curves differ in their magnitude for the undercooling at the constant level. Undercooling is a function of the number of nuclei, as we will see in the Figures 6.27 to 6.30.

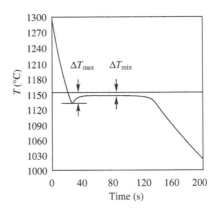

Figure 6.27 (a) Solidification curves of flake graphite and undercooled graphite in cast iron. Reproduced with permission from the Metallurgical Society.

For flake graphite and undercooled graphite the characteristic parameters ΔT_{max} and ΔT_{min} [Figure 6.27 (a)] are fairly independent of the number of nuclei [Figure 6.27 (b)]. A comparison between Figures 6.27 and 6.28 shows that the growths in nodular cast iron and in grey cast iron with flake graphite are very different. The undercooling is much more sensitive to the number of nuclei per unit volume in nodular iron than in flake graphite cast iron.

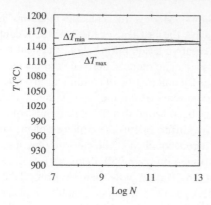

Figure 6.27 (b) Minimum and maximum undercooling as a function of the number of nuclei per unit volume of flake graphite and undercooled graphite in cast iron. Reproduced with permission from the Metallurgical Society.

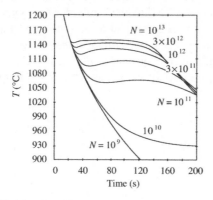

Figure 6.28 (a) Solidification curves for nodular graphite cast iron with varying numbers of nuclei per unit volume. Reproduced with permission from the Metallurgical Society.

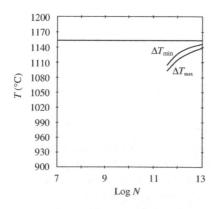

Figure 6.28 (b) Minimum and maximum undercooling as a function of the number of nuclei per unit volume of nodular graphite in cast iron. Reproduced with permission from the Metallurgical Society.

Corresponding curves for vermicular cast iron (Figure 6.29) show that the undercooling in this case is much greater than in the case of grey cast iron with flake graphite, but less than that of nodular cast iron. A comparison with the other curves above shows that the white structure in cast iron has the lowest undercooling of all the structures (Figure 6.30). It is less sensitive than the other ones to the influence of the number of nuclei.

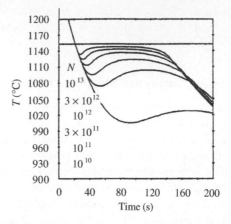

Figure 6.29 (a) Solidification curves for vermicular graphite in cast iron with a varying number of nuclei per unit volume. Reproduced with permission from John Wiley & Sons, Inc. and the Metallurgical Society.

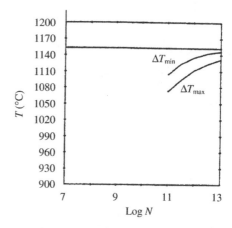

Figure 6.29 (b) Minimum and maximum undercooling as a function of the number of nuclei per unit volume of vermicular graphite in cast iron. Reproduced with permission from John Wiley & Sons, Inc. and the Metallurgical Society.

6.6 UNIDIRECTIONAL SOLIDIFICATION

In Chapters 4 and 5 we discussed solidification processes during unidirectional solidification, which means that the generated solidification heat is conducted away through the solidifying shell. This is the most common case of metal

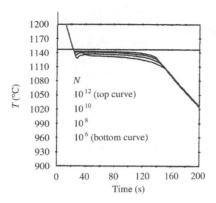

Figure 6.30 (a) Solidification curves of white structure in cast iron with a varying number of nuclei per unit volume. Reproduced with permission from the Metallurgical Society.

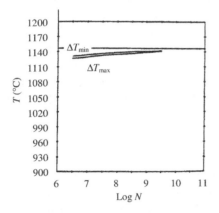

Figure 6.30 (b) Minimum and maximum undercooling as a function of the number of cementite plates per unit volume of white structure in cast iron. Reproduced with permission from the Metallurgical Society.

solidification. In this section we discuss the metal structures that arise during such a casting process.

6.6.1 Casting with the Aid of Unidirectional Solidification

During the last two decades of the Twentieth century, production of materials cast using unidirectional solidification started. One example is the casting of turbine blades used in jet engines, which is discussed in Section 6.6.2. The principle of the method used is described in Figure 6.31. A material cast with the aid of unidirectional solidification consists of columnar crystals or of a single crystal.

Casting using Unidirectional Solidification
A mould with melt is placed in an apparatus, the upper part of which consists of a furnace, the temperature of which exceeds the melting point of the metal. The lower part of the apparatus is cooled by air or water. The apparatus is

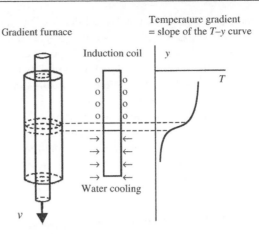

Figure 6.31 Apparatus for production of single crystals by controlled unidirectional solidification.

drawn upwards or, alternatively, the mould is moved downwards at a constant velocity. The rate of solidification of the metal melt will be the same as the relative velocity between the mould and the apparatus.

The advantages of unidirectional solidification, performed under careful temperature control, are (i) uniformity, and (ii) good mechanical properties of the cast materials. The cast structure is carefully controlled to get the same coarseness along the whole casting. The grain structure is also easy to control, due to competition in growth between crystals of different orientations. The advantage of unidirectional growth of single crystal materials is their superior mechanical strength as compared with ordinary casting materials.

Melt and Mould Temperatures during
Unidirectional Casting
The melt is normally cast in an inert atmosphere or in vacuum. During conventional casting the mould is kept at a temperature below that of the melt. The casting operation is finished within a short time (of about 1 minute). The temperature as a function of time for an ordinary turbine blade of a super-base alloy (steel, alloyed with Ni or Co) is illustrated in Figure 6.32.

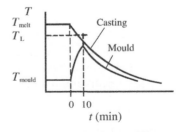

Figure 6.32 Melt and mould temperatures as functions of time during conventional component casting. The total solidification time and cooling time are much shorter for conventional casting than for unidirectional casting. Reproduced with permission from The Metals Society, The Institute of Materials.

In the case of unidirectional solidification, the mould is preheated above the solidus temperature to prevent thermal stresses during casting. The superheated alloy melt is poured into the preheated mould, which is then withdrawn from the furnace in such a way that the melt remains in contact with the chilled plate during the whole directional solidification process. The solidification and cooling time is of the magnitude of 1 to 10 hours (Figure 6.33). It can be seen from Figures 6.32 and 6.33 that the total solidification time from start to finish is much longer for unidirectional casting than for conventional component casting.

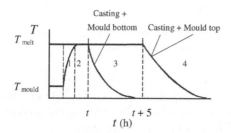

Figure 6.33   Melt and mould temperatures as functions of time for unidirectional solidification. Regions: (1) preheating of the mould; (2) casting time; (3) cooling of the bottom and solidification of the rest of the casting; (4) cooling of the casting and the mould at the top. Reproduced with permission from The Metals Society, The Institute of Materials.

6.6.2   Applications of Unidirectional Casting – Single Crystal Production

The Power-Down Process
By use of modern automatic control devices, the industrial equipment for directional solidification is much more sophisticated than the simple apparatus illustrated in Figure 6.31. One example of this is the so-called power-down process.

The power control of the electrical furnace is external and independent of the casting. The casting temperature can be monitored and the rate of power adjusted to the value required to give a constant solidification rate. The process is operated by use of induction heating, directly coupled to conventional investment moulds (wax-melting method) as illustrated in Figure 6.34. Quite regular and constant cooling rates can be obtained by programming the power input in different sections of the coil with careful design of the winding configuration.

Production of Turbine Blades
Turbine blades in jet engines or water power plants are exposed to very strong forces. Hence the demands on the mechanical strength of the blades are very high. The demands on turbine

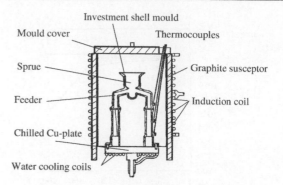

Figure 6.34   Schematic diagram of inductively heated mould apparatus for casting of unidirectionally solidified turbine blades using the power-down process. Reproduced with permission from The Metals Society, The Institute of Materials.

blades in jet engines are high because of the high temperatures (900–1100 °C) that arise when the fuel is burnt. Turbine blades in jet engines are cooled with the fuel or air. Below we briefly describe the production of turbine blades.

The first step is to make moulds of sufficient strength and that are resistant to high temperatures. The process is the same as that described on page 1 in Chapter 1. The moulds are produced by coating a wax model with several layers of ceramic powders (fine-grained $ZrSiO_4$, alumina Al_2O_3, or silica SiO_2), held together by a suitable binder, which can be either colloidal silica or ethyl silicate. Then the mould is dried and dewaxed before it is finally fired to increase its mechanical strength and remove the last traces of wax.

A turbine blade is complex, as it must be equipped with cooling channels. A *ceramic core*, which corresponds to the shape of the channels, has to be inserted into the mould. The core materials must be sufficiently stable to exclude interaction with the melt during casting and be capable of being removed from the channels after casting.

Single-Crystal Production
The method used for unidirectional solidification and illustrated in Figure 6.31 on pages 159–160 is also used in single-crystal production for various purposes. In practical cases the design of the equipment resembles that illustrated in Figure 6.34.

Constrictions can be inserted into the mould during the production of single-crystal components in order to select a single crystal for the top part of the casting. Such constrictions in various designs operate successfully. Below we discuss and analyse the formation of macrostructures in unidirectionally cast materials.

6.6.3   Crystal Growth in Unidirectional Solidification

Figure 6.35 shows a partly insulated melt in a mould that is in close contact with a strongly chilled copper plate. Heat is

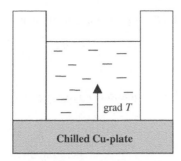

Figure 6.35 The temperature gradient at the bottom is directed from the cold bottom surface into the melt. The envelope surface of the melt is insulated.

removed from the melt and a temperature distribution is rapidly developed in the melt. The temperature gradient in the melt is vertical in this case and can be written:

$$|\text{grad } T| = \frac{dT}{dy}$$

The temperature is a scalar quantity. The temperature gradient is a vector that is directed from the lower to the higher temperature (compare page 61 in Chapter 4).

Small crystals having random orientations are nucleated in the melt when the melt at the bottom has reached the critical nucleation temperature. They start to grow in different directions under the influence of the temperature gradient, which has a constant direction. For this reason the process is called unidirectional solidification. During unidirectional solidification, the crystals are mainly oriented in the direction of the temperature gradient. The reason for this is so-called competitive growth.

With dendrite growth, the primary arms grow faster than the secondary and tertiary arms. Initially there is a random orientation of the crystals, which are formed by nucleation at the beginning of the solidification process. The crystals with primary arms in the solidification direction grow faster than all other crystals and conquer them in competition for the available space. When the slower primary arms of crystals with other orientations try to grow, the space in front of them is already filled with the network of primary, secondary and tertiary arms of the faster growing dendrites. After a short time the pattern with parallel crystals growing in the direction of the temperature gradient is distinctly established. We will illustrate competitive growth by a concrete example below.

Example 6.5

Two dendrites grow in a melt with a constant temperature gradient in the *y* direction. One of them grows in the

direction of the temperature gradient. The other one grows in a direction inclined at an angle of 45° to the gradient.

The growth rate of a dendrite can be written as a function of the undercooling of the melt, $v = \mu(T_L - T)^n$, where T_L is the liquidus temperature and T the temperature of the melt at the dendrite tip.

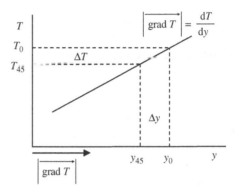

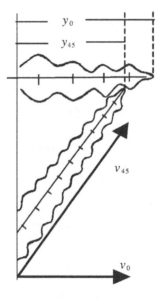

(a) Find the difference $\Delta y = y_0 - y_{45}$ in length in the *y* direction of the two dendrite tips as a function of the temperature gradient and the growth rate v_0 of the parallel-growing dendrite.

(b) Find the condition between Δy and the dendrite arm distance λ_{den} of the 'parallel' dendrite that enables the parallel dendrite to extend its secondary dendrite arms ahead of the primary dendrite arms of the 'inclined' dendrite.

Solution:

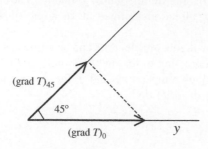

(a) The temperature gradient is a vector[1]. Its components in the parallel and inclined directions are given in the figure above. The ratio between the components is:

$$\frac{(\text{grad } T)_0}{(\text{grad } T)_{45}} = \sqrt{2} \tag{1'}$$

We want to find the ratio between the growth rates of the parallel-growing and the inclined-growing dendrites.

The relationship (dT/dt) can be written as:

$$\frac{dT}{dt} = \left(\frac{dT}{dy}\right)\left(\frac{dy}{dt}\right) = \text{grad } T \times v \tag{2'}$$

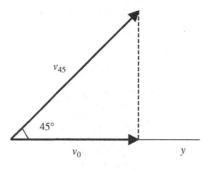

It is independent of direction. The product of the temperature gradient and the growth rate will thus be the same for the parallel and the inclined dendrites:

$$(\text{grad } T)_0 v_0 = (\text{grad } T)_{45} v_{45} \tag{3'}$$

or, in combination with Equation (1'):

$$\frac{(\text{grad } T)_0}{(\text{grad } T)_{45}} = \frac{v_{45}}{v_0} = \sqrt{2} \quad \Rightarrow \quad v_{45} = \sqrt{2} \times v_0 \tag{4'}$$

The next step is to calculate y_0 and y_{45}. The magnitude of the temperature gradient is constant in this case and can be written as:

$$|\text{grad } T| = \frac{dT}{dy} = \frac{\Delta T}{\Delta y} = \frac{T_0 - T_{45}}{y_0 - y_{45}} \tag{5'}$$

or

$$y_0 - y_{45} = \frac{T_0 - T_{45}}{\text{grad } T} \tag{6'}$$

or

$$y_0 - y_{45} = \frac{(T_L - T_{45}) - (T_L - T_0)}{\text{grad } T} \tag{7'}$$

---

[1] Here $\overline{\text{grad } T}$ is a vector in the positive y direction. Its magnitude is called grad T. The component of the gradient in the direction i is written $(\text{grad } T)_i$.

$(T_L - T_0)$ and $(T_L - T_{45})$ are solved from the equations below:

$$v_0 = \mu(T_L - T_0)^n \tag{8'}$$

$$v_{45} = \mu(T_L - T_{45})^n \tag{9'}$$

and introduced into Equation (7') together with Equation (4'):

$$\Delta y = y_0 - y_{45} = \frac{\left(\dfrac{\sqrt{2} \times v_0}{\mu}\right)^{\frac{1}{n}} - \left(\dfrac{v_0}{\mu}\right)^{\frac{1}{n}}}{|\text{grad } T|} \tag{10'}$$

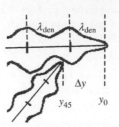

Obviously $y_0 > y_{45}$. The parallel dendrite is ahead of the inclined dendrite. This means that the parallel dendrite will stop the inclined dendrite as the space is already occupied by the parallel dendrite when the inclined dendrite arrives.

(b) The primary dendrite of the inclined crystal arrives after the second dendrite arm of the parallel dendrite if $\Delta y \geq \lambda_{\text{den}}$.

Answer:

(a) The desired difference Δy is given by Equation (10').

(b) The condition for competitive growth is $\Delta y \geq \lambda_{\text{den}}$.

6.7 MACROSTRUCTURES IN CAST MATERIALS

Eutectic growth and dendrite growth are discussed earlier in this chapter. Eutectic growth is a special case of solidification as it occurs only in alloys with eutectic compositions.

Solidification occurs in the majority of alloys and in pure metals by means of dendrite growth of nucleated crystals in the metal melts. This fact has been known for more than a century and will be applied to solidification processes of metal melts after casting.

Background

At the end of the Nineteenth century the Russian metallurgist Tschernoff published an epoch-making report concerning the solidification of steel ingots. He performed a detailed study of the crystal shapes of both the uncovered crystals that he found in the shrinkage cavities in steel ingots and the crystals tightly grown together, which he could observe in a microscope. He found that the macrostructure of a steel ingot could be divided into three distinct zones (Figure 6.36):

- a surface zone with small crystals of approximately equal size, *the surface crystal zone*;

Figure 6.36 Principle sketch of the macrostructure of the surface zone, the columnar zone and the equiaxed zone in the centre of an ingot. Medium casting temperature. Reproduced with permission from The Metals Society, The Institute of Materials.

- a zone with long columnar crystals, *the columnar zone*;
- a zone in the centre with relatively large equiaxed crystals, *the equiaxed crystal zone*.

The production of large steel ingots, which could be forged and rolled, started in the middle of the Nineteenth century, when new steel processes such as the Bessemer and Martin processes were developed. The knowledge of the metal structure was very diffuse, and since then much research on the structure of metals at solidification has resulted in greatly improved casting methods and properties of the final products.

Macrostructures of Unidirectionally Cast Materials

Tschernoff's observations have subsequently been confirmed experimentally in many ways and are the basis for the modern conception of the macrostructure of cast metals. If we look with the naked eye at the crystals in a macroetched sample we can see the macrostructure, i.e. the surface zone, the columnar zone and the central zone. The crystal region in the columnar zone has increased in accordance with the mechanism described in Example 6.5 on page 161 and in Figure 6.37.

Columnar crystal zone

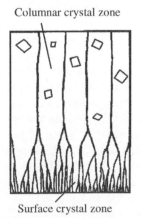

Surface crystal zone

Figure 6.37 Equiaxed crystals in the columnar crystal zone of an ingot.

In the columnar zone smaller and larger single crystals of different shapes, sizes and random orientation can be observed (Figure 6.37). They are equiaxed crystals of the same kind as the ones in the central zone, and their origin will be discussed on page 166.

The three basic zones occur in the final products of all types of casting process. Experimental evidence of the influence of various parameters that influence the macrostructure is discussed below. The formation of each of the three zones is then discussed separately in Sections 6.7.1, 6.7.2 and 6.7.3.

Influence of Casting Temperature and other Parameters on the Crystal Structure

In modern times much work has been done to explain the influence of various factors on the casting structure. The Swedish metallurgist Hultgren's series of publications, starting in the 1920s, constitutes a milestone within this field. The research has since continued for the whole of the Twentieth century and is still going on.

Hultgren showed that it is possible to vary the length of the columnar crystals by varying the casting temperature. An increase of the casting temperature leads to an increase of the columnar zone at the expense of the central zone [Figure 6.38 (a)]. A decrease in the casting temperature

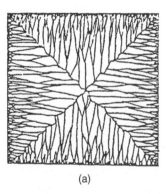

(a)

Figure 6.38 (a) Principle sketch of the macrostructure of the surface zone and the columnar zone in an ingot. High casting temperature.

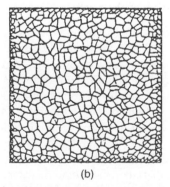

(b)

Figure 6.38 (b) Principle sketch of the macrostructure of an ingot. The equiaxed zone has grown at the expense of the columnar zone. Low casting temperature.

gives the structure illustrated earlier in Figure 6.36. At low temperature the columnar zone may be completely absent [Figure 6.38 (b)].

Hultgren also found that the structure was influenced by other factors such as stirring the melt during solidification, slow tapping into the mould and refilling during solidification. It has since been found that the structure can be changed similarly due to the properties of the mould, by addition of small amounts of foreign elements and by changing the composition of the alloy.

Experience shows that the surface crystal region is always small, while the shapes and relative sizes of the columnar and central zones vary considerably, depending on factors such as:

- casting temperature of the melt;
- casting method;
- growth rate;
- cooling rate.

Table 6.2 characterizes roughly the most common casting methods and relates the macrostructure of the metal to some of the most important factors listed in the table. The temperature distribution in the melt, solidified metal and mould results in the characteristic features of the different casting methods. Variation is achieved by change of cooling conditions.

TABLE 6.2 Influence of some parameters on the macrostructure of castings.

Casting method	Cooling rate	Growth rate	Columnar zone	Macrostructure
Continuous casting	Very strong	High	Long	Figure 6.38(a)
Ingot casting	Strong	Medium	Short	Figure 6.36
Sand mould	Weak	Low	Absent	Figure 6.38(b)

6.7.1 Formation of the Surface Crystal Zone

During permanent mould casting processes the melt is cast in close contact with a metal/mould surface, which is at room temperature or may be water cooled. The melt is rapidly cooled to the critical temperature T^* required for nucleation. A large number of nucleated, randomly oriented, small crystals is formed. The temperature gradient in the melt favours crystal growth in the direction of grad T (Example 6.5, page 161) at the lower surface. The structure is called the *surface crystal zone*.

We assume that the temperature distribution in the melt is given by the curve 1 in Figure 6.39 when the first nucleus of solid phase is formed. The surface of this growing crystal is initially small, which means that the generated solidifica-

tion heat per unit time is small even if the growth rate is high. This amount is not enough to balance the amount of heat that is carried away by cooling and the temperature in the melt decreases, as illustrated by curve 2 in Figure 6.39.

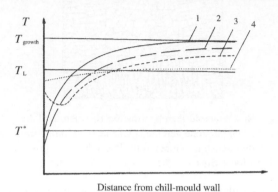

Figure 6.39 Temperature distribution in a metal melt at the initial stage of solidification as a function of time – one curve for each value of time.

Several nuclei may be formed, even within the very farthest layer. When the number of nuclei is large enough and there is a sufficiently large total surface, the generated solidification heat becomes so large that it more than balances the amount of heat carried away by cooling, if this latter is not too great. The temperature of the strongly undercooled zone increases and the temperature conditions are illustrated by curve 3 in Figure 6.39. No new nuclei are formed.

The temperature of the melt increases until the growth rate of the formed nuclei has decreased to the extent that the solidification heat balances the outer cooling and a relatively homogenous temperature of the melt is obtained – curve 4 in Figure 6.39.

6.7.2 Formation of the Columnar Crystal Zone

The whole initial solidification process occurs in connection with the growth of the nuclei of solid phase to crystal skeletons, dendrites. It has been shown that part of these crystal skeletons will be broken by strong convection, which is always present in the melt immediately after the casting. For this reason a large crystal multiplication appears in certain cases, which contributes considerably to the increase in the number of nuclei in the surface zone.

The reasoning above shows that formation of many nuclei is to be expected during the initial stage of solidification. These nuclei constitute the origin of the so-called surface crystal zone, which often is rather fine grained. After this initial stage the formation of nuclei normally ceases, which is caused by an increase of the temperature, as the curves 3 and 4 in Figure 6.39 show.

The continued solidification occurs almost entirely by growth of already nucleated crystals. Due to competitive growth, the crystals grow in the direction of the temperature gradient, i.e. inwards from the surface zone and towards the centre of the melt. Each crystal consists of several parallel primary dendrite arms, all of which have grown equally far into the melt. Dendrites are initially formed by growth of arms and branches in certain crystallographic directions. During a later stage these arms grow together and form distinct planes.

By making a cut through a columnar crystal from the surface and inwards towards the centre, it is possible to follow the extension of the individual dendrite crystals. Figure 6.40 shows a sketch of a columnar crystal.

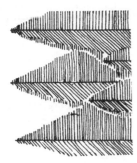

Figure 6.40 Parallel primary dendrite arms, which grow inwards in the melt, together form a columnar crystal. The three dendrite arms in the same horizontal plane.

As a consequence of the decrease in the growth rate at the solidification front with distance from the surface of the casting, the distance between the dendrite tips increases according to the relationship (6.9) on page 144. When the growth rate decreases, the structure becomes coarser, as is evident from Figures 6.43 and 6.45 on pages 168 and 169 respectively.

Example 6.6

In a foundry the dendrite arm distance of unidirectionally cast Al castings was studied as a function of the thickness of the casting. The measurements from the surface of the casting and inwards were plotted in a diagram. It was found that the dendrite arm distance was constant as a function of the casting thickness up to a certain critical thickness. The dendrite arm distance then increased parabolically with the thickness. Explain these results.

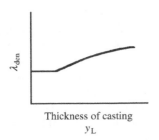

Solution and Answer:

The heat transport through the solidifying shell controls the growth rate, which in its turn controls the dendrite arm distance according to the relationship:

$$v_{\text{growth}} \, \lambda_{\text{den}}^2 = \text{const} \tag{1'}$$

which means that

$$\lambda_{\text{den}} = \frac{\text{const}}{\sqrt{v_{\text{growth}}}} \tag{2'}$$

We use the general expression for the solidification rate from Equation (4.46) on page 74 in Chapter 4:

$$v_{\text{growth}} = \frac{dy_L}{dt} = \frac{T_L - T_0}{\rho(-\Delta H)} \frac{h}{1 + \dfrac{h}{k} y_L} \tag{3'}$$

Case I: Thin castings, or at the start of the solidification when the solidified layer is thin, i.e. Nussel's number:

$$\text{Nu} = \frac{h}{k} \, y_L \ll 1$$

The term $\left(\frac{h}{k}\right) y_L$ can be neglected in comparison with 1 and Equation (3') can be reduced to:

$$v_{\text{growth}} = \frac{dy_L}{dt} = \frac{T_L - T_0}{\rho(-\Delta H)} h \tag{4'}$$

i.e. v_{growth} is constant. Consequently the dendrite arm distance is also constant at the beginning of the casting process, in agreement with the experimental results.

Case II: Thick castings, i.e. Nussel's number is not small compared with 1.

In this case Equation (3') can be written as:

$$v_{\text{growth}} = \frac{dy_L}{dt} = \frac{T_L - T_0}{\rho(-\Delta H)} \frac{h}{1 + \dfrac{h}{k} y_L} \tag{5'}$$

and we get:

$$\lambda_{\text{den}} = \frac{\text{const}}{\sqrt{\dfrac{dy_L}{dt}}} = \text{const} \sqrt{1 + \frac{h}{k} y_L} \tag{6'}$$

When $\left(\frac{h}{k}\right) y_L \gg 1$ the relationship (6') can be written as:

$$\lambda_{\text{den}} = \text{const} \sqrt{y_L} \tag{7'}$$

i.e. the dendrite arm distance increases parabolically with the casting thickness in agreement with the experimental results.

6.7.3 Formation of the Central Crystal Zone – Equiaxed Crystals of Random Orientation

Tschernoff was the first to discuss the formation of equiaxed crystals and a central zone in the scientific literature. His observations on ingots have been confirmed by microscopical studies and other convincing experimental evidence.

The crystals have a random orientation – all directions are equally frequent. The proper technical term to describe these crystals is *equiaxed crystals of random orientation.* The fact that the crystals have various orientations (Figure 6.41) when they precipitate from the melt shows that they are formed from separate nuclei. Sometimes during the precipitation process these crystals have floated freely in the melt. They are designated as free or freely floating crystals in the melt at this stage.

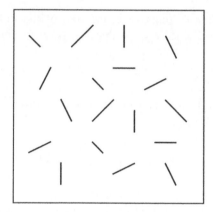

Figure 6.41 Illustration of the crystallographic directions of equiaxed crystals in the central zone, orientated at random.

Studies of the solidification process, based on rapid cooling while the reaction is going on, show that these crystals that float freely in the melt can grow to a considerable size.

Formation of Equiaxed Crystals of Random Orientation
There are several different theories about the nucleation of freely floating crystals. These theories have been applied to ingots in the text below but are valid for other types of castings as well.

One theory is that new nuclei are formed by *crystal multiplication* within the melt. This process is discussed earlier in Section 6.3.3. Both Hultgren and Southin have showed that there are crystals in the central zone that, in part, have the same structure as the crystals at the upper surface of the ingot. The explanation of this may be that dendrite fragments from the solid metal layer at the upper surface can be torn off, due to convection in the melt, and act as heterogeneities for nucleation of equiaxed crystals in the central zone (see Figure 6.37). Of course it is not necessary for crystal fragments to get torn off from the upper surface of the ingot in

particular. It may happen all over the ingot whenever the proper conditions for fractures of the crystal arms occur.

Another theory, presented by Howe, is that the concentration of segregated elements ahead of the solidification front may cause *undercooling of the melt* with the consequence that new crystals form at the lower temperature. Hultgren investigated this theory. He claimed that an undercooled zone may appear, due to the diffusion ahead of the solidification front, and that nucleation of new equiaxed crystals ahead of the front may occur there. Figure 6.42 shows how such an undercooled zone may arise. The left side of the figure shows the concentration profile of alloying elements in the melt in front of a growing dendrite tip. The right side of the figure shows how this profile can be transformed into a curve, which describes how the temperature must vary theoretically in order to cause every point in the melt to be at the liquidus temperature, the highest temperature at which solid phase can exist in the melt. If the real temperature profile is shallow enough, an undercooled zone may arise, which is the condition for formation of freely floating crystals.

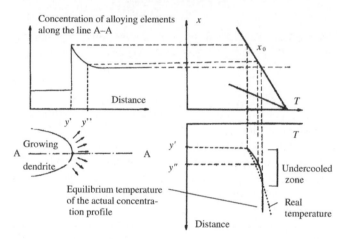

Figure 6.42 Formation of an undercooled zone in front of a growing dendrite tip in a metal.

In many cases a lot of impurities are present in the melt and new equiaxed crystals can easily be formed by nucleation on these heterogeneities. In addition, the number of growing crystals can be increased by crystal multiplication, enhanced by convection in the melt.

Formation of the Central Zone
When the number of freely floating crystals is large enough and the growing crystals have reached a certain critical size, they will effectively block the further growth of the columnar crystals. Then the central zone will replace the columnar zone. However, there will be no growth of the new crystals and no zone change unless the released heat of formation is transported away. As an example this topic is discussed for ingots on page 170.

Example 6.7

Find the relationship between the relative withdrawal velocity and the temperature gradient that must be valid in apparatus for controlled unidirectional solidification, if a sudden change from columnar crystals to equiaxed crystals of random orientation is to be avoided in a stellite alloy with the following properties:

(1) The relationship between the growth rate and the growth temperature T front is as follows:

$$v_{\text{growth}} - 10^{-4}(T_{\text{L}} \quad T_{\text{front}}) \text{ m/s}$$

where T_{L} is the liquidus temperature.

(2) The primary dendrite arm distance is described by:

$$v_{\text{growth}} \lambda_{\text{den}}^2 = 10^{-12} \text{ m}^3/\text{s}$$

Solution:

We have two growing objects: the solidification front and a nucleated crystal, which grows somewhere in the undercooled zone. The solidification front moves *upwards* with the growth rate v_{front} relative to the casting. Simultaneously the whole casting is moved *downwards* with the same velocity. The result is that the solidification front remains at rest relative to the surroundings. Thus the withdrawal velocity has the same magnitude as v_{front} but the opposite direction.

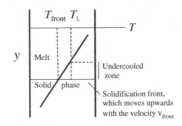

Motion of the Solidification Front:

We apply the condition 1 above to the solidification front of growing dendrites, where $T = T_{\text{front}}$, and get:

$$\frac{dy}{dt} = v_{\text{front}} = 10^{-4}(T_{\text{L}} - T_{\text{front}}) \qquad (1')$$

The temperature gradient close to solidification front is:

$$|\text{grad } T| = (dT/dy) \qquad (2')$$

Equation $(2')$ is multiplied by v_{front}, which is equal to dy/dt:

$$v_{\text{front}}|\text{grad } T| = \left(\frac{dy}{dt}\right)\left(\frac{dT}{dy}\right) = \frac{dT}{dt} \qquad (3')$$

Equation $(3')$ can be integrated. We want t_{max}, the time it takes for the solidification front to move (relative to the casting) from the position where $T = T_{\text{front}}$ to the position where $T = T_{\text{L}}$.

Both v_{front} and grad T are constant, which gives:

$$v_{\text{front}}|\text{grad } T| \int_0^{t_{\text{max}}} dt = \int_{T_{\text{front}}}^{T_{\text{L}}} dT$$

or, with the aid of Equation $(1')$:

$$t_{\text{max}} = \frac{T_{\text{L}} - T_{\text{front}}}{v_{\text{front}}|\text{grad } T|} = \frac{10^4}{|\text{grad } T|} \qquad (4')$$

Growth of Equiaxial Crystals

Next we will calculate the maximum size of a growing equiaxed crystal that has been nucleated and grown in the undercooled zone of the melt.

Provided that the critical temperature $T^* \approx T_{\text{L}}$ we get the maximum size of the crystal if it is nucleated at the upper end of the undercooled zone, where the temperature is T_{L} at $t = 0$. Then the temperature close to the crystal decreases when the casting is moved downwards and the crystal grows until it meets the solidification front. The growth temperature T_{crystal} drops linearly from T_{L} to T_{front}. The crystal growth follows the law:

$$\frac{dr}{dt} = v_{\text{crystal}} = 10^{-4}(T_{\text{L}} - T_{\text{crystal}})$$

which gives:

$$dr = 10^{-4}(T_{\text{L}} - T_{\text{crystal}})dt \qquad (5')$$

At $t = 0$ $T = T_{\text{L}}$ and $T_{\text{L}} - T = 0$

At $t = t$ $T = T_{\text{crystal}}$ and $T_{\text{L}} - T = T_{\text{L}} - T_{\text{crystal}}$

At $t = t_{\text{max}}$ $T = T_{\text{front}}$ and $T_{\text{L}} - T = T_{\text{L}} - T_{\text{front}}$

which gives:

$$\frac{t_{\text{max}}}{t} = \frac{T_{\text{L}} - T_{\text{front}}}{T_{\text{L}} - T_{\text{crystal}}} \quad \text{or} \quad T_{\text{L}} - T_{\text{crystal}} = \left(\frac{T_{\text{L}} - T_{\text{front}}}{t_{\text{max}}}\right)t \qquad (6')$$

This expression for $T_L - T_{crystal}$ is inserted into Equation (5′):

$$dr = 10^{-4} \left(\frac{T_L - T_{front}}{t_{max}} \right) t\, dt \qquad (7′)$$

Equation (7′) is integrated:

$$\int_0^{r_{max}} dr = 10^{-4} \left(\frac{T_L - T_{front}}{t_{max}} \right) \int_0^{t_{max}} t\, dt$$

which gives:

$$r_{max} = 10^{-4} \left(\frac{T_L - T_{front}}{t_{max}} \right) \frac{t_{max}^2}{2} \qquad (8′)$$

The value of t_{max} in Equation (4′) is introduced into Equation (8′):

$$r_{max} = 10^{-4} \left(\frac{T_L - T_{front}}{2} \right) \frac{10^4}{|grad\, T|} \qquad (9′)$$

Using Equation (1′) we get:

$$r_{max} = \frac{v_{front} \times 10^4}{2|grad\, T|} \qquad (10′)$$

To avoid a sudden change from columnar crystals to equiaxed crystals of random orientation the following condition must be fulfilled (compare text and figure at the end of Example 6.5 on page 162):

$$r_{max} < \lambda_{den} \qquad (11′)$$

Condition 2 in the example text gives:

$$\frac{v_{front} \times 10^4}{2|grad\, T|} < \sqrt{\frac{10^{-12}}{v_{crystal}}}$$

If we assume that the growth rate $v_{crystal}$ equals the withdrawal velocity $|v_{withdrawal}| = |v_{front}|$ we get the answer given below [in reality $v_{crystal}$ is somewhat lower, see Exercise 6.8 (c)].

Answer:
The desired relationship is $|grad\, T| > \frac{10^{10}}{2} |v_{withdrawal}|^{\frac{3}{2}}$.

Time for Change from Columnar Zone to Central Zone during the Solidification Process
General expressions for the time of the zone change and the length of the columnar zones can not be found as the heat flux through the solid shell varies with the casting method and the shapes and sizes of the castings. An example of

such calculations is given in Section 6.8 on page 171 for ingots when the convection in the melt is taken into account.

6.8 MACROSTRUCTURES IN INGOT CAST MATERIALS

In Section 6.7 a general discussion of the macrostructure in castings is given. In this and the following two sections some additional specific properties of the macrostructures of materials, cast using the main cast house processes, are discussed.

6.8.1 Columnar Zone in Ingots

In 1920 the English metallurgist Stead originally suggested the theory of competitive growth (Sections 6.6.3), which later was confirmed by Hultgren. Today it is generally accepted as the explanation of the columnar zone.

The columnar character of the crystal is illustrated in Figure 6.43. It arises by the simultaneous growth of several dendrite crystals side by side. The figure also shows that the cross-section area of the columnar crystal increases with the distance from the cooled surface. The reason for this is that there is an elimination of columnar crystals with less favourable crystal orientation, which results in an increase of the cross-section diameters of the remaining columnar crystals.

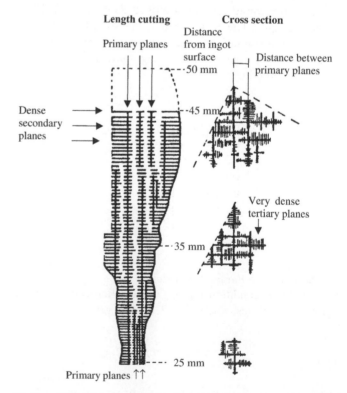

Figure 6.43 Cut through the length axis of a steel crystal. Reproduced with permission from the Scandinavian Journal of Metallurgy, Blackwell.

The crystal area, perpendicular to the mould surface, will thus increase during the development of the columnar crystal zone. Figure 6.44 shows the crystal area as a function of the distance from the ingot surface for two ingots of ball bearing steel. As a consequence of the fact that the growth rate of the solidification front decreases with the distance from the ingot surface, the distance between the dendrite tips increases. This is illustrated in Figure 6.45.

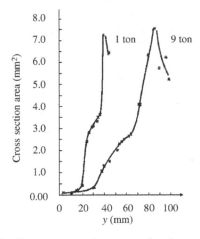

Figure 6.44 Transverse section area of columnar crystals in two ingots as a function of the distance from the ingot surface. Reproduced with permission from the Scandinavian Journal of Metallurgy, Blackwell.

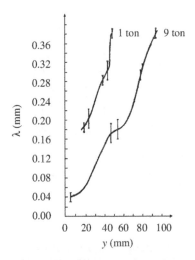

Figure 6.45 The spacing of the secondary plates in the columnar zone of the ingots in Figure 6.44 as a function of the distance from the ingot surface. Reproduced with permission from the Scandinavian Journal of Metallurgy, Blackwell.

Figure 6.45 does not correspond to the simple parabolic relationship, between the dendrite arm distance and the distance from the cooling surface [Equation (7′)] in Example 6.6 on page 16. The reason for this is that the growth conditions and consequently also the structure

morphologies are not constant. The dendrite arms and the crystal cross sections seem to grow in two or three different steps. For each step the relationship $v_{growth} \lambda_{den}^2 = const$ is valid, but with different contants for each step.

Close to the interface there is a subzone with cellular-like dendrite structure with weakly developed secondary arms (cellular crystals being crystals, with no secondary arms, that are formed at high growth rates and large temperature gradients). A second subzone of normal dendrites succeeds this zone. At larger distances from the ingot surface a third subzone of superdendrites can be identified (superdendrites are characterized by very large distances between their primary arms. They were first discussed by Bolling in 1968). The formation of superdendrites seems to be related to the superheating of the melt ahead of the growing dendrites.

6.8.2 Change from Columnar Zone to Central Zone

The question as to whether or not the change from columnar zone to central zone is coupled to equiaxed crystals has been debated for a long time. One of the very first scientists to discuss the formation process was the English metallurgist Stead. He claimed that the free crystals at the centre of an ingot might grow to a considerable size at the same time as there is a certain tendency of sedimentation. In this way he could explain another of his observations, namely that the ingot material is more pure in the lower than in the upper part (Figure 6.46 on page 170). The theory, that the zone of equiaxed crystals is built up from the bottom of the ingot by sedimentation, was further developed by among others Hultgren and is fully accepted today.

Various opinions concerning the ability of floating crystals to stop the growth of columnar crystals at the vertical solidification front of an ingot have been presented. To explain this change it was originally suggested that the columnar zone continues to grow as long as the remaining melt has a temperature that is sufficiently high to prevent the formation of floating crystals. However, Hensel showed that the melt is cooled surprisingly rapidly and often reached approximately the same temperature as the solidification front of the columnar crystals long before their growth ceased. He suggested convection in the melt as the reason for the rapid cooling process.

Nowadays the explanation of the change from columnar to equiaxed crystals in an ingot is the following. The floating crystals in the melt generally have a certain tendency to sedimentation, which increases gradually when the crystals grow. At last the crystals are so big that they can no longer float but sink to the bottom. On the way, some of the free crystals stick to the dendrite tips at the vertical solidification front and stop their growth. In those cases where the number of free crystals in the melt is large, this change will occur early,

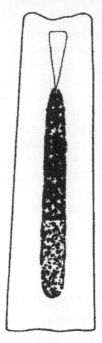

Figure 6.46 Sketch of the equiaxed crystal zone in the centre of an ingot according to Stead. The dark area is remaining melt with bright star-like equiaxed crystals. Reproduced with permission from the Scandanavian Journal of Metallurgy, Blackwell.

because the probability of the crystals sticking to the vertical solidification front increases with their number per unit volume. Numerous free crystals are nucleated in the melt if the casting temperature is low. At temperatures below the liquidus temperature a large number of free crystals are formed in the melt during the casting operation due to convection and crystal multiplication. In this case, the columnar crystal zone will end and the ingot structure will consist of small equiaxed crystals, provided that the heat flux from the ingot is sufficient.

A change from columnar crystals to equiaxed crystals of random orientation can occur only if the heat transfer to the surroundings is so rapid that the temperature gradient remains relatively unchanged by the heat released from the growing crystals.

6.8.3 Structures within the Central Zone

Different formation conditions for equiaxed crystals will result in differences in structure between the equiaxed zone, which is formed with the aid of sedimentation, and that which is formed at the vertical solidification front. This results in differences in material properties between the various parts of the ingot. From a practical point of view it may be important to realize that the zone of randomly oriented crystals in the centre of the ingot in reality consists of two different zones. The distinction

between them will be more pronounced the bigger the ingot is.

The crystals that stick to the vertical solidification front and stop its growth often grow somewhat themselves. Then they become larger than the crystals in the sediment zone and get a somewhat longish form. This is the reason why this zone often is designated as the *branched columnar zone* in the scientific literature.

The *globular dendrite zone* consists of sedimented freely growing crystals that grow in the melt and acquire a rounded shape before they settle. Thus the crystals in the sedimented zone often show a globular morphology.

In big ingots with long solidification times, the number of new free floating crystals will decrease during the solidification process and the central zone of branched columnar crystals will be extended over the whole cross section in the upper part of the ingot as shown in Figure 6.47.

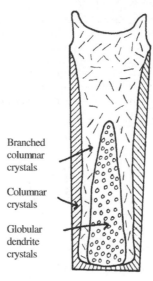

Branched columnar crystals

Columnar crystals

Globular dendrite crystals

Figure 6.47 Sketch of the extension of the zones in an ingot of 9 tons. Reproduced with permission from the Scandinavian Journal of Metallurgy, Blackwell.

6.8.4 Time for Change from Columnar to Central Zone in an Ingot during Solidification – Length of the Columnar Zone

The time for the change from columnar zone to central zone in an ingot is closely related to the temperature of the melt T_{melt} in the interior of the ingot as a function of time. This function is calculated in Section 5.3.2 in Chapter 5 for the case when the formation of equiaxed crystals or freely floating crystals is neglected. We will use these calculations and modify them by taking account of the influence of the formation of equiaxed crystals.

Heat Flux as a Function of Time with no Consideration to Formation of Freely Floating Crystals

In Section 5.3.2 in Chapter 5 we analysed the heat flow from the melt in the interior of an ingot to the surroundings and derived the temperatures T_{melt} of the cooling melt and the temperature T_{solid} as functions of the time t.

An analytical expression was found for T_{solid} while T_{melt} was presented graphically as the solution of the differential Equation (6.18) below. The heat flow from the cooling melt, through the solidifying shell, to the surroundings, was found to be:

$$\frac{dQ}{dt} = -c_{\text{p}}\rho abd\left(\frac{dT_{\text{melt}}}{dt}\right) \qquad (6.15)$$

This heat flow from the melt to the solid is transported by the aid of convection in the melt:

$$\frac{dQ}{dt} = \left(\frac{8kg^{\frac{1}{4}}}{3B}\right)ad^{\frac{3}{4}}[T_{\text{melt}} - T_{\text{solid}}]^{\frac{5}{4}} \qquad (6.16)$$

or

$$\frac{dQ}{dt} = \left(\frac{8k\,g^{\frac{1}{4}}}{3B}\right)ad^{\frac{3}{4}}\left[T_{\text{melt}} - T_{\text{L}} + \left(\frac{C}{2\mu\sqrt{t}}\right)^{\frac{1}{2}}\right]^{\frac{5}{4}} \qquad (6.17)$$

The heat balance requires that the expressions (6.15) and (6.17) are equal, which gives the differential equation:

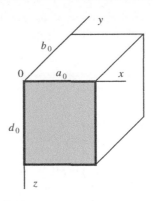

Figure 6.48 Definition of the dimensions of the mould used in Equation (6.18).

growing crystals has to be added. The total heat flow that is transported away from the melt in the central part of the ingot can be written as:

$$\frac{dQ}{dt} = -c_{\text{p}}\rho\,abd\left(\frac{dT_{\text{melt}}}{dt}\right)$$

Cooling heat flow
from the melt

$$+\ 4\pi r^2\left(\frac{dr}{dt}\right)fN\rho\,abd(-\Delta H) \qquad (6.19)$$

Solidication heat flow from
freely floating crystals

$$\frac{dT_{\text{melt}}}{dt} = \frac{-8k\,g^{\frac{1}{4}}}{3B(b_0 - 2C\sqrt{t})(d_0 - C\sqrt{t})^{\frac{1}{4}}\rho c_{\text{p}}}\left[T_{\text{melt}} - T_{\text{L}} + \left(\frac{C}{2\mu\sqrt{t}}\right)^{\frac{1}{2}}\right]^{\frac{5}{4}} \qquad (6.18)$$

where

c_{p} = heat capacitivity of the melt (J/kg)
T_{melt} = temperature of the melt in the interior of the ingot
B = constant in Equation (5.6) on page 96
C = constant in the equation $y_{\text{L}} = C\sqrt{t}$
g = gravitation constant
k = thermal conductivity of the melt
a_0, b_0, d_0 = dimensions of the mould (Figure 6.48).

In Equation (6.18) the influence of the thickness of the solidified shell has been taken into consideration. The derivation of Equation (6.18), step by step, is given on pages 98–99 in Chapter 5. No analytical solution of T_{melt} was presented.

Heat Flux as a Function of Time Taking into Account Formation of Freely Floating Crystals

When the growth of freely floating crystals is taken into consideration Equation (6.15) is no longer valid. A term that describes the heat flow due to the solidification heat of the

where

a, b, d = dimensions of the casting

r = radius of the freely floating crystals, which are assumed to be spherical
dr/dt = growth rate of the crystals
f = volume fraction solidified phase inside the crystals
$-\Delta H$ = heat of fusion (J/kg)
N = number of free crystals per unit volume in the melt.

The factor f has to be added because the crystals are not ideal solid spheres but consist in reality of solid dendrite arms surrounded by melt. Only the volume, which corresponds to the dendrite arms, has solidified.

A necessary condition for solidification is *undercooling*. According to Equation (6.11) on page 144, the growth rate of the crystals can be written:

$$\frac{dr}{dt} = \mu(T_L - T_{crystal})^n \qquad (6.20)$$

where μ = growth constant and $T_{crystal}$ = temperature on the surface of the floating crystals. We assume that μ and n have the same values both for the crystals and for the solidification front, i.e. for the growth of columnar crystals in the melt ($n = 2$). If we introduce the expression (6.20) into Equation (6.19) and assume that $T_{crystal} \sim T_{melt}$ we get:

$$\frac{dQ}{dt} = -c_p\rho\, abd \left(\frac{dT_{melt}}{dt}\right)$$
$$+ \mu(T_L - T_{melt})^2 4\pi r^2 f(-\Delta H) N\rho abd \qquad (6.21)$$

Calculation of T_{melt} as a Function of Time Taking into Account Formation of Freely Floating Crystals

Above we found the material balance (6.18) without taking into account the freely floating crystals by setting the expressions (6.15) and (6.17) equal. To get the corresponding differential equation for T_{melt}, when the solidification heat of the freely floating crystals is taken into consideration, we replace the expression (6.15) with expression (6.21) and make expression (6.21) equal to the expression (6.17).

$$- c_p\rho\, abd \left(\frac{dT_{melt}}{dt}\right) + \mu(T_L - T_{melt})^2 4\pi r^2 f(-\Delta H) N\rho abd$$
$$= \left(\frac{8k\, g^{\frac{1}{4}}}{3B}\right) ad^{\frac{3}{4}} \left[T_{melt} - T_L + \left(\frac{C}{2\mu\sqrt{t}}\right)^{\frac{1}{2}}\right]^{\frac{5}{4}}$$

which can be transformed into:

$$\frac{dT_{melt}}{dt} = \left(\frac{-8k\, g^{\frac{1}{4}}}{3B\cdot(b_0 - 2C\sqrt{t})(d_0 - C\sqrt{t})^{\frac{1}{4}}\rho c_p}\right)\left[T_{melt} - T_L + \left(\frac{C}{2\mu\sqrt{t}}\right)^{\frac{1}{2}}\right]^{\frac{5}{4}} + \left(\frac{\mu}{c_p}\right)(T_L - T_{melt})^2 4\pi r^2 f(-\Delta H) N \qquad (6.22)$$

Equation (6.22) is the differential equation for solution of T_{melt}. It is hard to find an exact solution, but the equation can be solved numerically, as illustrated in Figure 6.49. This is discussed in the next section.

Calculation of the Time for Change from Columnar Zone to Central Zone

The *first* term in Equation (6.22) is identical to the right hand-side of Equation (6.18) and to Equation (5.20) on page 99. The *second* term is the contribution that has to

be added due to the growth of the freely floating equiaxed crystals in the melt. As no crystal growth can occur until the melt is undercooled, the second term influences the solution of the differential equation only when T_{melt} is below the liquidus temperature T_L. The solidification heat from the growing equiaxed crystals results in the temperature increase of the melt, which is seen in the dotted part of the curve in Figure 6.49. It is the reason for the minimum of the dotted curve and the temperature increase after the minimum.

Initially the crystals are small and the effect of the generated solidification heat on the growth is thus small. The temperature of the melt increases and approaches the liquidus temperature when the crystals have grown to a certain size that depends on the number of crystals per unit volume. The convection increases, due to the temperature increase, and then the growth rate of the crystals increases. Simultaneously the probability increases that the free crystals will stick to the solidification front and retard the growth of the columnar crystals. If we could determine *when* the retardation becomes total, it would be possible to calculate the time for the change to equiaxed crystals.

The task of finding the exact time of change from the columnar zone to the central zone is difficult. Good agreement with experimental results is obtained, though, if we assume that *the change occurs when the melt in the interior of the ingot reaches its temperature minimum* (Figure 6.49).

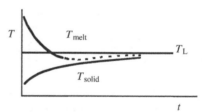

Figure 6.49 The temperature of the melt and the solid phase of an ingot as a function of time.

Length of the Columnar Zone

When the time for change to equiaxed crystals is known it is easy to calculate the length of the columnar zone. It can be calculated from Equation (5.16) on page 99 in Chapter 5:

$$y_L = C\sqrt{t} \qquad (6.23)$$

The length of the columnar zone decreases with an increased number of crystals per unit volume in the melt and with decreasing excess temperature $T_{melt} - T_L$. The reason for this is that an increased number of crystals per unit volume increases the probability of crystals sticking to the solidification front. At lower excess temperatures the temperature minimum is reached in shorter time, which gives a shorter length of the columnar crystal zone.

6.9 MACROSTRUCTURES IN CONTINUOUSLY CAST MATERIALS

The macrostructure of a continuously cast material resembles the macrostructure found in ingots. The formation mechanisms of the crystal types and crystal zones are the same in both cases, and the discrepancies that appear originate from the different casting conditions. Two cases of macrostructures in continuously cast materials are analysed and discussed below.

Figure 6.50 shows the macrostructure of a slab cast in a bent machine for continuous casting. The specimen consists of a small piece of the slab, cut along two parallel cross sections of the strand and then parted into two equal halves. The figure shows one of these halves. The width of the slab is 250 mm.

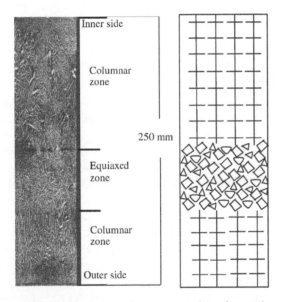

Figure 6.50 Macrostructure in a cross section of a continuously cast slab.

The upper part of the figure shows the structure formed on the inner radius side, and the lower part shows the structure on the opposite, outer side. The central part has a dark line along its length. This corresponds to a crack and segregated material, called centreline segregations. It has

nothing to do with the structure. Macrosegregation is discussed in Chapter 11. The structure can be characterized as a very fine surface zone that consists of a fine network of thin dendrite arms and *not*, as in ingots, of a great number of fine-grained crystals.

When a shell is formed in the chill-mould, close to the meniscus, some crystals grow along the surface in the withdrawal direction. However, cooling in the chill-mould is very strong and the dendrites become very thin. Due to the high temperature gradient the thin dendrite crystals grow inwards and form columnar crystals. The growth of columnar crystals is stopped by the formation of equiaxed crystals that have grown freely in the melt. The figure shows that the growth of the columnar crystal zone stopped *earlier* on the *outer* side (lower part of the figure) of the strand than at the inner side of the slab. The reason is sedimentation of the equiaxed crystals during their growth in the same way as was discussed for ingots.

The columnar crystals on the *inner* side grow towards the central part of the strand and sometimes even into the outer side (lower part of Figure 6.50). The reason is that the melt between the two growing zones does not contain enough equiaxed crystals to stop the columnar crystals effectively at the inner side.

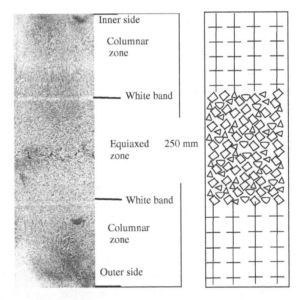

Figure 6.51 Macrostructure in a cross section of a continuously cast slab with electromagnetic stirring.

The pattern will be different if there is a great number of equiaxed crystals in the melt. Figure 6.51 shows the structure of such a continuously cast material in which measures have been taken to increase the number of freely growing equiaxed crystals. Electromagnetic stirring of the melt has increased the convection, and the convection caused formation of white bands (ghost lines), which are seen in Figure 6.51. In this way the number of crystals is enhanced by crystal multiplication at the same time as their opportunities to grow have increased. Again we disregard the appearance of the central part and the white bands. These effects have nothing to do with the macrostructure and are treated in a later chapter.

Figure 6.51 shows that the change from columnar zone to central zone occurs at the same distance from the cold wall for the inner and outer sides of the strand. The stirring eliminates the sedimentation and the asymmetry between the two sides that are present in Figure 6.50.

6.10 MACROSTRUCTURES IN NEAR NET SHAPE CAST MATERIALS

During the 1970's and later, different types of rapid solidification process were developed. Cast steel with better properties than those achieved using conventional casting methods was produced with the new rapid solidification methods. The effects of the higher cooling rates can be summarized as follows:

- refinement of grain size and more uniform dispersion of primary carbides;
- more uniform distribution of alloying elements in the matrix and reduction of alloy segregation, particularly in highly alloyed steels, where segregation can not be eliminated even after extensive working;
- increase of solid solubility of carbon and alloying elements;
- a better structure;
- formation of nonequilibrium crystalline phases or amorphous phases.

6.10.1 Structures of Rapidly Solidified Steel Strips

Figure 6.52 shows the structure of a steel alloy that consists of austenite crystals (γ-iron) and Cr–Fe–Mo eutectic carbides. Figure 6.52 (a) shows that the structure consists of two parts: a chill-region A and a cellular or dendrite region B.

Region A is restricted to the surface of the strip, where the contact between the strip and the wheel of the strip casting machine is good. This region consists of very fine-grained equiaxed crystals that are formed at the beginning of the solidification process. At some distance from the wheel surface the upper crystals in region A start to grow. Metallographic examination shows that these crystals grow with very fine cellular morphology and low microsegregation or even free from microsegregation.

Region B consists of either columnar or equiaxed crystals, depending on the local variation in the cooling rate. The mechanism behind the transition from the columnar to the

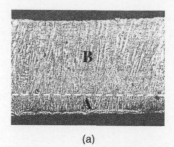

(a)

Figure 6.52 (a) Structure of a strip that consists of a mixture of γ-iron and Cr–Fe–Mo carbide. A = lower zone; B = upper zone. Reproduced with permission from Elsevier Science.

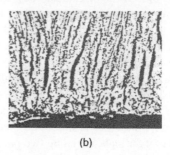

(b)

Figure 6.52 (b) Enlargement of the A-region, close to the wheel surface, at the bottom of Figure 6.52 (a). Reproduced with permission from Elsevier Science.

equiaxed zone is the same as that discussed earlier. The high cooling rate will increase the length of the columnar crystals.

6.10.2 Influence of Lattice Defects on Rapid Solidification Processes

In Section 4.3.4 in Chapter 4 we found that lattice defects are formed in the solid during solidification processes. This will be more and more pronounced the higher the cooling rate is. Crack formation is an indication of lattice defects. Such cracks can be seen in Figure 6.52 (b), where the long black, nearly vertical, stripes are elongated cracks.

Formation of lattice defects requires energy. When the defects are formed the energy of the solid increases proportionally to the amount of defects and varies with the defect type. The defects influence the latent heat of the material as was discussed in Section 4.3.4 in Chapter 4. The proportion of defects in the metal also influences the liquidus and solidus temperatures. This will thus affect the whole solidification process.

6.11 AMORPHOUS METALS

The kinetics of solidification can briefly be described by the constant μ in Equation (6.11) on page 144. Difficulties in transforming the system from an unordered structure in the melt to an ordered structure in the solid result in a small value of μ, mainly due to the diffusion process across the interface.

The more difficult the kinetics is, i.e. the slower the solidification process is, and the easier the nucleation of new equiaxed crystals occurs, the shorter will the columnar zone be. If the kinetics (the nucleation and growth processes) is slow enough, the crystallization process stops completely and the liquid becomes supercooled at a temperature where a so-called *glass transition* will occur.

The glass temperature is defined as the temperature at which the diffusion of the atoms is so low that no crystallization can occur. At the glass temperature T_{glass} an *amorphous phase* is formed.

6.11.1 Properties of Amorphous Phases

Amorphous phases have no structure. The atoms are in an unordered state, reminiscent of the liquid state. The lack of a regular or crystalline structure is confirmed by X-ray examination of amorphous materials. No sharp intense lines, characteristic of crystalline materials, are obtained, only diffuse and broad lines or rings that are typical of liquids.

At the glass transition temperature, sudden and drastic changes in specific volume, viscosity, and heat capacity are observed. Figure 6.53 illustrates the change in specific volume as a function of the temperature for an amorphous metal.

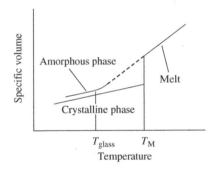

Figure 6.53 Specific volume as a function of the temperature for an amorphous metal. The dashed line indicates that no measurements have been possible at these temperatures.

The glass transition temperature is a material constant and is specific for each alloy. Amorphous metals have excellent mechanical properties. They are hard but very plastic, compared with crystalline metals. Amorphous metals show very good soft magnetic properties. They must not be heated above the glass transition temperature because of the risk of a transition to a crystalline phase.

6.11.2 Formation of Amorphous Phases

A necessary condition for formation of an amorphous phase is that *the metal melt has to be cooled quickly enough to avoid nucleation and crystallization.* The lower the temperature, the slower the solidification rate.

The critical cooling rates for metals are of the order 10^4 K/s or higher. In very special cases one can get amorphous metals at cooling rates of the magnitude 10^2 K/s. Experimental evidence shows that formation of amorphous metals will be enhanced if:

- the melting point of the alloy is *low*, compared with that of the pure metal;
- the glass transition temperature is *high* relative to the melting point of the alloy.

This is illustrated in a so-called temperature–time–transition diagram (TTT diagram) in Figure 6.54, where

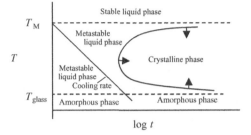

Figure 6.54 TTT diagram of an amorphous metal.

changes that promote formation of amorphous phases are marked with arrows.

The shape of the curve depends on (i) the melting point temperature T_M, (ii) the critical transition temperature T_{glass}, and (iii) the growth kinetics. Formation of amorphous material occurs in materials with a low melting point ($\downarrow$ in Figure 6.54), high glass transition temperature ($\uparrow$), and rapid nucleation and crystal growth ($\rightarrow$).

6.11.3 Casting Methods for Production of Amorphous Metals

The critical cooling rate, required for formation of an amorphous phase, is a material constant. It is the lowest possible cooling rate that can give an amorphous material and can be drawn in the TTT diagram as a tangent from the T_M point to the curve. This fact restricts the possible casting methods for production of amorphous metals.

Pressure die casting methods can be used at cooling rates of the order 12 K/s. At higher critical cooling rates spray casting methods are optimal. For continuous production of strips, the melt-drag, melt-spinning, and twin-roller technologies are used.

Example 6.8

During casting of thin metal strips using the melt-spinning process, a metal melt is sprayed onto a rotating copper wheel. The figure can be used to find the lowest possible cooling rate required to prevent crystalline solidification of a steel alloy. The curve represents the critical temperature T_{glass} as a function of time (logarithmic scale).

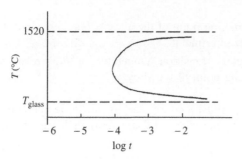

The critical cooling rate depends on the thickness of the strip. Calculate the maximum thickness the melt can have provided that it solidifies as an amorphous strip.

The heat transfer coefficient h between the wheel and the strip is $2.0 \times 10^4\,\text{W/m}^2\,\text{K}$. The excess temperature of the melt can be neglected.

Solution:
The heat transport across the interface wheel/metal is described by the equation

$$Ah(T_{\text{melt}} - T_0)\,\text{d}t = -A\delta\rho_{\text{L}}c_{\text{p}}^{\text{L}}\,\text{d}T \qquad (1')$$

heat transfer	change of 'heat content'
from the melt	from the melt when the
to the wheel	temperature is decreased
during time dt	by the amount d$T(<0)$.

The temperature of the melt is initially equal to the liquidus temperature. The strip thickness δ is solved from Equation (1'):

$$\delta = \frac{h(T_{\text{L}} - T_0)}{\rho_{\text{L}}c_{\text{p}}^{\text{L}}\left(-\dfrac{\text{d}T}{\text{d}t}\right)} \qquad (2')$$

We get the maximum thickness of the melt if we introduce the minimum cooling rate into Equation (2'). Starting at $t = 10^{-6}$ s when the temperature of the melt is 1520 °C, we draw the tangent to the curve and derive the time that corresponds to the intersection between the tangent and the horizontal log t axis.
From the figure we get:

$$\Delta t = 10^{-2.45} = 10^{0.55-3} \approx 3.55 \times 10^{-3}\,\text{s}$$
$$\Delta T = 0 - 1520 = -1520\,°\text{C}$$

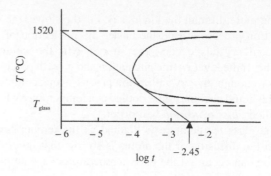

which gives:

$$\frac{\text{d}T}{\text{d}t} = \frac{\Delta T}{\Delta t} = -\frac{1520}{3.55 \times 10^{-3}} = -4.3 \times 10^5\,\text{K/s} \qquad (3')$$

As an approximate value of the temperature of the melt we use an average value of the liquidus temperature and the glass temperature:

$$\overline{T_{\text{melt}}} = \frac{T_{\text{L}} + T_{\text{glass}}}{2} \qquad (4')$$

The glass temperature is not known but can be described approximately by:

$$T_{\text{glass}} = T_{\text{L}} - t_{\text{glass}}\left(-\frac{\text{d}T}{\text{d}t}\right) \qquad (5')$$

Combination of Equations (4') and (5') gives:

$$T_{\text{melt}} = T_{\text{L}} - \frac{t_{\text{glass}}}{2}\left(-\frac{\text{d}T}{\text{d}t}\right) \qquad (6')$$

Equations (2') and (6') are combined, which gives the desired thickness:

$$\delta = \frac{h\left[(T_{\text{L}} - T_0) - \dfrac{t_{\text{glass}}}{2}\left(-\dfrac{\text{d}T}{\text{d}t}\right)\right]}{\rho_{\text{L}}c_{\text{p}}^{\text{L}}\left(-\dfrac{\text{d}T}{\text{d}t}\right)} \qquad (7')$$

From the figure we get $t_{\text{glass}} = 10^{-3} - 10^{-6} \approx 10^{-3}$ s.
By use of values from the text and calculated values from standard references and $T_0 = 20\,°\text{C}$ we get:

$$\delta = \frac{2.0 \times 10^4\left[(1520 - 20) - \left(\dfrac{10^{-3}}{2}\right)4.3 \times 10^5\right]}{6780 \times 650(4.3 \times 10^5)}$$

$$= 1.36 \times 10^{-5}\,\text{m}$$

Answer:
The maximum thickness of the melt for amorphous solidification is 13 μm.

6.11.4 Transition from Crystalline to Amorphous Solidification

The conclusion from Example 6.8 above is that it is necessary to cast very thin strips in order to achieve a cooling rate that is high enough to give an amorphous phase.

In practice a single-roller machine is used. The roller rotates with a peripheral velocity of the order of 4×10^3 m/s. A crucible with a narrow slit is located on top of the roller and the melt is pressed out of the crucible through the slit with the aid of gas pressure. The process is often called the *melt-spinning process*. The thickness of the strip is adjusted by changing the angular velocity of the roller.

Below we discuss two cases of strips with amorphous (lack of) structure. In the first case a transition from amorphous to crystalline phase has occurred. In the second case a transition from crystalline structure to amorphous phase has occurred.

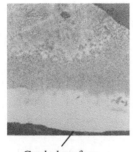

Cooled surface

Figure 6.55 Structure of a partially crystalline strip of an Al–Y alloy. The black area at the bottom of the figure is part of the sample holder, made of Bakelite.

Figure 6.55 shows the structure in a cross section of an Al–Y alloy, cast using the melt-spinning process. The alloy is a hypoeutectic alloy, where the first phase, formed at normal cooling rate, is Al $_3$Y. The amorphous phase is, in this case, formed close to the cooled surface. Later, a large number of small crystals are formed at greater distances from the cooled surface, when the conditions for formation of an amorphous phase are no longer fulfilled.

Figure 6.56 shows the structure of a strip, produced by the aid of the melt-spinning process. The cast alloy is a hypoeutectic Fe–B alloy. Solidification under normal

Figure 6.56 Structure of a partially crystalline strip of an Fe–B alloy. The black area at the bottom of the figure is part of the sample holder, made of Bakelite.

conditions results in a primary dendrite structure of Fe–B crystals. It has been observed that this structure becomes amorphous at higher cooling rates. The figure shows the structure of such a strip where Fe–B crystals have nucleated close to the cold surface of the roller. The crystals grow inwards (upwards in the Figure 6.56). The transition from crystalline to amorphous phase is clearly seen in the figure.

It is hard to analyse the formation of these different types of structure. However, we will make an attempt to interpret the structure in Figure 6.56. In this case we assume that the crystals are nucleated close to the liquidus temperature.

The crystals grow inwards (upwards in the figure) at an increasing growth rate. This is concluded from the fact that the dendrite arm spacing decreases with increased distance from the cooled surface ($v_{growth} \lambda_{den}^2 = $ const). The undercooling at the solidification front will therefore increase. It is reasonable to assume that the temperature of the strip is constant (Nu $\ll$ 1). This is a consequence of the fact that the strip is very thin.

Temperature Distribution and Growth Rate in the Strip
Figure 6.57 illustrates the temperature distributions in the strip in Figure 6.56 at two different times. Below we set up and discuss the heat balance for the heat transport in the strip. The strip temperature is constant and the heat flux is determined by the heat transfer between the strip and the Cu wheel.

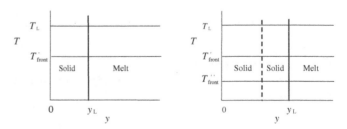

Figure 6.57 The temperature distribution in the strip in Figure 6.56 at two different times during the solidification process. The *y*-axis is perpendicular to the wheel and directed towards the melt.

When the solidification process in the strip starts, the heat transport occurs according to the following heat balance:

$$\underbrace{h(T_{i\,melt} - T_0)}_{\substack{\text{total heat flux across} \\ \text{the interface}}}$$

$$= \underbrace{\frac{dy_L}{dt}\rho(-\Delta H)}_{\substack{\text{solidification} \\ \text{heat flux}}} + \underbrace{\rho c_p y_L\left(\frac{-dT_{front}}{dt}\right)}_{\substack{\text{cooling} \\ \text{heat flux}}} \quad (6.24)$$

where

$T_{i\,melt}$ = temperature of the melt

T_{front} = temperature of the melt close to solidification front $T_{i\,melt}$

h = heat transfer coefficient across the interface melt/solid phase

T_0 = temperature of the wheel

y_L = thickness of the strip

dy_L/dt = growth rate of the solid phase

ΔH = heat of fusion (J/Kg)

ρ = density of the solid alloy

c_p = thermal capacity of the alloy.

The *first term* on the right-hand side of Equation (6.24) describes the heat flux caused by the heat release during the solidification process. The *second term* on the right-hand side represents the change in heat content of the cast strip. In Chapters 4 and 5 we disregarded this term in those cases where Nu $\ll$ 1 and assumed that the temperature in the strip or strips was constant and equal to the liquidus temperature. In this case the heat transport is so rapid and the cooling rate so high that the released heat of fusion is not sufficient to keep the temperature at the growth value. The temperature decreases in the solidified part of the strip as a consequence of the high cooling rate. Thus the second term cannot be neglected in strip casting processes with high cooling rates.

The *growth rate* in the strip is related to the solidification front temperature by [see Equation (6.11) on page 144]:

$$v_{growth} = \frac{dy_L}{dt} = \mu(T_L - T_{front})^n \quad (6.25)$$

where n is a number between 1 and 2.

By introduction of the expression (6.25) into Equation (6.24) we get:

$$h(T_{i\,melt} - T_0)$$
$$= \mu(T_L - T_{front})^n \rho(-\Delta H) + \rho c_p y_L\left(\frac{-dT_{front}}{dt}\right) \quad (6.26)$$

The solution of this differential equation gives the front temperature of the strip as a function of time. The transition from a crystalline to an amorphous phase will occur when T_{front} has reached the critical transition temperature T_{glass}.

The same analysis can be performed for the structure shown in Figure 6.55. However, in this case the homogeneous nucleation of new crystals occurs at a temperature *above* the glass transition temperature. The growth of the nucleated crystals is slow due to the low value of the kinetic coefficient μ. The melt between the crystals gets transferred to an amorphous state before the crystals have grown to full size.

SUMMARY

■ *Nucleation*

Solidification starts with formation of nuclei in the melt. The necessary condition for this process is that the melt is undercooled.

The nucleation is homogeneous if the nuclei are formed directly from the melt. At heterogeneous nucleation the nuclei are formed on foreign substances.

■ *Dendritic Growth*

Relationship between Growth Rate and Dendrite Arm Distance

$$v_{growth}\,\lambda_{den}^2 = \text{const}$$

Relationship between Growth Rate and Undercooling

$$v_{growth} = \mu(T_L - T)^n$$

■ *Eutectic Growth*

Precipitation of a eutectic alloy occurs at the constant eutectic temperature T_E.

Two phases with the compositions x_α and x_β are formed. A normal eutectic lamella structure arises when the two precipitated phases grow side by side in cooperation.

Relationship between Growth Rate and Lamella Distance

$$v_{growth}\,\lambda_{eut}^2 = \text{const}$$

In several casting processes, growth occurs radially from the centres. These eutectic cells grow until they collide and fill the whole volume.

Relationship between Growth Rate and Undercooling:

$$v_{\text{growth}} = \mu(T_E - T)^n$$

■ *Macrostructures in Sand Mould Cast Materials*

With casting in sand moulds, no columnar zone is normally formed. Usually only equiaxed crystals are formed.

Cast iron is an example of a eutectic alloy that is often cast in a sand mould. Cast iron solidifies into two eutectic structures: *grey cast iron* and *white cast iron.*

Cast iron solidifies by primary precipitation of austenite. When the eutectic point (carbon content 4.3 wt-% C) is reached complex eutectic reactions occur and graphite is precipitated.

The structure formation during the solidification process is reflected distinctly in the appearance of the solidification curve. The structure of cast iron is greatly diversified: four types of grey iron (A, B, C and D-graphite) and white iron.

- coarse flake graphite;
- nodular cast iron;
- vermicular cast iron;
- fine undercooled graphite;
- white cast iron, austenite/cementite.

Grey cast iron is easier to machine and has much better mechanical properties than does white iron. It is therefore desirable that cast iron solidifies as grey cast iron.

A *low* cooling rate and *low* growth rate promotes *grey iron* solidification.

White iron formation at solidification of cast iron is favoured by *all* factors that create high undercooling and high growth rate.

■ *Unidirectional Casting*

With the aid of a temperature gradient it is possible to achieve controlled solidification of a metal melt in the direction of the temperature gradient. Such solidification is called *unidirectional solidification.*

Crystal growth in other directions than the one of the temperature gradient is suppressed. Uniaxial solidification is used for single crystal production of components, for example turbine blades. The advantage of the single crystal technique is the superior mechanical strength of the material produced, compared with ordinary castings.

■ *Macrostructure in Cast Materials*

The macrostructures of all types of castings consist of three basic zones. The relative sizes of the zones depend on many factors, among them temperature of the melt, cooling rate, and thickness of the casting.

The *surface crystal zone* is formed when the melt is strongly undercooled. The nucleation stops when the temperature of the surface zone increases, due to the solidification heat emitted by the new crystals.

The *columnar crystal zone* is formed by dendrite growth of the already nucleated crystals in the surface zone. The growth occurs perpendicularly to the cooled surface in the direction of the temperature gradient.

The *equiaxed crystal zone* or *central zone* arises when randomly orientated crystals are nucleated and grow within the melt. The reason is assumed to be crystal multiplication – fragments of dendrite arms are broken and brought into the melt by convection and act as nuclei and/or undercooling of the melt in front of the dendrite tips.

A change from columnar zone to central zone with equiaxed crystals occurs when the latter have grown to a size that is sufficient to stop columnar growth. A large number of equiaxed crystals sediment in the melt, stick to the dendrite tips and stop their further growth.

Relationship between Casting Method and Structure

The factors in the headings of the table below are very important for the structure of the cast products.

Casting method	Cooling rate	Growth rate	Columnar zone	Macrostructure
Continuous casting	Very strong	High	Long	Figure 6.38(a)
Ingot casting	Strong	Medium	Short	Figure 6. 36
Sand mould	Weak	Low	Absent	Figure 6.38(b)

Macrostructures at Near Net Shape Casting

Strip casting is a special case of *rapid solidification*. The rapid solidification processes produce cast metals with better mechanical properties than are achieved using conventional casting methods. The most important quenching effects are:

- smaller grain size than normally;
- more uniform distribution of alloying elements, i.e. less microsegregation than normally;
- increase of solid solubility of alloying elements;
- formation of nonequilibrium crystalline phases or amorphous phases.

Amorphous Metals

If a metal melt is cooled fast enough to avoid nucleation and crystallization, an amorphous metal may be formed. The critical cooling rate is a material constant. One of the conditions is that the temperature is kept lower than the glass transition temperature, which is the temperature at which the diffusivity of the atoms is so low that no crystallization can occur.

Low melting point and high glass transition temperature of the metal and a very fast cooling rate under the critical value, promote formation of amorphous phases.

EXERCISES

6.1 Steel powder is added to a steel ingot in order to make the structure of the steel more fine grained. Calculate the minimum amount of steel powder that is required to achieve this effect. The melting point of the steel is 1470 °C and the excess temperature of the melt is 50 °C.

Hint A4

Material constants of steel:	
c_p^L	0.52 kJ/kg K
c_p^s	0.65 kJ/kg K
$-\Delta H$	272 kJ/kg

6.2 The dendrite arm distance in a cast material is in most cases determined by the growth rate and the relationship $v_{growth}\lambda_{den}^2 = \text{const}$, where v_{growth} is the growth rate and λ_{den} the dendrite arm distance. If the growth rate is measured in m/s and the dendrite arm distances in metres, the constant has the value $1.0 \times 10^{-10}\,\text{m}^3/\text{s}$ for low-carbon, iron-base alloys. Assume that the growth rate is described by:

$$\frac{dy}{dt} = \frac{1.5 \times 10^{-2}}{\sqrt{t}}\ \text{m/s}$$

Draw a diagram of λ_{den} as a function of the distance from the ingot surface.

Note that the limit of the growth rate is infinity for $t = 0$ according to the relationship above. For this reason start the diagram at a thickness of the solidifying layer equal to 1 mm. *Hint A9*

6.3 The dendrite arm distances in a cast material strongly influence the properties of the material. For an Al-base alloy, the dendrite arm distance varies with the solidification rate according to the following:

$$v_{growth}\ \lambda_{den}^2 = 1.0 \times 10^{-12}\,\text{m}^3/\text{s}$$

In a pressure casting process the solidification time is influenced by the pressure because the heat transfer number increases with increasing pressure. This can be described by $h = 400\,p$, where p is the pressure in atm and h is the heat transfer number measured in W/m² K. Calculate the dendrite arm distance as a function of the pressure. The temperature of the surroundings is 25 °C. The heat of fusion of the Al-base alloy is 398 kJ/kg. Other material constants are taken from standard tables.

Hint A11

6.4 During ingot casting of a Cu alloy a temperature–time curve in the centre of the ingot was measured during the solidification process. The curve is given below.

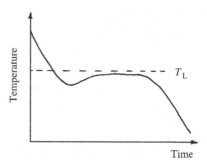

Divide the curve into sections and explain the appearance of the temperature–time curve. Relate the different sections of the curve to the macrostructure of the ingot.

Hint A5

6.5 Cast iron melts are often inoculated before casting in order to avoid solidification by precipitation of a cementite eutectic. At inoculation the number of grey cells in the structure increases. It has been found that the structure becomes coarser when the number of growing grey cells increases.

(a) Give a physical explanation of this phenomenon with the aid of simplified analytical relations.

Hint A52

(b) Give one or two reasons why one needs to inoculate the melt before casting, in spite of the knowledge that the mechanical properties will deteriorate.

Hint A280

6.6 Electromagnetic stirring is often used in various casting processes in order to reduce the length of the columnar zone and increase the central zone. One of the arguments has been that the stirring supplies so much heat to the solidification front that its further growth becomes inhibited and all further growth occurs by growth of freely floating heterogeneities in the melt.

Is this a correct argument? Answer this question by a discussion of the heat balance of the solidifying casting.

Hint A59

6.7 Rods of grey cast iron, which were cast by continuous casting, sometimes show a white rim zone. The reason for the white solidification is that the cooling rate has been too high. The limit between grey and white solidification is given by $v_{growth} = 4 \times 10^{-4}$ m/s.

Material constants

ρ_{Fe}^s	7.0×10^3 kg/m^3
$-\Delta H$	272 kJ/kg
T_L	1150 °C

If the growth rate exceeds this value there is a risk of white solidification. Make an estimation of the highest heat transfer number that can be allowed in order to avoid white solidification.

Hint A7

The temperature of the surroundings is 20 °C. You may assume that Nu ≪ 1.

6.8 Convection has a strong influence on the nucleation and growth of the equiaxed crystals in the central zone.

(a) Explain the influence of convection on the nucleation of the equiaxed crystals.

Hint A31

(b) Heat is transported by convection from the melt in the centre of a steel ingot via the solidification front to the surroundings. The heat transfer to the solidification front is described by:

$$\frac{dq}{dt} = h_{con} \, \Delta T_{melt}$$

where h_{con} is the average value of the heat transfer coefficient of convection and ΔT_{melt} is the difference

between the temperature of the melt in the interior of the ingot and the temperature of the melt in the boundary layer, close to the solidification front. The heat flux from the interior of the melt enables the growth of freely floating crystals in the melt. The growth rate of the free crystals as a function of the temperature of the steel melt is assumed to be:

$$v_{crystal} = \frac{dr}{dt} = \mu \left(T_L - T_{crystal} \right)$$

The same growth law is assumed to be valid for the solidification front $(\mu_{crystal} = \mu_{front} = \mu)$.

Given data

Dimensions of the ingot (height × width × length) = 1.50 × 0.40 × 0.60 m	
h_{con}	40×10^3 W/m^2 K
ΔT	0.5 K
μ	0.010 m/s K
ρ_{steel}	7.0×10^3 kg/m^3
$-\Delta H_{steel}$	272×10^3 J/kg

Calculate the average growth rate of these fragments as a function of their numbers per unit volume N. Use a shell thickness of 10 cm in your calculations and assume that the equiaxed crystals can be regarded as spheres with an average radius of 10 μm.

Hint A67

(c) Compare the calculated growth rate of the free crystals with the growth rate of the solidification front and discuss the possibility of a transition from columnar to equiaxed solidification.

Hint A42

6.9 In production of cast iron in a sand mould, wedge-shaped samples are often used to test the tendency of the cast iron to solidify as white iron. Consider a eutectic cast iron melt, which is to be tested.

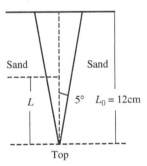

Cross section of the wedge-shaped sand mould for testing the cast iron melt. The melt is poured into the mould and is allowed to solidify and cool. The solid wedge is then cut and the structure of its cross-section is examined.

(a) Calculate the cooling rate along the central line as a function of the distance L from the wedge edge (see figure) in a wedge sample with a top angle of 10°. The initial excess temperature of the melt is 100 °C.

Hint A46

(b) Predict at what height L from the bottom of the wedge the change from white to grey iron will occur in a cast iron alloy that gives white solidification at a cooling rate of 60 °C/s.

Hint A250

Material constants for cast iron and for the sand mould and other constants are given below.

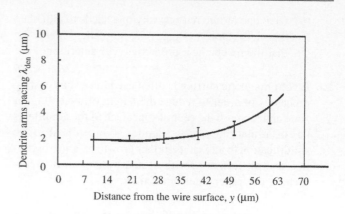

Material constants of cast iron		Other constants	
ρ^L	$7.0 \times 10^3 \, \text{kg/m}^3$	ρ_{mould}	$1.61 \times 10^3 \, \text{kg/m}^3$
c_P^L	$0.42 \, \text{kJ/kg K}$	k_{mould}	$0.63 \, \text{J/m s K}$
		c_p^{mould}	$1.05 \, \text{kJ/kg K}$
T_E	$1153 \, °C$	T_0	$25 \, °C$

Material and other constants	
ρ	$7.0 \times 10^3 \, \text{kg/m}^3$
$-\Delta H$	$280 \, \text{kJ/kg}$
T_L	$1450 \, °C$
T_0	$20 \, °C$

6.10 The figure below shows wire casting apparatus for the 'in-rotating water melt-spinning process'. A thin beam of melt from a nozzle is cast on the inside of a rotating drum. The drum is continuously sprayed with water to give rapid solidification of the wire.

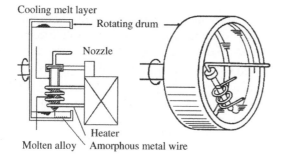

On one occasion the structure of the cast wire was analysed and the dendrite arm distance as a function of the distance y to the wire surface was found to vary as in the diagram above.

The casting rate was 10 m/s and the dendrite arm spacing was found to vary with the growth rate according to the relationship:

$$v_{\text{growth}} \, \lambda_{\text{den}}^2 = 1.0 \times 10^{-11} \, \text{m}^3/\text{s}$$

The melt had no excess temperature. Derive an expression for the heat transfer coefficient h as a function of the shell thickness $y = (r_0 - r)$ during the casting process and describe how h varied along the wire, i.e. with the distance z from the nozzle. The radius r_0 of the cast wire was 65 μm.

Hint A6

7 Microsegregation in Alloys – Peritectic Reactions and Transformations

7.1 INTRODUCTION

When an alloy solidifies by dendritic solidification the concentration of the alloying element will be unevenly distributed in the metal. This phenomenon is called *microsegregation*. In this chapter we will treat the causes of microsegregation, mathematical models for microsegregation, and discuss the factors that control its appearance. A short section about peritectic reactions and transformations is also included.

7.2 COOLING CURVES, DENDRITIC GROWTH, AND MICROSEGREGATION

Once a dendrite crystal has been formed, further solidification occurs by growth of solid phase on the existing dendrite arms. This process can be illustrated by a cooling curve and an analysis of the process that occurs in the melt during its cooling.

Such a curve is shown in Figure 7.1. The curve illustrates the temperature–time relationship for an Fe–C alloy that has cooled in a sand mould. The curve can in principle be divided into five different areas or time periods. Area 1 describes the cooling of a melt before the solidification starts. Area 2 corresponds to the process when a great number of dendrite crystals are formed in the melt and grow in a spherical or star pattern. Area 3 represents that part of the solidification process when the melt between the dendrite arms gradually solidifies. At the end of solidification the composition of the remaining melt becomes eutectic. During period 4 the remaining melt solidifies eutectically at constant temperature. Period

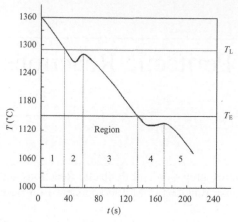

Figure 7.1 Cooling curve of an Fe–C alloy that has cooled in a sand mould. Primary precipitation of austenite is followed by a relatively short eutectic reaction.

5 corresponds to the cooling process of the solid phase.

Figure 7.2 shows a two-dimensional dendritic structure that contains 50 % solid phase.

The energy balance and the rate of heat removal control all solidification processes. The total enthalpy of a solidifying melt with growing spherical crystals of dendritic morphology can be written as:

$$Q = V_0 \rho_m c_p^m (T - T_{ref}) + V_0 \rho_m f N \frac{4}{3} \pi R^3 (-\Delta H_a) \quad (7.1)$$

heat content of the   heat of solidification of
melt and solid   the dendrite crystals

where

Q = total enthalpy of the system
V_0 = total volume of the system
ρ_m = density of the metal
c_p^m = thermal capacity of the metal
T = absolute temperature
T_{ref} = reference temperature
f = fraction of solid phase
N = number of crystals per unit volume
R = radius of the crystals
$-\Delta H_a$ = heat of fusion per atom of the metal.

When the melt solidifies, the quantities T, f and R change. If Equation (7.1) is derivatized with respect to time t we obtain:

Figure 7.2 Schematic two-dimensional picture of a dendritic structure containing 50 % solid phase.

The first term on the right-hand side of Equation (7.2) corresponds to the heat of cooling. The second term is the solidification heat emitted during crystal growth when the radius R increases. This term dominates during period 2. The third term describes the change in enthalpy due to change in the fraction of solid phase. This change is caused by temperature changes inside freshly formed crystals and dominates during period 3. During periods 1 and 5 the second and third terms in Equation (7.2) are zero.

In Chapters 4 and 5 we treat the heat transport caused by the first two terms. In the earlier treatment we assume that the fraction of solid phase in the second term is equal to 1 (at the analysis of the eutectic reaction at solidification in a sand mould) or 0.3 (at the change from columnar crystals to equiaxed crystals). The third term will be discussed in this section in order to calculate f and df/dt.

The derivative df/dt is determined by the phase diagram of the metal and by the rate of heat removal. This is seen from the equality:

$$\frac{df}{dt} = \frac{df}{dx}\frac{dx}{dT}\frac{dT}{dt} \quad (7.3)$$

where x is the mole fraction of the alloying element in the melt. dx/dT can be derived from the slope dT/dx of the liquidus line in the phase diagram. $(-dT/dt)$ is the cooling rate of the melt.

df/dx is an inverse measure of how the composition of the melt is changed due to a small change in the fraction of solid phase. Later in this chapter we will derive a relationship between the mole fraction of the alloying element x^L of the melt and the fraction f of the solid phase.

The phase diagram shows that the solid phase in most cases has a different composition to the melt, the concentration of the alloying element normally being lower in the

$$\frac{dQ}{dt} = V_0 \rho_m c_p^m \frac{dT}{dt} + V_0 \rho_m f N(-\Delta H_a) 4\pi R^2 \frac{dR}{dt} + V_0 \rho_m N(-\Delta H_a)\left(\frac{4\pi}{3}R^3\right)\frac{df}{dt} \quad (7.2)$$

solid phase than in the melt. This difference is described by the partition constant:

$$k_{part} = \frac{x^s}{x^L} \qquad (7.4)$$

where

k_{part} = partition coefficient or partition constant
x^s = mole fraction of alloying element in the solid phase
x^L = mole fraction of alloying element in the melt.

If the partition constant $k_{part} < 1$ the alloying element becomes concentrated in the remaining melt. If the diffusion rate in the solid phase is low, the *last* solidified parts will contain a *higher* concentration of the alloying element than the *first* solidified parts. If $k_{part} > 1$, the opposite is true. Such an uneven distribution of the alloying element in the solidified material is called *microsegregation.*

In both cases the microsegregation pattern indirectly reproduces the geometric shape of the crystals during their growth. Figure 7.3 (Top) illustrates this. It shows the nickel concentration close to a dendrite arm in an Fe–Ni–Cr

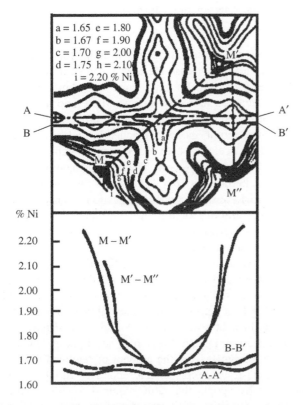

Figure 7.3 (Top) Isoconcentration curves close to a dendrite cross of the type illustrated in Figure 7.2. (Bottom) Measured concentration variations along some different measurement lines through the dendrite cross above. Reproduced by permission of The Minerals, Metals & Materials Society.

alloy. The basic information for the figure is obtained from microprobe measurements on the solidified material [Figure 7.3 (Bottom)].

7.3 SCHEIL'S SEGREGATION EQUATION – A MODEL OF MICROSEGREGATION

It is not possible to perform an exact calculation of df/dx in Equation (7.2) in a mathematical treatment of the solidification process of such a complicated geometrical shape as that in Figures 7.2 and 7.3. In order to analyse the influence of various factors on the distribution of the alloying element we have to simplify the geometry considerably. For a mathematical treatment of the phenomenon of microsegregation we choose the simplest possible geometry and consider the volume element given in Figure 7.4. We assume that this volume element represents a small interdendritic volume and make the following assumptions:

(1) The length of the element is equal to half the dendrite arm distance, $\lambda_{den}/2$.
(2) The volume element is so small that the temperature is the same within the element at any given moment.
(3) The solid and liquid phases have the same molar volume V_m.

If the last assumption is not fulfilled, the solidification results in a change in the volume of the material. Pores are formed or melt flows into the volume element, which might cause macrosegregation. In this chapter we will disregard such complications. Macrosegregation is treated in Chapter 11.

7.3.1 Scheil's Model of Microsegregation

In our first treatment of microsegregation we start with the following assumptions:

• The convection and diffusion in the melt are so violent and rapid that the melt at every moment has an even composition.
• The diffusion in the solid phase is so slow that it can be completely neglected.
• Local equilibrium exists between the solid phase and the melt. The equilibrium can be expressed by the partition constant:

$$k_{part} = \frac{x^s}{x^L} \qquad (7.4)$$

Consider Figure 7.4. The solidified material has reached a thickness y and grows by an amount dy during time dt. Solidification of the slice Ady requires a decrease in the concentration of the alloying element from x^L to x^s. The amount

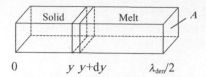

Figure 7.4 Solidifying volume element.

$(x^L - x^s)\, A dy/V_m$ of the alloying element has to be moved into the melt. Its concentration increases then by dx^L.

This amount of alloying metal is brought into the volume of the melt $A\left(\lambda_{den}/2 - y - dy\right)$ from the solidified slice $A dy$ and its concentration increases from x^L to $x^L + dx^L$. A material balance for the alloying element gives:

$$\frac{(x^L - x^s)\, A dy}{V_m} = \frac{A\left(\dfrac{\lambda_{den}}{2} - y - dy\right) dx^L}{V_m} \qquad (7.5)$$

By reduction and neglecting the product $dy dx^L$, Equation (7.5) can be simplified to:

$$\left(\frac{\lambda_{den}}{2} - y\right) dx^L = (x^L - x^s)\, dy \qquad (7.6)$$

By introduction of k_{part} we can eliminate x^s. Integration of Equation (7.6) from $y = 0$ to y and $x^L = x_0^L$ to x^L gives:

$$\int_{x_0^L}^{x^L} \frac{dx^L}{x^L - k_{part} x^L} = \int_0^y \frac{dy}{\left(\dfrac{\lambda_{den}}{2} - y\right)} \qquad (7.7)$$

where x_0^L is the initial concentration of the alloying element in the melt. After integration we get:

$$\frac{1}{1 - k_{part}} \ln \frac{x^L}{x_0^L} = -\ln \frac{\lambda_{den}/(2 - y)}{\lambda_{den}/2} \qquad (7.8)$$

Equation (7.8) is solved for x^L:

$$x^L = \frac{x^s}{k_{part}} = x_0^L \left(1 - \frac{2y}{\lambda_{den}}\right)^{-(1 - k_{part})} \qquad (7.9)$$

The derivatization of Equation (7.9) is to some extent associated with the geometrical shape of the volume element. The formula is valid for any geometrical shape if $2y/\lambda_{den}$ is replaced by the more general variable f, which represents the fraction of solidified material. f is called *degree of solidification* or *fraction of solid phase*. Equation (7.9) in its more general form be written as:

$$x^L = x_0^L \left(1 - f\right)^{-(1 - k_{part})} \qquad \text{Scheil's equation} \qquad (7.10)$$

Equation (7.10) is called *Scheil's segregation equation* after its originator. It is important to notice that x^L represents the

instant concentration of the alloying element in the melt while x^s represents the concentration of the alloying element in the last solidified material. Figure 7.5 shows how these two concentrations vary during the solidification process when f changes from 0 to 1 for the special case $k_{part} = 0.5$.

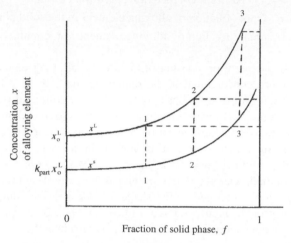

Figure 7.5 The concentrations of the alloying element in the melt and in the last solidified material during a solidification process for $k_{part} = 0.5$ as a function of the fraction of solid phase.

The dotted lines show the concentrations for three different cases. In the first case the melt has a composition that slightly exceeds x_0^L. In the second and third cases this concentration has increased exponentially and approaches infinity at the end of the solidification process.

Of course infinity is never reached in reality. A eutectic reaction may occur or there will be a homogenization during the solidification process. The homogenization process, which is known as *back diffusion*, is discussed in Section 7.5.1. Besides, Scheil's Equation is only valid for small concentrations of the alloying element because the partition coefficient k_{part} is constant only for low values of the concentration x.

In the treatment of microsegregation during the solidification process above we have disregarded the time and temperature aspects. The solidification process is controlled by the rate of heat removal from the volume element. Normally the heat of solidification dominates and the degree of solidification, i.e. the fraction of solid phase, can accurately be estimated from the amount of removed heat.

The temperature represents a secondary variable, which automatically adopts the value required for the condition that the instant value of the concentration x^L of the alloying element must lie on the liquidus line in the phase diagram of the alloy (Figure 7.6). However, there is an experimental method, called *controlled solidification*, where the temperature is determined from outside. It controls the solidification

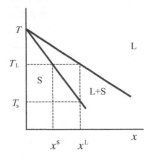

Figure 7.6 Phase diagram of a binary alloy.

process by varying the temperature of the material accord ing to a predecided temperature gradient (Chapter 6, pages 158–159).

Eutectic Solidification

When a molten binary alloy cools and starts to solidify, the compositions of both the melt and the solid phase change gradually and follow the liquidus and the solidus lines in the phase diagram. As long as these lines are fairly straight, the partition coefficient k_{part} is constant. This is the case for low concentrations of the alloying element.

We consider the case of when Scheil's equation [Equation (7.10)] is still valid at the eutectic temperature. At this temperature we get:

$$x_E^L = x_0^L \left(1 - f_E\right)^{-(1-k_{partE})} \tag{7.11}$$

where

x_E^L = concentration of the alloying element in the remaining melt at the eutectic temperature

x_0^L = initial concentration of the alloying element in the melt

k_{partE} = partition coefficient of the alloying element at the eutectic temperature

f_E = fraction of solid phase at the eutectic temperature.

The fraction of remaining liquid f_E^L at the eutectic temperature equals $(1 - f_E)$. It is solved using Equation (7.11):

$$\left(1 - f_E\right)^{-(1-k_{partE})} = \left(\frac{x_0^L}{x_E^L}\right)^{-1}$$

or

$$f_E^L = \left(1 - f_E\right) = \left(\frac{x_0^L}{x_E^L}\right)^{\frac{1}{1-k_{partE}}} \tag{7.12}$$

where f_E^L = the fraction of melt at the eutectic temperature, i.e. at the start of the eutectic reaction. It is reasonable to assume that the melt, which remains at the eutectic temperature, will solidify with eutectic composition and

structure. Thus the fraction $(1 - f_E)$ in Equation (7.12) also represents the fraction of the solid that has a eutectic composition when the molten alloy has solidified completely.

Example 7.1

As is seen from the phase diagram of the Al–Cu system, the solid Al phase has a maximum solubility of 2.50 at-% Cu at the eutectic temperature. The Cu concentration is 17.3 at-% at the eutectic temperature.

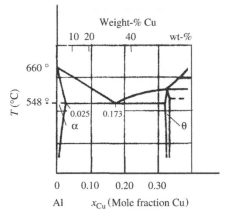

Part of the phase diagram of the system Al–Cu.

According to the phase diagram, a melt having an initial Cu concentration of 2.50 at-% would be able to solidify to a homogeneous Al phase with substituted Cu atoms if microsegregation, leading to an uneven distribution of the Cu atoms within the solid phase, could be disregarded. It cannot be disregarded, however. Take microsegregation into consideration by applying Scheil's equation and calculate the fraction of the material that solidifies with a eutectic composition and structure.

Solution:

It is reasonable to assume that the melt that is left when the temperature has dropped to the eutectic temperature will solidify eutectically. To calculate the desired fraction we only have to calculate the degree of solidification f_E from Scheil's equation when the melt has reached the eutectic composition. For application of Scheil's equation we have to know the partition constant k_{part}. An approximate value of this is:

$$k_{part} = \frac{x^s}{x^L} = \frac{2.50}{17.3} = 0.1445 \tag{1'}$$

Scheil's equation is applied for the eutectic melt:

$$x_E^L = x_0^L (1 - f_E)^{-(1-k_{part})}$$

or

$$0.173 = 0.0250 \, (1 - f_E)^{-(1-0.1445)}$$

which can be reduced to:

$$1 - f_E = 0.1445^{1/0.8555} = 0.1023 \qquad (2')$$

At the eutectic temperature the fraction f_E is solid. The rest, i.e. the fraction $(1 - f_E)$, will solidify with eutectic structure.

Answer:
10 % of the alloy will solidify with a eutectic composition.

7.3.2 Validity of Scheil's Segregation Equation – Convection in Melts. Diffusion in Melts and Solid Metals. The Lever Rule

One of the conditions for the derivation of Scheil's Equation was that the melt at every moment is homogeneous. This assumption is justified by the presence of convection and rapid diffusion in the melt. Both these processes are time dependent and the assumption is not fulfilled if the solidification process is rapid compared with the convection and diffusion in the remaining melt. In order to understand the factors that control microsegregation, we need to examine the magnitudes of convection and diffusion.

Convection
Convection of importance is supposed to occur only in volumes with a thickness exceeding 1 mm. Interdendritic volumes normally have a thickness less than 1 mm. For this reason we will neglect convection in connection with microsegregation here. The influence of interdendritic convection on macrosegregations is discussed in Chapter 11.

Diffusion of Alloying Elements in Metal Melts
In order to examine the influence of diffusion on microsegregation we will use Einstein's relationship for random walk:

$$l = \sqrt{2Dt} \qquad (7.13)$$

where
  l = average diffusion distance of an alloying atom during the time t
  D = diffusion constant (m^2/s)
  t = diffusion time.

Equation (7.13) can be used for an estimation of the concentration distribution of the alloying element within a dendrite. If the total solidification time of the dendrite is chosen as the time t and we make the reasonable assumption that

the diffusion constant is of the magnitude 10^{-10} m^2/s or larger, the average diffusion distance can be calculated. The calculations show that an atom during this time is able to move more than half the dendrite arm distance during the solidification process.

The diffusion constant of an alloying element in a metal melt is normally of the magnitude $10^{-9} - 10^{-8}$ m^2/s or $>10^{-10}$ m^2/s. The diffusion constants of carbon, nitrogen, hydrogen and other interstitially dissolved atoms (foreign atoms between the crystal lattice atoms) have the same magnitude in steel alloys, i.e. both in austenite and ferrite for example. The conclusion is that *no concentration gradients are likely to occur in the melt during the solidification process.*

The diffusion in the *melt* is rapid enough to prevent such gradients. There may be solute concentration differences of alloying elements in the *solid* phase, though. This topic is discussed below.

Diffusion of Alloying Elements in Solid Metals
When we derived Scheils equation we assumed that diffusion in the solid phase could be neglected. For the *interstitially* dissolved elements carbon, nitrogen and hydrogen in both austenite and ferrite in iron-base alloys, no concentration gradients have been found. The diffusion of interstitially dissolved elements in the solid phase of an alloy is *rapid* and gives no concentration gradients.

For *substitutionally* dissolved alloying elements (an alloying atom replaces a metal atom in the crystal lattice) there are concentration gradients in some cases. The conclusion is that for these elements the assumption that the rate of diffusion in the solid phase can be neglected is not always true and diffusion has to be taken into consideration.

In most metals with face-centred cube (FCC) structure (Figure 7.7) the diffusion constant is of the magnitude 10^{-13} m^2/s or less for substitutionally dissolved elements. Examples of metals with FCC-structure are: γ-Fe (austenite), Cu, Al and Pb. In these cases there is probably a certain diffusion in the solid phase at the end of the solidification process. This occurs when the concentration gradient has become large. The assumption that there is no diffusion in the solid phase is consequently comparatively valid in the cases of FCC-metals.

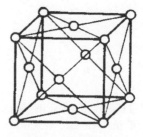

Figure 7.7 Face-centred cube (FCC) structure. There is an atom in the centre of each lateral surface.

The assumption is more uncertain if the solid phase has body-centred cube (BCC) structure (Figure 7.8), as is the case for ferrite in steel and for the β-phase in copper alloys, for example. In ferrite the diffusion constant for alloying metals is about 10^{-11} m²/s at 1400 °C. For sulfur in ferrite the diffusion constant is 10^{-10} m²/s.

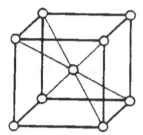

Figure 7.8 Body-centred cube (BCC) structure. There is an atom in the centre of the unit cell.

Examples of metals with BCC-structure are: δ-Fe (ferrite), β-brass, Li and V. It is evident from the values of the diffusion constants given above that there are cases when diffusion is rapid enough to smooth out the differences in composition, i.e. the concentration gradient, in the solid phase. If this occurs before the melt has disappeared completely, there will be an exchange of the alloying element between the melt and the already solidified material. This phenomenon is called *back diffusion*. If back diffusion occurs after solidification has been completed, the phenomenon is called *homogenization*. In the presence of back diffusion, Scheil's Equation (7.10) on page 186 does not describe the microsegregation properly but has to be modified. This is treated on page 197.

The Lever Rule

In the case of very rapid diffusion of the alloying element in the solid phase Scheil's equation can not be used at all. Rapid diffusion of the alloying element results in an even distribution of the alloying element in each phase, i.e. even compositions in both the solid phase and in the melt.

Consider a mass element $\mathrm{d}m$. The fraction f of the mass element has solidified and the rest, fraction $(1 - f)$, is molten. Instead of an even concentration x_0^{L} of the alloying element the concentration is x^{L} everywhere in the melt and x^{s} everywhere in the solid phase (Figure 7.9). A material balance of the alloying element will be:

$$f\,\mathrm{d}m(x_0^{\mathrm{L}} - x^{\mathrm{s}}) = (1 - f)\,\mathrm{d}m(x^{\mathrm{L}} - x_0^{\mathrm{L}}) \qquad (7.14)$$

amount of the alloying	amount of the alloying
element removed	element transferred to
from the solid phase	to the melt

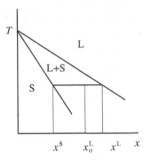

Figure 7.9 Phase diagram of a binary alloy.

which gives, using Equation (7.4) on page 185:

$$x^{\mathrm{L}} = \frac{x^{\mathrm{s}}}{k_{\mathrm{part}}} = \frac{x_0^{\mathrm{L}}}{1 - f(1 - k_{\mathrm{part}})} \qquad \text{Lever rule} \qquad (7.15)$$

where

x_0^{L} = initial concentration of the alloying element

f = fraction solid phase in the mass element

k_{part} = partition coefficient $(x^{\mathrm{s}}/x^{\mathrm{L}})$ of the alloying element.

Equation (7.15) is called the *lever rule*. It is valid during rapid diffusion of alloying elements both in the solid phase and in the solidifying melt. Most practical cases are likely to be a hybrid between this extreme (rapid diffusion in the solid phase) and the extreme represented by Scheil's equation (no diffusion at all in the solid phase).

7.4 SOLIDIFICATION PROCESSES IN ALLOYS

Scheil's segregation equation [Equation (7.10) on page 186] is independent of the cooling rate of the melt. As was pointed out on page 186, *time* has not been involved in the treatment of microsegregation so far. For a more accurate treatment *time* and *diffusion rates* in the solid phase must be taken into consideration. The coarseness of the alloy structure is related to the cooling rate, and this determines the time available for diffusion. Thus the coarseness of the structure is indirectly involved.

Before we can discuss back diffusion and modify Scheil's model of microsegregation, it is necessary to analyse the solidification process for alloys with wide solidification intervals.

7.4.1 Solidification Intervals of Alloys

In Chapters 4 and 5 we treated the solidification process in an alloy by solving the general thermal conduction equation under the assumption that the solidification front was planar. However, in most alloys solidification occurs over a certain temperature range, the so-called *solidification interval* (Figure 7.10). The solidification front is described not by one but by *two* planes enclosing a two-phase region that contains both melt and solid phases.

Figure 7.10 Solidification interval T_L–T_s.

In those cases where a well-defined two-phase region exists, it is limited by two solidification fronts given by the solidus temperature T_s, and the liquidus temperature T_L. It is possible to calculate (i) the width of the two-phase region, expressed as a temperature difference or a distance, and (ii) the solidification time.

When a well-defined two-phase region is missing, for example in a melt with equiaxed crystals, only the solidification time can be calculated. These calculations involve the *rate of diffusion in the solid phase*, and the *cooling rate of the melt*.

The numerical calculations are normally performed using computers but this procedure gives little information of the real process. For the sake of better understanding, we will discuss two alternative calculation methods below. One of them is based on Scheil's equation and the assumptions associated with this equation. The other one uses the lever rule as a basic equation and the conditions valid for this case.

7.4.2 Calculation of Widths of Solidification Intervals and Solidification Times of Alloys

Figure 7.11 describes alloys with a solidification interval and the outlines of the temperature distribution, concentration distribution, and fraction of solid phase close to and around the solidification front. If the temperature distribution and the phase diagram are known, then the concentration distribution and the fraction of solid phase can be calculated.

In order to calculate the temperature distribution within the solidification interval we have to solve the general thermal conduction equation by analogy with calculations in Chapter 4. With the same designations as in Chapter 4 the equation, valid within the solidification interval, can be written as:

$$\left(\rho c_p + \rho \left(-\Delta H \right) \frac{df}{dT} \right) \frac{\partial T}{\partial t} = k \, \frac{\partial^2 T}{\partial y^2} \qquad (7.16)$$

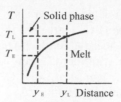

Figure 7.11 (a) Temperature distribution in and around the solidification front as a function of the distance from the cooling surface for the alloy in Figure 7.11 (d). Reproduced with permission from Merton C. Flemings.

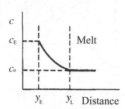

Figure 7.11 (b) Concentration distribution in and around the solidification front as a function of the distance from the cooling surface for the alloy in Figure 7.11 (d). Reproduced with permission from Merton C. Flemings.

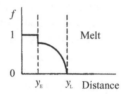

Figure 7.11 (c) Fraction of solid phase f in and around the solidification front as a function of the distance from the cooling surface for the alloy in Figure 7.11 (d). Reproduced with permission from Merton C. Flemings.

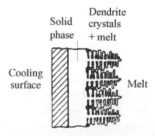

Figure 7.11 (d) Solidifying alloy with a solidification interval. The region with dendrite crystals also contains melt and is a two-phase region. Reproduced with permission from Merton C. Flemings.

where $k =$ thermal conductivity of the solid metal. It should be noted that the thermal conductivity k must be clearly distinguished from the partition coefficient k_{part}.

The key to the solutions in the two cases is the quantity df/dT in the heat Equation (7.16). Below we concentrate on finding expressions for this term in two different cases.

Calculation of $\frac{df}{dT}$ as a Function of T

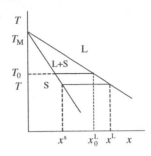

The figure gives a relationship between the coordinates x^L and T:

$$\frac{x^L}{T - T_M} = \frac{1}{m_L} \qquad (m_L < 0) \qquad (1')$$

where T_M is the melting point of the pure metal. T_0 and x_0^L are constants while x^L and T are variables that change along the liquidus line. T_0 is the temperature where the alloy with the initial composition x_0^L starts to solidify.

Equation $(1')$ is applied to the initial concentration x_0^L of the alloying element and to x^L:

$$x_0^L = \frac{1}{m_L}(T_0 - T_M) \qquad (2')$$

$$x^L = \frac{1}{m_L}(T - T_M) \qquad (3')$$

These expressions for x^L and x_0^L are inserted into Equation (7.10):

$$T_M - T = (T_M - T_0)(1 - f)^{-(1 - k_{part})} \qquad (4')$$

The equation can be transformed into:

$$(1 - f) = \left(\frac{T_M - T}{T_M - T_0}\right)^{\frac{-1}{1 - k_{part}}} \qquad (5')$$

The fraction of solid f as a function of temperature for the Al–2.5 at-% Cu alloy in Example 7.1 (page 187) is given in Figure 7.12 (a).

The easiest way to find df/dT is to take the logarithm of Equation $(5')$ and differentiat the new equation with respect to T:

$$\ln(1 - f) = \frac{-1}{1 - k_{part}}\left[\ln(T_M - T) - \ln(T_M - T_0)\right] \qquad (6')$$

After reduction, the derived equation can be written as:

$$\frac{df}{dT} = \frac{-(1 - f)}{(1 - k_{part})(T_M - T)} \qquad (7')$$

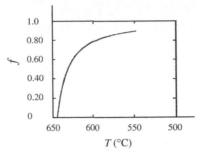

Figure 7.12 (a) Fraction of solid as a function of temperature when Scheil's equation is valid.

These expressions are then introduced into Equation (7.16), which gives the condition for the solution in the two cases. To keep the main thread as clear as possible parts of the calculations below have been enclosed in boxes.

Alternative I: *Scheil's equation is valid.*
When Scheil's equation (Equation 7.10) is valid, it can be used for calculation of df/dT. x^L has to be expressed in terms of temperature. This can easily be done by using that phase diagram where the liquidus line of the alloy has a constant slope m_L. The derivation is given in the box.

The factor $(1 - f)$ in Equation $(5')$ is inserted into Equation $(7')$ to give the desired relation in alternative I:

$$\frac{df}{dT} = \frac{-1}{1 - k_{part}}(T_M - T)^{-\left(\frac{2 - k_{part}}{1 - k_{part}}\right)}(T_M - T_0)^{\frac{1}{1 - k_{part}}} \qquad (7.17)$$

The function (7.17) is illustrated in Figure 7.12 (b) for the Al–2.5 at-% Cu alloy in Example 7.1 on pages 187–188.

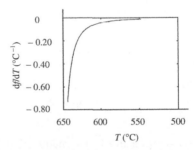

Figure 7.12 (b) Fraction of solid per degree df/dT as a function of temperature when Scheil's equation is valid.

It can be seen from the figure that the fraction formation per degree df/dT is largest at the beginning of the solidification process. Note that the function (7.17) is only valid down to the eutectic temperature. At this temperature a 10 % eutectic structure is precipitated, in accordance with Example 7.1.

Alternative II: *The lever rule is valid.*

When the lever rule [Equation (7.15) on page 189] is valid, it can be used for calculation of df/dT.

$$x^{L} = \frac{x^{s}}{k_{part}} = \frac{x_{0}^{L}}{1 - f(1 - k)} \qquad (7.15)$$

x^{L} must be expressed in terms of temperature. The derivation of df/dT is given in the box below.

Calculation of $\frac{df}{dT}$ as a Function of T

If the general expression in Equations (2′) and (3′) in the box on page 191:

$$x_{0}^{L} = \frac{T_{0} - T_{M}}{m_{L}} \qquad (1')$$

and

$$x^{L} = \frac{T - T_{M}}{m_{L}} \qquad (2')$$

is introduced into the lever rule [Equation (7.15)] we get:

$$\frac{T_{M} - T}{m_{L}} = \frac{T_{M} - T_{0}}{m_{L} \left[1 - f(1 - k_{part})\right]}$$

which can be written as:

$$1 - f\left(1 - k_{part}\right) = \frac{T_{M} - T_{0}}{T_{M} - T} \qquad (3')$$

or

$$f = \frac{1}{1 - k_{part}} - \frac{T_{M} - T_{0}}{\left(1 - k_{part}\right)\left(T_{M} - T\right)} \qquad (4')$$

The function (4′) is illustrated in Figure 7.13 (a) for an iron-base alloy with 0.70 wt-% carbon. The same alloy occurs in Example 7.3 on page 194.

Differentiation of Equation (4′) with respect to the variable temperature T gives Equation (7.18) below.

The desired relationship in alternative II will therefore be:

$$\frac{df}{dT} = \frac{-(T_{M} - T_{0})}{\left(1 - k_{part}\right)\left(T_{M} - T\right)^{2}} \qquad (7.18)$$

The function (7.18) is illustrated by Figure 7.13 (b) for an iron-base alloy with 0.70 wt-% carbon as in Example 7.3 on pages 194–197. Figure 7.13 (b) shows that the solid formation

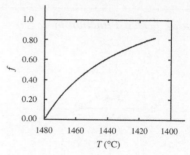

Figure 7.13 (a) Fraction f of solid as a function of temperature when the lever rule is valid.

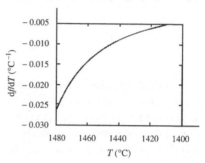

Figure 7.13 (b) Fraction of solid formed per degree df/dT as a function of temperature when the lever rule is valid.

per degree is greatest at the beginning of the solidification process (small values of f).

When the function df/dT is known, it is introduced into the general heat equation [Equation (7.16)]. The differential equation has to be solved using boundary conditions. It is convenient to use computer programs for the numerical calculations.

In the examples below we calculate the solidification interval of one alloy and derive the solidification time for another one. We will use a simpler method of calculation than the general heat equation by applying the analytical expressions that were derived in Chapter 4, in combination with the phase diagram of the alloy.

Calculation of the Solidification Interval of an Alloy with a Wide Solidification Interval

Example 7.2 below illustrates the method of calculating the width of the solidification interval, i.e. the width of the two-phase region, when an alloy solidifies.

Example 7.2

Calculate the solidification interval, i.e. the width of the two-phase region, as a function of the heat transfer coefficient in an Al alloy containing 2.5 at-% Cu, which solidifies in contact with a cooled Cu plate. Perform the calculation for the three different cases, namely when the eutectic solidification front has advanced to the depths 10 mm, 50 mm and 100 mm respectively, into the cast metal. The temperature profile is shown in the figure.

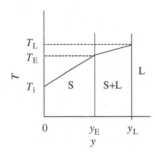

From Example 7.1 on page 187 it is known that up to a fraction of $f_E = 0.90$ solidifies with a variable composition, due to microsegregation, and the remainder $(1 - f_E = 0.10)$ solidifies with eutectic composition. The thermal conductivity of the alloy is 202 W/m K. The phase diagram of Al–Cu is given on page 187. The temperature of the surroundings is 20 °C.

Solution:

During the solidification process, solidification heat is created simultaneously over the whole two-phase region [Figure 7.11 (d) on page 190 and Figure 4.10 on page 66]. In order to simplify the calculations we assume that:

- when an arbitrary volume element has solidified completely, the fraction f_E has solidified at the dendrite tips at temperature T_L and the remainder, fraction $(1 - f_E)$, has solidified at the eutectic solidification front at temperature T_E, and
- the heat of solidification of the alloy is the same at the two solidification fronts: $(-\Delta H_E) = (-\Delta H_{dendrite}) = (-\Delta H)$.

Starting with these assumptions we will set up two heat transport equations, one for the dendrite tips and the other one for the eutectic front.

For the dendritic solidification the heat flux can be written as:

$$\frac{dq_{den}}{dt} = -(1 - f_E)\rho(-\Delta H)\frac{dy_L}{dt} = -k\frac{T_L - T_E}{y_L - y_E} \quad (1')$$

Dendrite solidification   Heat flux through
heat flux   the two-phase region

The total heat flux dq_E/dt through the eutectic solidification front is the sum of the dendrite and eutectic solidification

fluxes. This can be written as:

$$-k\frac{T_L - T_E}{y_L - y_E} - f_E\,\rho(-\Delta H)\frac{dy_E}{dt} = -h\,(T_i - T_0) \quad (2')$$

Dendrite   Eutectic   Total heat flux
heat flux   heat flux   across the interface
                            solid alloy/mould

where

$h =$ heat transfer coefficient between the Cu plate and the solid phase of the alloy
$k =$ thermal conductivity of the two-phase region
$y_E, y_L =$ coordinates of the two solidification fronts
$T_E =$ eutectic temperature of the alloy
$T_L =$ liquidus temperature of the alloy
$T_0 =$ temperature of the surroundings
$f_E =$ fraction of solid phase at the eutectic temperature
$\rho =$ density of the alloy.

The positions y_E and y_L of the solidification fronts are both functions of time. The thermal capacity in the solid alloy is small in comparison with the solidification heat $(-\Delta H)$. Thus it is reasonable to neglect the cooling heat as compared with the solidification heat. In this case Equations (4.45) and (4.46) on page 74 are valid at the eutectic solidification front:

$$T_i = \frac{T_E - T_0}{1 + \dfrac{h}{k}y_E} + T_0 \quad (3')$$

$$\frac{dy_E}{dt} = \frac{T_E - T_0}{\rho(-\Delta H)}\frac{h}{1 + \left(\dfrac{h}{k}\right)y_E} \quad (4')$$

The expression of dy_E/dt [Equation (4')] is introduced into Equation (2') on the left-hand side. The expression of T_i [Equation (3')] is introduced into Equation (2') on the right-hand side to give:

$$k\left(\frac{T_L - T_E}{y_L - y_E}\right) + f_E\rho(-\Delta H)\left(\frac{T_E - T_0}{(-\Delta H)}\right)\frac{h}{1 + \dfrac{h}{k}y_E}$$

$$= h\frac{T_E - T_0}{1 + \dfrac{h}{k}y_E} \quad (5')$$

The width of the two-phase region in the Al–Cu alloy is $y_L - y_E$. We can solve $(y_L - y_E)$ from Equation (5') to get after reduction:

$$y_L - y_E = \frac{k + hy_E}{h(1 - f_E)}\frac{T_L - T_E}{T_E - T_0} \quad (6')$$

We want $(y_L - y_E)$ as a function of h. All quantities on the right-hand side of Equation (6') are known except h, which is the independent variable. By inserting the values listed below and choosing convenient values of h $(200 \text{ W/m}^2 \text{ K} < h < 2000 \text{ W/m}^2 \text{ K})$ we can calculate the corresponding values of $(y_L - y_E)$ and plot the desired function graphically.

$y_E = 0.010$ m, 0.050 m, 0.100 m (text)
$f_E = 0.90$ (text)
$k = 202$ W/m K (text)
$T_0 = 20\,°C$ (text)
$T_L = 660\,°C$ (phase diagram of Al–Cu on page 187)
$T_E = 548\,°C$ (phase diagram of Al–Cu on page 187).

Answer:
The solidification interval is $T_L - T_E = 112\,°C$.
The width of the two-phase region $= y_L - y_E$
The desired function is $y_L - y_E = \dfrac{k + hy_E}{h(1 - f_E)} \dfrac{T_L - T_E}{T_E - T_0}$
The function is plotted in the figure below for the three values of y_L as given in the text.

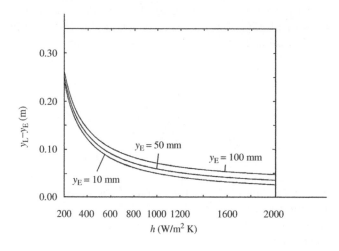

The figure in the answer in Example 7.2 illustrates that the widths of the two-phase region decrease very rapidly with increasing cooling rate, which corresponds to an increase in h. At low cooling rates the material is isothermal with a two-phase region that covers the whole casting. This situation is discussed below.

Calculation of the Solidification Time of Castings Made of Alloys with a Wide Solidification Interval

The solidification time of a casting can be derived from a temperature–time curve like that in Figure 7.1 on page 184 in this chapter. The solidification time for a casting that solidifies in a sand mould, for example, can be read directly from the curve.

A second method is to use Chvorinov's rule to calculate the solidification time for castings cast in sand moulds (e.g. in Example 4.4 on page 81).

A third method for thin castings is mentioned in Section 4.4.5 in Chapter 4. Thin castings are isothermal during the solidification and cooling processes. This is also valid for thin castings made of alloys having wide solidification intervals. The solidification time can be calculated using a simple formula [Equation (4.85) on page 86].

Solidification Time of a Casting Made of an Alloy with a Wide Solidification Interval

A general method for calculating solidification times is to solve Equation (7.2) on page 184 in this chapter. In Chapter 4 this equation was solved under the assumption that the last term on the right-hand side could be neglected.

In alloys with a large solidification interval we can instead neglect the middle term on the right-hand side of Equation (7.2). The physical significance of this assumption is that we can assume that the dendrite network is formed instantly as very thin arms and that the whole solidification process occurs by secondary solidification on these arms. In reality, we made the same assumption when we derived Scheil's equation.

If we neglect the second term on the right-hand side of Equation (7.2) and combine it with Equation (7.3) on page 184 we get the following differential equation:

$$\frac{dQ}{dt} = V_0 \rho_{\text{metal}} \left[c_p^{\text{metal}} \frac{dT}{dt} + (-\Delta H) \frac{df}{dT} \right] \qquad (7.19)$$

where $(-\Delta H)$ is the solidification heat of the metal (J/kg). By integrating Equation (7.19) the solidification time can be calculated. A concrete example of such a calculation is given in Example 7.3 below.

Example 7.3
An iron-base alloy containing 0.70 wt-% C is cast in a sand mould to a plate with a thickness of 20 mm at a temperature of 1580 °C.

(a) Calculate the cooling time before solidification and the solidification time of the casting.

(b) Calculate and plot the temperature–time curve for the corresponding cooling and solidification processes.

The material constants are found in general reference tables. The phase diagram of the system Fe–C is given below.

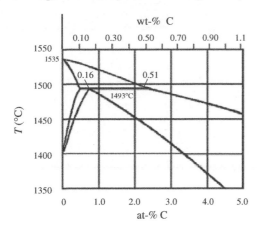

Solution:

In order to plot the temperature–time curve we have to find two functions:

(1) The T–t relationship during cooling from $1580\,^\circ$C down to the liquidus temperature T_L of the alloy when solidification starts.

(2) The T–t relationship during solidification when the temperature decreases from T_L down to the solidus temperature T_s of the alloy.

Cooling Curve before Solidification

The heat flux from a solidifying alloy in a sand mould can be written [see Equation (4.70) on page 79]:

$$\frac{dq}{dt} = \sqrt{\frac{k_{mould}\;\rho_{mould}\;c_p^{mould}}{\pi t}}(T_i - T_0) \qquad (1')$$

and also:

$$\frac{dq}{dt} = \rho_{melt}\;c_p^{melt}\;l\left(-\frac{dT}{dt}\right) \qquad (2')$$

where l is the thickness of the plate. The surface temperature T_i is assumed to be constant and equal to the liquidus temperature T_L [Equation (4.60) on page 78 in Chapter 4]. Equation $(1')$ is valid only if $T_i = T_L$.

The reason for the minus sign in Equation $(2')$ is that a *positive* heat flux causes a temperature *decrease*. The two expressions for the heat flux are equal; this gives a third equation:

$$\int_0^t \frac{dt}{\sqrt{t}} = \int_{1580}^{T} \rho_{melt}\;c_p^{melt}\;l\sqrt{\frac{\pi}{k_{mould}\;\rho_{mould}\;c_p^{mould}}} \times \frac{-dT}{(T_L - T_0)} \qquad (3')$$

which is integrated to give:

$$\sqrt{t} = \frac{\rho_{melt}\;c_p^{melt}\;l}{2}\sqrt{\frac{\pi}{k_{mould}\;\rho_{mould}\;c_p^{mould}}}\left(\frac{1580 - T}{T_L - T_0}\right)$$

This can be transformed into:

$$T = 1580 - \frac{2\sqrt{k_{mould}\;\rho_{mould}\;c_p^{mould}}\;(T_L - T_0)}{\rho_{melt}\;c_p^{melt}\;l\sqrt{\pi}} \times \sqrt{t} \quad (4')$$

Equation $(4')$ is the relationship that represents the temperature–time curve before solidification starts. The cooling interval starts at $1580\,^\circ$C and finishes when $T = T_L$.

Calculations:

The calculations below have been performed using the following material constants:

$k_{mould} = 0.60\,\text{W/mK}$	$l = 0.020\,\text{m}$
	$\rho_{Fe} = 7.3 \times 10^3\,\text{kg/m}^3$
$\rho_{mould} = 1.5 \times 10^3\,\text{kg/m}^3$	$T_0 = 20\,^\circ$C
	$c_p^{Fe} = 6.70 \times 10^2\,\text{J/kg K}$
$c_p^{mould} = 1.13 \times 10^3\,\text{J/kg K}$	$T_L = 1480\,^\circ$C
	(phase diagram Fe-C)

The resulting function is:

$$T = 1580 - 17t^{0.5} \qquad (5')$$

t (s)	$T = 1580 - 17\,t^{0.5}$ ($^\circ$C)
0	1580
5	1542
10	1526
15	1514
20	1508
25	1495
30	1487
35	1480

The calculated values are listed in the table.
$T = T_L = 1480\,^\circ$C is inserted in Equation $(5')$, which gives:

$$t_{cool} = 34.6\,\text{s} \approx 35\,\text{s}.$$

Solidification Curve

Equation (7.19) on page 194 should be applied during the solidification process instead of Equation $(2')$.

Equation (7.19) can be transformed into the right-hand side of Equation (6′) by using Equation (1′) and by introducing the volume $V_0 = Al$ and the heat per unit area $q = Q/A$.

$$\frac{dq}{dt} = \sqrt{\frac{k_{\text{mould}}\ \rho_{\text{mould}}\ c_p^{\text{mould}}}{\pi t}}(T_i - T_0)$$

$$= l\rho_{\text{metal}}\left(-\frac{dT}{dt}\right)\left[c_p^{\text{metal}} + (-\Delta H)\frac{df}{dT}\right] \quad (6')$$

Carbon is a comparatively small atom compared with Fe and the carbon diffusion in the solid alloy has to be taken into consideration. We assume that *the lever rule* is valid, and in this case Equation (7.18) on page 192 can be applied.

$$\frac{df}{dT} = \frac{-(T_M - T_L)}{(1 - k_{\text{part}})(T_M - T)^2} \quad (7')$$

During casting in a sand mould the surface temperature T_i is equal to T_L (Equation (4.60) on page 78 in Chapter 4). If the expression (7′) is introduced into Equation (6′) we get, after separation of the variables t and T and integrating:

$$\left(\sqrt{\frac{k_{\text{mould}}\ \rho_{\text{mould}}\ c_p^{\text{mould}}}{\pi}}\right)\int_{t_{\text{cool}}}^{t}\frac{dt}{\sqrt{t}}$$

$$= -\int_{T_L}^{T}\frac{l\rho_{\text{metal}}\left[c_p^{\text{metal}} + (-\Delta H)\dfrac{-(T_M - T_L)}{(1 - k_{\text{part}})(T_M - T)^2}\right]}{(T_L - T_0)}dT \quad (8')$$

where

T_M = melting point of pure metal (iron)
T_L = initial solidification temperature
T_s = final solidification temperature
T_0 = temperature of the surroundings.

Equation (8′) is valid over the interval $T_L \geq T \geq T_s$. We separate the equation into three integrals I_1, I_2 and I_3.

After interpration they are:

$$I_1 = \sqrt{\frac{k_{\text{mould}}\ \rho_{\text{mould}}\ c_p^{\text{mould}}}{\pi}}(2\sqrt{t} - 2\sqrt{t_{\text{cool}}}) \quad (9')$$

$$I_2 = \frac{l\rho_{\text{metal}}\ c_p^{\text{metal}}}{(T_L - T_0)}(T_L - T) \quad (10')$$

$$I_3 = \left[-\frac{l\rho_{\text{metal}}(-\Delta H)}{(1 - k_{\text{part}})}\right]\frac{-(T_M - T_L)}{(T_L - T_0)}\frac{-1}{(T_M - T)}\Big|_{T_L}^{T} \quad (11')$$

or

$$I_3 = \left[\frac{l\rho_{\text{metal}}(-\Delta H)}{(1 - k_{\text{part}})}\right]\frac{(T_M - T_L)}{(T_L - T_0)}\left[\frac{1}{(T_M - T_L)} - \frac{1}{(T_M - T)}\right]$$

or

$$I_3 = \frac{l\rho_{\text{metal}}(-\Delta H)}{(1 - k_{\text{part}})}\frac{T_L - T}{(T_L - T_0)(T_M - T)} \quad (12')$$

We combine the three integrals and solve $\sqrt{t}$:

$$\left(2\sqrt{\frac{k_{\text{mould}}\ \rho_{\text{mould}}\ c_p^{\text{mould}}}{\pi}}\right)(\sqrt{t} - \sqrt{t_{\text{cool}}})$$

$$= \frac{l\rho_{\text{metal}}\ c_p^{\text{metal}}}{(T_L - T_0)}(T_L - T) + \frac{l\rho_{\text{metal}}(-\Delta H)}{(T_L - T_0)}\frac{T_L - T}{(1 - k_{\text{part}})(T_M - T)}$$

or

$$\sqrt{t} = \sqrt{t_{\text{cool}}} + \left(\frac{\sqrt{\pi}\ l\rho_{\text{metal}}}{2\sqrt{k_{\text{mould}}\ \rho_{\text{mould}}\ c_p^{\text{mould}}}}\right)\frac{(T_L - T)}{(T_L - T_0)}\left[c_p^{\text{metal}} + \frac{-\Delta H}{1 - k_{\text{part}}}\frac{1}{(T_M - T)}\right] \quad (13')$$

Equation (13′) is the relationship that represents the temperature–time curve during the solidification process. The solidification period starts at $T = T_L$ and is finished when $T = T_s$.

Calculations

In addition to the material constants used for calculation of the cooling curve we will use $t_{\text{cool}} = 35s$ (calculated above),

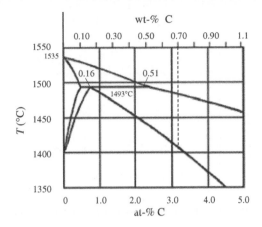

$-\Delta H = 272 \times 10^3$ J/kg and $T_s \approx 1410\,^\circ$C (read from the phase diagram of Fe–C for $c_0 = 0.70$ wt %). In order to calculate the partition coefficient k_{part} we use, for example, the peritectic line $T = 1493\,^\circ$C (Section 7.7 on page 207–208) in the phase diagram above and read the corresponding weight percentages for the intersections between this line and the solidus and liquidus lines:

$$k_{part} = \frac{c^s}{c^L} = \frac{0.16}{0.51} = 0.314$$

Equation (13′) is an equation of second degree in T but it is not worthwhile solving T as a function of t. Instead we insert the calculated values and the other known quantities into Equation (13′) to give:

$$\sqrt{t} = \sqrt{t_{cool}} + \frac{\sqrt{\pi}\, l\rho_{metal}}{2\sqrt{k_{mould}\, \rho_{mould}\, c_p^{mould}}} \frac{T_L - T}{T_L - T_0}$$

$$\left[c_p^{metal} + \left(\frac{-\Delta H}{l - k_{part}} \right) \frac{l}{(T_M - T)} \right]$$

and get

$$\sqrt{t} = \sqrt{34.6} + \frac{\sqrt{\pi} \times 0.020 \times (7.3 \times 10^3)}{2\sqrt{0.60(1.5 \times 10^3)(1.13 \times 10^3)}} \frac{1480 - T}{1480 - 20}$$

$$\left[6.7 \times 10^2 + \left(\frac{272 \times 10^3}{1 - 0.314} \right) \frac{1}{(1535 - T)} \right]$$

or

$$t \approx \left[5.88 + (8.86 \times 10^{-5})(1480 - T) \right.$$

$$\left. \left(6.7 \times 10^2 + \frac{3.965 \times 10^5}{(1535 - T)} \right) \right]^2 \qquad (14')$$

Equation (14′) is the basis for plotting the temperature–time curve during solidification time. A number of given T-values can be inserted into Equation (14′) and the corresponding t-values are calculated. $T = T_s = 1410\,^\circ$C gives $t = 890\,\text{s} = 14.8\,\text{min}$.

By plotting these pairs of values in a diagram, the temperature–time curve can be drawn. The calculated cooling and solidification curves are shown in the figure below.

Answer:

(a) The cooling process before solidification requires 35 s.

(b) The cooling and solidification time is $\approx$15 min.

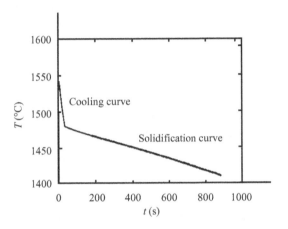

7.5 INFLUENCE OF BACK DIFFUSION IN THE SOLID PHASE ON MICROSEGREGATION OF ALLOYS

When we derived Scheil's equation and the lever rule, we did not consider the geometrical effects on the distribution of alloying elements at solidification. As is seen from Figure 6.4 on page 143, the dendrite solidification structure is very complex. Primary and secondary arms grow simultaneously, and during the growth process some of the secondary arms grow faster than others. The concentration distribution of the alloying elements in the solid alloy will thus contain a three-dimensional pattern of numerous concentration maxima and minima.

For this reason it is difficult to specify a particular diffusion distance and calculate the back diffusion. Below we discuss a method of handling this problem and derive an improved and successful model for microsegregation that includes the effects of back diffusion. It was primarily derived by Flemings and is frequently used in metallurgical literature.

7.5.1 Scheil's Modified Segregation Equation

We consider the same volume element as in Section 7.3.1 (Figure 7.4 on page 186) when we derived Scheil's equation. If back diffusion is taken into consideration the basic material balance Equation (7.6) on page 186 has to be completed on the right-hand side by a term that represents the amount of alloying element transported from the melt back into the solid phase.

$$\text{Amount of alloying element} = \int_0^y \Delta x^s dy \qquad (7.20)$$

If this term is incorporated into Equation (7.6) (page 186) and we assume that the molar volumes of the solid phase and the melt are equal we get, after reduction:

$$\left(\frac{\lambda}{2} - y\right)dx^L = (x^L - x^s)dy - \int_0^y \Delta x^s dy \qquad (7.21)$$

Increase of alloying element in the melt	Addition of alloying element from the solidfied volume	Return of alloying element from melt to solid (back diffusion)

Equation (7.21) should be compared with Equation (7.6). The shaded area in Figure 7.14 represents the last term on the right-hand side in Equation (7.21), i.e. the effect of back diffusion.

We have to find an expression for the concentration gradient in the solid phase close to the interface. This, in principle, requires application of Fick's second law to the whole solid phase and calculation of its concentration distribution. This is a complicated procedure and has not been done analytically so far. An approximate calculation can be performed if we assume that:

$$dx^s = k_{part}dx^L \qquad (7.22)$$

In addition, we set up a heat balance equation for the solidifying volume element. The amount of heat emitted from the volume element can be written as:

$$\frac{dQ}{dt} = A\rho(-\Delta H)\frac{dy}{dt} + A\rho c_p\left(-\frac{dT}{dt}\right)\left(\frac{\lambda_{den}}{2}\right) \qquad (7.23)$$

where $(-dT/dt)$ is the cooling rate during the solidification process. The first term in Equation (7.23) represents the heat of fusion and the second one is the amount of heat associated with the temperature decrease. The last term is

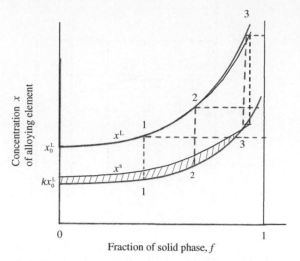

Figure 7.14 The concentrations of the alloying element in the melt and in the last solidified material during a solidification process for $k = 0.5$ as a function of the fraction solid with and without considering back diffusion. The marked area represents the effect of back diffusion.

often small compared with the former. Thus the second term on the right-hand side in Equation (7.23) can be neglected.

We further assume that the amount of heat that is removed per unit time (dQ/dt) is constant. As is seen from Equation (7.23), dy/dt also becomes constant if we make these assumptions. We can replace dy/dt by $\lambda_{den}/2\theta$, where λ_{den} is the dendrite arm distance and θ is the *total solidification time*. This value is inserted into Equation (7.21) to give:

$$D_s \frac{2\theta}{\lambda_{den}}k_{part}dx^L + \left(\frac{\lambda_{den}}{2} - y\right)dx^L = (x^L - x^s)dy \qquad (7.24)$$

where

D_s is the diffusion constant of the alloying element in the solid (see page 188).

Integrating and using the relationship $x^s = k_{part}x^L$ gives:

$$\int_{x_0^L}^{x^L} \frac{dx^L}{x^L(1 - k_{part})} = \int_0^y \frac{dy}{D_s \dfrac{2\theta}{\lambda_{den}}k_{part} + \dfrac{\lambda_{den}}{2} - y}$$

$$\frac{1}{1 - k_{part}} \ln \frac{x^L}{x_0^L} = -\ln \frac{D_s \dfrac{2\theta}{\lambda_{den}}k_{part} + \dfrac{\lambda_{den}}{2} - y}{D_s \dfrac{2\theta}{\lambda_{den}}k_{part} + \dfrac{\lambda_{den}}{2}}$$

which can be reduced to:

$$\left(\frac{x^L}{x_0^L}\right)^{\frac{-1}{1 - k_{part}}} = \frac{D_s \dfrac{2\theta}{\lambda_{den}}k_{part} + \dfrac{\lambda_{den}}{2} - y}{D_s \dfrac{2\theta}{\lambda_{den}}k_{part} + \dfrac{\lambda_{den}}{2}}$$

or

$$x^{L} = x_0^{L} \left(\frac{D_s \dfrac{2\theta}{\lambda_{\text{den}}} k_{\text{part}} + \dfrac{\lambda_{\text{den}}}{2} - y}{D_s \dfrac{2\theta}{\lambda_{\text{den}}} k_{\text{part}} + \dfrac{\lambda_{\text{den}}}{2}} \right)^{-(1-k_{\text{part}})}$$

which can be transformed into:

$$x^{L} = x_0^{L} \left(1 - \frac{\dfrac{2y}{\lambda_{\text{den}}}}{1 + D_s \dfrac{4\theta}{\lambda_{\text{den}}^2} k_{\text{part}}} \right)^{-(1-k_{\text{part}})}$$

$$= x_0^{L} \left(1 - \frac{f}{1 + D_s \dfrac{4\theta}{\lambda_{\text{den}}^2} k_{\text{part}}} \right)^{-(1-k_{\text{part}})} \qquad (7.25)$$

Equation (7.25) is a modified form of Scheil's segregation equation. A comparison between Equations (7.25) and (7.10) shows that they differ by a correction term B in the denominator:

$$B = \frac{4D_s \theta k_{\text{part}}}{\lambda_{\text{den}}^2} \qquad (7.26)$$

This correction term is caused by the back diffusion. The correction term is unimportant as long as it is $\ll 1$, but it can be seen from Equation (7.25) that back diffusion becomes more and more important at the end of the solidification process, when f approaches the value 1.

To carry out more detailed calculations, we require knowledge of the value of the parameter B, which contains the solidification time, directly and indirectly, because the dendrite arm distance depends on the cooling rate. The parameter B contains the ratio $4\theta/\lambda_{\text{den}}^2$ and the dendrite arm distance decreases with increasing cooling rate. The total solidification time θ also decreases with increasing cooling rate. Thus the factors λ_{den} and θ counteract each other.

Example 7.4

The figure shows the dendrite arm distance as a function of composition of the alloy when the cooling rate is known for a number of Al–Cu alloys. Use this information to decide whether the importance of back diffusion in Al–Cu alloys increases or decreases with the cooling rate.

Solution:

It is reasonable to assume that the heat flux is constant during cooling and solidification, which gives:

$$c_p^{L} \left(-\frac{dT}{dt} \right) = (-\Delta H) \frac{df}{dt} \qquad (1')$$

The total solidification time corresponds to $f = 1$. Hence $df/dt = (1 - 0)/(\theta - 0) = 1/\theta$. We insert this value into Equation (1') and solve θ.

$$\theta = \frac{-\Delta H}{c_p^{L} \left(-\dfrac{dT}{dt} \right)} \qquad (2')$$

Next we compare the correction terms required to account for the back diffusion at two different cooling rates:

$$B_1 = \frac{4D_s \theta_1 k_{\text{part}}}{\lambda_1^2} = \frac{4D_s k_{\text{part}}(-\Delta H)}{\lambda_1^2 c_p^{L} \left(-\dfrac{dT_1}{dt} \right)} \qquad (3')$$

and

$$B_2 = \frac{4D_s \theta_2 k_{\text{part}}}{\lambda_2^2} = \frac{4D_s k_{\text{part}}(-\Delta H)}{\lambda_2^2 c_p^{L} \left(-\dfrac{dT_2}{dt} \right)} \qquad (4')$$

Equations (3') and (4') are divided by each other to give:

$$\frac{B_1}{B_2} = \left(-\frac{dT_2}{dt} \right) \lambda_2^2 \Big/ \left(-\frac{dT_1}{dt} \right) \lambda_1^2 , \qquad (5')$$

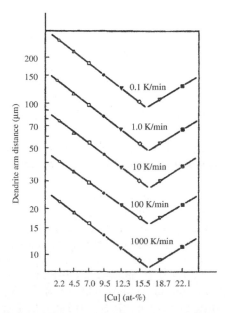

The dendrite arm distance as a function of the concentration of the alloying element and the cooling rate (parameter) for a number of Al–Cu alloys. A eutectic reaction occurs at 17.3 at-% Cu. At higher Cu-concentrations than 17.3 at-%, θ-phase is precipitated, which results in the curves given in the figure.

The figure above shows that the faster the cooling rate is, the smaller will the dendrite arm distance be. Quantitatively it can be seen that an increase in the cooling rate of a factor of 104 will lead to a decrease of the dendrite arm distance by a factor of about 12, regardless of the composition of the alloy.

The values $\left(-\frac{dT_2}{dt}\right)\Big/\left(-\frac{dT_1}{dt}\right) = 10^4$ and $\frac{\lambda_2^2}{\lambda_1^2} = \frac{1}{12^2}$, both obtained from the figure, are inserted into Equation (5′) to give:

$$\frac{B_1}{B_2} = \frac{10^4}{12^2} \quad \text{or} \quad B_2 = B_1 \times 0.014$$

The parameter B decreases when the cooling rate increases.

Answer:
The importance of back diffusion in Al–Cu alloys decreases when the cooling rate increases.

If $\lambda_{den} = \theta^{0.5}$ [see Equation (6.10) on page 144] the back diffusion in Al–Cu alloys will be independent of the cooling rate in accordance with Example 7.4 above. In reality the exponent is lower than 0.5. The less the back diffusion is, the greater will the microsegregation be. Thus we can conclude from Example 7.4 that microsegregation will be greater the more rapid the cooling rate is. Fast homogenization after solidification can often compensate for microsegregation. We come back to this in Section 7.6 and in Chapter 8.

We integrated Equation (7.24) on the assumption that the parameter B was constant. This condition is well fulfilled in the many cases when the alloy has a narrow solidification interval. However, with wide solidification intervals the temperature dependence of D^s will be considerable. The removal of cooling heat must also be taken into consideration, as this leads to a temperature decrease in the volume element.

Scheil's segregation Equation (7.10) on page 186 shows that the first solidified material gets and retains the composition $k_{part}x_0^L$ if back diffusion is small. In cases of strong back diffusion, the initially solidified material will even out in composition during continued solidification. In these cases there are no simple approximate solutions but numerical calculations have been performed for special cases. One such case is the Al–Cu alloy described in Figure 7.15. Figure 7.15 shows the result of such a numerical calculation for an Al alloy with 4.5 wt-% Cu. The calculated partition constant is $k_{part} = 0.136$. The first material will therefore have the composition:

$$c^s = k_{part}c_0^L = 0.136 \times 4.5 = 0.61\,\text{wt-% Cu}$$

when it is formed. This value corresponds to the lowest point in Figure 7.15. When the fraction of solid phase is 0.1 the Cu concentration in the first solidified material has increased. This is illustrated by the short curve that

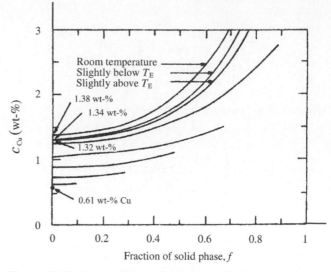

Figure 7.15 Concentration distribution of Cu in an Al–4.5 wt-% Cu alloy as a function of degree of solidification. Reproduced by permission of the Minerals, Metals & Materials Society.

ends at $f = 0.1$. The upper curves show how the solidified layer grows and how its composition changes. The curves in Figure 7.15 were calculated by using a B value of $6.1 \times 10^6 D_s$. The diffusion constant was assumed to vary exponentially with the temperature.

7.5.2 Choice of Method to Calculate Alloy Composition

Calculation of the concentration of the alloying element as a function of the fraction of solid phase $f = 2y/\lambda_{den}$ is performed as follows:

- At *high* diffusion rates the lever rule is valid [Equation (7.15) on page 189].
- At *low* diffusion rates Scheil's modified segregation equation is valid [Equation (7.25) on page 199].
- When $D_s = 10^{-11}\,\text{m}^2/\text{s}$ it is difficult to decide which relation one should use. The cooling rate is allowed to decide.
- At *low* cooling rates the lever rule is valid.
- At *high* cooling rates Sheil's modified equation agrees best with reality.

7.5.3 Degree of Microsegregation

Microsegregation in a solidified material can be determined by using a microprobe. A great number of such investigations have been performed on steel alloys. These have shown that in most cases the concentration of the alloying elements at the centre of a dendrite arm is constant during the whole solidification process, provided the material solidifies as austenite. The reason for this is that homogenization (page 189) occurs to a very small extent during solidification. The concentration distribution in the interdendritic areas is, in such a case, described by Scheil's modified segregation equation [Equation (7.25) on page 199] when the primary precipitation is austenite.

Various alloying elements have very different partition constants and different tendencies of segregation. In order to describe the tendency of segregation the concept of *degree of segregation* has been introduced:

The degree of microsegregation is the ratio between the highest and the lowest measured values of the concentration of the alloying element in a dendrite crystal aggregate.

$$S = \frac{x_{max}^{s}}{x_{min}^{s}} \qquad (7.27)$$

Because the relationship $x^{s} = k_{part}x^{L}$ is always valid, Scheil's modified segregation equation [Equation (7.25)] can be used for calculating the degree of segregation. The ratios of x_{max}^{L}, when the segregation is highest $(2y/\lambda_{den} = f = 1)$, and x_{min}^{L}, when the segregation is lowest $(f = 0)$, are formed:

$$x_{max}^{L} = x_{0}^{L} \left[\left(1 - \frac{\frac{2y}{\lambda_{den}}}{D_{s} \left(\frac{4\theta}{\lambda_{den}^{2}} \right) k_{part} + 1} \right)^{-(1-k_{part})} \right]_{f = \frac{2y}{\lambda_{den}} = 1}$$

$$= x_{0}^{L} \left(1 - \frac{1}{B+1} \right)^{-(1-k_{part})} = x_{0}^{L} \left(\frac{B}{1+B} \right)^{-(1-k_{part})}$$

and

$$x_{min}^{L} = x_{0}^{L} \left[\left(1 - \frac{\frac{2y}{\lambda_{den}}}{D_{s} \left(\frac{4\theta}{\lambda_{den}^{2}} \right) k_{part} + 1} \right)^{-(1-k_{part})} \right]_{f = \frac{2y}{\lambda_{den}} = 0} = x_{0}^{L}$$

which gives:

$$S = \frac{x_{max}^{s}}{x_{min}^{s}} = \frac{k_{part} \cdot x_{max}^{L}}{k_{part} \cdot x_{min}^{L}} = \frac{x_{max}^{L}}{x_{min}^{L}} = \left(\frac{B}{1+B} \right)^{-(1-k_{part})} \qquad (7.28)$$

It has been known for a long time that an increase of the carbon content in ternary Fe–Cr–C alloys increases the degree of segregation of chromium. The reason for this increase is that the partition constant of chromium between austenite and melt decreases with increasing carbon

content, with the result that the segregation increases. In steel alloys there are alloying elements or impurities that have partition constants between austenite and melt close to unity or slightly smaller, and also substances that have very small partition constants.

Because the degree of segregation S depends on the partition constant k_{part} as in Equation (7.28) directly and indirectly (k_{part} is involved in B too), the degree of segregation can be expressed as a function of the partition constant. The result of such calculations for steel alloys which solidify as austenite is given in Figure 7.16. The calculations have been performed for two different cooling rates, which simulate the cooling rates close to the centre and close to the surface zone in a 9-ton ingot.

As expected, low k_{part} values correspond to very high values for the degree of segregation. For $k_{part} \geq 0.90$ the material becomes practically homogeneous. In Figure 7.16 the approximate values for the partition constants of the most common alloying elements are plotted: phosphorus shows the highest degree of segregation.

The cooling rate is faster in the surface zone (dotted line) than in the centre. Figure 7.16 therefore shows that the degree of segregation increases with increasing cooling rate. This is compensated for by the fine structure. The smaller the diffusion distances are, i.e. the dendrite arm distances in the solid, the more rapid will the subsequent homogenization during cooling and heat treatment be.

In Section 7.6 microsegregation and solidification processes in iron-base alloys are discussed in more detail because of their technical importance.

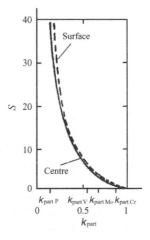

Figure 7.16 Theoretical calculations of the degree of segregation S of steel alloys as a function of the partition constant k_{part}. The continuous line in the figure corresponds to the centre of a 9-ton ingot. The dotted line illustrates the degree of segregation close to the surface zone of the same ingot.

7.6 SOLIDIFICATION PROCESSES AND MICROSEGREGATION IN IRON-BASE ALLOYS

The solidification process in iron-base alloys starts with dendrite solidification and precipitation of either ferrite (δ) or austenite (γ) as seen in Figure 7.17.

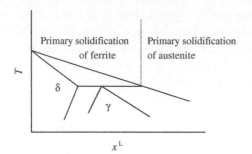

Figure 7.17 Solidification regions for precipitation of ferrite and austenite.

The microsegregation in the two cases is completely different. This difference depends on two factors:

- the difference between the diffusion rates of the alloying element in ferrite and austenite;
- the difference between the partition coefficients of the alloying element in ferrite and austenite.

7.6.1 Microsegregation in Primary Precipitation of Ferrite

During primary precipitation of ferrite, microsegregation is low for most alloying elements. The reason for this may be that the partition coefficient k_{part} is close to 1 or that back diffusion is very large. This results in a homogenization process after solidification, i.e. the concentration of the alloying element is levelled out between the solid phase and the melt.

The phase diagram above shows that the partition coefficient is *not* equal to 1 during primary precipitation of ferrite. Ferrite has BCC structure (page 189). It is known that the diffusion rates for both substitutionally and, especially, for interstitially dissolved atoms are comparatively high for BCC structures. Consequently it is the diffusion rate that is of crucial importance for the microsegregation in primary precipitation of ferrite.

In many cases the lever rule gives a good description of the solidification process and the distribution of the alloying elements because the diffusion rate is often high. However, there are exceptions. Some alloying elements have low diffusion rate in ferrite. In these cases, especially if the cooling rate is high, the lever rule gives a poor description of reality. Instead we start from the mathematical model given below to calculate the degree of segregation S as a function of the

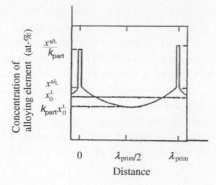

Figure 7.18 The concentration distribution of the alloying element across a dendrite arm within the interval λ_{prim}. Reproduced with permission from The Metal Society, Institute of Materials, Minerals & Mining.

partition coefficient, the diffusion rate in the solid phase and the cooling rate.

We assume that the primary dendrite arms are situated in parallel planes and that the concentration of the alloying element in the solid phase across the dendrite arms after complete solidification can be described by a sine function (Figure 7.18). We call the distance between the parallel primary planes λ_{prim}, which we may call the primary dendrite arm distance, and emphasize that it is *not* equal to the secondary dendrite arm distance λ_{den}. λ_{den} is much smaller than λ_{prim}.

The sine function has a minimum in the centre of the dendrite arms (first solidified parts) and a maximum in the interdendritic areas half way between the arms (last solidified parts of the melt). The concentration of the alloying element in the solid phase is a function of time t and position y. We assume that it can be described by the function:

$$x^s(y,t) = x^s(t) - \left[x^s(t) - k_{part}x_0^L\right]\exp\left(\frac{-\pi^2 D_s t}{\lambda_{prim}^2}\right)\sin\frac{\pi y}{\lambda_{prim}}$$

$$(7.29)$$

where

$$t = \text{time}$$
$$\lambda_{prim} = \text{distance between the parallel primary planes}$$
$$y = \text{distance from origin within the interval } 0-\lambda_{prim}$$
$$(\text{see Figure 7.18})$$
$$k_{part} = x^s/x^L = \text{partition coefficient of alloying element between ferrite and melt}$$
$$x_0^L = \text{initial concentration of alloying element in the the melt}$$
$$x^s(t) = x^s(0,t) = \text{concentration of alloying element at } y = 0$$
$$D_s = \text{diffusion coefficient of alloying element in the solid phase.}$$

To be able to use Equation (7.29) we must find an expression for the concentration of the alloying element $x^s(t)$. We consider a solidification process within the interval λ_{prim} and set up a material balance for the amount of alloying element:

$$x_0^L \lambda_{prim} = \int_0^\lambda \left[x^s(t) - \left(x^s(t) - k_{part} x_0^L \right) \exp\left(\frac{-\pi^2 D_s t}{\lambda_{prim}^2} \right) \sin\frac{\pi y}{\lambda_{prim}} \right] dy \quad (7.30)$$

After integration we solve $x^s(t)$:

$$x^s(t) = x_0^L \left[\frac{1 - \left(\frac{2k_{part}}{\pi} \right) \exp\left(-\frac{\pi^2 D_s t}{\lambda_{prim}^2} \right)}{1 - \left(\frac{2}{\pi} \right) \exp\left(-\frac{\pi^2 D_s t}{\lambda_{prim}^2} \right)} \right] \quad (7.31)$$

[compare Equation (2′) on page 199]:

$$t = \theta = \frac{-\Delta H}{c_p \left(-\dfrac{dT}{dt} \right)} \quad (7.33)$$

$$\lambda_{prim} = A \left(-\frac{dT}{dt} \right)^n \quad (7.34)$$

where A and n are constants.

If we introduce these expressions into Equation (7.32) we get a final expression for S that depends on the phase diagram (partition coefficient and solidification interval), diffusion rate and cooling rate.

Using Equation (7.35) the degree of segregation can be calculated for different alloying elements in steel. We will

$$S = \left\{ 1 - \left[1 - \frac{k_{part} \left(1 - \left(\frac{2}{\pi} \right) \exp\left[\frac{-\pi^2}{A^2 c_p \left(-\frac{dT}{dt} \right)^{1+2n}} D_s(-\Delta H) \right] \right)}{1 - \left(\frac{2k_{part}}{\pi} \right) \exp\left[-\frac{\pi^2}{A^2 c_p \left(-\frac{dT}{dt} \right)^{1+2n}} D_s(-\Delta H) \right]} \right] \exp\left[-\frac{\pi^2}{A^2 c_p \left(-\frac{dT}{dt} \right)^{1+2n}} D_s(-\Delta H) \right] \right\}^{-1} \quad (7.35)$$

We introduce the expression (7.31) into Equation (7.29) and get a useful relationship for x^s as a function of y and t. This function has a maximum when the sine function is equal to zero and a minimum when the sine function is equal to 1.

We also want an expression for the degree of segregation S [Equation (7.27) on page 201]. By inserting the maximum and minimum values of the function x^s discussed above, we get the following expression for S as a function of t:

analyse how the k_{part} *value* and the *diffusion rate* influence the degree of segregation for primary precipitation of ferrite. For this purpose we choose the three alloying elements Nb, C, and Cr. Their phase diagrams are given in the Figures 7.19, 7.20 and 7.21.

As is seen from Figures 7.19 and 7.21 the calculation of the $k_{part}^{\delta/L}$ values is straightforward for <u>Nb</u> and <u>C</u>[1]. Readings at the eutectic and peritectic lines are used

$$S = \left\{ 1 - \left[1 - \frac{k_{part} \left(1 - \frac{2}{\pi} \exp\left(\frac{-\pi^2}{\lambda_{prim}^2} D_s t \right) \right)}{1 - \left(\frac{2k_{part}}{\pi} \right) \exp\left(-\frac{\pi^2}{\lambda_{prim}^2} D_s t \right)} \right] \exp\left(-\frac{\pi^2}{\lambda_{prim}^2} D_s t \right) \right\}^{-1} \quad (7.32)$$

Equation (7.32) will be more useful for calculations if we replace t and λ_{prim} with terms containing the cooling rate

[1] Underlined alloying atoms indicate that they are dissolved in a liquid or solid phase. This notation is often used in the rest of this chapter and in Chapters 6 and 9.

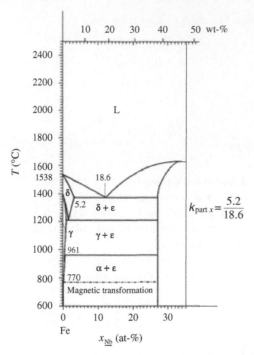

Figure 7.19 Part of the phase diagram of the system Fe–Nb.

$$k_{\text{part }x} = \frac{5.2}{18.6}$$

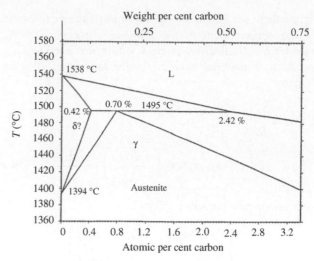

Figure 7.21 Simplified phase diagram of the system Fe–C: $k_{\text{part}_C}^{\delta/L} = 0.42/2.42$. Reproduced with permission from the American Society for Metals.

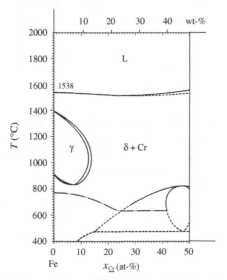

Figure 7.20 Part of the phase diagram of the system Fe–Cr. Reproduced with permission from the American Society for Metals.

and the ratios of x^s/x^L when the melt solidifies as ferrite (δ-phase) are calculated. In the case of Fe–Cr, the solidus and liquidus lines ($T \sim 1535\,°C$) are very close and not resolved in Figure 7.20. In a diagram with better resolution $k_{\text{part}}{}^{\delta/L}$ for Cr can be derived, and this is found to have a value below but close to 1.

Influence of the Partition Coefficient and Diffusion Rate on the Microsegregation of the Alloying Element in Case of Ferrite Solidification

Influence of the Partition Coefficient on the Degree of Segregation

The degree of segregation has been studied as a function of the cooling rate for three alloys Fe–Nb, Fe–C, and Fe–Cr. The result is illustrated in Figure 7.22.

Cr and Nb have nearly the same diffusion rate in ferrite but different k_{part} values, which can be seen from their phase diagrams. The k_{part} values can either be derived from Figures 7.19 and 7.20 or read from Figure 7.16 on page

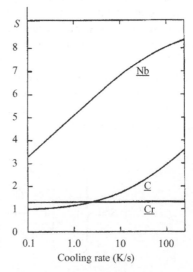

Figure 7.22 Degree of segregation for Nb, C, and Cr as functions of the cooling rate in primary precipitation of ferrite. Reproduced with permission from The Metal Society, Institute of Materials, Minerals & Mining.

201. The degree of segregation is then calculated using Equation (7.35). Figure 7.22 shows that the *closer to* 1 the k_{part} value is, the *lower* will the degree of segregation be.

Influence of Diffusion Rate on the Degree of Segregation

Nb and C have approximately equal k_{part} values but very different diffusion rates. The Nb atoms, which are substitutionally dissolved, have a low diffusion rate. The small, interstitially dissolved C atoms diffuse rapidly through the ferrite. Figure 7.22 shows that the *higher* the diffusion rate is, the *lower* will the degree of segregation be. Figure 7.22 also shows that the degree of segregation S is close to 1 for C and Cr unless the cooling rate is very high. For Nb the degree of segregation is large both for high and low cooling rates.

Figure 7.22 describes the conditions at the time of complete solidification, i.e. at the end of the solidification process. During the subsequent cooling, the homogenization process goes on and at room temperature there is probably very low segregation in the material.

When the *lever rule* is valid there is no microsegregation in the material and $S = 1$. Figure 7.22 also shows that the lever rule gives a relatively good approximation for the concentration of the alloying element in ferrite at a *low* cooling rate during the solidification process. It should be added that this is true especially for interstitially dissolved elements such as C, N, and H (rapid diffusion).

7.6.2 Microsegregation in Primary Precipitation of Austenite

Most alloys that solidify by primary precipitation of austenite show a completely different segregation pattern than that which appears in primary precipitation of ferrite. The main reason for this is that the diffusion rate in general is much lower in austenite (FCC structure) than in ferrite (BCC structure). The back diffusion is therefore less important in primary precipitation of austenite than with ferrite but it cannot be neglected completely.

The mathematical model that describes the segregation pattern during the solidification process in primary precipitation of austenite best is Scheil's modified equation:

$$x^L = x_0^L \left(1 - \frac{\frac{2y}{\lambda_{den}}}{1 + D_s \left(\frac{4\theta}{\lambda_{den}^2} \right) k_{part}} \right)^{-(1-k_{part})}$$

$$= x_0^L \left(1 - \frac{f}{1 + B} \right)^{-(1-k_{part})} \tag{7.36}$$

where

x^L = concentration of the alloying element of the melt

x_0^L = initial concentration of the alloying element in the melt

f = fraction of solid phase

$k_{part} = x^s / x^L$ = partition coefficient between austenite and melt.

The parameter B is defined as:

$$B = \frac{4D^s \theta}{\lambda_{den}^2} k_{part} \tag{7.37}$$

where

θ = total solidification time

D^s = diffusion rate of the alloying element in the solid phase

λ_{den} = dendrite arm distance.

B is a correction term in Scheil's equation [Equation (7.10) on page 186]. Equation (7.36) shows that as long as $B \ll 1$, Scheil's simple equation satisfactorily describes the concentration distribution of the alloying element. When $(1 - f)$ decreases at the end of the solidification process, then back diffusion increases. This is a consequence of the very large concentration gradients that arise at the end of the solidification process between the austenite and the remaining melt.

From Equation (7.28) on page 201 we have:

$$S = \frac{x_{max}^s}{x_{min}^s} = \left(\frac{B}{1 + B} \right)^{-(1-k_{part})} \tag{7.28}$$

In the same way as with primary precipitation of ferrite (page 203) we introduce solidification time θ and cooling rate into the expression for the degree of segregation. Equation (7.28) can then be written as:

$$S = \left(\frac{1 + B}{B} \right)^{(1-k_{part})} = \left(\frac{1 + \left(\frac{4D_s \theta}{\lambda_{den}^2} \right) k_{part}}{\left(\frac{4D_s \theta}{\lambda_{den}^2} \right) k_{part}} \right)^{(1-k_{part})} \tag{7.38}$$

where

$$t = \theta = \frac{-\Delta H}{c_p \left(-\frac{dT}{dt} \right)} \tag{7.33}$$

and

$$\lambda_{den} = A \left(-\frac{dT}{dt} \right)^n \tag{7.34}$$

Inserting these terms for θ and λ_{den} into Equation (7.38) gives:

$$S = \left(k_{\text{part}} 1 + \frac{4k_{\text{part}} D_{\text{s}}(-\Delta H)}{A^2 c_{\text{p}} \left(-\frac{\text{d}T}{\text{d}t} \right)^{1+2n}} \Bigg/ \frac{4k_{\text{part}} D_{\text{s}}(-\Delta H)}{A^2 c_{\text{p}} \left(-\frac{\text{d}T}{\text{d}t} \right)^{1+2n}} \right)^{(1-k_{\text{part}})} \tag{7.39}$$

where n and A are constants. Equation (7.39) gives the degree of segregation S as a function of the cooling rate for primary precipitation of austenite.

Using Equation (7.39) the degree of segregation as a function of the phase diagram of the alloy (partition constant and solidification interval), diffusion rate and cooling rate can be calculated. The results of such calculations for Fe–P, Fe–Mo, Fe–V, and Fe–Cr are given in the following section. The phase diagram for the system Fe–Cr is given on page 204. The phase diagrams for the systems Fe–P, Fe–Mo, and Fe–V are given in Figures 7.23, 7.24, and 7.25, respectively. It should be noted that the k_{part} values are *not* the same when an iron-base alloy solidifies as ferrite or austenite. One example is chromium: $k_{\text{part}\underline{\text{Cr}}}^{\delta/\text{L}} \neq k_{\text{part}\underline{\text{Cr}}}^{\gamma/\text{L}}$. Below we will describe how $k_{\text{part}\underline{\text{Cr}}}^{\gamma/\text{L}}$ can be derived from the phase diagram of the system Fe–Cr.

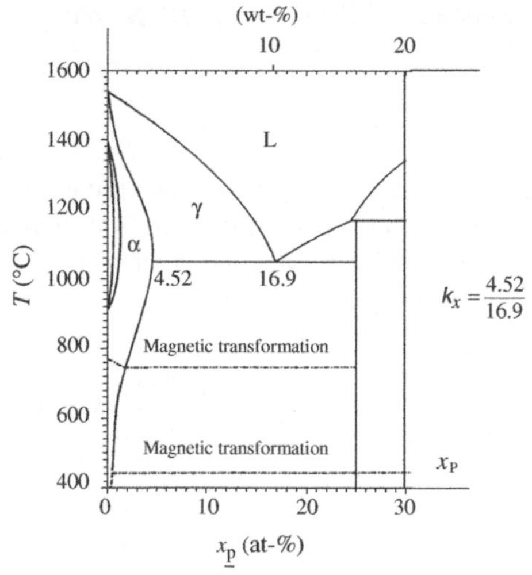

Figure 7.23 Part of the phase diagram of the system Fe–P. Reproduced with permission from the American Society for Metals.

$k_{\text{part}\underline{\text{Cr}}}^{\gamma/\text{L}}$ can not be derived directly because the phases γ and L are nowhere adjacent with a two-phase region in between. The partition constant $k_{\text{part}\underline{\text{Cr}}}^{\gamma/\text{L}}$ is derived indirectly in two steps. Initially $k_{\text{part}\underline{\text{Cr}}}^{\delta/\text{L}}$ and $k_{\text{part}\underline{\text{Cr}}}^{\gamma/\delta}$ have to be calculated.

The derivation of the partition constant $k_{\text{part}\underline{\text{Cr}}}^{\delta/\text{L}}$ is described on page 204. $k_{\text{part}\underline{\text{Cr}}}^{\gamma/\delta}$ can be derived from the phase diagram of Fe–Cr. The x^{γ} and x^{δ} values are read

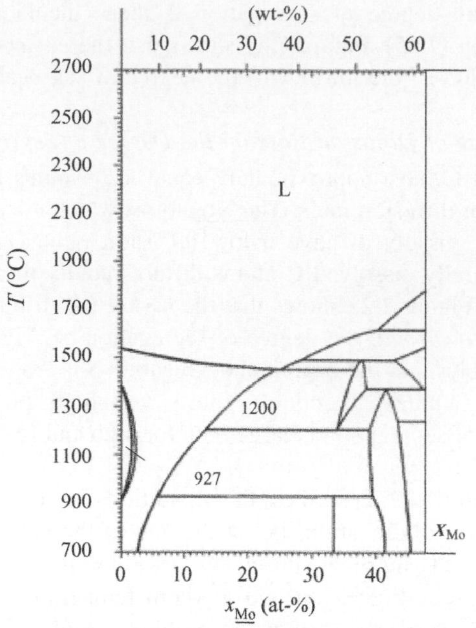

Figure 7.24 Part of the phase diagram of the system Fe–Mo. Reproduced with permission from the American Society for Metals.

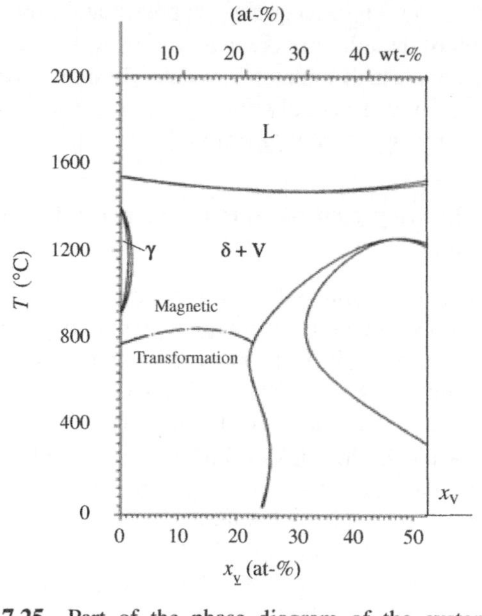

Figure 7.25 Part of the phase diagram of the system Fe–V. Reproduced with permission from the American Society for Metals.

from the loop at $T \approx 1000\,°\text{C}$ (see Figure 204, page) and their ratio is calculated.

Going back to the definition of k_{part}, one realizes that the relationship $k_{\text{part}\underline{\text{Cr}}}^{\gamma/\text{L}} = k_{\text{part}\underline{\text{Cr}}}^{\gamma/\delta} k_{\underline{\text{Cr}}}^{\delta/\text{L}}$, which gives $k_{\text{part}\underline{\text{Cr}}}^{\gamma/\text{L}}$ when the other two $k^{\gamma/\text{L}}$ values are known. As is seen from Figure 7.23, calculation of $k_{\text{part}\underline{\text{P}}}^{\gamma/\text{L}}$ is straightforward. The phase diagrams of Fe–Mo and Fe–V are of the same type as that of Fe–Cr, and $k_{\text{part}\underline{\text{Mo}}}^{\gamma/\text{L}}$ and $k_{\text{part}\underline{\text{V}}}^{\gamma/\text{L}}$ are derived in the same way as $k_{\text{part}\underline{\text{Cr}}}^{\gamma/\text{L}}$.

Influence of the Partition Coefficient and Diffusion Rate on the Microsegregation of the Alloying Element in the Case of Austenite Solidification

The k_{part} values of $\underline{P}$, $\underline{Mo}$, $\underline{V}$, and $\underline{Cr}$ are marked in Figure 7.16 on page 201 and can be read there. Figure 7.26 shows clearly that the degree of segregation *increases* with *decreasing* k_{part} *values* and *increasing cooling rate*. Interstitially dissolved elements diffuse more easily than do substitutionally dissolved elements. The former have a higher diffusion rate and lower degree of segregation than the latter, which are part of the crystal lattice.

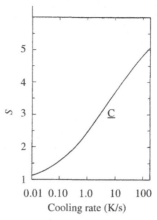

Figure 7.27 The degree of segregation of carbon as a function of the cooling rate during primary precipitation of austenite. The temperature gradient is constant during the cooling and equal to 6×10^3 K/m. For the calculations the same values for the constants n and A [Equation (7.34) on page 203] as in Figure 7.22 have been used. Reproduced with permission from The Metal Society, Institute of Materials, Minerals & Mining.

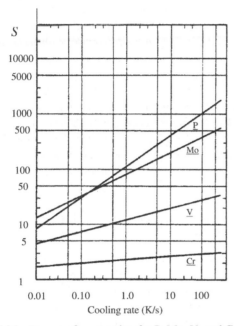

Figure 7.26 Degree of segregation for $\underline{P}$, $\underline{Mo}$, $\underline{V}$, and $\underline{Cr}$ as functions of the cooling rate in primary precipitation of austenite. The temperature gradient is constant during the cooling and equal to 6×10^3 K/m. For the calculations the same values for the constants n and A as in Figure 7.22 have been used. Reproduced with permission from The Metal Society, Institute of Materials, Minerals & Mining.

The Degree of Segregation of Carbon and Other Interstitially Dissolved Elements

Fe–C is an example of an alloy with interstitially dissolved alloying atoms. Carbon has, consequently, a comparatively high diffusion rate and Equation (7.39) on page 206 *cannot* be used for calculating the degree of segregation. Instead Equation (7.35) on page 203 can be used. The result of such a calculation is illustrated in Figure 7.27. The phase diagram of the system Fe–C is given in Figure 7.21 on page 204.

When $S = 1$ the lever rule is valid. The *lever rule* gives a good approximation when calculating the carbon concentration in austenite at *low* cooling rates during the solidification process. Corresponding arguments are valid

for other interstitially dissolved elements, for example $\underline{N}$ and $\underline{H}$.

7.7 PERITECTIC REACTIONS AND TRANSFORMATIONS IN BINARY IRON-BASE ALLOYS

In addition to microsegregation, so-called *peritectic reactions* and *transformations* may occur during the solidification process in some cases. They also change the structure and composition of the alloy. This is the case for many technically important alloys, for example, Fe–C alloys.

A *peritectic reaction* is a reaction that occurs between a primary solid phase α and the melt of an alloy with the alloying element B.

$$L + \alpha \rightarrow \beta$$

A secondary phase β is formed. This peritectic reaction is followed by a *peritectic transformation* where one of the solid phases is transformed into the other one. The reaction can be described by the phase diagram of the alloy. Figure 7.28 shows a schematic phase diagram with a primary phase α and a B-rich β-phase. The peritectic reaction $L + \alpha \rightarrow \beta$ starts at the peritectic temperature T_P, or slightly below with a minor undercooling.

At equilibrium all alloys with a composition less than the concentration corresponding to the vertical line I solidify primarily as α-phase. All alloys with a composition greater than that corresponding to the vertical line III will primarily solidify as β-phase.

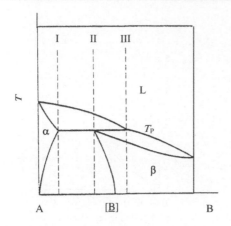

Figure 7.28 Phase diagram with the peritectic reaction L + α → β.

For alloys with a composition within the region between the lines II and III, the melt will primarily solidify as α-phase, which is later transformed into stable β-phase. Alloys with a composition within the region between the lines I and II will also primarily solidify as α-phase, which later is partially transformed into β-phase.

There is a distinction between a *peritectic reaction* and a *peritectic transformation*. In a *peritectic reaction* all three phases α, β and melt are normally in direct contact with each other. Figure 7.29 shows the principle of the growth. B atoms diffuse through the melt to the B-depleted α-phase and a layer of β-phase is formed around the α-phase. The diffusion rate in the melt is fast. The lateral growth rate and the surface tension determine the thickness and extension of the secondary β-phase.

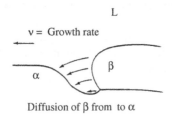

Figure 7.29 Peritectic reaction, in which the secondary β-phase grows along the surface of the primary α-phase. Reproduced with permission from The Metal Society, Institute of Materials, Minerals & Mining.

In a *peritectic transformation*, the melt and the primary α-phase are separated by the secondary β-phase (Figure 7.30). The transformation occurs at the α/β interphase. Diffusion of the alloying element B through the secondary β-phase makes the transformation [α + B → β] possible. The thickness of the β-layer grows during cooling, due to growth both at the β/L and the α/β interfaces.

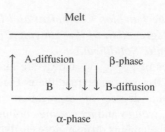

Figure 7.30 Peritectic transformation, in which the secondary β-phase insulates the melt from the primary α-phase.

The precipitation of β-phase directly from the melt at the β/L interface depends on the shape of the phase diagram and the cooling rate. The diffusion through the β-layer of B atoms and precipitation of β-phase at the α/β interface depends on the diffusion rate, the shape of the phase diagram, and the cooling rate.

Some alloys have phase diagrams that are *mirror-inverted* compared with that in Figure 7.28. Then the primary α-phase is more B-rich than the secondary β-phase. In these cases the β-phase is transformed into α-phase during the peritectic transformation. Later we give some concrete examples of peritectic reactions and transformations.

7.7.1 Peritectic Reactions and Transformations in the Fe–C System

Because peritectic reactions and transformations occur rapidly (C is interstitially dissolved in Fe) careful experimental investigations are very difficult to perform in this field. One example of such experiments is Stjerndahl's study of peritectic reactions for the alloy Fe–0.3 % C.

On the right-hand side of Figure 7.31 the structure consists of dendrites with ferrite (grey or white) and

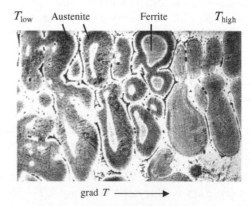

Figure 7.31 Peritectic reaction and transformation in Fe–0.3% C: L + δ → γ. The specimen has solidified in a temperature field having a temperature gradient. The temperature is higher to the right than to the left in the figure. The γ-phase (austenite) that surrounds the δ-grains (ferrite) gets thicker during cooling.

solidified interdendritic melt (bright area between the grains). The peritectic reaction starts with formation of a layer of austenite (dark grey) around the ferrite dendrite arms.

The austenite layer increases successively in thickness by transformation of ferrite (δ) and melt into austenite (γ). Finally the reaction has proceeded so far that the ferrite areas disappear completely (not visible in the figure). The transformation is very rapid and occurs within a temperature interval of a few degrees celsius.

Theory of Peritectic Transformation

Using a simple mathematical model it is possible to calculate the *peritectic transformation rate* in the Fe–C system. The transformation of ferrite into austenite is very rapid, and the driving force of the peritectic reaction is a small undercooling ($T_P - T$).

Calculation of the Growth Rate of the Austenite Layer

For the calculations we assume that the dendrites consist of parallel plates and that the distance between neighbouring plates is equal to the dendrite distance λ_{den} (Figure 7.32).

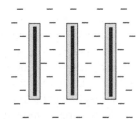

Figure 7.32 Parallel plates as a model of parallel dendrite arms at equal distances λ_{den}. A transformation of ferrite (black areas) into surrounding austenite (grey areas) is going on.

Initially the interior of each plate consists of ferrite (δ) surrounded by a layer of austenite (γ). The ferrite is thus insulated from the melt. We also assume that:

(i) $\underline{C}$-atoms diffuse from the melt through the austenite layer to δ-phase. The driving force is the concentration difference between the phases.

(ii) The $\underline{C}$-concentration decreases linearly from the interface γ/L to the interface γ/δ.

(iii) The phase diagram for the system Fe–C determines the equilibrium concentrations in the L-, γ- and δ-phases. A schematic phase diagram and the concentration distribution of $\underline{C}$ are given in Figure 7.33.

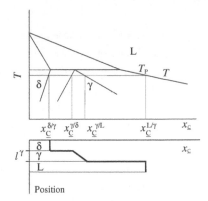

Figure 7.33 Concentration distribution of $\underline{C}$ at a peritectic reaction and transformation. Undercooling $= T_P - T$.

Two material balances are set up, one at the interface γ/δ and one at the interface γ/L (Figure 7.34):[2]

$$
\delta \rightarrow \gamma : \underbrace{\frac{AD_{\underline{C}}^{\gamma}}{V_m^{\gamma}} \frac{x_{\underline{C}}^{\gamma/L} - x_{\underline{C}}^{\gamma/\delta}}{l^{\gamma}}}_{\substack{\text{Added amount of } \underline{C}\text{-atom} \\ \text{per unit times, due to} \\ \text{diffusion}}}
$$
$$
= \underbrace{A \frac{dl^{\gamma/\delta}}{dt} \frac{x_{\underline{C}}^{\gamma/\delta} - x_{\underline{C}}^{\delta/\gamma}}{V_m^{\delta}}}_{\substack{\text{Added amounts of } \underline{C}\text{-atoms per} \\ \text{unit times in the new volume} \\ \text{element of the } \gamma\text{-phase}}} \qquad (7.40)
$$

$$
L \rightarrow \gamma : \frac{AD_{\underline{C}}^{\gamma}}{V_m^{\gamma}} \frac{x_{\underline{C}}^{\gamma/L} - x_{\underline{C}}^{\gamma/\delta}}{l^{\gamma}}
$$
$$
= A \frac{dl^{\gamma/L}}{dt} \frac{x_{\underline{C}}^{L/\gamma} - x_{\underline{C}}^{\gamma/L}}{V_m^{L}} \qquad (7.41)
$$

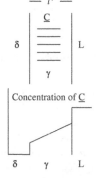

Figure 7.34 Carbon atoms diffuse from the melt through the austenite to the ferrite on the left-hand side.

[2] Consider a boundary between two phases α and β of a binary alloy. The concentration in at-% of the alloying element in phase α, close to the boundary, is designated $x^{\alpha/\beta}$. The symbol $x^{\beta/\alpha}$ means the concentration in at-% of the alloying element in phase β close to the boundary.

where

$$\mathrm{d}l^{\gamma/\delta}/\mathrm{d}t = \text{growth rate of } \gamma\text{-phase at the interface } \gamma/\delta$$

$$\mathrm{d}l^{\gamma/L}/\mathrm{d}t = \text{growth rate of } \gamma\text{-phase at the interface } \gamma/L$$

$$D_{\underline{C}}^{\gamma} = \text{diffusion constant of } \underline{C}\text{-atoms in the } \gamma\text{-phase}$$

$$l^{\gamma} = \text{thickness of the } \gamma\text{-phase}$$

$$V_{m}^{\gamma}, V_{m}^{\delta}, V_{m}^{L} = \text{molar volumes of austenite, ferrite, and melt.}$$

The total growth rate of the γ-layer is the sum of the growth rates at the two interfaces:

$$\frac{\mathrm{d}l^{\gamma}}{\mathrm{d}t} = \frac{\mathrm{d}l^{\gamma/\delta}}{\mathrm{d}t} + \frac{\mathrm{d}l^{\gamma/L}}{\mathrm{d}t} \tag{7.42}$$

or

$$\frac{\mathrm{d}l^{\gamma}}{\mathrm{d}t} = \frac{D_{\underline{C}}^{\gamma}}{l^{\gamma}} \frac{x^{\gamma/L} - x^{\gamma/\delta}}{V_{m}^{\gamma}} \left[\frac{V_{m}^{\delta}}{x^{\gamma/\delta} - x^{\delta/\gamma}} + \frac{V_{m}^{L}}{x^{L/\gamma} - x^{\gamma/L}} \right] \tag{7.43}$$

The expression (7.43) is the total growth rate of the γ-layer.

It can be seen from Figure 7.33 that the expression $(x^{\gamma/L} - x^{\gamma/\delta})$ is zero at the peritectic temperature and increases strongly with increasing undercooling. The second factor on the right-hand side in Equation (7.43) is positive and also increases with increasing undercooling for the Fe–C system. In addition, Equation (7.43) shows that the growth rate depends heavily on the diffusion constant of carbon in the austenite layer. Interstitially dissolved alloying elements often have diffusion constants that are very high and of the magnitude $10^{-9} \, \mathrm{m^2/s}$. The carbon atoms are interstitially dissolved in austenite, which explains why the diffusion of the $\underline{C}$ atoms has such a great influence on the growth rate of the austenite layer.

Calculation of the Temperature Interval of the Peritectic Transformation

Equation (7.43) is valid at a constant temperature T slightly below the peritectic temperature T_P. For the derivation we assumed that the peritectic transformation occurs at constant cooling rate. In this case the temperature interval within which the peritectic transformation occurs can be calculated. We assume that:

$$x^{\gamma/L} - x^{\gamma/\delta} = \text{const}(T_p - T) \tag{7.44}$$

where

$$T_P = \text{peritectic temperature}$$

$$T = \text{temperature of the melt}$$

$$(T_P - T) = \text{undercooling.}$$

The expression (7.44) is inserted into Equation (7.43) and $\mathrm{d}l^{\gamma}/\mathrm{d}t$ is replaced by:

$$\frac{\mathrm{d}l^{\gamma}}{\mathrm{d}t} = \frac{\mathrm{d}l^{\gamma}}{\mathrm{d}T} \times \frac{\mathrm{d}T}{\mathrm{d}t} \tag{7.45}$$

The resulting equation is integrated. The integration is not given here but the final result can be written as:

$$l^{\gamma} = \text{const} \times \Delta T \tag{7.46}$$

where ΔT is the temperature interval within which the peritectic transformation occurs. The value of the constant depends on the cooling rate and the quantities that are involved in Equation (7.43). With reasonable values for the concentration differences and other quantities in Equation (7.46), calculation of ΔT shows that the peritectic transformation in most cases is very rapid and is completed within 6–10 °C below the peritectic temperature, depending on the cooling rate.

7.7.2 Peritectic Reactions and Transformations in Fe–M Systems

Peritectic reactions also occur in alloys where the alloying element M is substitutionally dissolved. Examples of such binary alloys are Fe–Ni and Fe–Mn alloys.

Experimental investigations into stainless steel and Fe–Ni–S alloys show that a layer of austenite surrounds the primary dendrite crystals. The thickness of the austenite layer grows with decreasing temperature. The γ-layer grows both inwards by a reaction with ferrite and outwards by precipitation of austenite from the melt. The growth of the austenite layer by transformation, due to diffusion of the alloying element through the austenite, is almost completely missing. The reason is that the diffusion rate of substitutionally dissolved elements is very low in austenite.

The fundamental difference between Fe–C and Fe–M alloys is that $\underline{C}$ atoms that are dissolved interstitially have *high* diffusion rates in the crystal lattice while, for example, $\underline{Ni}$, $\underline{Cr}$, and $\underline{Mo}$ atoms are substitutionally dissolved and have *very low* diffusion rates. Very small amounts of these types of atom are able to pass the austenite layer from the melt to the dendrites of ferrite.

Growth of the Austenite Layer

The mathematical model for calculating the concentration of the alloying element as a function of position in a

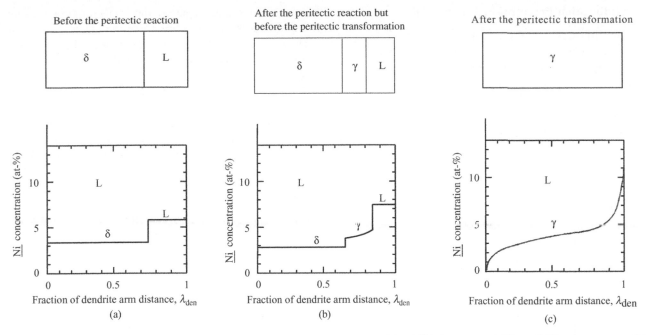

Figure 7.35 <u>Ni</u> distribution in an Fe–4at-% Ni alloy. A temperature gradient of 6×10^3 K/m was used. The solidification rate was 0.01 cm/min. Ni distribution (a) before the peritectic reaction, (b) after the peritectic reaction but before the transformation, and (c) after the peritectic transformation. Reproduced with permission from The Metal Society, Institute of Materials, Minerals & Mining.

peritectic transformation is described in the preceding section. It *cannot* be used for calculating the distribution of M atoms in the austenite layer because the lever rule is assumed to be valid, i.e. the diffusion rate of the M atoms must be very high in the solid phase, which is not the case in reality. Instead, the simple model, briefly described below, can be used.

With the lever rule a rough estimate is made of the amount of ferrite that is precipitated *before* the peritectic reaction starts. When it starts, austenite is precipitated between the ferrite and the melt and surrounds the ferrite completely. In the following peritectic transformation the austenite layer grows both at the austenite/ferrite interface and at the austenite/melt interface by precipitation of austenite.

The concentration of the alloying element in the austenite as a function of position can be calculated with good approximation for both interface reactions by using Scheil's equation, provided that the diffusion rate of the alloying element in austenite is zero. Figure 7.35 illustrates an example of such calculations for the alloy Fe–4 % Ni, performed by Fredriksson and Stjerndahl. Figure 7.35 (a) shows the Ni distribution before the peritectic reaction has started. Figure 7.35 (b) represents the Ni distribution immediately after the peritectic reaction but before the peritectic transformation has started.

Figure 7.35 (c) illustrates the Ni distribution at the end of the peritectic transformation.

It is known that addition of a second alloying element often influences the microsegregation of the first alloying element considerably. However, very few quantitative studies have been performed on how the interaction between different alloying elements affects the microsegregation in multicomponent systems. This topic will be discussed briefly in Section 7.9 on pages 217–219.

7.8 MICROSEGREGATION IN MULTICOMPONENT ALLOYS

So far we have only discussed the microsegregation of binary alloys in this chapter. However, most technically interesting alloys consist of more than two components. It is reasonably easy to illustrate the solidification process in ternary systems but harder in more complex alloys having more than two alloying elements. In the latter case, computer calculations are necessary but in most cases these only give qualitative results.

We will restrict discussion of this topic to a review of the basic principles of microsegregation in ternary alloys and illustrate this with a couple of important ternary alloys such as Al–Si–Cu and Fe–Cr–C (Section 7.9).

7.8.1 Solidification Processes in Ternary Alloys

Figure 7.36 illustrates a three-dimensional model of a ternary system with the components A–B–C. It involves the phase diagrams of the three binary systems AB, AC, and BC.

The three binary phase diagrams are projections on the walls of the three-dimensional phase diagram of the ternary alloy. In Figure 7.36 (b) they have been turned down into

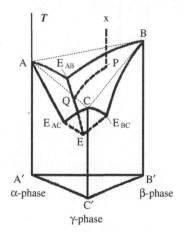

Figure 7.36 (a) When a melt with a composition corresponding to the point x cools, it starts to solidify at point P, its composition changes, due to microsegregation, along the line PQ. The composition point then moves along the eutectic line down the ternary eutectic point E. This motion describes the solidification process.

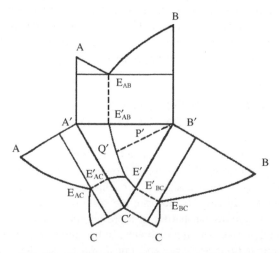

Figure 7.36 (b) The binary phase diagrams and the bottom plane A'B'C' are situated in the same plane when the three planes A'ABB', C'CBB' and A'ACC' are turned down into the paper plane. The eutectic lines E'_{AB}–E', E'_{AC}–E' and E'_{BC}–E' meet at E', the projection of the ternary eutectic point E. At point E the melt and the phases α, β and γ are in equilibrium with each other. The eutectic temperature T_E is the lowest possible temperature of the melt.

the same horizontal plane as the concentration triangle. The three phase diagrams are all of the eutectic type with a low solubility of the alloying element in the solid state (Figure 7.37). This is the simplest and most common type of ternary alloy.

The primary solid solutions, formed when the temperature decreases, are called α, β, and γ. Eutectic valleys (lines instead of points) run downwards in the ternary diagram and meet at the ternary eutectic point E. E corresponds to the lowest possible temperature at which the alloy exists as a liquid. At this temperature only one composition is possible. Figure 7.36 (b) illustrates the projections of the eutectic valleys and the three-dimensional phase diagram on the bottom plane. The differences in temperature of the points on the lines cannot be seen in this representation.

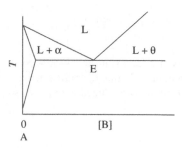

Figure 7.37 Typical binary eutectic phase diagram. The concentration (solubility) of the alloying element in the solid α-phase can be read on the horizontal axis.

For an alloy with an initial composition, corresponding to an arbitrary point P in the two-phase region in Figure 7.36 (a), the solidification starts with a primary precipitation of pure component A, normally as α-dendrites. During continued solidification the composition of both the precipitated α-phase and the liquid change gradually until the point Q on the eutectic valley line E_{AB}–E in Figure 7.36 (a) is reached. Then the eutectic reaction $L \rightarrow \alpha + \beta$ follows. The process is analogous to that for binary alloys as described in Section 6.4 on pages 145–147.

The eutectic reaction goes on until the composition of the melt corresponds to point E. The composition change of the melt during this process can be described as a motion along the eutectic line E_{AB}–E from point Q to the ternary eutectic point E. At the ternary eutectic point E the eutectic reaction changes to $L \rightarrow \alpha + \beta + \gamma$. The newly precipitated solid phases together form the ternary eutectic composition, which corresponds to point E. Analogous processes occur in alloys with other initial compositions, where the precipitation starts with either β- or γ-precipitation.

7.8.2 Simple Model of Microsegregation in Ternary Alloys

The pathway in a ternary system can be calculated using the same expressions as for binary alloys. In order to simplify the calculations and make them more illustrative we will assume that there is no back diffusion in the primarily precipitated phases (α-, β- or γ-dendrites) during solidification. The basic assumptions, which result in Scheil's segregation equation for the microsegregation of binary alloys, are listed on page 185.

Composition of the Melt during Primary Dendrite Solidification

To describe microsegregation during the primary precipitation of a solid phase in a ternary system, for example, for components A and B along the path PQ in Figure 7.36 (b), we use Scheil's equation [Equation (7.10) on page 186]. It can be applied simultaneously to the two components B and C as follows.

On the way from P to Q only α-phase is precipitated: L $\rightarrow$ α. For this reason, the partition coefficients $k_{\text{part}}^{\alpha/L}$ for the elements B and C occur in Equations (7.47) and (7.48).

$$x_B^L = x_B^{0L}(1 - f_\alpha)^{-(1 - k_{\text{part}B}^{\alpha/L})} \tag{7.47}$$

$$x_C^L = x_C^{0L}(1 - f_\alpha)^{-(1 - k_{\text{part}C}^{\alpha/L})} \tag{7.48}$$

where

$x_{B,C}^L$ = concentration of alloying elements B and C in the melt at an arbitrary point along the line PQ

$x_{B,C}^{0L}$ = initial concentration of elements B and C in the melt

f_α = volume fraction of the α-phase

$k_{\text{part}B,C}^{\alpha/L}$ = partition coefficient for elements B and C between the α-phase and the melt.

As the fraction f is the same for both components at a given temperature, f_α is the same in both equations. This gives the following relationship between f_α and the concentrations of the two components:

$$1 - f_\alpha = \left(\frac{x_B^L}{x_B^{0L}}\right)^{-\frac{1}{1 - k_{\text{part}B}^{\alpha/L}}} = \left(\frac{x_C^L}{x_C^{0L}}\right)^{-\frac{1}{1 - k_{\text{part}C}^{\alpha/L}}} \tag{7.49}$$

The Equations (7.49) are derived by solving $(1 - f_\alpha)$ from the Equations (7.47) and (7.48). When the temperature decreases, f_α increases and controls the concentrations of the three components, using Equations (7.47) and (7.48) and the relationship:

$$x_A^L = 1 - x_B^L - x_C^L \tag{7.50}$$

The relationships, which are valid between the fraction f_α and the concentrations of the alloying elements during the reaction, mean that the solidification process follows the line PQ in Figure 7.36 (a). The α-precipitation lasts until point Q is reached.

If the solidification occurs under equilibrium conditions and back diffusion can be neglected, the relationships (7.49) prove to be a good description of the concentrations of the components B and C in the melt during the solidification of the melt. These relationships give no information about the reaction temperatures. The temperatures are read for each set of concentrations of the alloying elements from the phase diagram or can be derived by using the thermodynamic relationships, which are used for calculation of the phase diagram.

Composition of the Melt during the Binary Eutectic Solidification

At point Q in Figure 7.36 (a), the precipitation of α-phase stops and is replaced by precipitation of a *eutectic* phase $\alpha + \beta$ when the temperature decreases. The composition of the melt changes during the eutectic reaction, due to segregation, and follows the eutectic line from point Q down to the ternary eutectic point E in Figure 7.36 (a). The α- and β-phases are precipitated simultaneously along the line QE.

If the volume fraction of each phase in the eutectic structure and the corresponding partition coefficient are known, the following relationships describe the segregation path during the eutectic reaction:

$$x_B^{L\,\text{bin}} = x_B^{LQ}(1 - f_{\text{bin eut}}^{\alpha+\beta})^{-\left[(1 - k_{\text{part}B}^{\alpha/L})f_{\text{bin eut}}^\alpha + (1 - k_{\text{part}B}^{\beta/L})f_{\text{bin eut}}^\beta\right]} \tag{7.51}$$

$$x_C^{L\,\text{bin}} = x_C^{LQ}(1 - f_{\text{bin eut}}^{\alpha+\beta})^{-\left[(1 - k_{\text{part}C}^{\alpha/L})f_{\text{bin eut}}^\alpha + (1 - k_{\text{part}C}^{\beta/L})f_{\text{bin eut}}^\beta\right]} \tag{7.52}$$

where

$x_{B,C}^{L\,\text{bin}}$ = concentration of elements B and C in the melt at an arbitrary point on the line QE during the binary eutectic reaction

$x_{B,C}^{LQ}$ = concentration of elements B and C in the melt at the start of the binary eutectic reaction (point Q)

$f_{\text{bin eut}}^{\alpha+\beta}$ = volume fraction of solidified binary eutectic, i.e. the ($\alpha + \beta$) structure, in the melt

$f_{\text{bin eut}}^{\alpha,\beta}$ = volume fractions of α-phase and β-phase in the binary eutectic structure.

It is reasonable to assume that the partition coefficients as well as the volume fractions in the eutectic structure are almost constant. If we assume that they are constant, the k_{part} and f values in Equations (7.51) and (7.52) can be evaluated from the binary phase diagrams. The k_{part} values are derived by the aid of the solidus and liquidus lines. The volume fractions are found by using the lever rule at the eutectic temperature in the binary phase diagrams. The values are assumed to be the same in the ternary phase diagram.

Ternary Eutectic Solidification

At the ternary eutectic point the composition of the melt remains constant during the rest of the solidification process. At point E the remaining melt solidifies at the constant temperature T_E in accordance with the ternary eutectic reaction: $L \rightarrow \alpha + \beta + \gamma$.

Along path PQ the fraction f_α of the initial volume solidifies. Along path QE the fraction $f_{\text{bin eut}}^{\alpha+\beta}$ of the *remaining* melt, i.e. the volume fraction $(1 - f_\alpha)$, solidifies. Thus the total fraction of solid f_E of the *initial* volume at point E will be:

$$f_E = f_\alpha + f_{\text{bineut}}^{\alpha+\beta}(1 - f_\alpha) \tag{7.53}$$

The remaining fraction $(1 - f_E)$ of melt solidifies with the constant ternary eutectic structure $(\alpha + \beta + \gamma)$. This can be calculated by using Equations (7.47), (7.48), (7.51), (7.52) and the binary phase diagrams of the component systems. This is illustrated in Example 7.5 below.

The formulae above have been derived for concentrations expressed as mole fractions. These equations are valid for weight per cent as well *provided that the partition coefficients $k_{\text{part}}x$ are replaced by $k_{\text{part}}c$.*

Application of the Model for Microsegregation

As an application of the simple model for microsegregation in ternary alloys we will analyse the solidification pattern of the alloy Al–10 wt-% Cu–6.0 wt-% Si. Figure 7.38(a) shows the microstructure of a solidified sample of the alloy.

- The large white regions, called Type I in Figure 7.38 (a), consist of pure aluminium. They form the matrix of the alloy and are formed during primary precipitation of α-phase. This corresponds to the reaction $L \rightarrow \alpha$;
- The mixture of white phase and facetted grey crystals, called Type II in Figure 7.38 (a), is the result of the eutectic reaction $L \rightarrow Al + Si$;

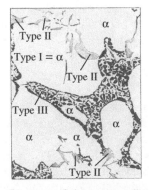

Figure 7.38 (a) Structure of the ternary alloy Al–10 wt-% Cu–6 wt-% Si. White areas = α-phase = Al; grey areas = β-phase = Si; black areas = θ-phase = Al$_2$Cu. Type I regions arise from the reaction $L \rightarrow \alpha$; Type II regions arise from the reaction $L \rightarrow \alpha + \beta$; and Type III regions arise from the reaction $L \rightarrow \alpha + \beta + \theta$.

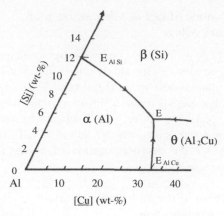

Figure 7.38 (b) The Al-rich corner of the Al–Cu–Si ternary system. The figure shows the eutectic lines $E_{AlSi} - E$ and $E_{AlCu} - E$ and the ternary eutectic point E.

- The mixture of white, grey and black phases, called Type III in Figure 7.38 (a), is the result of the eutectic reaction $L \rightarrow Al + Si + Al_2Cu$.

Figure 7.38 (b) shows a projection on the bottom plane of the Al-rich corner of the ternary system Al–Si–Cu. The phase diagrams of the binary systems Al–Si and Al–Cu, given in Example 7.5 below, confirm that there are two possible eutectic reactions:

$$L \rightarrow Al + Si \quad (\alpha + \beta) \quad \text{and} \quad L \rightarrow Al + Al_2Cu \quad (\alpha + \theta)$$

The initial composition of the alloy determines which one of the eutectic reactions will occur after precipitation of the α-phase. In Figure 7.38 (b) the eutectic lines $E_{AlSi} - E$ and $E_{AlCu} - E$ meet at the point E. A mixture of two solid phases, either $(\alpha + \beta)$ or $(\alpha + \theta)$, with segregated composition is precipitated. The reaction occurs at decreasing temperature until the ternary eutectic point E is reached. At point E the binary eutectic reaction stops and is replaced by the ternary eutectic reaction:

$$L \rightarrow Al + Si + Al_2Cu$$

This reaction results in precipitation of a mixture of three solid phases with a constant ternary eutectic composition. All the remaining melt at point E solidifies with this composition.

Example 7.5

When a molten Al–10 wt-% Cu–6.0 wt-% Si alloy has solidified three types of structures in the solid material have been identified: pure Al-phase, a mixed structure of (Al + Si)-phases and a ternary eutectic structure, which consists of a mixture of Al + Si + Al$_2$Cu phases.

Copper is extremely insoluble in silicon. The binary phase diagrams of AlSi and AlCu and the Al corner of the ternary phase diagram of the Al–Si–Cu system are given below.

(a) Calculate the composition of the melt when the binary eutectic reaction L → Al + Si starts.

(b) Calculate the volume fractions of the three different structures, Type I (Al), Type II (Al + Si), and Type III (Al+ Si + Al₂Cu), in the alloy after complete solidification.

Solution:

(a): Primary precipitation of Al phase and eutectic Al + Si phase:

During the primary precipitation of the Al phase the concentrations in the melt of the alloying elements Si and Cu are described by the Equations (7.47) and (7.48) on page 213. If we assume that B = Si and C = Cu we get (all concentrations in wt-%, which is correct if proper $k_{part}c$-values instead of $k_{part}x$-values are used):

$$c_{\underline{Si}}^{L} = c_{\underline{Si}}^{0L}(1-f_{Al})^{-(1-k_{part\underline{Si}}^{\alpha/L})} \quad (1')$$

and

$$c_{\underline{Cu}}^{L} = c_{\underline{Cu}}^{0L}(1-f_{Al})^{-(1-k_{part\underline{Cu}}^{\alpha/L})} \quad (2')$$

where the initial values $c_{\underline{Si}}^{0L} = 6.0$ wt-% and $c_{\underline{Cu}}^{0L} = 10$ wt-% are given in the text and marked by the point P in the ternary phase diagram in the figure below. Equations (1') and (2') are valid along the pathway PQ [Figure 7.36 (a) on page 212].

We want to calculate the fraction f_{Al} of the Al-phase. It can be obtained if we (i) derive the partition coefficients for Si and Cu, and (ii) calculate the coordinates $(c_{\underline{Cu}}^{LQ}; c_{\underline{Si}}^{LQ})$ of point Q.

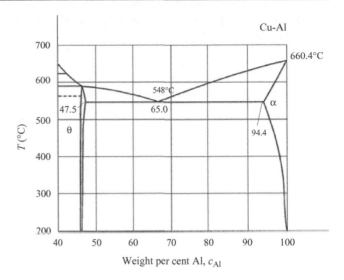

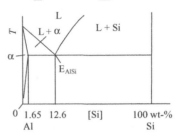

Calculation of $k_{part\underline{Si}}^{\alpha/L}$ and $k_{part\underline{Cu}}^{\alpha/L}$

The liquidus and solidus lines in the phase diagrams are approximately straight lines. In this case the partition coefficients can easily be derived from the readings at the eutectic lines in the two phase diagrams above.

$$k_{part\underline{Si}}^{Al/L} \approx \frac{1.65}{12.6} = 0.13 \quad \text{and} \quad k_{part\underline{Cu}}^{Al/L} \approx \frac{100-94.4}{100-65} = 0.16$$

Calculation of $(c_{\underline{Cu}}^{LQ}; c_{\underline{Si}}^{LQ})$

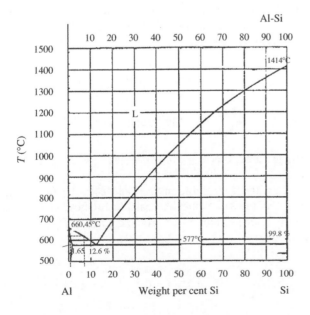

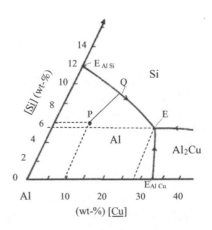

Coordinates, given in the text, calculated, or read from the phase diagram

Point P:	(10;6)	wt-%
Point E_{AlSi}:	(0;126)	wt-%
Point Q:	(14;9)	wt-%
Point E:	(28;5.5)	wt-%

We assume that the line $E_{AlSi}E$ in the ternary phase diagram is straight. It is determined by the two points $E_{AlSi} = (0; 12.6)$ and $E = (28; 5.5)$, which are read from the ternary phase diagram. The equation of the line is found to be:

$$c_{\underline{Si}}^{L} = 12.6 - 0.25\, c_{\underline{Cu}}^{L} \qquad (3')$$

The point Q lies on the line $E_{AlSi}E$. Thus equation $(3')$ is valid for point Q. Point Q also lies on the line PQ. We apply Equations $(1')$ and $(2')$ to point Q and solve $(1 - f_\alpha)$. The fraction f_α is the same for both components, which gives:

$$1 - f_\alpha = \left(\frac{c_{\underline{Si}}^{LQ}}{c_{\underline{Si}}^{0L}}\right)^{-\frac{1}{1-k_{\text{part}\underline{Si}}^{\alpha/L}}} = \left(\frac{c_{\underline{Cu}}^{LQ}}{c_{\underline{Cu}}^{0L}}\right)^{-\frac{1}{1-k_{\text{part}\underline{Cu}}^{\alpha/L}}} \qquad (4')$$

The initial concentrations of Si and Cu are known and the values of the partitions coefficients have been derived from the binary phase diagrams of the Al–Si and Al–Cu systems on page 215. The coordinates of point Q can be solved from Equations $(3')$ and $(4')$. The Si concentration in Equation $(3')$ is introduced into Equation $(4')$:

$$\left(\frac{12.6 - (0.25 \times c_{\underline{Cu}}^{LQ})}{6.0}\right)^{-\frac{1}{1-0.13}} = \left(\frac{c_{\underline{Cu}}^{LQ}}{10.0}\right)^{-\frac{1}{1-0.16}}$$

The solution of equation $(4')$ is: $\quad c_{\underline{Cu}}^{LQ} \approx 14 \text{ wt-}\%$

Equation $(3')$ gives: $\quad c_{\underline{Si}}^{LQ} = 12.6 - 0.25 \times 14 \approx 9.1 \text{ wt-}\%$

(b): Fraction of Al phase and (Al + Si) phase
We want the volume fractions of Types I, II and III. If we can calculate and $f_\alpha = f_{Al}$ and $f_{\text{bin eut}}^{Al+Si}$, then f_E can easily be found using Equation (7.53) on page 214 and the volume fractions can be derived.

Calculation of f_{Al}
When the coordinates of point Q are known the value of f_{Al} can be calculated from Equation $(4')$ using the Si or

Cu values:

$$1 - f_{Al} = \left(\frac{c_{\underline{Si}}^{LQ}}{c_{\underline{Si}}^{0L}}\right)^{-\frac{1}{1-k_{\text{part}\underline{Si}}^{\alpha/L}}} = \left(\frac{9.1}{6.0}\right)^{-\frac{1}{1-0.13}} = 0.620$$

$$1 - f_{Al} = \left(\frac{c_{\underline{Cu}}^{LQ}}{c_{\underline{Cu}}^{0L}}\right)^{-\frac{1}{1-k_{\text{part}\underline{Cu}}^{\alpha/L}}} = \left(\frac{14}{10}\right)^{-\frac{1}{1-0.16}} = 0.670$$

We choose the average value:

$$f_{Al} = 1 - 0.645 = 0.355 \qquad (5')$$

Calculation of $f_{\text{bin eut}}^{Al+Si}$
During the binary eutectic reaction $L \rightarrow Al + Si$, Equation (7.51) on page 213 is valid. The reaction starts at point Q and ends at the ternary eutectic point E. At point E we have:

$$c_{\underline{Si}}^{LE} = c_{\underline{Si}}^{LQ}(1 - f_{\text{bin eut}}^{Al+Si})^{-\left[(1-k_{\text{part}\underline{Si}}^{Al/L})f_E^{Al}+(1-k_{\text{part}\underline{Si}}^{Si/L})f_E^{Si}\right]} \qquad (6')$$

where f_E^{Al} and f_E^{Si} are the volume fractions of the Al and Si phases at the eutectic point E.

We assume that the volume fractions f^{Al} and f^{Si} are the same in the binary and ternary systems. Thus we get from the binary Al–Si phase diagram by using the lever rule:

$$f_{\text{bin eut}}^{Al} = f_E^{Al} = \frac{100 - 12.6}{100 - 1.65} = 0.89$$

and

$$f_{\text{bin eut}}^{Si} = f_E^{Si} = 1 - 0.89 = 0.11$$

The Si coordinate of the ternary eutectic point E was determined in (a) and found to be 5.5 wt-%. The Si coordinate of point Q was calculated in (a) and found to be 9.1 wt-%. We calculate the $k_{\text{part}\underline{Si}}^{\alpha/L}$ and $k_{\text{part}\underline{Si}}^{Si/L}$ values for the points Q and E and use their average values.

$$Q: k_{\text{part}\underline{Si}}^{Al/L} \approx \frac{1.65}{9.1} = 0.18 \quad \text{and} \quad k_{\text{part}\underline{Si}}^{Si/L} = \frac{100}{9.1} = 11$$

$$E: k_{\text{part}\underline{Si}}^{Al/L} = \frac{1.65}{5.5} = 0.30 \quad \text{and} \quad k_{\text{part}\underline{Si}}^{Si/L} = \frac{100}{5.5} = 18$$

Average values are $k_{\text{part}\underline{Si}}^{\alpha/L} = 0.24$ and $k_{\text{part}\underline{Si}}^{Si/L} = 14.5$

We insert the concentrations $c_{\underline{Si}}^{LE}$ and $c_{\underline{Si}}^{LQ}$, the f_E^{Al} and f_E^{Si} values and the k_{part} values (all calculated above) into Equation $(6')$ and the value of the volume fraction $f_{\text{bin eut}}^{Al+Si}$ can be calculated:

$$f_{\text{bin eut}}^{Al+Si} = 1 - \left(\frac{5.5}{9.1}\right)^{\frac{-1}{(1-0.24)0.89+(1-14.5)0.11}} = 0.455$$

Now we have all the information we need to calculate the volume fractions of the three structures.

Calculation of Volume Fractions of the Structure Types

At point Q the fraction of solid Al phase (Type I structure) is f_{Al}. The fraction of the remaining melt is $(1 - f_{Al})$.

At point E the fraction of solid mixture of the $(Al + Si)$-phases of the remaining melt is $f_{bin\ eut}^{Al+Si}$. The corresponding fraction of Type II structure related to the initial volume is $f_{bin\ eut}^{Al+Si}(1 - f_{Al})$. The fraction of the total solid structure at the ternary eutectic point E is:

$$f_E = f_{Al} + f_{bin\ eut}^{Al}(1 - f_{Al}) = 0.355 + 0.455(1 - 0.355) = 0.648 \tag{7'}$$

The fraction of remaining melt at point E is:

$$1 - f_E = 1 - 0.648 = 0.352 \tag{8'}$$

This fraction solidifies with ternary eutectic structure.

Volume fraction of Type I structure: $f_{Al} = 0.355$
Volume fraction of Type II structure:

$$f_{bin\ eut}^{Al+Si}(1 - f_{Al}) = 0.455 \times 0.645 = 0.293$$

Volume fraction of Type III structure:

$$1 - f_{Al} - f_{bin\ eut}^{Al+Si}(1 - f_{Al}) = 1 - 0.355 - 0.455 \times 0.645 = 0.352$$

Answer:

(a) The composition of the melt when the binary eutectic reaction starts is 9 wt-% Si, 14 wt-% Cu and 77 wt-% Al.

(b) The volume fraction of Type I structure is 36 %. The volume fraction of Type II structure is 29 %. The volume fraction of Type III structure is 35 %.

If the initial composition is different, the precipitated phases and the structure of the solid alloy will be different. If the precipitation of primary α-phase ends in a point on the $E_{AlCu} - E$ line then a mixture of α-phase and Al_2Cu will precipitate instead of α and Si. Analogous calculations can be performed for this case.

The total volume fraction f_E of solid, independent of type, at point E can be calculated in a simpler way than above by going from point P straight to the ternary eutectic point E in a ternary phase diagram (Figure 7.39).

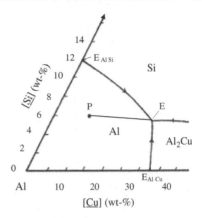

Figure 7.39 The ternary phase diagram of Al–Cu–Si with the path from point P to the ternary eutectic point E drawn in the figure.

7.9 MICROSEGREGATION AND PERITECTIC REACTIONS AND TRANSFORMATIONS IN MULTICOMPONENT IRON-BASE ALLOYS

In Section 7.7 it was mentioned that both peritectic reactions and transformations and microsegregation during cooling change the structures and compositions of alloys. Microsegregation is treated extensively for binary alloys and briefly for ternary alloys in the preceding sections of this chapter. Peritectic reactions in binary iron-base alloys of technical importance are treated in Section 7.7. Iron-base alloys often contain more than two components and for this reason we have added a section on microsegregation in some ternary iron-base alloys in this chapter.

7.9.1 Microsegregation and Peritectic Reactions and Transformations in Ternary Iron-Base Alloys

As an example of microsegregation in ternary iron-base alloys we will discuss Fe–Cr–C alloys. They belong to a group of commercially important alloys that has been investigated quantitatively.

As the behavior of the alloys is closely related to their ternary phase diagrams we will start with a description of the phase diagram of the Fe–Cr–C system as a basis for further discussion.

Phase Diagram of the System **Fe–Cr–C**

In the scientific literature there are a number of investigations on the phase diagram of Fe–Cr–C. Bungart and his coworkers performed one of the most detailed studies. A summary of his results is shown in Figure 7.40.

The phase diagram in Figure 7.40 looks different as compared with the ternary phase diagram we saw in Section 7.8. It is a projection on a horizontal plane and hence there is no temperature axis. It illustrates the Fe corner of the ternary phase diagram Fe–Cr–C but does not use the concentration

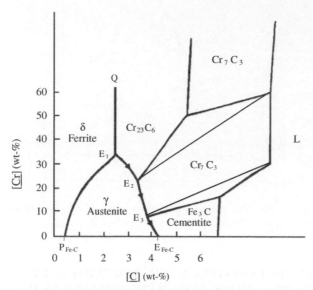

Figure 7.40 Simplified Fe corner of the ternary phase diagram of the system Fe–Cr–C according to Bungart. Reproduced by permission of Verlag Stahleisen GmbH, Düsseldorf, Germany. © 1958 Verlag Stahleisen GmbH, Düsseldorf, Germany.

triangle that is the normal representation [Figure 7.38 (b) on page 214] of component concentrations. The difference is not especially sensational, though. The concentrations of $\underline{Cr}$ and $\underline{C}$ in Figure 7.40 are plotted along ordinary orthogonal axes instead of axes along two triangle sides.

Eutectic and Peritectic Reaction in Solidifying* Fe–Cr–c *Melts

Two types of metal carbides are formal and dissolved in solidifying Fe–M–c melt during cooling:

23-carbide $= M_{23}C_6$
7-carbide $= M_7C_3$

where M is a metal atom, in this case Cr.

If the alloy melt has a composition with a comparatively low carbon concentration, the primary precipitation when the melt cools will be ferrite. At high $\underline{C}$ concentration and moderate $\underline{Cr}$ concentrations, primary precipitation of austenite will occur. In this case no precipitation of ferrite will ever occur. The peritectic line P_{Fe-C}–E_1 in Figure 7.40 is the intersection between the liquidus surfaces L/δ and L/γ. P_{Fe-C} is the peritectic point in the binary Fe–C phase diagram.

The liquidus surface of the δ-phase meets the liquidus surface of the so-called 23-carbide at high $\underline{Cr}$ concentration and a eutectic valley is obtained. Along the valley Q–E_1 the

reaction L $\rightarrow$ δ + $M_{23}C_6$ occurs, which results in δ-phase and 23-carbide during the cooling. The valley ends at point E_1 where the four phases L, δ, γ and 23-carbide are at equilibrium with each other. At point E_1 the δ-phase is dissolved and transformed into γ-phase by a peritectic reaction and transformation.

During further cooling the composition of the melt follows the eutectic valley E_1–E_2. The eutectic reaction results in the precipitation of a eutectic structure of γ-phase and 23-carbide. At the eutectic point E_2, four phases, L, γ, 23-carbide and 7-carbide, are in equilibrium with each other. At the eutectic point E_2, the 23-carbide is dissolved and transformed during cooling into so-called 7-carbide, by a peritectic reaction and transformation.

As the cooling process continues, the composition of the melt follows the eutectic valley E_2–E_3. Precipitation of a eutectic structure of γ-phase and 7-carbide occurs. At the eutectic point E_3, four phases, L, γ, 7-carbide and iron carbide, Fe_3C, called cementite, are in equilibrium with each other. At point E_3 the 7-carbide is dissolved and transformed during cooling into cementite by a peritectic reaction and transformation.

Further cooling results in precipitation of a eutectic structure of γ-phase and cementite. This process goes on until solidification is complete. The Fe–Cr–C system has no ternary eutectic point. The point E_{Fe-C} is the eutectic point in the binary Fe–C phase diagram. The cooling process, described above, is summarized in Table 7.1.

The phase diagram will be used to explain the influence of carbon concentration on the microsegregation of chromium and how the precipitation of carbides in iron-base alloys can be controlled.

Influence of Carbon on Microsegregation of* $\underline{Cr}$ *in* Fe–Cr–C *Alloys

The degree of segregation of $\underline{Cr}$ as a function of the $\underline{C}$ concentration in a series of Fe–1.5 % Cr steels with increasing $\underline{C}$ concentrations has been examined experimentally. The result is illustrated in Figure 7.41. The figure shows that in a binary Fe–Cr alloy there is no segregation at all ($S = 1$ for the $\underline{C}$ concentration $= 0$). The degree of segregation of $\underline{Cr}$ increases initially with increasing $\underline{C}$ concentration but decreases again when the addition of C increases. At 1.6 wt-% $\underline{C}$, the degree of segregation of $\underline{Cr}$ reaches its maximum $S = 5$.

The reason for the influence of the $\underline{C}$ concentration on the $\underline{Cr}$ segregation has been investigated thoroughly in the scientific literature.

The increase in the degree of $\underline{Cr}$ segregation at low $\underline{C}$ concentrations can be explained by the decrease of the partition coefficient k_{part}, caused by the interaction between $\underline{C}$ and $\underline{Cr}$ atoms in austenite and in the melt. The phase diagram of the ternary system Fe–Cr–C in Figure 7.40

TABLE 7.1 **Survey of peritectic and eutectic reactions in a solidifying Fe–Cr–C melt.**

Reaction peritectic reaction = (Pe) eutectic reaction = (Eu)		Point/line in the phase diagram	Temperature (°C)	Cr concentration (wt-%)	C concentration (wt-%)	Comments
$L + \delta \rightarrow \gamma$	(Pe)	E_1	1275	34	2.4	δ dissolves, γ enters
$L \rightarrow \gamma + Cr_{23}C_6$	(Eu)	$E_1 \; E_2$				Precipitation of γ and $Cr_{23}C_6$
$L + Cr_{23}C_6 \rightarrow Cr_7C_3$	(Pe)	E_2	1255	23	3.5	$Cr_{23}C_6$ dissolves, Cr_7C_3 enters
$L \rightarrow \gamma + Cr_7C_3$	(Eu)	$E_2 \; E_3$				Precipitation of γ and Cr_7C_3
$L + Cr_7C_3 \rightarrow Fe_3C$	(Pe)	E_3	1175	8.0	3.8	Cr_7C_3 dissolves, Fe_3 C enters
$L \rightarrow \gamma + Fe_3C$	(Eu)	$E_3 \; E_{Fe-C}$				Precipitation of γ and Fe_3C

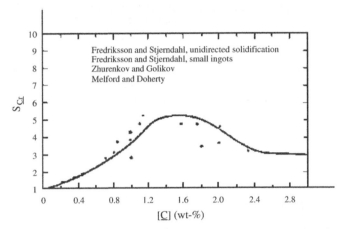

Figure 7.41 The degree of segregation of C as a function of the C concentration (in wt-%) in a series of Fe–Cr–C alloys with constant Cr concentration. The values, which are plotted in the diagram, originate from four different and independent investigations. Reproduced with permission from the Scandinavian Journal of Metallurgy, Blackwell.

can be used to explain the decrease of the degree *S* of segregation at increasing carbon concentration, which is illustrated in Figure 7.41. It can be seen from the phase diagram that alloys with increasing C content and a constant Cr content meet the eutectic line $E_1 \; E_2 \; E_3$ at a lower Cr content when the carbon content increases due to the negative slope of the eutectic line.

Peritectic Reactions and Other Solidification Processes in Cr-Bearing Steels

In connection with the discussion of the ternary phase diagram of the system Fe–Cr–C we found that different carbides are formed and disappear during the solidification process. Figure 7.42 shows the microstructure of a 7-carbide eutectic mixture. The single-phase regions of austenite-dendrite arms in cross section with the eutectic structure $\gamma + M_7C_3$ between the arms can be observed.

This topic is further discussed in the shape of the solved example below, which concentrates on avoiding unwanted carbides in the solid material after complete solidification.

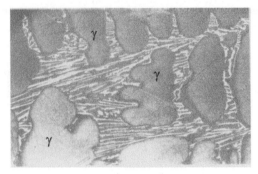

Figure 7.42 Broom-shaped morphology of a 7-carbide eutectic structure. The white lines in the figure are carbide plates, and the dark areas between the plates consist of γ-phase.

Example 7.6

When a steel melt, which contains chromium, solidifies there is a risk of precipitation of 7-carbide, $Cr_7 \; C_3$.

(a) Suggest a method for calculating the relationship between the maximum carbon concentration $c_{\underline{C}}^0$ and the maximum chromium concentration $c_{\underline{Cr}}^0$ that the steel can be allowed to contain if Cr_7C_3 precipitation is to be avoided.

(b) What is the maximum chromium concentration, expressed in wt-%, that can be allowed if the carbon concentration in the steel is 2.0 wt-% and $Cr_7 \; C_3$ precipitation is to be avoided?

The phase diagram on page 218 may be used. The partition constants of chromium and carbon are $k_{part_{\underline{Cr}}} = 0.88$ and $k_{part_{\underline{C}}} = 0.42$ respectively.

Solution:

From the phase diagram we can read the solubility product of Cr_7C_3:

$$c_{\underline{Cr}}^7 \times c_{\underline{C}}^3 = const \qquad (1')$$

The value of the constant is obtained simply by using data from Table 7.1 on this page for the eutectic point E_3 in the

Fe–Cr–C phase diagram. At this point the melt is in equilibrium with the 7-carbide.

$$\text{const} = c_{\underline{Cr}}^7 \times c_{\underline{C}}^3 = 8.0^7 \times 3.8^3 \ (\text{wt-}\%)^{10} \qquad (2')$$

where the Cr and C concentrations at the eutectic point are 8.0 wt-% and 3.8 wt-% respectively. When the steel melt solidifies its composition is gradually changed by microsegregation. In order to be able to calculate the C and Cr concentrations in the melt at various solidification fractions f we will use segregation equations.

The diffusion of the Cr atoms in the solid phase is *slow* and it is therefore reasonable to assume that Scheil's equation (page 186) is valid for Cr concentration in the melt:

$$c_{\underline{Cr}}^L = \frac{c_{\underline{Cr}}^s}{k_{\text{part}\,\underline{Cr}}} = c_{\underline{Cr}}^0 (1 - f)^{-(1 - k_{\text{part}\,\underline{Cr}})} \qquad (3')$$

On the other hand the diffusion of the C atoms is *rapid* and thus the C concentration in the melt is best described by the aid of the lever rule (page 189):

$$c_{\underline{C}}^L = \frac{c_{\underline{C}}^s}{k_{\text{part}\,\underline{Cr}}} = \frac{c_{\underline{C}}^0}{1 - f(1 - k_{\text{part}\,\underline{Cr}})} \qquad (4')$$

(a): First of all we calculate the fraction of solid phase in the eutectic composition, when the C concentration reaches its maximum value, $c_{\underline{C}}^L = 3.8$ wt-%. The initial C concentration in the melt equals the average value of the C concentration in the melt (compare Example 7.1 on pages 187–188). The partition constants are known and it is, in principle, possible to solve f from Equation (4'). By inserting the known value of f and the known Cr concentration at the eutectic composition, $c_{\underline{Cr}}^L = 8.0$ wt-%, into Equation (3') we can calculate the maximum Cr concentration in the melt before it starts to solidify. This concentration is the desired maximum Cr concentration that can be allowed.

(b): We introduce the known values into Equation (4'):

$$c_{\underline{C}}^L = \frac{c_{\underline{C}}^0}{1 - f(1 - k_{\text{part}\,\underline{Cr}})} \qquad (4')$$

and get:

$$3.8 = \frac{2.0}{1 - f(1 - 0.42)}$$

which gives $f = 0.817$. This value is inserted into Equation (3') together with the other known quantities:

$$c_{\underline{Cr}}^L = \frac{c_{\underline{Cr}}^s}{k_{\text{part}\,\underline{Cr}}} = c_{\underline{Cr}}^0 (1 - f)^{-(1 - k_{\text{part}\,\underline{Cr}})}$$

$$\Rightarrow 8.0 = c_{\underline{Cr}}^0 (1 - 0.817)^{-(1 - 0.88)}$$

which gives $c_{\underline{Cr}}^0 = 6.52$ wt-%

Answer:

(a) See the description above.

(b) If the C concentration is 2.0 wt-%, the Cr concentration must be below 6.5 wt-% if 7-carbide precipitation is to be avoided.

SUMMARY

■ *Microsegregation*

When an alloy solidifies by dendritic growth the concentration of the alloying element will be distributed unevenly in the material. This phenomenon is called microsegregation.

If the partition constant $k_{\text{part}} = x^s / x^L < 1$ the alloying element will be concentrated in the remaining melt during the solidification process. If the diffusion rate in the solid phase is *low* the last parts to solidify will have a higher content of the alloying element than the earlier solidified parts.

If $k_{\text{part}} > 1$ the opposite is true.

■ *Scheil's Segregation Equation*

Provided that:

- the convection and diffusion of the alloying element in the melt at each moment is so violent that the melt has an equal composition x^L everywhere;
- the diffusion of the alloying element in the solid phase is so *slow* that it can be completely neglected;
- a local equilibrium exists at the interface between the solid phase and the melt.

The equilibrium can be described by a partition constant, then Scheil's segregation equation is valid:

$$x^L = \frac{x^s}{k_{\text{part}}} = x_0^L (1 - f)^{-(1 - k_{\text{part}})}$$

A special case is when the fraction that solidifies with eutectic structure f_E has a concentration for the alloying element of:

$$x_E^L = x_0^L (1 - f_E)^{-(1 - k_{\text{part}})}$$

■ *Lever rule*

If the diffusion is *rapid* in the solid phase then the lever rule is valid:

$$x^L = \frac{x^s}{k_{\text{part}}} = \frac{x_0^L}{1 - f(1 - k_{\text{part}})}$$

■ Solidification Processes in Alloys

Solidification Process

Alloys have two solidification fronts. The alloy starts to solidify at the first solidification front. Its temperature is the liquidus temperature T_L. Solidification is complete at the second solidification front. Its temperature is the solidus temperature T_s. Between the two solidification fronts there is a two-phase region with solid phase and melt.

The basis for calculation of temperatures, concentration of alloying element, and fraction of solid phase is the heat equation:

$$\left(\rho c_p + \rho(-\Delta H)\frac{df}{dT} \right)\frac{\partial T}{\partial t} = k\left(\frac{\partial^2 T}{\partial y^2} \right)$$

which is integrated after introduction of a convenient expression for df/dT. Two alternatives for df/dT are:
If Scheil's equation is valid:

$$\frac{df}{dT} = \frac{-1}{1 - k_{part}}(T_M - T)^{-\frac{2 - k_{part}}{1 - k_{part}}}(T_M - T_0)^{\frac{1}{1 - k_{part}}}$$

If the lever rule is valid:

$$\frac{df}{dT} = \frac{-(T_M - T_0)}{(1 - k_{part})(T_M - T)^2}$$

A simple alternative to solving the heat equation is calculation of desired quantities with the aid of basic equations of heat transport.

Solidification Interval and Solidification Time of Alloys

Solidification Interval:
$$[\Delta T = T_L - T_s]$$

Solidification Time:
In alloys with a wide solidification interval the solidification time can be calculated by integration of the differential equation:

$$\frac{dQ}{dt} = V_0 \rho_{metal}\left[c_p^{metal}\frac{dT}{dt} + (-\Delta H)\frac{df}{dT} \right]$$

If the heat flux is constant during cooling and solidification then:

$$c_p^L\left(-\frac{dT}{dt} \right) = (-\Delta H)\frac{df}{dt}$$

The total solidification time corresponds to $f = 1$.

$$\theta = \frac{-\Delta H}{c_p^L\left(-\dfrac{dT}{dt} \right)}$$

■ Scheil's Modified Segregation Equation

Back Diffusion

Rapid diffusion in the solid phase eliminates or reduces the concentration gradient of the alloying element. This leads to an exchange of atoms between the liquid and solid phases. The return of alloying atoms from the liquid to the solid is called back diffusion.

Scheil's Modified Segregation Equation

If the back diffusion in the solid phase is considered, Scheil's modified equation is valid:

$$x^L = x_0^L\left(1 - \frac{f}{1 + D_s\left(\dfrac{4\theta}{\lambda_{den}^2} \right)k_{part}} \right)^{-(1 - k_{part})}$$

$$= x_0^L\left(1 - \frac{f}{1 + B} \right)^{-(1 - k_{part})}$$

where $B = \dfrac{4 D_s \theta k_{part}}{\lambda_{den}^2}$ is a correction term, which depends on the back diffusion in the solid phase.

■ Calculation of the Concentration of the Alloying Element as a Function of the Fraction of Solid Phase

- At *high* diffusion rates *the lever rule* is valid.
- At *low* diffusion rates *Scheil's modified segregation equation* is valid.
- When $D_s = 10^{-11}\ \mathrm{m^2/s}$ it is difficult to decide which relation one should use. The cooling rate is allowed to decide.
- At *low* cooling rates *the lever rule* is valid.
- At *high* cooling rates *Scheil's modified equation* agrees best with reality.

■ Degree of Microsegregation

$$S = \frac{x_{max}^s}{x_{min}^s} = \left(\frac{B}{1 + B} \right)^{-(1 - k_{part})} \quad \text{where} \quad B = \frac{4 D_s \theta k_{part}}{\lambda_{den}^2}$$

- The *higher* the k_{part} value is, the *lower* will be the S value.
- The *higher* the diffusion rate is the *lower* will be the S value.
- The *higher* the cooling rate is the *higher* will be the S value.

Interstitially dissolved elements diffuse more easily through a crystal lattice than do substitutionally dissolved alloying elements, which are parts of the lattice. Hence the former have a higher diffusion rate and consequently a lower degree of segregation than do the latter.

■ *Solidification Processes and Microsegregation in Iron-Base Alloys*

The solidification process in iron-base alloys starts with dendritic solidification and precipitation of either ferrite (δ) or austenite (γ).

The microsegregation in the two cases is different, depending on two factors:

- differences in diffusion rate of the alloying element in ferrite and in austenite.
- differences in partition constants of the alloying element in δ/L and γ/L.

The degree of segregation S can be calculated using the phase diagram, the diffusion rate of the alloying elements, and the cooling rate.

Microsegregation in Primary Precipitation of Ferrite (BCC)

The diffusion rates of the alloying elements are generally comparatively high and thus the microsegregation is relatively small.

- The *higher* the k_{part}-value is the *smaller* will S be.
- The *higher* the diffusion rate is the *smaller* will S be.

When the lever rule is valid and the cooling rate is low, then $S = 1$. This is valid especially for interstitially dissolved elements such as $\underline{C}$, $\underline{N}$, and $\underline{H}$ (rapid diffusion).

■ *Microsegregation in Primary Precipitation of Austenite (FCC)*

Microsegregation in austenite is greater than in ferrite because the diffusion rate is generally much lower in austenite than in ferrite, due to the structure. However, interstitially dissolved elements, for example $\underline{C}$, have high diffusion rates even in austenite and their degree of segregation is small.

Peritectic Reactions and Transformations

A *peritectic reaction* is a reaction that occurs when a melt reacts with a primarily formed phase α and a secondary phase β is formed:

$$L + \alpha = \beta$$

The peritectic reaction is followed by *a peritectic transformation*. There is a distinction between a peritectic reaction and a peritectic transformation.

In a *peritectic reaction*, the three phases α, β and melt are normally in direct contact with each other. The alloying element B diffuses from the secondary phase β through the melt to the primary phase α. A layer of β is formed around the α-phase.

In a *peritectic transformation*, the melt and the primary α-phase are separated by the secondary β-phase. B atoms diffuse through the secondary β-phase to the α-phase and the α-phase is transformed into B-rich β-phase. The thickness of the β-layer increases during the cooling. If the phase diagram is mirror-inverted the β-phase becomes transformed into α-phase.

Peritectic Reactions and Transformations in **Fe–C**

Peritectic reactions and transformations in Fe–C alloys occur rapidly, i.e. within a temperature interval ΔT of a few degrees celsius.

$\underline{C}$ diffuses through the austenite layer around dendrites of ferrite. By setting up material balances at the interfaces it is possible to find an expression for the *peritectic transformation rate* as a function of the diffusion rate of $\underline{C}$ in austenite and the $\underline{C}$ concentrations at the interfaces. The thickness of the austenite layer can be written

$$l^{\gamma} = \text{const}(\Delta T)$$

Peritectic Reactions and Transformations in **Fe–M** *Alloys*

Peritectic reactions also occur in alloys where the alloying element is substitutionally dissolved. Examples of such binary alloys are Fe–Ni and Fe–Mn alloys.

The fundamental difference between Fe–C and Fe–M alloys is that the $\underline{C}$ atoms that are interstitially dissolved have high diffusion rates in the crystal lattice while, for example, the $\underline{M}$ atoms are substitutionally dissolved and have very low diffusion rates. Very small amounts of these atoms can pass through the austenite layer from the melt to the dendrites of ferrite. The M concentrations in the austenite layer can be calculated using Scheil's equation because the diffusion rates of M in the solid are very small.

■ *Microsegregation in Multicomponent Alloys*

Most technically interesting alloys contain two or several alloying elements. The microsegregation of an alloying element is influenced by the other alloying elements.

Solidification Process in a Ternary Alloy

The projections on the walls of the three-dimensional phase diagram are the three binary phase diagrams of the systems A–B, A–C, and B–C. The diagram contains three eutectic valleys, which run downwards and end in the ternary eutectic point E.

At high temperatures the alloy is molten. The solidification starts with a primary precipitation of α-, β-, or γ-phase. Solidification of a binary eutectic occurs along one of the eutectic valleys and ends in point E. E corresponds to the only temperature and composition at which the three components and the melt can exist simultaneously. The rest of the alloy solidifies as a ternary eutectic mixture of three phases.

■ *Microsegregation in Ternary Alloys*

Primarily Precipitated Structure (Type I)
Scheil's equation is applied to the two alloying elements simultaneously. The equations are valid during the primary precipitation of, for example, α-phase (type I structure):

$$x_B^L = x_B^{0L}(1 - f_\alpha)^{-(1 - k_{partB}^{\alpha/L})}$$
$$x_C^L = x_C^{0L}(1 - f_\alpha)^{-(1 - k_{partC}^{\alpha/L})}$$

The initial composition of the alloy determines which phase precipitates (α-, β-, or γ-phase).

Binary Eutectic Structure along Eutectic Valley (Type II)

$$x_B^{Lbin} = x_B^{LQ}(1 - f_{bin\ eut}^{\alpha+\beta})^{-\left[(1 - k_{partB}^{\alpha/L})f_{bin\ eut}^\alpha + (1 - k_{partB}^{\beta/L})f_{bin\ eut}^\beta\right]}$$
$$x_C^{Lbin} = x_C^{LQ}(1 - f_{bin\ eut}^{\alpha+\beta})^{-\left[(1 - k_{partC}^{\alpha/L})f_{bin\ eut}^\alpha + (1 - k_{partC}^{\beta/L})f_{bin\ eut}^\beta\right]}$$

Ternary Eutectic Structure at Point E (Type III)
At the ternary eutectic point, the remaining melt solidifies at constant temperature and a eutectic mixture of the three phases α, β, and γ is precipitated.

Volume fraction of Type I structure $= f_\alpha$
Volume fraction of Type II structure $= f_{bin\ eut}^{\alpha+\beta}(1 - f_\alpha)$
Volume fraction of Type III structure $= 1 - f_\alpha - \left[f_{bineut}^{\alpha+\beta}(1 - f_\alpha)\right]$

Total fraction of precipitated structures at point E:

$$f_E = f_\alpha + \left[f_{bin\ eut}^{\alpha+\beta}(1 - f_\alpha)\right]$$

Fraction of ternary eutectic structure: after total solidification

$$1 - f_E = 1 - f_\alpha - \left[f_{bin\ eut}^{\alpha+\beta}(1 - f_\alpha)\right]$$

Microsegregation and Peritectic Reactions in Iron-Base Ternary Alloys

Phase diagram of the ternary system Fe–Cr–C
When an Fe–Cr–C alloy solidifies, the primarily precipitated phase is either ferrite or austenite, depending on the composition of the alloy. The phase diagram contains eutectic valleys and a peritectic line. Two solid Cr carbides are precipitated and dissolved during the solidification process (Figure 7.40). The system has no ternary eutectic point. The final stable phases are austenite and Fe_3C (cementite).

Carbides in Cr-bearing steel
Using the phase diagram for Fe–Cr–C the influence of C on the microsegregation of Cr in Cr-bearing steel can be explained. At higher carbon concentration in Fe–Cr–C alloys iron carbide is formed by a eutectic reaction to give α-phase + iron carbide. The negative slope of the eutectic line $E_3E_{Fe–C}$ (Figure 7.40) in the ternary phase diagram of the system Fe–Cr–C is the reason why the degree of segregation of Cr decreases at higher carbon concentrations.

EXERCISES

7.1 An Al–Zr alloy with 20 at-% Al is to be cast. Calculate the fraction of eutectic structure that is formed at 1350 °C.

Hint B1

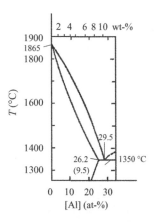

The phase diagram for the system Al–Zr is given.

7.2 An Al–Mg alloy with 40 at-% Al is to be cast. The circumstances are such that it is reasonable to assume that Scheil's equation is valid. The phase diagram of the system Al–Mg is given below.
Calculate the fraction of eutectic structure formed at solidification of the molten alloy.

Hint B12

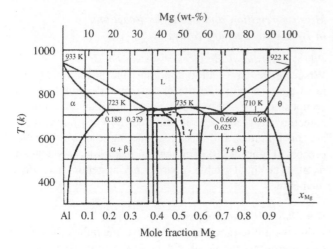

Mg (wt-%)

Mole fraction Mg

The values required for the calculations can be obtained from the text or read from the phase diagram.

7.3 In the technical literature the concept of degree of segregation is often mentioned. It is defined as the maximum concentration divided by the minimum concentration of the alloying element over a dendrite distance in the solid phase. It is easy to see that if the back diffusion is large the degree of segregation will be $x^s_{max}/x^s_{min} = 1$.

(a) Calculate the degree of segregation of an alloying element A at three different diffusion coefficients: (i) $D^s = 10^{-15}$ m^2/s; (ii) $D^s = 10^{-12}$ m^2/s, and (iii) $D^s = 10^{-9}$ m^2/s

Hint B28

known date:
initial concentration of the alloying
  element in the melt = 17 at-%
cooling rate = 5.0 °C/min;
temperature interval = 40 °C;
dendrite arm distance = 150 μm;
partition coefficient = 0.85.

(b) What is your conclusion from the results in (a)?

Hint B52

7.4 An Fe–Ni alloy containing 15 % Ni is cast in a mould that allows unidirectional solidification. It was found from temperature measurements during the solidification process that the isotherms move from the surface towards the last solidified parts of the melt according to the parabolic growth law $y = 0.50\sqrt{t}$, where the shell thickness y is measured in mm and the time in seconds. The dendrite arm distance λ_{den} as a function of the solidification time θ is given by:

$$\lambda_{den} = 2.0 \times 10^{-6} \times \theta^{0.4} \qquad \text{m}$$

The diffusion constant for Ni is:

$$D_{Ni} = (11 \times 10^{-4})e^{-\frac{38062}{T}} \qquad \text{m}^2/\text{s}$$

where T is the absolute temperature. The solidification temperature is 1470 °C. The partition coefficient $k_{\text{part Ni}}^{\gamma/L}$ is 0.73.
Calculate and show in a diagram how the degree of segregation varies with the distance from the surface of the casting.

Hint B8

7.5 You want to make an estimation of the segregation effect of Cr, Ni, and Mo in a stainless austenite steel. You intend to cast a very big casting and you are interested to know the microsegregation of these alloying elements in the centre of the casting. However, you cannot extract a specimen and make measurements but are restricted to theoretical calculations.

You know that the cooling rate in the centre is 5 °C/min and that the solidification interval of the alloy is 40 °C. In addition, you have found by measurements that the dendrite arm distance is equal to 150 μm at your cooling rate. The partition coefficients for Ni, Cr, and Mo at the interface austenite/melt are 0.90, 0.85 and 0.65, respectively. You also know that your colleague has measured the diffusion constants at the melting point of the alloy and they are:

$$D_{Ni}^{\gamma} = 2.0 \times 10^{-13} \qquad \text{m}^2/\text{s}$$

$$D_{Cr}^{\gamma} = 7.5 \times 10^{-13} \qquad \text{m}^2/\text{s}$$

$$D_{Mo}^{\gamma} = 7.5 \times 10^{-13} \qquad \text{m}^2/\text{s}$$

Perform calculations for the degree of segregation for Ni, Cr and Mo, defined as the maximum concentration divided by the minimum concentration over a dendrite arm distance, for a steel containing 13 % Ni, 17 % Cr and 2 % Mo.

Hint B44

7.6 A cast house is developing a new Fe–Ni–V–C steel. The Ni concentration, which controls the austenite–ferrite transformation, has been chosen at a high enough value for austenite solidification. The C concentration has been chosen as 3.0 at-%. The remaining problem is to choose the vanadium concentration. The V concentration is partly restricted by the amount of eutectic structure of the type L → γ + VC, which is precipitated during the solidification process.
Improve the basis for deciding the V concentration by calculation of the fraction of precipitated eutectic structure as a function of the V concentration. Plot the result in a diagram.

Hint B18

The solubility product for VC precipitation and the partition coefficients are derived from the ternary phase diagram. They were found to be:

$$x_{\underline{C}}^{L} \times x_{\underline{V}}^{L} = 2.5 \times 10^{-4} \quad \text{(mole fraction)}^2$$

and

$$k_{\text{part}\underline{C}}^{\gamma/L} = k_{\text{part}\underline{V}}^{\gamma/L} = 0.40$$

7.7 It is intended to produce a cast alloy of aluminium base, with the alloying elements Cu and Si, which is precipitation hardenable. It is desirable for the concentrations of the alloying elements to be as high as possible. For machining-technical and mechanical reasons, the two eutectic reactions:

$$L \rightarrow Al(\alpha) + Al_2Cu \quad \text{and} \quad L \rightarrow Si(\alpha) + Al_2Cu$$

have to be avoided.

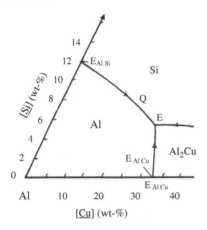

On the other hand precipitation of the three-phase eutectic:

$$L \rightarrow Al(\alpha) + Al_2Cu + Si$$

can be accepted to a certain extent. No more than 5.0 volume-% of this structure can be allowed if precipitation of the binary structures is to be avoided. What maximum concentrations of Si and Cu in the Al–Cu–Si alloy are in agreement with the conditions given above?

Hint B4

The phase diagrams of the systems Al–Si and Al–Cu and the ternary diagram of Al–Cu–Si are given in Example 7.5 on page 215.

7.8 Cast iron often contains small amounts of phosphorus, which has a very low solubility in the solid phases, both in austenite and in graphite. The consequence is that phosphorus segregates strongly. The solidification process often ends with precipitation of a ternary eutectic phase after the eutectic reaction:

$$L \rightarrow Fe(\gamma) + C(\text{graphite}) + Fe_3P$$

In the ternary eutectic Fe–C–P system this reaction occurs at 960 °C and at a composition of 3.5 wt-% $\underline{C}$ and 7.2 wt-% $\underline{P}$.

(a) Calculate the volume fraction of the ternary eutectic phase $Fe(\gamma) + C(\text{graphite}) + Fe_3P$ as a function of the initial P concentration.

Hint B23

(b) What will the volume fraction of ternary eutectic phase be if the initial $\underline{P}$ concentration is 0.1 wt-%?

Hint B137

7.9 Figure 7.41 on page 219 shows the segregation ratio of chromium as a function of the carbon concentration for a series of steel alloys that contain 1.5 wt-% $\underline{Cr}$ and various $\underline{C}$ concentrations. The segregation ratio S_{Cr} of chromium shows a maximum at around 1.6 wt-% $\underline{C}$ and then decreases gradually with increasing $\underline{C}$ concentration.

The decrease of the ratio S at higher $\underline{C}$ concentrations is supposed to depend on an earlier start for the eutectic reaction $L \rightarrow Fe(\gamma) + Fe_3C$ at high $\underline{C}$ concentrations than at low $\underline{C}$ concentrations and the fact that the $\underline{Cr}$ concentration in the eutectic melt decreases with increasing $\underline{C}$ concentration.

In order to check these statements calculate:

(a) the $\underline{Cr}$-concentration in the melt;

Hint B231

(b) the degree of $\underline{Cr}$-segregation S_{Cr} in the eutectic melt,

Hint B175

as functions of the initial carbon concentration in the melt for the interval 1.5 wt-% $\leq c_{\underline{C}}^{0} \leq$ 2.8 wt-% in steps of 0.1 wt-%. Plot the function in (b) in a diagram.

For your calculations you may assume that:

$$k_{\text{part}\underline{C}}^{\gamma/L} = k_{\text{part}\underline{C}} = 0.50 \quad \text{and} \quad k_{\text{part}\underline{Cr}}^{\gamma/L} = k_{\text{part}\underline{Cr}} = 0.73.$$

The partition constants are assumed to be constant at high $\underline{C}$ concentrations, and there is practically no back diffusion of $\underline{Cr}$ at high $\underline{C}$ concentrations. The phase diagram for the Fe corner of the system Fe–Cr–C is to be found in Figure 7.40 on page 218.

8 Heat Treatment and Plastic Forming

8.1 INTRODUCTION

It is well known that steel alloys and cast iron undergo several phase transformations during the cooling process after casting. These transformations are used in different types of heat treatment in order to improve the properties of the alloys.

Well-known methods include annealing, normalizing quench-hardening, and isothermal treatment. Precipitation hardening is another method that is used to improve the properties of both Al-base and Ni-base alloys.

Homogenization is a very important process, which eliminates or reduces the effects of the inevitable microsegregation

formed during solidification of castings and improves the properties of the alloys.

Some alloys contain secondary phases, which are not desired. Heat treatment at a constant temperature for a certain time results in dissolution of the secondary phases. In fact, dissolution treatment is also a sort of homogenization.

Homogenization and dissolution of secondary phases during isothermal heat treatment will be discussed extensively in this chapter. We shall also discuss various types of treatments which are especially designed for castings and ingots, and also heat treatment at both high temperature and high pressure in order to remove pores from the castings. In addition, we shall discuss the effect of phase transformations on the alloy structure and the influence of plastic deformation on the homogenization process.

8.2 HOMOGENIZATION

During the discussion of the origin of microsegregations in Chapter 7 (pages 197–198), we found that back-diffusion in the solid phase counteracts and lowers the degree of microsegregation. The effect of back diffusion during solidification depends on:

- the value of the diffusion constant of the alloying element;
- the solidification rate;
- the dendrite arm distance in the solid phase.

The conclusion from the third point is that the effect of diffusion in the solid phase can be strengthened if the cooling rate is lowered during or after the solidification. The effect will be even stronger if the cooling is interrupted and the temperature of the material is kept constant, i.e. the material undergoes an isothermal treatment. This treatment may also be performed by reheating after cooling to room temperature and/or possible plastic machining of the material.

Such an isothermal treatment has a homogenizing effect on the concentration of the alloying element. Initially the mathematically simplest case is treated, when the material contains only *one* phase during the whole heat treatment. In this case, the process is called *homogenization*.

Throughout earlier chapters we have denoted the dendrite arm distance λ_{den}. In this chapter, we shall describe concentration distributions with the aid of sine functions and introduce a mathematical *wavelength* λ. As we call the dendrite arm distance λ_{den}, there will be no difficulty distinguishing these two quantities.

8.2.1 Mathematical Model of the Distribution of the Alloying Element in the Structure as a Function of Position and Time

Dendritic structure has been treated in Section 6.3 (pages 143–145) and in Section 7.3 (pages 185–186 and 197–198) concerning microsegregation. Scheils equation reflects the corresponding microsegregation in the solid dendrite arms and predicts an infinite concentration of solute in the last solidified melt ($f = 1$).

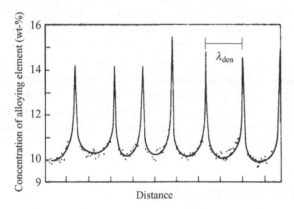

Figure 8.1 Concentration distribution of Mn across the secondary dendrite arms in a high-Mn steel. Reproduced with permission from M. Hove, McMaster University.

Figure 8.1 shows the concentration of manganese in a cast high-Mn steel. There are no measurements of the solute in the last solidified parts where two dendrite arms meet at complete solidification. Back diffusion limits the concentration to finite maximum values. Similar concentration changes of the alloying element to that in Figure 8.1 are observed in most cast binary alloys.

During isothermal heat treatment, the solute atoms diffuse from higher to lower concentration in the dendrite arms and the inhomogeneous solute concentration becomes smoothed out.

This process is the same for primary and secondary dendrite arms. As the diffusion distances, i.e. the dendrite arm distances, are much longer for primary than for secondary arms, the time required for homogenization is much longer for primary arms than for secondary arms.

The theory is the same in both cases. The only difference is that the primary arm spacings are up to 30 times larger than those for the secondary arms. This can be seen from Figure 6.5 (page 144). The corresponding homogenization time for the primary arms exceeds the time for the secondary arms by a factor of magnitude 10^3. The reason is that the homogenization time is a parabolic function of the dendrite arm distance [Equation (8.10), page 231].

Mathematical Model of Homogenization

Figure 8.1 shows that the concentration distribution of the alloying element across the secondary dendrite arms before the homogenization is not a simple function. In order to find a mathematical description of the change in the concentration distribution as a function of time during the heat treatment, it is necessary to:

1. find a simple mathematical model, which describes the initial concentration distribution x in the structure as a function of position y at $t = 0$;
2. simulate the homogenization process by calculation of the concentration of the alloying element as a function of position and time with the aid of Fick's second law.

$$\frac{\partial c}{\partial t} = D \left(\frac{\partial^2 c}{\partial x^2} + \frac{\partial^2 c}{\partial y^2} + \frac{\partial^2 c}{\partial z^2} \right)$$

[Fick's second law]

Initial Concentration Distribution

In order to describe the concentration distribution of an alloying element as a function of position, it is necessary to find a model that is as simple as possible.

The first approximation is to consider the concentration as a function of only *one* coordinate, i.e. the problem is treated one-dimensionally. This is a serious restriction and its consequences have to be carefully analysed. We have used such an approximation on several earlier occasions, for example in Chapter 6 when we discussed unidirectional solidification and in Chapter 7 when we treated microsegregation.

In order to simplify the problem further, the concentration distribution is in most cases described with the aid of a sine function. Unfortunately, the concentration profile has seldom or never the shape of a sine curve in reality. This difficulty can be overcome by application of Fourier analysis. Every arbitrary profile, for example a rectangular wave (Figure 8.2), can be described by a function of sine and cosine terms of multiple frequencies. The adaption to the profile in question is done by determination of the coefficients of the sine and cosine terms.

During the homogenization process, the amplitude of the concentration profile decreases, but not uniformly. The initial profile consists of the fundamental tone in combination with its overtones, which changes the profile. The latter fade away much more rapidly than the fundamental tone, which will be explained in the box on page 230. Normally only the fundamental tone remains after a short time, independently of the initial shape of the concentration profile.

In fact, the true initial concentration profile is not so important for the study of the homogenization process. Hence it is possible to approximate the concentration

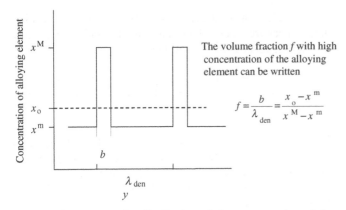

Figure 8.2 Rectangular distribution of the concentration of the alloying element as a function of position. The amount of the alloying element is proportional to the area under the curve. $x_0 =$ Average concentration.

profile by, for example, a rectangular wave (Figure 8.2) or a simple sine wave (Figure 8.3). These two alternatives are often useful approximations and will be treated separately below.

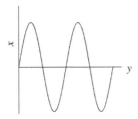

Figure 8.3 Simple sine distribution of the solute concentration as a function of position.

Rectangular Concentration Profile
A rectangular wave is an acceptable approximation in many cases, for example for the concentration distribution in Figure 8.1. Other examples will be given later in this chapter.

The true concentration profile of the square wave is of minor importance as the 'overtones' disappear very quickly and only the fundamental tone remains. A bold approximation, which works well, is to use a sine wave from the beginning. The problem is to find the amplitude of this fundamental wave.

The solute concentration cannot be lower than the minimum value x^m anywhere in the specimen. The material balance requires that the average concentration distribution is the same as the real one when we use the fundamental wave as concentration distribution.

For this reason, the fundamental wave must be a simple sine wave with the amplitude $x_0 - x^m$ and an average concentration equal to x_0. It differs considerably from reality but the approximation has been verified experimentally and it works. This model will be used in Example 8.2 (page 232).

Sine-shaped Concentration Profile
The concentration of the solute distribution in a dendrite is illustrated in two dimensions in Figure 7.3b (page 185). This distribution is illustrated by the continuous curve in Figure 8.4. In this case, a rectangular wave is not a satisfactory approximation whereas a simple sine function is suitable as the fundamental wave. The additional Fourier terms help to give the initial true concentration profile.

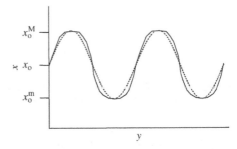

Figure 8.4 Sketch of the solute profile in a dendrite arm immediately after solidification. The dotted curve represents the fundamental tone.

The dotted sine wave in Figure 8.4 can be used to describe the concentration profile as a function of position instead of the initial true profile. Figure 8.5 illustrates schematically the concentration in two adjacent dendrites just before total solidification. The minimum value is found in the centres and the highest concentrations appear in the last solidified parts of the dendrites.

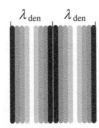

Figure 8.5 Schematic solute distribution in two dendrites immediately before total solidification.

The minimum value at the centre of the secondary dendrites in Figure 8.4 is denoted x_0^m and the maximum value x_0^M. The maximum *amplitude* of the sine wave before the homogenization has started will obviously be:

$$\text{amplitude} = \frac{x_0^M - x_0^m}{2} \tag{8.1}$$

The solute concentration x as a function of an arbitrary distance y from origin of the coordinate system in Figure 8.4 can be written as:

$$x = x_0 + \frac{x_0^M - x_0^m}{2}\sin\frac{2\pi y}{\lambda} \tag{8.2}$$

The volume fraction f with high concentration of the alloying element can be written

$$f = \frac{b}{\lambda_{den}} = \frac{x_0 - x^m}{x^M - x^m}$$

where x_0 is the average solute concentration in the dendrite. The wavelength λ of the sine wave for the first Fourier component (fundamental tone) is equal to λ_{den} in this case. The multiple frequencies of the overtones can be described by the relation

$$\lambda_n = \frac{\lambda}{n} = \frac{\lambda_{den}}{n} \qquad (8.3)$$

where n is an integer.

Concentration Distribution as a Function of Time and Position

Fick's second law controls the diffusion in the solid dendrite, i.e. the homogenization process. For the one-dimensional case, Fick's second law can be written as:

$$\frac{\partial x}{\partial t} = D \frac{\partial^2 x}{\partial y^2} \qquad (8.4)$$

where

D = diffusion constant
x = concentration of alloying element (mole fraction)
y = position coordinate.

The solute distribution x as a function of position y is given in Equation (8.2). This expression is introduced into Equation (8.4). By solving the partial differential Equation (8.4) we obtain the desired expression of the concentration of the

alloying element as a function of both time and position during heat treatment. The solution of Fick's second law for the fundamental wave will be:

$$x = x_0 + \frac{x_0^M - x_0^m}{2} \sin\left(\frac{2\pi y}{\lambda_{den}}\right) \exp\left(-\frac{4\pi^2 D}{\lambda_{den}^2} t\right) \qquad (8.5)$$

The solution can be verified by differentiating Equation (8.5) twice with respect to y and once with respect to t and inserting the derivatives into Equation (8.4). An additional check is that Equation (8.5) becomes identical with Equation (8.2) when $t = 0$ is inserted.

It is important to observe that the diffusion constant D is strongly temperature dependent and has a much higher value during the heat treatment than at room temperature.

Solution of Fick's Second Law for Homogenization

Each Fourier component has its own solution of Equation (8.4) The solution depends on the equation that is valid for the Fourier component instead of Equation (8.2). The solution for overtone n can be written as:

$$x_n = A + B \sin\frac{2\pi y}{\lambda_n} \exp\left(-\frac{4\pi^2 D}{\lambda_n^2} t\right) = A + B \sin\frac{2\pi n y}{\lambda} \exp\left(-\frac{4\pi^2 n^2 D}{\lambda^2} t\right) \qquad (8.6)$$

where A and B are integration constants. In the present case, the quantity λ is equal to the dendrite arm distance λ_{den} for the fundamental wave ($n = 1$). It can be concluded from Equation (8.6) that the exponential factor equals 1 for $t = 0$.

The solutions (8.6) of Equation (8.4) have the following properties:

1. The solutions describe what happens to each component independently of the others (superposition principle).
2. The solutions show that an arbitrary sine-shaped component will keep its sine shape while its amplitude decreases with time.
3. The rate of amplitude decrease for component n depends strongly on the wavelength owing to the factor n^2 in the exponential expression.

The multiple Fourier components or the overtones have short 'wavelengths' ($\lambda_n = \lambda/n, n \geq 2$) and disappear rapidly because the exponential function decreases very rapidly towards zero. Hence the overtones can in most cases be neglected in comparison with the fundamental tone with wavelength λ, which in this case is equal to λ_{den}.

The general solution for all the Fourier components is discussed in the above box. Point 3 in the box implies that the segregation pattern during heat treatment decays from the initial shape illustrated in Figure 8.4 to the shape of a sine curve. The amplitude decreases with time. The decrease makes it possible to calculate the homogenization time.

8.2.2 Homogenization Time

At time $t = 0$, the solidification of the dendrites is complete and the homogenization has not started. At this time the amplitude of the fundamental wave (the solute concentration) at position y can be written as [Equation (8.5) with $t = 0$]:

$$\text{amplitude}(t = 0) = \frac{x_0^M - x_0^m}{2}\sin\left(\frac{2\pi y}{\lambda_{\text{den}}}\right)\exp\left(-\frac{4\pi^2 D}{\lambda_{\text{den}}^2}\cdot 0\right)$$

$$(8.7a)$$

At time t, the amplitude of the wave at position y has decreased to:

$$\text{amplitude}(t) = \frac{x_0^M - x_0^m}{2}\sin\left(\frac{2\pi y}{\lambda_{\text{den}}}\right)\exp\left(-\frac{4\pi^2 D}{\lambda_{\text{den}}^2}t\right) \quad (8.7b)$$

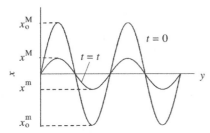

Figure 8.6 Concentration of the alloying element as a function of time and position in a dendrite arm.

The amplitudes of the damped wave can be written with the aid of the concentrations x^M and x^m, which are defined in Figure 8.6.

$$\text{amplitude}(t) = \frac{x^M - x^m}{2}\sin\left(\frac{2\pi y}{\lambda_{\text{den}}}\right) \quad (8.8)$$

Identity of Equations (8.7b) and (8.8) gives:

$$\frac{x^M - x^m}{x_0^M - x_0^m} = \exp\left(-\frac{4\pi^2 D}{\lambda_{\text{den}}^2}t\right) \quad (8.9)$$

Equation (8.9) shows that the homogenization never becomes complete because that would require an infinite homogenization time. The time for reduction of the amplitude to 10 % of the initial value is often used as a measure of the homogenization time.

The homogenization time can be solved from Equation (8.9):

$$t = \frac{\lambda_{\text{den}}^2}{4\pi^2 D}\ln\left(\frac{x_0^M - x_0^m}{x^M - x^m}\right) \quad (8.10)$$

It can be seen that the treatment time is proportional to the square of the dendrite arm distance and inversely proportional to the diffusion constant.

Equation (8.10) gives an idea of the effect of a heat treatment of the cast material and can be used for calculation of the homogenization time when the ratio of the amplitudes is given. The method is illustrated by Example 8.1 below.

Example 8.1

The distribution of an alloying element in a direction perpendicular to the secondary dendrite arms (y-direction) in a casting has the shape of a sine wave. The diffusion constant at the given temperature is known and is denoted D. The dendrite arm spacing is equal to λ_{den}.

The casting is heat-treated. How long a treatment time is required to reduce the difference between the maximum and minimum concentrations of the alloying element to one-tenth of the initial value?

Solution:
See Figure 8.6.
From the text we know that:

$$x^M - x^m = (x_0^M - x_0^m)/10$$

When this ratio is known, we obtain the homogenization time from Equation (8.10):

$$t = \frac{\lambda_{\text{den}}^2}{4\pi^2 D}\ln\frac{x_0^M - x_0^m}{x^M - x^m} = \frac{\lambda_{\text{den}}^2}{4\pi^2 D}\ln\frac{1}{0.1}$$

or

$$t = \frac{\lambda_{\text{den}}^2}{4\pi^2 D}\ln\frac{1}{0.1} = \frac{\ln 10}{4\pi^2}\frac{\lambda_{\text{den}}^2}{D} = \frac{2.3}{4\pi^2}\frac{\lambda_{\text{den}}^2}{D}$$

Answer:
The homogenization time $t = 0.058\lambda_{\text{den}}^2/D$.

From Example 8.1, we can conclude that a decay of the amplitude by a factor of 100 requires a heat treatment time only twice as long as that required to reduce the amplitude by a factor of 10.

In the general case, it can be seen from Equation (8.10) that the decay factor has a moderate influence on the heat treatment time. The treatment time is much more sensitive to the value of the dendrite arm distance λ_{den} and especially to the absolute temperature, which changes the diffusion constant D exponentially.

Example 8.2

The concentration of the alloying element across the dendrite arms in a casting shows a distribution that can approximately be described by a rectangular wave, reminiscent of that in Figure 8.2, with the value

$$\frac{b}{\lambda_{den}} = \frac{x_0 - x^m}{x^M - x^m} = 0.09$$

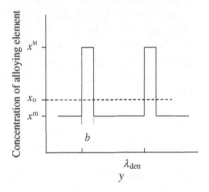

The casting is heat-treated at a constant temperature. The dendrite arm distance λ_{den} and the diffusion constant D of the material at the temperature in question are known.

How long a time will it take to lower the amplitude of the concentration of the alloying element to one-tenth of the initial concentration?

Solution:

The rectangular wave can be described as the sum of a fundamental tone and a number of 'overtones' with multiple frequencies of the fundamental frequency v:

$$v_n = nv \qquad (1')$$

The wavelengths λ_n of the overtones correspond to integer parts of the wavelength of the fundamental wave $\lambda = \lambda_{den}$ [Equation (8.3)]:

$$\lambda_n = \lambda_{den}/n \qquad (2')$$

We assume that the overtones decay within a very short time. In agreement with the discussion on page 229, we neglect the real concentration distribution and assume that the initial concentration distribution can be described by a fundamental sine wave with amplitude $x_0 - x^m$ at the start of the homogenization process.

At the end of the heat treatment, it is realistic to assume that the true concentration distribution corresponds to a sine wave with lower amplitude than the initial value. It is equal to the amplitude of the initial sine wave multiplied by the exponential decay factor.

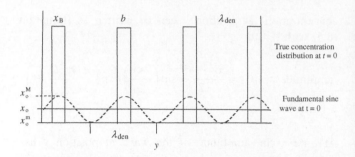

True concentration distribution at $t = 0$

Fundamental sine wave at $t = 0$

According to the text, the amplitude of the concentration distribution has dropped to 10 % of its initial value after the heat treatment time. This condition gives another expression of the amplitude after time t. The two expressions, which both are based on sine waves, must be equal. This condition gives:

$$(x_0 - x^m) \exp\left(-\frac{4\pi^2 D}{\lambda_{den}^2} t\right) = 0.10(x^M - x_0) \qquad (3')$$

For the sine wave we have $x_0^L - x^m = x^M - x_0^L$, which gives:

$$-\frac{4\pi^2 D}{\lambda_{den}^2} t = \ln 0.10 \qquad (4')$$

or

$$t = \frac{\lambda_{den}^2}{4\pi^2 D} \ln\left(\frac{1}{0.1}\right) = \frac{\ln 10}{4\pi^2} \frac{\lambda_{den}^2}{D} = \frac{2.3}{4\pi^2} \frac{\lambda_{den}^2}{D} = 0.058 \times \frac{\lambda_{den}^2}{D} \qquad (5')$$

Answer:

The amplitude of the concentration of the alloying element is reduced to one-tenth of the initial concentration after the time $t = 0.058\lambda_{den}^2/D$.

It is important to note that the heat treatment time is independent of the value of b/λ_{den}. This ratio is *not* involved in the calculations in either example 1 or example 2. The ratio influences the average solute concentration but not the homogenization time.

Example 8.3

Consider the casting in Example 2 above and calculate the distribution of the alloying element for the purpose in (a) below.

(a) How long a time does it take before the difference between the maximum and the minimum concentration has been reduced to 75 % of the original value?

(b) Compare the results of Examples 2 and 3.

Solution:

(a) In Example 8.2 above we assumed that the 'overtones' had decayed and that we could approximate both the initial and final states with merely the first term of the Fourier series in the calculations. In the present case we can *not* neglect the additional terms. It would be complicated to solve the problem by use of the Fourier method. We will instead solve the problem with the aid of a completely different method, which is more approximate but simple and fast.

The high concentration of the alloying element is originally restricted to a narrow range with width *b*. The rectangular shape of the distribution of the alloying element is *not* maintained during the heat treatment. We make the assumption that the concentration initially decays linearly with the distance from the high-concentration region. The distribution can then be described by two triangles before it finally is approximated by a sine curve when the overtones have decayed completely. The triangle shape appears owing to diffusion of the alloying element at the beginning of the heat treatment.

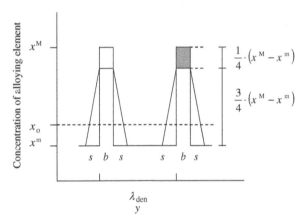

The area under the curve is proportional to the total amount of alloying element and hence constant, independent of its shape. Heat treatment means that alloying atoms from the regions *b* diffuse sideways. The concentration of the alloying element is assumed to decay linearly at the beginning of the heat treatment. The shaded area = the sum of the areas of the two triangles.

By a material balance, the base *s* of each triangle can be calculated:

$$Cb\frac{x^M - x^m}{4} = C \times 2 \times \frac{1}{2} \times s\frac{3(x^M - x^m)}{4} \qquad (1')$$

where *C* is the proportional constant between the amount of alloying element and the area under the curve. The solution is

$$s = b/3 \qquad (2')$$

During the time *t*, the atoms diffuse an average distance *s'* in a material with diffusion constant *D*. According to the

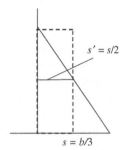

The area of the dashed rectangle equals the area of the triangle.

law of random walk, the following relation between these quantities is valid:

$$(s')^2 = 2Dt \qquad (3')$$

The average diffusion distance *s'* outside the high-concentration region equals half the triangle base *s*:

$$s' = s/2 = b/6 \qquad (4')$$

We insert *s'* into Equation (3') and obtain:

$$\left(\frac{b}{6}\right)^2 = 2Dt$$

which gives the following simple expression:

$$t = b^2/72D \qquad (5')$$

(b) In order to be able to compare the time above with the result in Example 8.2, we have to introduce λ_{den} instead of *b* into Equation (5') above. According to the text of Example 8.2, we have:

$$b/\lambda_{den} = 0.09 \qquad (6')$$

If we insert the Equation (6') into Equation (5'), we obtain:

$$t = \frac{b^2}{72D} = \frac{(0.09\lambda_{den})^2}{72D} = 0.00011 \times \frac{\lambda_{den}^2}{D} \qquad (7')$$

The results in Examples 8.2 and 8.3 can now be compared directly. It does not matter whether Example 8.2 deals with the difference between the maximum and the average concentrations or with the difference between the maximum and minimum concentrations because the former is proportional to the latter. The time will be the same in both cases.

The ratio between the times will be = 0.0058/0.00011 = 527.

Answer:

(a) The difference between the maximum and the average concentrations is reduced to 75 % of the original value in the time $t = 0.00011\lambda_{den}^2/D$.

(b) It takes more than 500 times as much time to reduce the difference between the maximum and minimum concentrations to one-tenth instead of three-quarters of the original value.

The assumption that the 'the overtones' disappear completely at the end of the heat treatment time is obviously justified.

Experimental examinations show that the properties of a cast material, especially the ductility, are steadily improved with increasing degree of homogenization.

The examples above show that if a homogeneous material is desired in the shortest possible time, it is important to start with a material with *small* dendrite arm distances [compare Equation (8.10) on page 231]. Short dendrite arm distances are obtained at high solidification and cooling rates.

It is obviously a significant advantage to cast thin sections by continuous casting rather than to cast large ingots, which later are rolled to thin dimensions.

8.3 DISSOLUTION OF SECONDARY PHASES

Precipitated secondary phases often have a complicated structure. To be able to calculate the heat treatment time required to dissolve the precipitated secondary phases, it is necessary to know the geometry of the areas in question and to express them in terms of mathematics.

Often radical simplifications have to be made in the description of the geometry of the structure. We will restrict the discussion below to a one-dimensional case, where the regions are approximated by narrow plates of high concentration of the alloying element, surrounded by a matrix of low concentration.

8.3.1 Simple Model for Heat Treatment of One-Dimensional Secondary Phases

We consider a binary alloy, for example an Al–Cu alloy. Figure 8.7 a shows the phase diagram of this system.

When a melt with a composition x_0, which corresponds to the cross in Figure 8.7a, solidifies and cools, the solid phase will initially consist of α-phase with a gradually increasing concentration of the alloying element B, due to microsegregation. A eutectic structure of α-phase with low B concentration and θ-phase with high B concentration precipitates when the temperature has decreased to the eutectic temperature at the

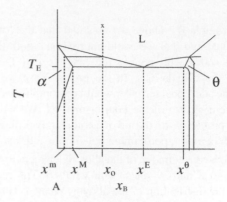

Figure 8.7 Part of the phase diagram of the Al–Cu system.

end of the solidification process The θ-phase is a precipitated secondary phase, which is not desired.

If the composition x_0 of the melt is known, it is possible to calculate the amount of the precipitated eutectic structure with the aid of Scheil's segregation equation (Chapter 7, page 186) and hence also the amount of θ-phase (Example 7.1, page 187).

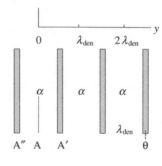

Matrix: α Secondary phase: θ

Figure 8.8 Schematic diagram of precipitated secondary phases in a casting structure. Reproduced with permission from McGraw-Hill.

We want to calculate the heat treatment time required to dissolve the whole θ-phase at a temperature close to the eutectic temperature. We will start with a one-dimensional model and describe the precipitated θ-phase regions as parallel plates divided by regular intervals of matrix (Figure 8.8). The concentration of the alloying element as a function of the ratio $2y/\lambda_{den}$ is shown in Figure 8.9. Two cases have to be distinguished:

1. The concentration fields from each plate overlap.
2. The concentration fields from each plate do not overlap.

Case I: The Concentration Fields Overlap
The diffusion fields (the concentration as a function of distance) from adjacent plates overlap outside the plates.

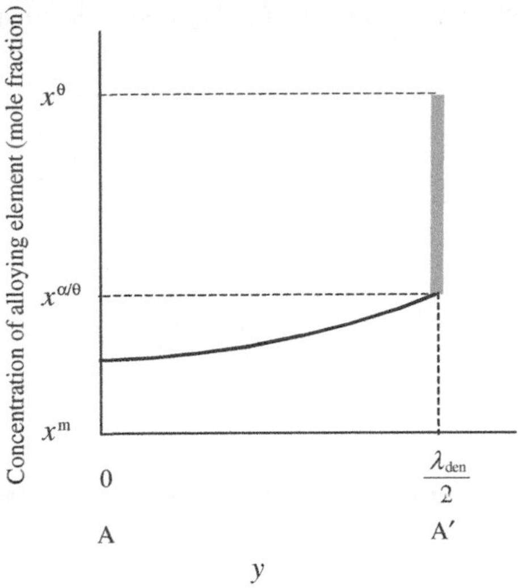

Figure 8.9 Concentration of alloying element in the casting structure as a function of position. Reproduced with permission from McGraw-Hill.

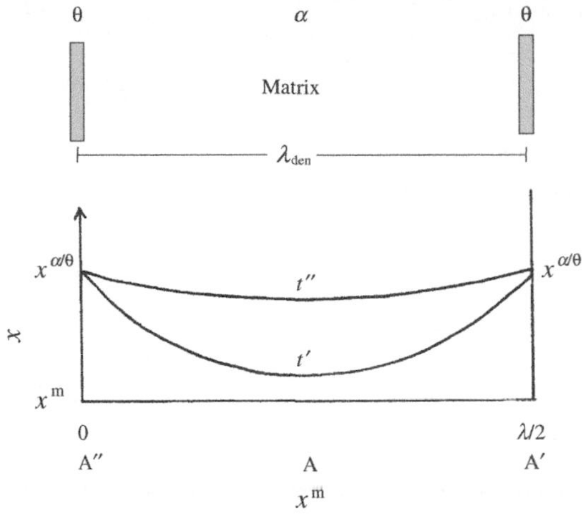

Figure 8.10 Case I. The concentration of the alloying element as a function of position in the matrix when the diffusion fields from adjacent plates overlap. $t'' > t'$. In the present case only *half* a sine wave is included within the distance λ_{den}: $\lambda_{den} = \lambda/2$.

The concentration in each point is the sum of two terms, caused by the diffusion from each plate (the superposition principle).

The overlap leads to an effect similar to the so-called 'impingement' on precipitation, i.e. the two precipitations compete with each other for the same alloying atoms. The result is a decrease in the dissolution rate on heat treatment because the concentration gradient decreases. The smaller the difference in concentration of the alloying element between the plates and their surroundings, the slower will the diffusion be.

We assume that the width of the plates is small compared with the distance between them. For this reason, the interface between the matrix and the plates can be regarded as approximately stationary. As long as the plates are not completely dissolved, the concentration of the alloying element at this interface equals the equilibrium composition $x^{\alpha/\theta}$ of the solid phase [Figures 8.10 and (later) 8.14].

The same reasoning that we used on page 229 can also be used in this case. The solution of Fick's law [Equation (8.6) in the box on page 230] contains overtones in the Fourier series, which rapidly decays during the heat treatment. In the end only the fundamental tone is left, i.e. the first sine term.

When the concentration profile has become sine-shaped [in analogy with Equation (8.7)], we can write (see Figure 8.10):

$$x = x^{\alpha/\theta} - B\exp\left(-\frac{4\pi^2 D}{\lambda^2}t\right)\sin\left(\frac{2\pi y}{\lambda}\right) \tag{8.11}$$

where

$x^{\alpha/\theta}$ = equilibrium composition at the α/θ interface

B = a constant.

As is seen in Figure 8.10, there is *half* a sine wave with wavelength λ within the distance λ_{den} and we obtain:

$$\lambda_{den} = \frac{\lambda}{2} \tag{8.12}$$

We insert $\lambda = 2\,\lambda_{den}$ into Equation (8.11). The amount m of alloying element which has been added to the matrix by diffusion from the plates at the time t can then be written as [C is a proportionality constant between the amount of alloying element and the area represented by the integral (compare page 233)]:

$$m = C\int_0^{\lambda_{den}}(x - x^m)\,dy = C\int_0^{\lambda_{den}}\left[x^{\alpha/\theta} - B\exp\left(-\frac{4\pi^2 Dt}{4\lambda_{den}^2}\right)\sin\left(\frac{2\pi y}{2\lambda_{den}}\right) - x^m\right]dy \tag{8.13}$$

which can be written as:

$$m = C(x^{\alpha/\theta} - x^m)\lambda_{den} - BC\exp\left(-\frac{\pi^2 Dt}{\lambda_{den}^2}\right)\int_0^{\lambda_{den}}\sin\left(\frac{\pi y}{\lambda_{den}}\right)dy \tag{8.14}$$

Integration gives:

$$
m = C(x^{\alpha/\theta} - x^{\mathrm{m}})\lambda_{\mathrm{den}} - BC
$$

$$
\times \exp\left(-\frac{\pi^2 Dt}{\lambda_{\mathrm{den}}^2}\right)\left[-\cos\left(\frac{\pi y}{\lambda_{\mathrm{den}}}\right)\frac{\lambda_{\mathrm{den}}}{\pi}\right]_0^{\lambda_{\mathrm{den}}} \tag{8.15}
$$

or

$$
m = C(x^{\alpha/\theta} - x^{\mathrm{m}})\,\lambda_{\mathrm{den}} - BC\exp\left(-\frac{\pi^2 Dt}{\lambda_{\mathrm{den}}^2}\right)\frac{\lambda_{\mathrm{den}}}{\pi}\times 2 \tag{8.16}
$$

The constant B is determined by the boundary condition $m = 0$ at $t = 0$, which gives:

$$
B = (x^{\alpha/\theta} - x^{\mathrm{m}})\frac{\pi}{2} \tag{8.17}
$$

Inserting of the value of B into equation (8.16) gives:

$$
m = C(x^{\alpha/\theta} - x^{\mathrm{m}})\lambda_{\mathrm{den}} - C(x^{\alpha/\theta} - x^{\mathrm{m}})\frac{\pi}{2}\exp\left(-\frac{\pi^2 Dt}{\lambda_{\mathrm{den}}^2}\right)\frac{2\lambda_{\mathrm{den}}}{\pi} \tag{8.18}
$$

or

$$
m = C(x^{\alpha/\theta} - x^{\mathrm{m}})\lambda_{\mathrm{den}}\left[1 - \exp\left(-\frac{\pi^2 Dt}{\lambda_{\mathrm{den}}^2}\right)\right] \tag{8.19}
$$

The condition for complete dissolution of the plates is that m equals the content of the alloying element in a plate above the level of the matrix:

$$
m = C(x^\theta - x^{\mathrm{m}})b \tag{8.20}
$$

The original volume fraction f_0^θ of the θ-phase can be written as (compare Figure 8.2 on page 229):

$$
f_0^\theta = \frac{b}{\lambda_{\mathrm{den}}} = \frac{x_0 - x^{\mathrm{m}}}{x^\theta - x^{\mathrm{m}}} \tag{8.21}
$$

or

$$
b = \lambda_{\mathrm{den}}\frac{x_0 - x^{\mathrm{m}}}{x^\theta - x^{\mathrm{m}}} \tag{8.22}
$$

The two expressions for m in Equations (8.19) and (8.20) are equal:

$$
m = C(x^\theta - x^{\mathrm{m}})b = C(x^{\alpha/\theta} - x^{\mathrm{m}})\lambda_{\mathrm{den}}\left[1 - \exp\left(-\frac{\pi^2 Dt}{\lambda_{\mathrm{den}}^2}\right)\right] \tag{8.23}
$$

The expression $b = f_0^\theta\lambda_{\mathrm{den}}$ [Equation (8.21)] is inserted into Equation (8.23), which gives:

$$
(x^\theta - x^{\mathrm{m}})f_0^\theta\lambda_{\mathrm{den}} = (x^{\alpha/\theta} - x^{\mathrm{m}})\lambda_{\mathrm{den}}\left[1 - \exp\left(-\frac{\pi^2 Dt}{\lambda_{\mathrm{den}}^2}\right)\right] \tag{8.24}
$$

which can be transformed into:

$$
f_0^\theta = \frac{x^{\alpha/\theta} - x^{\mathrm{m}}}{x^\theta - x^{\mathrm{m}}}\left[1 - \exp\left(-\frac{\pi^2 Dt}{\lambda_{\mathrm{den}}^2}\right)\right] \tag{8.25}
$$

where f_0^θ, $x^{\alpha/\theta}$, x^{m}, x^θ, λ_{den}, and D are known quantities. From Equation (8.25), it is therefore easy to calculate how long a time it will take to dissolve the secondary phase completely:

$$
t = -\frac{\lambda_{\mathrm{den}}^2}{\pi^2 D}\ln\left[1 - \frac{f_0^\theta(x^\theta - x^{\mathrm{m}})}{x^{\alpha/\theta} - x^{\mathrm{m}}}\right] \tag{8.26}
$$

It is also possible to calculate what fraction g of the solid phase is dissolved after a given time t_g. The condition for partial dissolution is that m equals the content of alloying element in the fraction of the plate which has been dissolved [compare Equation (8.20) above]:

$$
m = C(x^\theta - x^{\mathrm{m}})gb \tag{8.27}
$$

The expression for b in Equation (8.22) is inserted into Equation (8.27). The expression for m is inserted into Equation (8.19) and we obtain

$$
m = C(x^\theta - x^{\mathrm{m}})g\lambda_{\mathrm{den}}\frac{x_0^{\mathrm{L}} - x^{\mathrm{m}}}{x^\theta - x^{\mathrm{m}}}
$$

$$
= C(x^{\alpha/\theta} - x^{\mathrm{m}})\lambda_{\mathrm{den}}\left[1 - \exp\left(-\frac{\pi^2 Dt_g}{\lambda_{\mathrm{den}}^2}\right)\right] \tag{8.28}
$$

which can be simplified to:

$$g = \frac{x^{\alpha/\theta} - x^{\mathrm{m}}}{x_0^{\mathrm{L}} - x^{\mathrm{m}}}\left[1 - \exp\left(-\frac{\pi^2 D t_g}{\lambda_{\mathrm{den}}^2}\right)\right] \quad (8.29\mathrm{a})$$

If we introduce $b = f_0^{\theta}\lambda_{\mathrm{den}}$ [Equation (8.21)] into Equation (8.27) and combine the new equation with Equation (8.19), we obtain:

$$m = C(x^{\theta} - x^{\mathrm{m}})g f_0^{\theta}\lambda_{\mathrm{den}}$$

$$= C(x^{\alpha/\theta} - x^{\mathrm{m}})\lambda_{\mathrm{den}}\left[1 - \exp\left(-\frac{\pi^2 D t_g}{\lambda_{\mathrm{den}}^2}\right)\right] \quad (8.29\mathrm{b})$$

which can be simplified to:

$$g = \frac{x^{\alpha/\theta} - x^{\mathrm{m}}}{f_0^{\theta}(x^{\theta} - x^{\mathrm{m}})}\left[1 - \exp\left(-\frac{\pi^2 D t_g}{\lambda_{\mathrm{den}}^2}\right)\right] \quad (8.29\mathrm{c})$$

For $g = 1$, Equation (8.29c) will of course be identical with Equation (8.25). From Equation (8.29c) we can calculate how long a time t_g it takes to dissolve a given fraction g of the original volume fraction f_0^{θ} when the other quantities are known:

$$t_g = -\frac{\lambda_{\mathrm{den}}^2}{\pi^2 D}\ln\left[1 - \frac{g f_0^{\theta}(x^{\theta} - x^{\mathrm{m}})}{x^{\alpha/\theta} - x^{\mathrm{m}}}\right] \quad (8.30)$$

In Figure 8.11, the fraction g is plotted as a function of the time t_g divided by the square of the distance λ_{den} between the plates.

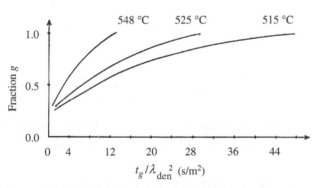

Figure 8.11 The fraction g of secondary phase as a function of the time t_g divided by the square of the distance λ_{den} between the stripes. Reproduced with permission from McGraw-Hill.

Figure 8.11 shows that the lower the temperature, the longer the dissolution process will take. The explanation of this is evident from Figure 8.12.

The *larger* the concentration difference between the average concentration x_0 and the equilibrium concentration $x^{\alpha/\theta}$, the *faster* will be the diffusion and the dissolution rate. The *lower* the temperature, the *smaller* will be the concentration difference and the *slower* will be the dissolution process. If the temperature equals the equilibrium limit (T'''') it will theoretically require an infinite time to dissolve the secondary phase completely.

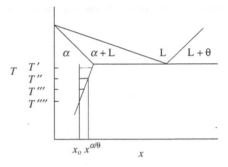

Figure 8.12 Part of the phase diagram of the Al–Cu system.

Example 8.4

An Al–Cu melt contains 2.5 at-% Cu. When the melt solidifies, two phases are precipitated, θ-phase and solid phase, with gradually changing composition (microsegregation) until the temperature has decreased to the eutectic temperature. Subsequently the rest of the melt solidifies with eutectic composition.

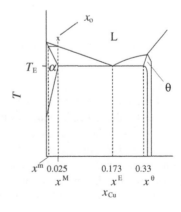

The diffusion constant of Cu atoms in Al at temperatures close to the eutectic temperature is assumed to be

1.0×10^{-13} m^2/s. The cooling rate of the melt is 5 K/min. The dendrite arm distance can be taken from the figure on page 199 in Chapter 7.

(a) How much eutectic phase (%) does the solid piece of metal contain?

(b) How much θ-phase (%) does the solid piece of metal contain?

(c) It is desirable to dissolve the θ-phase by heat treatment at a temperature above but close to the eutectic temperature. For how long a time at least must the heat treatment last if all the θ-phase is to be dissolved?

(d) What part of the θ-phase has dissolved after a treatment time of 1 h?

Solution:

(a) The fraction of eutectic phase can be calculated with the aid of Scheil's equation in the same way as in Example 7.1 on page 187.

The remaining melt which is left when the temperature has decreased to the eutectic temperature will solidify with a eutectic composition. Hence we can simply calculate the fraction f_E from Scheil's equation when the melt has reached the eutectic composition. The partition coefficient k_{part} can be derived from the phase diagram:

$$k_{part} = \frac{x^s}{x^L} = \frac{2.50}{17.3} = 0.1445 \qquad (1')$$

Scheil's equation, $x^L = x_0^L (1 - f_E)^{-(1-k_{part})}$, gives:

$$0.173 = 0.025(1 - f_E)^{-(1-0.1445)}$$

or

$$1 - f_E = 0.1445^{1/0.855} = 0.1023$$

which gives $f_E = 0.90$.

Some 90 % of the melt has solidified when the temperature has decreased to the eutectic temperature. The rest, 10 %, solidifies with a eutectic composition.

(b) Some 10 % of the solid metal piece consists of eutectic structure, i.e. a mixture of α-phase and θ-phase. We want to know how much θ-phase (%) it contains. With the aid of the phase diagram in the text and the lever rule, we obtain:

$$f_0^\theta = (1 - f_E) \frac{x^E - x^M}{x^\theta - x^M} \qquad (2')$$

The concentration of the alloying element x^M is given in the phase diagram in the text, $x^M = 0.025$, and we obtain the fraction of the θ-phase before heat treatment:

$$f^\theta = f_0^\theta = (1 - f_E) \frac{x^E - x^M}{x^\theta - x^M} = 0.10 \times \frac{0.173 - 0.025}{0.33 - 0.025} = 0.049$$

(c) For calculation of the dissolution time we must know the concentration x^m and the dendrite arm distance.

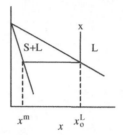

Calculation of x^m:

The solid phase has the lowest concentration of the alloying element. This phase is precipitated when the melt with concentration x_0^L of the alloying element starts to solidify. With the aid of the figure shown and the known value of the partition coefficient [Equation (1')], we obtain:

$$x^m = x_0^L k_{part} = 0.025 \times 0.1445 = 0.0036 \quad \text{mole fraction}$$

Calculation of λ_{den}:

In the figure in Example 7.4 on page 199 we plot an interpolated curve for the cooling rate 5 K/min and read λ_{den} for 2.5 at-% Cu as 90×10^{-6} m.

Total dissolution time:

The time for total dissolution ($g = 1$) of the secondary phase at the eutectic temperature can be calculated from Equation (8.26):

$$t = -\frac{\lambda_{den}^2}{\pi^2 D} \ln\left[1 - \frac{f_0^\theta (x^\theta - x^m)}{x^{\alpha/\theta} - x^m}\right] \qquad (3')$$

At the eutectic temperature $x^{\alpha/\theta} = x^M = 0.025$. If we insert the numerical values into Equation (3'), we obtain:

$$t = -\frac{(90 \times 10^{-6})^2}{\pi^2 \times 1.0 \times 10^{-13}} \ln\left[1 - \frac{0.049(0.33 - 0.004)}{0.025 - 0.004}\right]$$

$$= 12 \times 10^3 \text{ s} = 200 \text{ min}$$

(d) We insert $t = 3600$ s into Equation (8.29c):

$$g = \frac{x^{\alpha/\theta} - x^{m}}{f_0^\theta(x^\theta - x^{m})}\left[1 - \exp\left(-\frac{\pi^2 Dt}{\lambda_{den}^2}\right)\right]$$

and calculate g:

$$g = \frac{0.025 - 0.004}{0.049(0.33 - 0.004)}\left\{1 - \exp\left[-\frac{\pi^2 \times 1.0 \times 10^{-13} \times 3600}{(90 \times 10^{-6})^2}\right]\right\} = 0.47$$

Answer:

(a) The metal piece contains 10 % eutectic phase.
(b) The metal piece contains about 5 % θ-phase.
(c) After $\sim$3.5 h the whole θ-phase is dissolved.
(d) About 50 % of the θ-phase is dissolved after 1 h.

Case II: The Concentration Fields Do Not Overlap

The diffusion fields (the concentration profile as a function of distance) of adjacent plates of the secondary phase do *not* overlap (Figure 8.13). This is an extreme case, which occurs when the concentration of the alloying element in the matrix is so far from saturation that the diffusion fields do not overlap until the dissolution is finished.

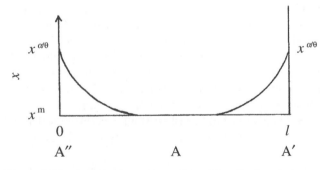

Figure 8.13 Case II. The concentration of the alloying element as a function of position in the matrix with no overlap between the diffusion fields of adjacent bands.

In this case the dissolution rate of the plates with the alloy composition x^θ can be calculated with the aid of the same laws that are valid for the growth of such particles on precipitation. The initial concentration of the alloying element in the matrix is x^{m}. At the interface between the plate and the matrix, the alloy composition is $x^{\alpha/\theta}$ (see Figure 8.14) as long as the θ-phase is not completely dissolved.

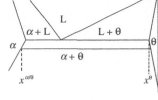

Figure 8.14 Phase diagram of the system α–θ.

We use Fick's first law (with constant C, due to units):

$$\frac{dm}{dt} = -D\frac{dx}{dy}C \tag{8.31}$$

and assume that the concentration of the alloying element that has diffused into the solid phase decreases linearly with distance y. With the aid of Figure 8.15, we obtain:

$$\frac{dx}{dy} = \frac{-(x^{\alpha/\theta} - x^{m})}{s} \tag{8.32}$$

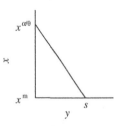

Figure 8.15 The alloy composition in the matrix as a function of the distance y from the plate with the alloy composition x^θ.

We can determine the distance s with the aid of a material balance. The amount of the alloying element that has diffused into the surroundings can be written in two ways (see Figure 8.16):

$$m = \underbrace{C(x^\theta - x^{m})gb}_{} = C\int_0^s x\,dy = \underbrace{C\frac{2(x^{\alpha/\theta} - x^{m})s}{2}}_{} \tag{8.33}$$

| Amount of alloying element that has diffused to the surroundings, which has the alloy composition x^{m} when the fraction g has been dissolved | Amount of alloying element that has diffused into the matrix = the area under the curve in Figure 8.15 = the sum of two triangle areas |

It is necessary to consider that the alloying element diffuses in *two* directions. The factor 2 in the numerator on

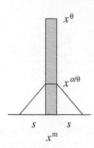

Figure 8.16 Concentration profile of a plate before and after heat treatment. The shaded area illustrates the θ distribution before the heat treatment and the triangles at the bottom represent the θ distribution after heat treatment.

the right-hand side of Equation (8.33) is due to *two* triangles, one on either side of the particle.

We differentiate Equation (8.27) on page 236 with respect to time:

$$\frac{dm}{dt} = C(x^\theta - x^m) b \frac{dg}{dt} \qquad (8.34)$$

s is solved from Equation (8.33):

$$s = \frac{(x^\theta - x^m) g b}{x^{\alpha/\theta} - x^m} \qquad (8.35)$$

and inserted into Equation (8.32):

$$\frac{dx}{dy} = \frac{-(x^{\alpha/\theta} - x^m)}{s} = \frac{-(x^{\alpha/\theta} - x^m)^2}{(x^\theta - x^m) g b} \qquad (8.36)$$

This value of dx/dy and the expression dm/dt [Equation (8.34)] are inserted into Equation (8.31), which gives:

$$\frac{dm}{dt} = C(x^\theta - x^m) b \frac{dg}{dt} = -DC \frac{-(x^{\alpha/\theta} - x^m)^2}{(x^\theta - x^m) g b} \qquad (8.37)$$

which can be transformed to:

$$b^2 \int_0^1 g\,dg = \int_0^t D \frac{(x^{\alpha/\theta} - x^m)^2}{(x^\theta - x^m)^2}\,dt \qquad (8.38)$$

After integration, we obtain:

$$\frac{b^2}{2} = D t \frac{(x^{\alpha/\theta} - x^m)^2}{(x^\theta - x^m)^2} \qquad (8.39)$$

or

$$t = \frac{b^2}{2D} \frac{(x^\theta - x^m)^2}{(x^{\alpha/\theta} - x^m)^2} \qquad (8.40)$$

The time of complete dissolution can be calculated from Equation (8.40). If we change the upper integration limit in Equation (8.38) to g on the left-hand side and t_g on the right-hand side, we obtain a relation between the dissolving fraction and the time of partial dissolution. The result is:

$$\frac{b^2}{2} = \frac{D t_g}{g^2} \frac{(x^{\alpha/\theta} - x^m)^2}{(x^\theta - x^m)^2} \qquad (8.41)$$

If we introduce λ_{den} instead of b with the aid of Equation (8.21) ($b = f_0^\theta \lambda_{den}$), we obtain the two expressions:

$$g = \frac{\sqrt{2D t_g}}{f_0^\theta \lambda_{den}} \frac{(x^{\alpha/\theta} - x^m)}{(x^\theta - x^m)} \qquad (8.42)$$

and

$$t_g = \frac{(g f_0^\theta \lambda_{den})^2}{2D} \frac{(x^\theta - x^m)^2}{(x^{\alpha/\theta} - x^m)^2} \qquad (8.43)$$

The total dissolution time is obtained if $g = 1$ is introduced into Equation (8.43).

8.4 CHANGE OF CASTING STRUCTURE DURING COOLING AND PLASTIC DEFORMATION

It can be seen from the phase diagrams in Chapter 7, for example the phase diagram of the Fe–C system, that steel alloys experience several phase transformations during the cooling process. Many stainless-steel alloys, and also low-carbon steel alloys, solidify primarily as ferrite (δ-iron) and transform later into austenite (Sections 7.6 and 7.7). The low-carbon steel alloys then transform again into ferrite and perlite (Figure 8.17). The microsegregation pattern and the coarseness of the structure in addition to the cooling conditions influence the phase transformations.

An ingot or a continuously cast alloy normally undergoes a plastic deformation process, which influences the

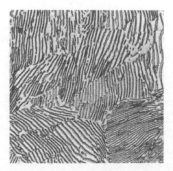

Figure 8.17 Perlite is a low-carbon steel (~0.8 wt-% C). It has a lamella structure. The layers consist of ferrite and cementite. Reproduced from the American Society for Metals (ASM).

homogenization process and/or the structure. The material becomes refined and interdendritic voids shut. These effects of plastic deformation will be briefly discussed later in this chapter.

Plastic Deformation

All materials are more or less elastic and expand and contract due to tensile and compression forces, respectively. When the forces are removed, the material returns to its original shape. On the other hand, a *plastic deformation* is a *permanent deformation*, which does *not* disappear when the load is withdrawn.

Plastic deformation can be performed at low temperature (cold working) or at high temperature (hot working). Plastic working refines the treated material and improves its mechanical properties. Plastic working includes various processes such as rolling, forging, extruding, stamping, drawing and pressing.

When a metal is exposed to plastic deformation, the segregation pattern will generally change. On plastic deformation, areas with different concentration of alloying elements are stretched to bands in the deformation direction. The structure becomes banded or, in technical language, lamella-shaped.

In low-alloyed steels, this lamella structure may appear as bands of ferrite or perlite. Figures 8.18 and 8.19 give examples of the micro- and respectively, macro-structure in a worked low-carbon steel.

In a material exposed to plastic deformation, the mechanical properties of the material will be different in different directions relative to the deformation direction. Generally it

Figure 8.18 Microstructure of a worked low-carbon steel. Reproduced with permission from John Wiley & Sons, Inc.

| (a) | (b) |

Figure 8.19 Macrostructure in a low-carbon steel before and after working.

is true that extension, yield point limit, and notch value become lower *across* than *along* the deformation direction as a consequence of the lamella structure.

8.4.1 Homogenization of Ingots and Continuously Cast Materials by Deformation

It is not always possible to achieve a completely homogeneous material in spite of strong deformation of ingots. This is illustrated below by the analysis of a simple example.

It is possible to stretch the ingot by a machining operation, for example, rolling extrusion. It is difficult to describe the result of such a working operation in detail, owing to the complex geometry of the dendrites, but one may gain a better understanding if a cubic element of the material is considered. The plastic deformation in a die is illustrated in two dimensions in Figure 8.20.

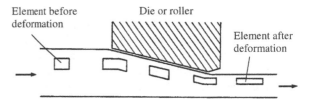

Figure 8.20 Plastic deformation of metal elements which are exposed to mainly compressive forces but also to pulling tensile forces. Reproduced with permission from the Institute of Materials.

Plastic working deforms and elongates the material, as illustrated in Figure 8.20. In addition to the geometric changes of its shape, plastic deformation also leads to changes in the distribution of alloying elements.

Consider a cube made of an alloy with a simple cubic crystalline structure and an inhomogeneous distribution of the alloying element before plastic deformation. The deviations of the solute concentration from the average value in various regions are marked by + and − signs. There are two cases, depending on the positions of the maximum and minimum concentrations during the deformation process. The two cases are illustrated on the next page.

Plastic Deformation of Cast Materials and Products

The purpose of plastic deformation is two-fold:

- to change of shape and thickness of the material as a step in the chain to manufacture finished products;
- to close cracks and pores, i.e. to improve the properties and quality of the material.

Rolling and forging deform cast materials. The effect of the plastic deformation depends on the direction of the rolling. To illustrate this, we will consider the effect of rolling a cube made of a dendritic alloy in two perpendicular directions.

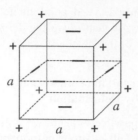

Figure 8.21 Alloy cube with normal dendritic distribution of the alloying element. The distribution is symmetrical in the three perpendicular directions. Reproduced with permission from the American Society for Metals.

Figure 7.3 (a) (page 185) illustrates the inhomogeneous distribution of the alloying element in a dendritic material. The isoconcentration curves represent regions of maximum concentration. Between them there are regions with minimum concentration.

The inhomogeneous distribution of alloying elements in the cube in Figure 8.21 is illustrated by + and − signs. The signs have nothing to do with electrical charge. The corners have maximum concentration and the sides correspond to minimum concentration. The distances between consecutive + signs represent the dendrite arm distances.

We will consider how they change when the material is rolled. On deformation *the volume of the cube remains constant.*

Consider the cube in Figure 8.21 before it is rolled in the direction illustrated in Figure 8.22 (a). This can be done in many different ways and we will discuss two of them [Figure 8.22 (b)].

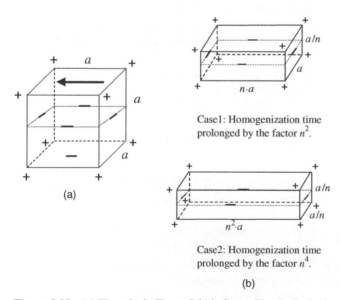

Case1: Homogenization time prolonged by the factor n^2.

Case2: Homogenization time prolonged by the factor n^4.

(b)

Figure 8.22 (a) The cube in Figure 8.21 before rolling in the indicated direction. (b) The deformed cube after rolling in the two extreme cases. For the sake of clearness, the minus signs in the top and bottom surfaces have been omitted. Reproduced with permission from the American Society for Metals.

In the *first* case, the deformation is done in such a way that the cross-sectional area of the cube becomes rectangular and its height is reduced by a factor $1/n$ whereas the other side remains constant. As the volume of the body is constant, the new length of the side in the deformation direction will be na.

In the *second* case, the cross-sectional area remains square and its sides are both reduced by a factor $1/n$. The length of the side in the deformation direction will be extended from a to n^2a.

When the side in the rolling direction is elongated, all other distances, even the dendrite arm distances, will be elongated by the same factor in this direction.

After hot pressing, the material is heat-treated to achieve homogenization. Equation (8.10) on page 231 shows that *the homogenization time is proportional to the square of the dendrite arm distance.* For this reason, the homogenization time will be extended by a factor of n^2 and n^4, compared with the unchanged cube, in the two extreme cases above. The material is in reality deformed to a state between the two cases above. The normal procedure is to compromise.

The material is deformed to an intermediate state between the two extreme cases above. A strong deformation gives an initial homogenization, but it is inevitable that the final state will be strongly delayed or may even fail to be achieved within the available homogenization time.

8.4.2 Secondary Phase Structures in Ingots

Columnar crystals in ingots are not always formed by primary dendrite growth directly from the melt. This is illustrated by Figures 8.23a and b. The figures show the structures of a normal 18/8 stainless steel.

Figure 8.23a shows the primary structure formed by crystals grown directly from the melt. The letters a, b and c represent the different zones, defined in Figure 6.47 (page 170).

Figure 8.23b illustrates very large columnar bent crystals. The crystals are formed when ferrite is transformed into austenite (see the phase diagram on page 150) in the solid state. The process is illustrated schematically in Figure 8.24b.

Figures 8.24a and b show a corner of an ingot. The liquidus isotherm (dotted curve) and the 'equidistant' transition isotherm are assumed to have the shapes illustrated.

Figure 8.24b illustrates the formation of γ-Fe from δ-Fe. The figure illustrates the growth of elongated γ-crystals. The growth of these crystals is not restricted to any particular crystallographic direction.

However, during the solidification process, the primary growth of the dendrite crystals is restricted to certain crystallographic directions. In this case, the time to adjust the dendrite tips in the directions of the isotherms is too long and the crystals will retain their directions.

The formation of austenite from ferrite, $\delta \rightarrow \gamma$, does not affect the transition from columnar to equiaxed crystals. The

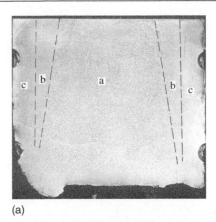

(a)

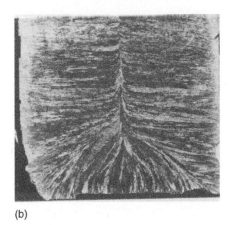

(b)

Figure 8.23 (a) Primary structure of crystals grown directly from the melt of an 18/8 stainless steel. Different zones of the structure: (a) globular dendrite crystals; (b) branched columnar crystals; (c) columnar crystals.

(b) Large columnar austenite grains, formed by transformation of ferrite into austenite by recrystallization or secondary crystallization in the solid state: $\delta \rightarrow \gamma$. The process is illustrated schematically in Figure 8.24b. Reproduced with permission from Merton C. Flemings.

austenite crystals easily override the equiaxed crystals in the central part of the ingot, i.e. the austenite crystals grow unaffected by the borders between the crystal zones. The microsegregation pattern is not affected by the $\delta \rightarrow \gamma$ transformation.

8.4.3 Structure Changes of Secondary Phases by Plastic Deformation

The appearance and distribution of precipitated secondary phases are also influenced by plastic deformation. In these cases, the final state is affected by the plastic properties of the particles in the secondary phase relative to the matrix.

There are three different cases:

1. The particles are *harder* and *stronger* than the matrix; no effect of the plastic deformation.
2. The particles are *softer* than the matrix; they become elongated in the direction of deformation.

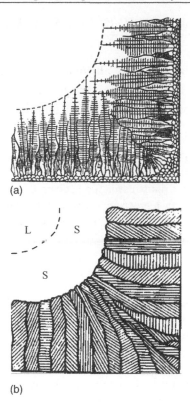

(a)

(b)

Figure 8.24 (a) Primary crystallization of a corner in a steel ingot. The dotted curve at the dendrite tips represents the isotherm $T = T_L$. (b) Secondary crystallization (recrystallization in the solid state) of the corner in (a). The dotted curve represents the isotherm $T = T_L$ and the later profile corresponds to a secondary isotherm with lower temperature, where the transformation $\delta \rightarrow \gamma$ occurs. Reproduced with permission from Jernkontorets Annaler.

3. The particles are *brittle*; they break during the deformation process and distribute along lines parallel to the deformation direction.

The three cases are illustrated in Figure 8.25. It can be seen that the distances between the particles decrease on deformation if the particles are brittle or soft. This means that the heat-treatment time for dissolution becomes shorter after deformation of the material with these types of inclusions than during the corresponding treatment without prior deformation.

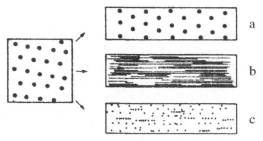

Figure 8.25 The effect of deformation on secondary phase particles. (a) Hard particles, no change; (b) soft particles, plastic deformation; (c) brittle particles, the particles break. Reproduced with permission from McGraw-Hill.

8.5 COOLING SHRINKAGE AND STRESS RELIEF HEAT TREATMENT

8.5.1 Origin of Casting Shrinkage

The cooling shrinkage from the solidification temperature down to room temperature starts immediately when an ingot or casting has solidified. The casting shrinkage generates problems of several kinds. If the shrinkage is *obstructed* in any part of the chill-mould during the solidification and cooling process, mechanical stress will arise.

However, the main reason for mechanical stress in castings is a *non-uniform temperature distribution*, which results in uneven shrinkage. The uneven shrinkage causes:

- plastic deformation at higher temperatures;
- mechanical stress in the cast material.

Plastic deformation causes deformation of cast components. The remaining mechanical stress results in unacceptably low mechanical strength of the castings and generates a risk of cold cracks. Stress in the cast material may lead to spontaneous fractures in the material even before or shortly after the cooling. If the component breaks at a light hammer-stroke or when it is dropped it is useless.

Another negative consequence of the cooling shrinkage is the difficulty of predicting the dimensions of the final cast component in cases when the utmost accuracy is required.

8.5.2 Plastic and Elastic Shape Deformations

The origin of stress depends on the fact that the deformations during the cooling initially are *plastic*. Shape changes caused by, for example, stretching remain without causing any stress in the material. When the temperature has decreased below a certain value, characteristic for each alloy, the deformations become *elastic* and cause mechanical stress in the material. For cast iron, for example, the boundary temperature between plastic and elastic deformations is approximately 650 °C.

With the aid of a so-called 'shrinkage harp', it is possible to measure the deformations and calculate the generated stress in different materials.

The profile shown in Figure 8.26 (a) is cast. The two thin outer rods cool most rapidly. When their elasticity limit is exceeded, the shrinkage of the thicker middle rod is still plastic and does not cause any mechanical stress until the temperature has dropped below the plasticity limit. Therefore, it becomes shorter than the thin rods. When this happens, the temperature difference between the thick middle rod and the thin outer rods can be up to 200 °C.

After cooling, the profile adopts the appearance shown in Figure 8.26 (b). When the middle rod is cut across its axis, a gap is formed. From the magnitude of this gap, it is possible to calculate the earlier mechanical strain in the material.

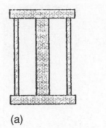

(a)   (b)

Figure 8.26 Shrinkage harp: (a) at a temperature above the elasticity limit and (b) at room temperature. Reproduced with permission from Karlebo.

8.5.3 Methods to Avoid Stress in the Material

To avoid or reduce the problem of stress in cast materials, efforts must be made to make *the temperature in the different parts of the casting decrease uniformly.*

The shape and thickness of the product must be designed in such a way that this condition is fulfilled as far as possible. Projecting thin parts ought to be avoided because they cool much more rapidly than more compact parts. Other simple arrangements are also possible.

If a circular plate is to be cast, for example, it will cool primarily at the periphery and finally in the centre if its thickness is even (Figures 8.27 and 8.28). This causes remaining stress in the material. If the future use of the product does not prevent it, one can cast a ring instead, i.e. remove the middle part, and in addition make the ring thicker at its outer parts. Both of these measures lead to a more even temperature distribution and less stress in the material than when the plate is compact and has a uniform thickness.

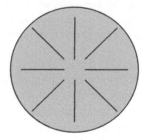

Figure 8.27 Circular plate casting with uniform thickness.

In other cases, uneven cooling may compensate for differences in material thickness and maintain a more even temperature distribution in the casting.

When a tube, i.e. a cylindrical shell, is to be cast, special precautions have to be taken. The massive sand core in the centre is cold and is warmed by the hot melt. The sand cylinder tends to expand. Simultaneously, the melt in the mould solidifies and cools, which results in shrinkage of the space inside the tube. These opposing effects lead to strong mechanical stress in the tube. The result may be that the tube cracks and has to be discarded.

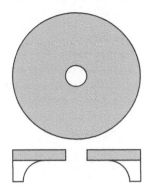

Figure 8.28 Ring-shaped casting which is thicker at the periphery than at the centre. The casting is seen from above and from the side.

Crack formation in the tube can be avoided if a less durable binding agent is used in the inner part of the sand mould. It breaks owing to the mechanical pressure and solidification and cooling shrinkage can occur without any restrictions (Figure 8.29).

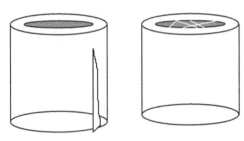

Figure 8.29 Two tube castings, one with a rigid core and the other with a more fragile core, after cooling. The former has cracked during the cooling process, owing to strong mechanical stress.

Stresses depend on the shape and size of the casting. They can be reduced or even eliminated by an optimal design and/or choice of a suitable material. Cast iron has a lower expansion coefficient than other casting alloys and is therefore to be preferred from this point of view.

The residual stresses will cause shape changes of the metal object or even severe defects such as cracks. Even if measures have been taken to avoid stress on cooling shrinkage, it is inevitable that cast components will often contain residual stresses. The stresses can be eliminated or at least strongly reduced by so-called stress relief treatment.

8.5.4 Stress Relief Heat Treatment

Stress relief heat treatment is normally performed in the following way. The casting is heated to a temperature at which plastic deformation is easily achieved in the metal and kept at this temperature for a certain time.

After the heat treatment, *the cooling rate* of the casting often has to be *low* prevent new stresses from developing in the casting. Normally the casting is left to cool inside the furnace when it is switched off.

Choice of Temperature and Time

The choices of time and temperature are important in stress relief heat treatment. The higher the temperature the better, if the aim is preferably stress relief. On the other hand, it cannot be chosen so high that the structure of the metal changes. Structure changes in cast iron start, for example, around 600 °C. A disadvantage with high temperature is that the mechanical strength of the alloy will be reduced. The choice of temperature has to be a compromise between different interests. The choice of time depends on how rapidly the stress decreases with time and on how much residual stress one is willing to accept.

Choice of Alloy

The required temperatures and times for stress relief depend on the type of alloy. Copper-base castings do not require stress-relieving treatment because their good thermal conductivity leads to low temperature gradients in the material and little risk of stress. Aluminium objects, cast in sand moulds, have low residual stress levels after the heat treatment for the same reason.

Aluminium- and magnesium-base alloys are recommended for applications where severe stress is to be expected. Heat relief treatment is normally required for cast iron and steel products. Table 8.1 gives some examples, based on experience, of suitable times and temperatures.

In all cases of isothermal heat treatment at high temperature, the cooling rate must be low otherwise new thermal stresses are built up in the materials. Some rule of thumb temperatures and times are given in Table 8.1.

TABLE 8.1   Rule of thumb for stress-relief heat treatment of iron-base alloys.

Material	Temperature (°C)	Treatment time (hours/cm)
Grey iron	480–590	2–0.4
Carbon steel	590–680	0.4
Cr–Mo steel	730–760	0.8–1.2
18-8 Cr–Ni steel	815	0.8

8.6   HOT ISOSTATIC PRESSING

Hot isostatic pressing (HIP) is a method in which the castings are heat-treated at high temperatures and simultaneously exposed to high pressure. It was commercially developed and was available during the last two decades of the 20th century, primarily for Al- and Ni-base alloys.

The castings treated in this way acquire properties similar to those of forged products.

8.6.1 Principles of HIP

The process is performed in a high-pressure chamber, normally in an argon atmosphere at a pressure up to 10^4 atm. The temperature in the chamber is chosen to be as high as possible, just as in stress relief heat treatment and for the same reasons (plastic deformation, page 244). The high pressure in combination with the high temperature forces pores and cracks in the casting to be welded or sintered together.

Figure 8.30 illustrates how the porosity (volume fraction of pores in the casting) of an Al casting, which is heat-treated at 500 °C, decreases with time. The decrease in the pore volume is initially very fast.

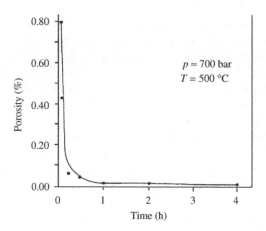

Figure 8.30 Porosity (volume fraction of gas pores) as a function of time for a cast Al alloy which is exposed to a high isostatic press. Reproduced with permission from Bodycote Powdermet.

Thanks to the high pressure, the pore volume is strongly reduced, as can be seen in the example below. The high temperature counteracts the decrease in volume slightly, but permits a permanent volume reduction due to plastic deformation.

Example 8.5
A pore with initial volume V_0 contains hydrogen gas at pressure 1 bar at room temperature (20 °C). The pore is included in an alloy, which is HIP treated at 500 °C and 10 kbar. Calculate the volume of the pore at this temperature and pressure.

Solution:
We assume that the general gas law is valid for the pore during the HIP treatment:

$$pV = nRT \qquad (1')$$

where
p = pressure of the pore
V = volume of the gas pore
n = number of kilomoles of the hydrogen gas
T = absolute temperature of the gas pore.

The number of kilomoles in the gas pore is the same at the two locations. Hence we obtain:

$$n = \frac{p_0 V_0}{RT_0} = \frac{pV}{RT} \qquad (2')$$

which gives:

$$V = V_0 \frac{p_0 T}{pT_0} = V_0 \frac{1 \times (500 + 273)}{10^4 \times (20 + 273)} \approx \frac{V_0}{3.8 \times 10^3} \qquad (3')$$

Answer:
The volume of the gas pore is reduced by a factor of 3800.

The above example explains the drastic decrease in the pore volume at the beginning of the process illustrated in Figure 8.30. It is initially due to reversible temperature and pressure changes of the pore volume. In order to reduce the pore volume permanently, it is necessary to reduce the amount of gas included in the pore. The mechanism behind this process is diffusion of gas from the pore through the solid to the surroundings. The process is illustrated in the example below.

Example 8.6
An Al casting is exposed to hot isostatic pressing in order to remove gas pores in the material. The HIP treatment is performed at 500 °C in an argon atmosphere at a pressure of 10 kbar.

(a) Calculate the radius of a spherical hydrogen pore with an initial volume (at a pressure of 1 bar and room temperature, 20 °C) equal to 4.2×10^{-3} mm^3 as a function of time and distance from the pore to the surface during the treatment time. Plot the radius of the pore as a function of time in a diagram for some different values of the distance between the pore and the surface.

(b) Calculate the time required to empty the pore ($V_{\text{pore}} = 0$) if it is situated at a distance of 10 mm from the surface of the casting.

Hydrogen is dissolved in the Al casting as $\underline{H}$ atoms. At equilibrium, the hydrogen mole fraction $x_{\underline{H}}$ in the casting, close to the pore, is proportional to the pressure of the included hydrogen gas inside the pore:

$$x_{\underline{H}} = C\sqrt{p_{H_2}} = 5 \times 10^{-8}\sqrt{p_{H_2}}$$

where $x_{\underline{H}}$ is measured in mole fraction and the pressure in bar. The diffusion of $\underline{H}$ atoms through the Al casting can be described with the aid of Fick's diffusion law:

$$\frac{dn}{dt} = -D_{\underline{H}} \frac{A}{V_m^{\underline{H}}} \cdot \frac{dx_{\underline{H}}}{dy}$$

The diffusion constant for $\underline{H}$ atoms in the Al alloy is 1.0×10^{-7} m²/s. The molar volume of hydrogen atoms in the Al alloy is assumed to be 0.010 m³/kmol.

Solution:
(a) If we apply Fick's law on the diffusion of the $\underline{H}$-atoms through the alloy and introduce the disappearance of H_2 molecules from the pore, we obtain:

$$\frac{dn_{H_2}}{dt} = \frac{1}{2}\frac{dn_{\underline{H}}}{dt} = -\frac{1}{2}D_{\underline{H}}\frac{A}{V_m^{\underline{H}}}\frac{dx_{\underline{H}}}{dy} \qquad (1')$$

where

n_{H_2} = number of kilomoles of H_2 molecules in the pore
$n_{\underline{H}}$ = corresponding number of kilomoles of $\underline{H}$ atoms (not in the gas, but when they successively dissolve in the casting and diffuse from the pore through the casting to the surroundings)
$D_{\underline{H}}$ = diffusion constant of $\underline{H}$ atoms, which diffuse through the casting
A = area of the pore
$V_m^{\underline{H}}$ = molar volume of $\underline{H}$ atoms in the casting
$x_{\underline{H}}$ = concentration of $\underline{H}$ atoms dissolved in the casting
y = coordinate.

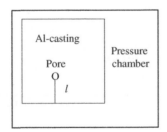

The dissolved $\underline{H}$ atoms will diffuse through the alloy from the pore to the surface (surroundings). In this case we obtain:

$$\frac{dn_{H_2}}{dt} = \frac{1}{2}\frac{dn_{\underline{H}}}{dt} = -\frac{1}{2}D_{\underline{H}}\frac{A}{V_m^{\underline{H}}}\frac{x_{\underline{H}}^{\text{pore}} - x_{\underline{H}}^{\text{surface}}}{l} \qquad (2')$$

where l is the distance between the pore and the outer surface of the casting, as illustrated. We want the radius r_{pore} as a function of time. The pore volume changes from the initial volume V_0 to zero. The volume is not involved in Equation (2').

If we apply the general gas law on the hydrogen gas (H_2) included in the pore, we obtain:

$$pV = n_{H_2}RT \qquad (3')$$

This equation is differentiated with respect to time. As p and T are constant in the pore during the treatment time, we obtain:

$$\frac{dn_{H_2}}{dt} = \frac{p}{RT}\frac{dV}{dt} \qquad (4')$$

According to the first equation in the example text (page 246) the $\underline{H}$ concentration in the casting, close to a pore, is proportional to the H_2 pressure inside the pore. It is reasonable to assume that the pressure inside the pore is constant when the pore volume decreases.

The $\underline{H}$ concentration close to the surface of the casting is proportional to the H_2 pressure in the pressure chamber, which is practically zero as the surrounding gas is argon:

$$x_{\underline{H}}^{\text{pore}} = C\sqrt{p_{H_2}^{\text{pore}}} \qquad (5')$$

$$x_{\underline{H}}^{\text{surface}} = C\sqrt{p_{H_2}^{\text{surface}}} = 0 \qquad (6')$$

The derivative of n_{H_2} in Equation (4') is introduced into Equation (2') together with the $\underline{H}$ concentrations in equations (5') and (6'). Then we obtain:

$$\frac{p}{RT}\frac{dV}{dt} = -\frac{1}{2}D_{\underline{H}}\frac{A}{V_m^{\underline{H}}}\frac{C\sqrt{p_{H_2}^{\text{pore}}} - 0}{l} \qquad (7')$$

Both V and A depend on the radius. After variable separation, Equation (7') is integrated, which gives the volume of the pore as a function of time. In this case we obtain:

$$\frac{p}{RT}\int_{V_0}^{V_{\text{pore}}}\frac{dV}{A} = -\frac{1}{2}D_{\underline{H}}\frac{1}{V_m^{\underline{H}}}\frac{C\sqrt{p_{H_2}^{\text{pore}}}}{l} \cdot \int_0^t dt$$

We introduce the radius instead of the volume and area of the pore. The integral can then be written as:

$$\frac{p}{RT}\int_{r_0}^{r_{\text{pore}}}\frac{4\pi r^2 dr}{4\pi r^2} = -\frac{1}{2}D_{\underline{H}}\frac{1}{V_m^{\underline{H}}}\frac{C\sqrt{p_{H_2}^{\text{pore}}}}{l}\int_0^t dt$$

The desired function is:

$$r_{\text{pore}} = r_0 - \frac{RTD_{\underline{H}}C\sqrt{p_{H_2}^{\text{pore}}}}{2pV_m^{\underline{H}}}\frac{t}{l} \qquad (8')$$

Next we have to derive the value of r_0. We use the result from Example 8.5 that the volume changes from V_0 to $V_0/3800$ when the pressure and temperature are increased to HIP conditions. Hence we obtain:

$$\frac{4\pi}{3}r_0^3 = \frac{4.2 \times 10^{-3}}{3800} \times 10^{-9}\,\text{m}^3 \quad \Rightarrow \quad r_0 = 6.41 \times 10^{-6}\,\text{m}$$

Inserting numerical values into Equation (8'), we obtain:

$$r_{\text{pore}} = 6.41 \times 10^{-6} - \frac{8.31 \times 10^3 \times (500+273) \times 1.0 \times 10^{-7} \times 5 \times 10^{-8}}{2 \times 10^4 \times 10^5 \times 0.010}\frac{\sqrt{10^4}}{l}t$$

or

$$r_{\text{pore}} = 6.41 \times 10^{-6} - 1.6 \times 10^{-13}\frac{t}{l} \qquad (9')$$

(b) The desired time is obtained if $r_{\text{pore}} = 0$ and $l = 0.010\,\text{m}$ are inserted into Equation (9'):

$$t_{\text{pore}} = \frac{6.41 \times 10^{-6}l}{1.6 \times 10^{-13}} = \frac{6.41 \times 10^{-6} \times 0.010}{1.6 \times 10^{-13}}$$

$$= 4.0 \times 10^5\,\text{s} = 1.1 \times 10^2\,\text{h}$$

Answer:

(a) The pore radius as a function of time (time unit = hour) is:

$$r_{\text{pore}} = r_0 - \frac{RTD_{\underline{\text{H}}}C\sqrt{p_{\text{H}_2}^{\text{pore}}}}{2pV_{\text{m}}^{\underline{\text{H}}}}\frac{t}{l} \times 3600$$

or with the given numerical values:

$$r_{\text{pore}} = 6.4 \times 10^{-6} - \frac{1.6 \times 10^{-13} \times 3600}{l}t$$

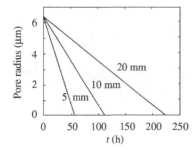

(b) The time required to empty the pore is about 110 h.

HIP treatment requires a long time, which makes HIP-treated alloys expensive. They are used solely for special purposes where the highest possible quality is needed, for instance in the aircraft industry.

The diffusion process is not reversible. The extremely high pressure during the HIP treatment presses the pore walls together and results, after many hours, in complete metallurgical fusion of the earlier pore or crack walls.

The time will be shortest if the pore consists of H_2 because the diffusion rate for hydrogen is very high. For pores which contain N_2 or O_2, the time will be much longer unless nitrides or oxides are precipitated in the matrix. Pores which contain Ar will never disappear.

In addition, it can be noted that the microstructure of the alloy is practically unchanged. This is a general property of HIP. No general deformations of the material occur as the pressure is isostatic, i.e. the same in all directions. Figures 8.31 (a) and (b) illustrate this effect of HIP. The figures show the same alloy before and after HIP treatment. The dark pores in Figure 8.31 (a) have disappeared completely after HIP treatment.

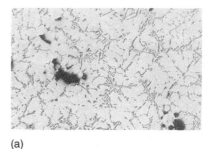

(a)

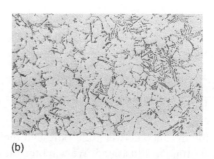

(b)

Figure 8.31 Microstructure of an Al–Si alloy (a) before and (b) after HIP treatment. Reproduced with permission from Bodycote Powdermet.

SUMMARY

■ Homogenization

The concentration of alloying elements is always unevenly distributed in cast materials, mainly owing to microsegregation.

By heat treatment at a constant temperature, the concentration differences can be smoothed out. Homogenization of the materials can be performed by heat treatment and/or plastic deformation.

■ Homogenization of Single-Phase Materials

Single-phase materials are materials in which the concentrations of alloying elements depend only on *one* position coordinate.

A one-dimensional concentration distribution of an alloying element is a relatively good approximation for rolled materials. The approximation is more doubtful for other cast materials.

(i) A simple model for the initial concentration distribution in the alloy as a function of position is set up.

(ii) Then the concentration of an alloying element as a function of time can be calculated as a solution of Fick's second law. The solution describes the homogenization process during heat treatment.

The concentration profile of the solute over a dendrite arm is a periodic function of position. An arbitrary profile can be described by a Fourier series of sine and cosine functions of a fundamental frequency and its multiple frequencies. The amplitudes of the overtones decay rapidly and only the fundamental tone with $\lambda = \lambda_{den}$ remains. We can write:

$$x = x_0 + \frac{x_0^M - x_0^m}{2} \sin\left(\frac{2\pi y}{\lambda_{den}}\right) \exp\left(-\frac{4\pi^2 D}{\lambda_{den}^2} t\right)$$

Homogenization Time

The heat treatment time required to reduce the amplitude of the fundamental wave from x_0^M to x^M is given by:

$$\frac{x^M - x^m}{x_0^M - x_0^m} = \exp\left(-\frac{4\pi^2 D}{\lambda_{den}^2} t\right) \quad \text{or} \quad t = \frac{\lambda_{den}^2}{4\pi^2 D} \ln\left(\frac{x_0^M - x_0^m}{x^M - x^m}\right)$$

The heat treatment time is often calculated for a decay ratio = 1/10. The decay ratio has a moderate influence on the homogenization time. The treatment time is much more sensitive to the dendrite arm distance and especially to the absolute temperature, which changes the diffusion constant D exponentially.

In order to obtain a homogenous material in the shortest possible time, it is important to start with a material which has *small* dendrite arm distances. Short dendrite arm distances are obtained at high solidification and cooling rates.

It is a significant advantage to cast thin sections by continuous casting rather than large ingots that later are rolled to thin dimensions.

■ Dissolution of Secondary Phases

When a binary alloy cools, a primary α-phase is precipitated. When T has decreased to the eutectic temperature, both an α-phase and a θ-phase, with a high concentration of the alloying element, are precipitated.

The non-desired θ-phase is a secondary phase, which can be dissolved by heat treatment. The one-dimensional case, i.e. when the secondary phase consists of parallel plates at regular distances from each other, is discussed.

■ Conditions for Dissolution of Secondary Phases

Case I: The Concentration Fields Overlap
The diffusion fields (the concentration field as a function of distance) from adjacent plates overlap. The concentration at each point between two plates is the sum of two terms, caused by the diffusion from each plate.

The time for complete dissolution of the secondary phase is:

$$t = -\frac{\lambda_{den}^2}{\pi^2 D} \ln\left[1 - \frac{f_0^\theta (x^\theta - x^m)}{x^{\alpha/\theta} - x^m}\right]$$

The fraction g of the plates dissolved during the treatment time t_g is:

$$g = \frac{x^{\alpha/\theta} - x^m}{f_0^\theta (x^\theta - x^m)}\left[1 - \exp\left(-\frac{\pi^2 D t_g}{\lambda_{den}^2}\right)\right]$$

The time required to dissolve the fraction g of the plates is:

$$t_g = -\frac{\lambda_{den}^2}{\pi^2 D} \ln\left[1 - \frac{g f_0^\theta (x^\theta - x^m)}{x^{\alpha/\theta} - x^m}\right]$$

The dissolution time is proportional to the square of the dendrite arm distance. The *larger* λ_{den} is, the *longer* will be the dissolution time. The *higher* the treatment temperature is, the *shorter* will be the dissolution time.

Case II: The Concentration Fields do not Overlap
The diffusion fields (the concentration profile as a function of the distance) of adjacent bands of the secondary phase do *not* overlap.

This is an extreme case, which occurs when the concentration of the alloying element in the matrix is so strongly subsaturated that the diffusion fields do not overlap until the dissolution is finished.

The time for complete dissolution of the secondary phase is

$$t = \frac{b^2}{2D} \frac{(x^\theta - x^m)^2}{(x^{\alpha/\theta} - x^m)^2}$$

The fraction g of the bands dissolved during the treatment time t_g is:

$$g = \frac{\sqrt{2Dt_g}}{f_0^\theta \lambda_{den}} \frac{(x^{\alpha/\theta} - x^m)^2}{(x^\theta - x^m)^2}$$

The time required to dissolve the fraction g of the bands is:

$$t_g = \frac{(gf_0^\theta \lambda_{den})^2}{2D} \frac{(x^\theta - x^m)^2}{(x^{\alpha/\theta} - x^m)^2}$$

■ Change of Casting Structure During Cooling and Plastic Deformation

When a metal is exposed to plastic deformation, the segregation pattern will generally change. On plastic deformation, areas with different concentrations of alloying elements are stretched into bands in the deformation direction. A lamella structure arises.

In low-alloyed steels, this lamella structure may appear as bands of ferrite or perlite.

In a plastically deformed material, the mechanical properties of the material will be different in different directions relative to the deformation direction.

Plastic Deformation of Cast Materials and Products

The purpose of plastic deformation is:

(i)   to change the shape and thickness of the material as a step in the chain to manufacture finished products;
(ii) to close cracks and pores, i.e. to improve the properties and quality of the material.

Rolling and forging deform cast materials. The volume of the material remains constant during the process.

The effect of the plastic deformation depends on the direction of the rolling. When the side in the rolling direction is elongated, all other distances, even the dendrite arm distances, will be elongated by the same factor in this direction.

There are two extreme cases. The homogenization time is proportional to the square of the dendrite arm distance.

The material is in reality deformed to a state between the two cases above. The normal procedure is to compromise.

Structure Changes of Secondary Phases by Plastic Deformation

The appearance and distribution of precipitated secondary phases are also influenced by plastic deformation. In these cases the final state is affected by the plastic properties of particles in the secondary phase relative to the matrix:

1. The particles are *harder* and *stronger* than the matrix: there is no effect of the plastic deformation.
2. The particles are *softer* than the matrix: they become elongated in the direction of deformation.
3. The particles are *brittle*: they break during the deformation process and distribute along lines parallel to the deformation direction.

■ Cooling Shrinkage

The main reason for mechanical stress in castings is a non-uniform temperature distribution, which in turn results in uneven shrinkage. The uneven shrinkage causes

(i)   plastic deformation at higher temperatures;
(ii) remaining and destructive mechanical stress.

It is important to make the temperature in the different parts of the casting decrease uniformly. A low cooling rate is essential to avoid thermal stresses.

Mechanical stress in materials can be measured with the aid of a shrinkage harp. If stress cannot be avoided, it can be eliminated or at least reduced by heat treatment.

■ Stress Relief Heat Treatment

The temperature for heat treatment is chosen as high as possible, but not so high that the structure of the material changes.

The treatment time depends on the type of casting and the shape of the casting. Cu-base alloys need no heat treatment.

■ High Isostatic Pressure

Castings are treated at high temperature (500–600 °C) and high pressure (magnitude 10 kbar) for several hours.

The extremely high pressure during the HIP treatment presses the pore walls together and results, after many hours, in complete metallurgical fusion of the earlier pore or crack walls. The microstructure of the alloy remains practically unchanged.

EXERCISES

8.1 Eutectic alloys often form a lamella structure, which consists of two solid phases. On heating, a melt is formed between the two phases. Calculate the time required for homogenization of such a melt as a function of the lamella distance, λ_{eut}, and the diffusion constant, D. The phase diagram is given below.

Hint B3

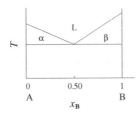

8.2 It has been suggested that dissolution of secondary phases would be faster if they were heat-treated at a temperature above the melting point of the secondary phase. See Figure 8.10 on page 235 and that illustrated here.

Test this suggestion by calculating the dissolution time of a eutectic melt at a temperature 15 °C above the eutectic temperature for the Al–Cu alloy with 2.5 at-% Cu, which was discussed in Example 8.4 (page 237). The coarseness of the structure depends on the cooling rate. You may use the diagram in Example 7.4 (page 199) to find the relation in this case.

Is it a good suggestion?

Hint B22

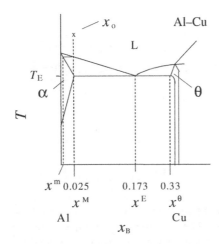

8.3 An old principle, which has been applied, for example, in the homogenization of high-speed steel, is to heat treat the material at a temperature above the eutectic temperature. High-speed steel contains about 10 % eutectic carbide. The eutectic phase melts during the heat treatment. During the homogenization, the melt solidifies slowly while the alloying elements diffuse into the solid phase.

This principle can also be applied at liquid-phase sintering. In this case, two powders with high and low carbon concentrations are mixed.

Consider a low-carbon steel which is mixed with eutectic cast iron. The powder mixture is heated to a temperature just above the eutectic temperature of Fe–C. At an excess temperature of 10 °C the eutectic iron powder melts.

A thin film of eutectic melt is formed around the low-carbon steel grains during the melting process. The diffusion process of $\underline{C}$ atoms from the melt into the low-carbon steel can be treated as if it were one-dimensional.

Plot the dissolution time for the molten cast iron powder as a function of the volume fraction of eutectic

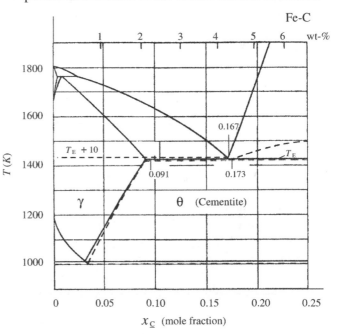

Reproduced by Permission of ASM International.

iron powder within the interval 0–20 % in a diagram.

Hint B17

The diffusion constant $D_{\underline{C}}^{\gamma} = 1.0 \times 10^{-10}$ m^2/s. The dendrite arm distance is $10\overline{0}$ μm. Other constants can be derived from the phase diagram of the Fe–C system.

8.4 Cast Al–Cu-base alloys are precipitation hardenable. In most cases it is desirable to have a Cu concentration as high as possible in these alloys in order to give them the highest possible hardness. This frequently results in the precipitation of θ-phase during the solidification process. This phase is then dissolved by heat treatment.

(a) Give an account of the possible advantages with heat treatment above and under the eutectic temperature.

Hint B33

(b) Discuss with the aid of the phase diagram below the optimal choice of dissolution temperature.

Hint B46

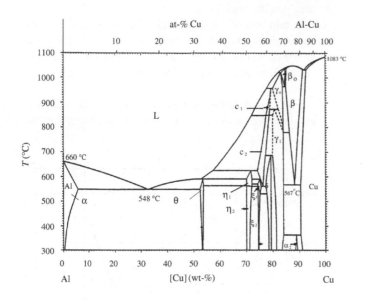

8.5 Graphite powder is sometimes used as an inoculation means in cast iron. Make an estimation of the time during which the graphite has an inoculation effect by calculation of:

(a) the dissolution time of the graphite particles;

Hint B9

(b) the homogenization time required to smooth out the carbon variations after dissolution.

Hint B39

The inoculation effect is considered to have disappeared when the variation of the carbon concentration in the melt is <0.10 wt-% $\underline{C}$. Assume that the cast iron contains 4.3 wt-% $\underline{C}$ and that the graphite is added at 1300 °C.

The graphite particles are disk-shaped and have a thickness of 0.10 mm. The diffusion constant $D_{\underline{C}} = 1.0 \times 10^{-9}$ m²/s. The phase diagram of Fe–C is given below.

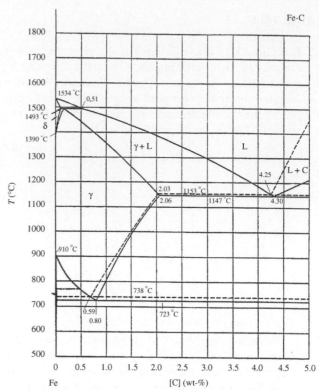

Reproduced with Permission from McGraw-Hill.

8.6 An Al–Cu alloy, which contains 3 wt-% Cu, has been cast. A metallographic examination of the casting structure shows that a θ-phase has been precipitated. The amount of θ-phase is estimated as 2 vol.-% of the material.

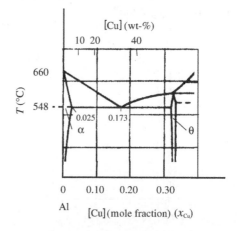

The θ-phase can be dissolved by heat treatment of the casting. The heat-treatment temperature is 500 °C. The dendrite arm distance is 50 µm and the diffusion coefficient of Cu at the given temperature is 4.8×10^{-14} m²/s. How long a time is required to dissolve the θ-phase completely?

Hint B7

8.7 In Cr-alloyed steel, chromium carbides are often obtained during casting processes. These carbides are undesired in most cases and have to be dissolved after the solidification process. The coarseness of the structure varies with the cooling rate and thickness of the casting.

Find an expression for the dissolution time as a function of the thickness y of the ingot and discuss with the aid of your result how the shortest possible dissolution time is obtained. The structure of the ingot is assumed to consist entirely of columnar crystals.

Hint B25

8.8 The aim of the so-called direct casting methods is to produce products which are as close as possible to their final dimensions. In the ideal case, one wants to avoid all sorts of hot working and perform only cold working and heat treatment.

An Fe–1 wt-% Cr–C alloy is cast in a twin-roller machine to a slab of thickness 1.0 mm. The heat transfer coefficient between the slab and the rollers is 1.0×10^3 W/m^2 K. The diffusion constant can be written as:

$$D = D_0 \exp \frac{Q_D}{RT}$$

The activation energy Q_D for diffusion of Cr atoms in γ-iron is 292×10^6 J/kmol. $D_0 = 10.8 \times 10^{-4}$ m^2/s.

The relation between the dendrite arm distance λ_{den} and the total solidification time θ_{sol} is:

$$\lambda_{den} = 2.0 \times 10^{-3} \theta_{sol}^{0.5}$$

Both quantities are measured in SI units.

Calculate the time required to eliminate the microsegregation in the cast plate by heat treatment at 1000 °C. Material constants of iron are given in the table.

Material constants:	
ρ	$= 7.2 \times 10^3$ kg/m^3
$-\Delta H$	$= 272 \times 10^3$ J/kg
T_L	$= 1500$ °C

Hint B11

8.9 An alternative method of homogenization to ordinary heat treatment is to combine heat treatment and deformation. A typical example is given below.

The Cr-bearing alloy in Exercise 8.8 is cast to a slab of thickness 0.20 m and then rolled to a thickness of 1.0 mm in two steps. The slab is cast with the aid of continuous casting with an average heat transfer coefficient of 0.5×10^3 W/m^2 K. The slab is hot-rolled from a thickness of 200 mm down to 2.0 mm and then cold-rolled down to a thickness of 1.0 mm and finally heat-treated. The dendrite arm distance and the solidification rate in the slab depend on each other according to the relation:

$$\lambda_{den}^2 v_{growth} = 1.0 \times 10^{-10} \text{ m}^2/\text{s}$$

The heat treatment is performed at the same temperature as in Exercise 8.8. The thermal conductivity of the alloy is 30 W/m K. Calculate the homogenization time.

Hint B40

The material constants and other relations which are valid for the alloy in Exercise 8.8 are the same in this case and can be used here also.

8.10 On casting and solidification of a steel with 0.50 at-% <u>Mn</u>, precipitation of FeS has been obtained in the shape of thin films with an average thickness of 3.0 μm. These films make the material brittle.

By heat treatment of the casting, the films can be transformed into MnS, which results in a reduction of the fragility of the material. The <u>Mn</u> concentration in the melt close to the films is 0.25 at-%. This value is assumed to be constant during the whole heat treatment. The <u>Mn</u> atoms come from the melt, which surrounds the FeS films.

Calculate the time for the transformation if the heat treatment is performed at 1150 °C. The diffusion constant D_{Mn}^γ for <u>Mn</u> atoms in the steel melt at this temperature is 1.0×10^{-12} m^2/s.

Hint B16

9 Precipitation of Pores and Slag Inclusions during Casting Processes

Materials Processing during Casting   H. Fredriksson and U. Åkerlind
Copyright © 2006 John Wiley & Sons, Ltd.

9.1 INTRODUCTION

The production of ingots and manufacturing of products by casting are controlled by determined requirements of the product taking its future use into account. In the production of many components, for example turbine blades, there are great demands on homogeneity, elastic properties, and mechanical strength of the cast material.

Chemical reactions between the hot melt and the surrounding materials are inevitable at the high temperatures of metal casting. The reaction products may be so-called *secondary phases*, i.e. precipitation of gas pores or slag inclusions at solidification and cooling of the casting.

On melting of the metal, and to a certain extent also during the casting, gases in the surrounding atmosphere dissolve in the melt, which may result in cavities, so-called *pores*, in the castings. Chemical reactions between the melt and the dissolved gases at high temperature may alternatively result in secondary solid phases, so-called *slag* products, in the castings after solidification and cooling. Such solid secondary phases may also form after chemical reactions between the melt and non-metallic mould materials in contact with the melt, when it solidifies and cools.

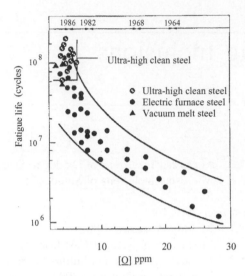

Figure 9.1 Relation between fatigue life and oxygen content in ball bearing steels. Reproduced with permission from The Iron & Steel Institute of Japan.

shows the negative effect of oxygen on the fatigue properties of ball bearing steels. The impact resistance of the metals also becomes lowered and cracks may occur, especially for metals with high mechanical strength. Pores and slag inclusions must therefore be avoided as much as possible.

In this chapter, the general principles and the reaction mechanisms behind gas and non-metallic slag precipitations will primarily be discussed. The general laws will then be applied to different types of alloys, such as aluminium alloys, copper alloys, steel and iron alloys, cast iron, and nickel-base alloys.

The problems connected with the solubilities of the most common gases in metal melts and the most important types of nonmetallic inclusions will be discussed separately for each group of alloys. The reasons for various types of gas precipitation will be penetrated. The possible ways to prevent or eliminate macroslag inclusions in the alloys will be discussed. Different methods to neutralize the negative effects of pores and slag inclusions will be suggested for each group of alloys.

Depending on which gas has reacted with the melt, metal oxides and nitrides may form and precipitate as solid secondary phases in the shape of slag inclusions during the solidification process. Examples of other types of such inclusions, formed from impurities in the melt, are sulfides in steel alloys and iron aluminates in aluminium-base alloys.

Both pores and solid secondary phases lower the quality of the casting. An example is given in Figure 9.1, which

9.2 UNITS AND LAWS

In order to facilitate the discussions of solubilities and the risk of precipitation of various gases in various alloys, we will need some special concepts and units. These are listed in Table 9.1. Table 9.2 reviews the most important physical and chemical laws, which will be applied further on. Example 9.1 illustrates some common types of unit transformations.

TABLE 9.1 **Units and symbols used in the text.**

Quantity	Symbol	Unit	Comments
Pressure	p	N/m^2, atm	
Volume	V	m^3	
Mole	n	kmol	Mass (kg) of molar weight
Absolute temperature	T	K	
Concentration	[ ]		General designation
		ppm	(weight) parts per million
	c	weight-%	Subscript 0 = initial value
		(wt-%)	Subscript L = liquid phase
		kmol/l	Subscript s = solid phase
	x	mole fraction	Number of kmole of the
		atomic-%	substance/total number of
		(at-%)	kmol. Subscripts as above
Partition coefficient Solid phase/liquid phase	k_{part}, k_{partx}, k_{partc}		
Solubility of a gas in a melt or a solid	S	ppm	(weight) parts per million
		cm^3/100 g	cm^3 (NTP)/per 100 g
		wt-%	Subscripts as above

TABLE 9.2 Some basic physical and chemical laws.

- $pV = nRT$ *General gas law*
- The saturation value of the concentration of a gas dissolved in a melt is proportional to the square root of the pressure of the gas:

$$c = \text{constant} \times \sqrt{p} \qquad \textit{Sievert's law}$$

 where $p =$ partial pressure of the gas.

- *Condition for precipitation of a gas in a melt*

$$p_{\text{gas}} = p_a + p_h + 2\sigma/r$$

 where

   $p_{\text{gas}} =$ total pressure inside the gas pore
   $p_a \;\;=$ atmospheric pressure (*not* the partial pressure of the gas)
   $p_h \;\;= \rho g h =$ hydrostatic pressure
   $\rho \;\;\;=$ density of the melt
   $h \;\;\;=$ distance from the gas pore to the free surface of the melt
   $\sigma \;\;\;=$ surface energy per unit area of the melt
   $r \;\;\;=$ radius of the gas pore.

- $p + \rho g h + \dfrac{\rho v^2}{2} = \text{constant}$ *Bernoulli's equation*
 where

   $p \;=$ pressure in a liquid or a gas
   $h \;=$ height over an arbitrarily chosen zero level
   $v \;=$ velocity of the particles in the streaming fluid.

- For the reaction $aA + bB \rightleftharpoons cC + dD$ we have

$$\frac{[C]^c \times [D]^d}{[A]^a \times [B]^b} = \text{constant} \qquad \textit{Guldberg--Waage's law}$$

  Endothermic reaction   Energy has to be supplied
  Exothermic reaction   Energy is emitted.

The value of the constant depends on the concentration units and on the temperature. The equilibrium is altered when the temperature is changed.

- $\dfrac{dm}{dt} = -DA\dfrac{dc}{dy}$ *Fick's first law*

When Fick's law is applied one *must not* simply use dimensionless units for concentration c, for example wt-%, ppm, mole fraction, or at-%.
See page 303 for a special analysis of this topic.

Example 9.1
A binary alloy has a composition which is described by the masses and molar weights m_A, M_A and m_B, M_B, respectively. Express its composition in ppm, weight-% (wt-%), mole fraction, and atomic-% (at-%). In addition, transfer c (wt-%) to x (at-%) and x to c.

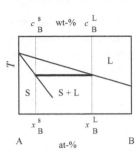

Solution and Answers

$$c_B = \frac{m_B}{m_A + m_B} \times 10^6 \ \text{ppm} = \frac{m_B}{m_A + m_B} \times 100 \ \text{wt-\%}$$

$$c_B = \frac{x_B M_B}{x_A M_A + x_B M_B} \times 100 \ \text{wt-\%}$$

$$x_B = \frac{\dfrac{m_B}{M_B}}{\dfrac{m_A}{M_A} + \dfrac{m_B}{M_B}} \ (\text{mole fraction}) = \frac{\dfrac{m_B}{M_B}}{\dfrac{m_A}{M_A} + \dfrac{m_B}{M_B}} \times 100 \ \text{at-\%}$$

$$x_B = \frac{\dfrac{c_B}{M_B}}{\dfrac{c_A}{M_A} + \dfrac{c_B}{M_B}} \times 100 \ \text{at-\%}$$

The following relations are also valid for a binary system:
Weight percent:

$$c_A + c_B = 100$$

Mole fractions:

$$x_A + x_B = 1$$

9.3 PRECIPITATION OF GASES IN METAL MELTS

9.3.1 Gas Reactions and Gas Precipitation in Metal Melts

Precipitation of gas is one of the most severe problems that arise at casting of iron and steel and aluminium and copper alloys. Hence it is urgent to analyse the mechanisms which control pore formation during casting.

The Most Common Gases in Metal Melts
During melting, the metal is for a long time in contact with the surrounding atmosphere, which normally is air. *Oxygen* and *nitrogen* are therefore always present and may in some cases dissolve in the melt. Because air always contains water vapour, one must also take *hydrogen* into account.

All steel and iron alloys contain carbon. The possibility exists in these cases that carbon and oxygen, both dissolved in the melt, may generate *carbon monoxide*.

During casting, which lasts for a relatively short time, the melt may carry away air mechanically. By a suitable design of the casting process and the mould, this complication can be prevented.

Other sources of gases in contact with the melt are chemical reactions between the melt and the material of the mould. Some reaction products may be gaseous, including *hydrogen, oxygen, carbon monoxide, carbon dioxide,* and *sulfur dioxide*. An important example is moisture in sand moulds, which leads to water vapour in contact with the melt. The gas dissociates and generates *hydrogen*. This hydrogen source is difficult to eliminate.

Reason and Conditions for Gas Precipitation

The solubility of gases is much higher in the melt than in the solid phase for all common metals. Pressure, temperature and composition of the melt influence the solubility of a gas, i.e. its concentration in the melt. Examples of this in concrete cases will be given later.

When the melt starts to solidify, the gas concentrates gradually in the melt. When the solubility reaches its saturation value, one of two alternatives will occur.

In the absence of suitable condensation nuclei and/or if the solidification process is faster than the time necessary for reaching equilibrium between the dissolved phase and the gas phase, a *supersaturated solution* is formed and gas precipitation fails to occur. Under opposite conditions, *gas pores* are nucleated and grow in the melt. The pressure inside the pores (Figure 9.2) is controlled by the equilibrium equation:

$$p_{gas} = p_{atm} + p_h + 2\sigma/r \qquad (9.1)$$

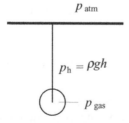

p_{atm}

$p_h = \rho g h$

p_{gas}

Figure 9.2 Pore in a melt at depth h below the surface.

where
p_{gas} = total pressure inside the gas pore
p_{atm} = atmospheric pressure (*not* the partial pressure of the gas)
$p_h = \rho g h$ = hydrostatic pressure
ρ = density of the melt

h = distance from the gas pore to the free surface of the melt
σ = surface energy per unit area of the melt
r = radius of the gas pore.

Depending on the actual circumstances, the bubbles may have various shapes, sizes and distribution after solidification. We will discuss this later in this chapter.

Equation (9.1) is a necessary but not sufficient condition for gas precipitation. Gas precipitation is a particularly complicated process, which depends on many factors, among others:

- concentration of the dissolved gas in the melt;
- saturation concentration of the gas in the alloy;
- composition of the alloy;
- rates of the chemical reactions;
- diffusion rates of the gas in the melt and the solid phase;
- access to suitable condensation nuclei;
- surface tension of the melt.

We will consider these factors more closely.

9.3.2 Solubility of Gases in Metals – Sievert's Law

Sievert's Law

The difference between the solubility of gases in a metal melt and in the solid phase after solidification is the basic reason for gas precipitation in metals at casting.

The solubility of a gas in a metal melt depends on its partial pressure in the surrounding atmosphere. Most common gases are diatomic at room temperature. At high temperature (temperature of the metal melt), a dissociation of the gas G_2 occurs at the metal surface in the presence of metal atoms M:

$$M + G_2 \rightleftharpoons M + 2\underline{G}$$

A chemical equilibrium between the dissolved gas atoms $\underline{G}$ and the surrounding gas is established. Guldberg–Waage's law controls the equilibrium:

$$\frac{[\underline{G}]^2}{p_{G_2}} = \text{constant} \qquad (9.2)$$

where the value of the constant depends on the temperature. The gas atoms then diffuse into the melt and/or the solid phase and stay there in the shape of a solution. We denote the concentration in the solution with [$\underline{G}$], where the underlined letter means that the gas is dissolved in the metal as G atoms and not G_2 molecules.

At equilibrium, the concentration of the dissolved gas atoms in the melt or in the solid phase is proportional to the square root of the partial pressure of the gas in the surrounding atmosphere. This statement can be written as:

$$[\underline{G}] = \text{constant} \times \sqrt{p_{G_2}} \qquad \text{Sievert's law} \qquad (9.3)$$

Sievert's law follows directly from Equation (9.2).

The solubility of a gas in metals is often given in the literature at the standard pressure 1 atm. Sievert's law is extraordinarily useful for calculation of the solubility [$\underline{G}$] in equilibrium with a known partial pressure in the surrounding atmosphere. [$\underline{G}$] is normally expressed as $c_{\underline{G}}$ (wt-%) or $x_{\underline{G}}$ (mole fraction).

Partition Constant

Sievert's law gives the concentration of dissolved gas in equilibrium with the surrounding atmosphere. Analogously there exists an equilibrium between the concentrations of dissolved gas in the melt and in the solid phase. At equilibrium their ratio is constant:

$$\frac{x^s}{x^L} = k_{\text{part }x} \quad \text{or} \quad \frac{c^s}{c^L} = k_{\text{part }c} \qquad (9.4)$$

The constant k_{part} is the *partition constant* or *partition coefficient*. This important concept will often be used throughout this chapter.

The k_{part} value depends on the concentration units used.

Solubility Diagram

Figure 9.3 shows a typical case of the solubility of a gas in a pure metal as a function of temperature.

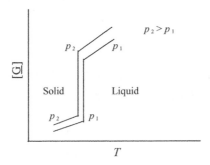

Figure 9.3 Solubility of a gas in a metal as a function of temperature for the case when the solubility process is *endothermic*. The solubility of the gas also depends on its pressure.

It can be seen that:

- the solubility of the gas is much greater in the melt than in the solid phase.
- the solubility (concentration) of the gas increases with temperature both in the melt and in the solid phase.

This latter information can be included in Equation (9.3) by replacing it by Equation (9.5):

$$[\underline{G}] = \text{constant} \times \exp\left(\frac{-\Delta H_S}{RT}\right) \times \sqrt{p_{G_2}} \qquad (9.5)$$

where

$$
\begin{aligned}
[\underline{G}] &= \text{solubility of the gas} \\
-\Delta H_S &= \text{heat of solution per kmol of the gas} \\
R &= \text{gas constant} \\
T &= \text{absolute temperature.}
\end{aligned}
$$

When hydrogen dissolves in the melt energy must be supplied. The reaction is *endothermic* and $-\Delta H_S$ is *positive*. If we differentiate Equation (9.5) with respect to T we obtain:

$$\frac{\mathrm{d}[\underline{G}]}{\mathrm{d}T} = \text{constant} \times \exp\left(\frac{-\Delta H_s}{RT}\right) \times \frac{-(-\Delta H_s)}{R} \times \frac{-1}{T^2} \times \sqrt{p_{G_2}}$$

or

$$\frac{\mathrm{d}[\underline{G}]}{\mathrm{d}T} = \text{constant} \times \exp\left(\frac{-\Delta H_s}{RT}\right) \times \frac{-\Delta H_s}{RT^2} \times \sqrt{p_{G_2}} \qquad (9.6)$$

The derivatives, i.e. the slopes of the curves, are positive if $-\Delta H_S$ is positive. This is in agreement with Figure 9.3 and is valid for hydrogen in most common metals such as Fe, Al, Mg, and Cu.

When hydrogen dissolves in, for example, Ti or Zr, energy is released, i.e. the reaction is *exothermic*. The solubility curves have *negative* slopes in this case (Figure 9.4).

The melt generally contains less gas than the saturation value. This means that there is no risk of gas precipitation during the solidification process in this case.

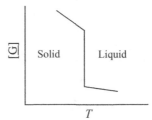

Figure 9.4 The solubility of a gas in a metal as a function of temperature when the solution process is *exothermic*.

9.3.3 Reaction Mechanisms and Reaction Kinetics

Absorption of a gas in a metal melt occurs in three steps:

1. diffusion of the surrounding gas to the surface;
2. chemical reaction at the surface;
3. atomic diffusion of one or more of the reaction products into the melt.

The slowest of these steps controls the rate of the gas absorption.

Step 1 is normally fast and there is no need to discuss it further. Most gases in question are diatomic, for example H_2, N_2, and O_2.

Step 2 is often a dissociation, which occurs at the metal surface, for example:

$$M + H_2 \rightleftharpoons 2\underline{H} + M$$

where M is the chemical symbol for the molten metal, which is not consumed in the reaction. This step is in many cases the slowest one and hence it determines the reaction rate.

Step 3 implies that the atoms diffuse into the melt and stay there as a solution. As mentioned above, an underlined atom symbol indicates a monoatomic gas atom dissolved in a melt or a solid phase, for example $\underline{H}$ or $\underline{N}$. Often the diffusion process is supported by the convection in the melt, which increases the rate of this step considerably.

Step 2 or 3 determines how rapidly a gas dissolves in the metal melt. The same is true in reverse order in gas precipitation. Step 3 controls the pore growth during the solidification process. In the case of no convection in the melt, diffusion occurs in the solid–liquid two-phase region during the solidification process.

Gas Absorption

During casting from ladle to mould or chill-mould, the temperature of the melt decreases owing to emission of heat from the metal jet stream to the surroundings. Simultaneously, strong absorption of oxygen, hydrogen, and nitrogen occurs because a large area is exposed to the atmosphere. It contains water vapour, which reacts with the metal and results in metal oxide or dissolved oxygen and monoatomic hydrogen.

If the stream is split up, its reaction area is enlarged many times. Experience shows that the length and shape of the stream are very important.

Example 9.2 shows that the hydrogen concentration in steel increases during the casting process. The hydrogen concentration in steel is very important for pore formation during the solidification process.

Example 9.2

An ingot of 10 tons will be cast. The casting time is 2 min. During the casting from ladle to mould, it is to be expected from experience that the temperature of the steel melt will decrease by 20 °C (from 1520 to 1500 °C). A chemical

analysis shows that the steel contains 4 ppm of hydrogen before the casting. In the present case, the third step (see above), diffusion of H atoms through a boundary layer into the melt, is the slowest process and determines the absorption rate. This boundary layer is assumed to have an average thickness of 10 μm.

Make a rough calculation to find the increase in hydrogen concentration in the steel during the casting operation.

From the literature the following information has been extracted:

Diffusion rate of hydrogen in a steel melt $= 1.0 \times 10^{-9}$ m^2/s. Maximum solubility of hydrogen in steel at the present temperature and moisture concentration in the air $= 20$ ppm. Thermal capacity of steel $= 0.76$ kJ/kg K.

The constant in the Stefan–Boltzmann's law is 5.67×10^{-8} J/sm^2 K^4. The steel melt is no ideal black body and a reduction factor, the emissitivity ε, has to be introduced. Its value can in this case be estimated as 0.6.

Solution:

Step 1:
From the information that the melt cools by 20 °C, we can calculate the effective area A of the jet stream. The heat loss is equal to the emitted energy of the jet stream.

$$c_p M \Delta T = \varepsilon \sigma (T^4 - T_0^4) A t \qquad (1')$$

where
- M = mass of the ingot
- ΔT = temperature decrease of the melt
- T = temperature of the steel melt $= 1520\,°C = 1793$ K
- T_o = temperature of the surroundings $= 20\,°C = 293$ K
- t = casting time $= 2 \times 60 = 120$ s.

T_0^4 can be neglected compared with T^4 and we obtain the following equality, which gives $A = 3.58$ m^2.

$$0.76 \times 10^3 \times 10 \times 10^3 \times 20$$
$$= 0.6 \times 5.67 \times 10^{-8} \times 1793^4 \times A \times 120$$

Step 2:
The monoatomic hydrogen diffuses into the steel melt. With the aid of Fick's diffusion law we can calculate the hydrogen concentration of the ingot at the end of the casting time.

Fick's first law can be written as:

$$\frac{dm}{dt} = -DA \frac{dc}{dy} \qquad (2')$$

where

m = mass of the diffused hydrogen
D = diffusion coefficient of hydrogen in steel
A = total effective area of the stream
c = concentration (wt-%) multiplied by the density of steel (see page 303)
y = diffusion distance perpendicular to the surface.

At stationary flow, Equation (2') can be written as:

$$\frac{dm}{dt} = -DA \frac{\rho_{Fe} \Delta c}{\Delta y} = -DA \frac{\rho_{Fe}(c - c_{eq})}{\delta} \qquad (3')$$

where δ is the thickness of the boundary layer and c_{eq} is the solubility limit for hydrogen in steel at the given temperature.

The mass dm of hydrogen, which is transported into the steel melt during the time dt, increases the hydrogen concentration in the steel melt with the amount dc and we obtain the equality $dm = dcM$ and Equation (3') can be written as:

$$\frac{Mdc}{dt} = -DA\rho_{Fe} \frac{c - c_{eq}}{\delta} \qquad (4')$$

If we integrate Equation (4'), we obtain the hydrogen concentration in the ingot as a function of time. From this relation we can obtain the desired concentration of hydrogen by inserting $t = 120\,s$; $1\,ppm = 10^{-4}\,wt\text{-}\%$.

$$\int_{4\times10^{-4}}^{c} \frac{dc}{c - c_{eq}} = -\frac{DA\rho_{Fe}}{M\delta} \int_{0}^{c} dt \qquad (5')$$

or

$$\left[\ln(c - c_{eq}) \right]_{4\times10^{-4}}^{c} = -\frac{DA\rho_{Fe}}{M\delta} t$$

which can be transformed into:

$$\frac{c - 20 \times 10^{-4}}{4 \times 10^{-4} - 20 \times 10^{-4}} = \exp\left(-\frac{DA\rho_{Fe}}{M\delta} t\right)$$

where c is given in wt-%; c is solved, giving:

$$c = 20 \times 10^{-4} - 16 \times 10^{-4} \times \exp\left(-\frac{DA\rho_{Fe}}{M\delta} t\right) \qquad (6')$$

The value of c can be calculated by inserting known values in Equation (6'). In addition to earlier given values, we have

$\rho_{Fe} = 7.8 \times 10^3$ kg/m^3 (from standard reference table) and $\delta = 1.0 \times 10^{-5}$ m (from text), which gives:

$$c = 20 \times 10^{-4} - 16 \times 10^{-4} \exp(-0.33)$$
$$= 8.5 \times 10^{-4}\,\text{wt-%} = 8.5\,\text{ppm}$$

Answer:
The hydrogen concentration will be doubled during the casting operation.

From the ladle, the stream falls into the mould. The stream causes motion and flow in the metal melt in the mould. When the stream hits the surface of the melt in the mould, air bubbles are carried down into the steel bath (Figure 9.5).

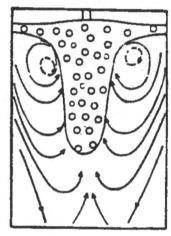

Figure 9.5 Flow in a steel melt when the stream from the ladle hits the mould.

If the flow is laminar, air is carried into the space between the jet stream and the melt and accompanies the melt into the mould. If the flow is turbulent, air bubbles, trapped in the jet stream, follow the melt into the mould. When the air bubbles in the mould rise and leave the surface they cause a strong motion in the steel melt. The rising bubbles also decrease the penetration depth of the jet stream.

The trapped gas in the bubbles reacts chemically with the melt. If the gas is air it causes slag inclusions. For this reason, it is important that the atmosphere is as inert as possible.

In downhill casting in a mould, this can be achieved by covering the inside of the mould with tar, which is evaporated during the casting. In continuous casting without a casting tube (Chapter 10) rape oil can be added to protect the surface of the melt in the mould from reacting with air.

9.3.4 Formation of Pores

In casting, two main types of cavities or pores in castings and ingots may appear: *shrinkage pores* and *gas pores*. The classification is made with respect to the completely different origins and appearances of the pores.

It is well known that the solid phase generally has a higher density than the melt. The solidification shrinkage is normally compensated by the fact that new melt is sucked into the two-phase area during the solidification process. In many cases, especially when the solidification process is rapid, it is very difficult to compensate for the whole solidification shrinkage, and instead pores are formed.

Shrinkage pores arise by the formation of cavities in the remaining melt in the interdendritic areas, owing to the resistance that the melt offers to the transport through the narrow channels. The resistance is caused partly by the viscosity of the melt and partly by the dendrites around the channels. For this reason, the transport channels of the melt are normally curved. In this case the surface of the pore becomes thorny and uneven. Shrinkage pores will be treated further in Chapter 10.

Gas pores are characterized by the fact that their surface is smooth and even. Absorbed gas in a molten metal is in most cases more soluble in the melt than in the solid phase. Solidification of the metal results in gas precipitation in the interdendritic regions. The released gas has a large volume, compared with the melt, which in most cases causes the formation of large pores because the melt is squeezed out of the two-phase areas. The sizes of these pores are determined by the concentration of the dissolved gas and by the difference in solubility of the gas in the melt and in the solid phase.

The gas pores can be divided into two subgroups: spherical pores and elongated pores. They will be discussed separately below.

When the precipitation of a eutectic structure at the end of the solidification process was discussed in Chapter 7, we used a phase diagram in order to facilitate our understanding of the process. This method can also be used when precipitation of gas pores is concerned.

Precipitation of Gas Pores

Consider a gas with a solubility curve similar to that in Figure 9.3. In this case we exchange the axes and obtain a figure like that in Figure 9.6. It is the phase diagram of the metal–gas system.

When solidus and liquidus lines have been added, the diagram will be even more like a 'eutectic' phase diagram. As can be seen from Figure 9.6, the position of the eutectic point E is very sensitive to the pressure *p*.

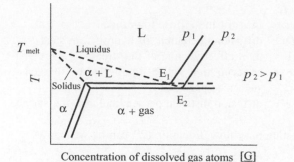

Concentration of dissolved gas atoms [G]

Figure 9.6 Phase diagram of a metal–gas system.

Spherical Pores

Spherical pores in the melt are formed when the gas concentration exceeds the solubility of the gas at the current temperature. This occurs when the gas concentration is higher than the eutectic concentration of the gas in the melt as is illustrated in Figure 9.6.

Pores are formed preferably at the upper parts of an ingot. In this case the surface tension forces and the pressure forces in the melt determine the pressure inside the pores according to Equation (9.1).

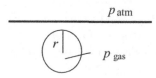

Figure 9.7 Spherical pore at a depth *h* below the surface of a metal melt.

In order to analyse the pore growth, i.e. the pore radius as a function of time, we consider a pore at a depth *h* under the surface of the melt (Figure 9.7) and the terms $P_{atm} + \rho g h$ in Equation (9.1) can be replaced by P_{total}, giving:

$$p_{gas} = p_{total} + \frac{2\sigma}{r} \qquad (9.7)$$

where

p_{gas} = total pressure inside the pore
p_{total} = atmospheric pressure (*not* the partial pressure of the gas) + the hydrotatic pressure
σ = surface energy per unit area of the melt
r = radius of the gas pore.

In addition, we assume that the gas is diatomic and dissolved as atoms in the melt. The pore grows when the following reaction occurs at the melt/pore interface:

$$2\underline{G} \rightarrow G_2$$

The gas is precipitated as diatomic molecules inside the pore. With the aid of the general gas law, the number (kmol) of diatomic gas molecules in the pore can be calculated:

$$n_{G_2} = \frac{p_{gas} V_{pore}}{RT} \tag{9.8}$$

where

n_{G_2} = kmol of the diatomic gas in the pore
p_{gas} = pressure inside the pore
V_{pore} = volume of the pore
R = the general gas constant
T = temperature of the melt.

Equation (9.8) if combined with Equation (9.7) gives the number n_G of the gas atoms (kmol) dissolved in the melt, which are required to form the pore, as a function of the radius r of the pore. The number of gas atoms is twice the number of gas molecules:

$$n_{\underline{G}} = 2n_{G_2} = \frac{2\left(p_{total} + \frac{2\sigma}{r}\right)\frac{4\pi r^3}{3}}{RT} \tag{9.9}$$

If we differentiate Equation (9.9) with respect to time we obtain a relation between the *decrease* in the number of dissolved gas atoms (kmole) per unit time in the pore and the *increase* in the pore radius per unit time:

$$-\frac{dn_{\underline{G}}}{dt} = \frac{2dn_{G_2}}{dt} = \left(\frac{2dn_{G_2}}{dr} = \frac{8\pi}{3R}\right)\frac{dr}{dt} = \frac{(3p_{total}r^2 + 4\sigma r)}{T}\frac{dr}{dt} \tag{9.10}$$

We can obtain another expression of the number of gas atoms (kmol) per unit time by considering the diffusion of gas atoms from the melt into the melt/pore interface. Fick's first law (see Table 9.2 on page 257) will be applied. The increase in mass per unit time is proportional to the area and the concentration gradient.

The growth rate of the pore is proportional to the concentration gradient. We assume that the concentration of dissolved gas atoms increases linearly from $x_{\underline{G}}^{L/pore}$ at the pore surface to the gas concentration in the melt at the distance r from the pore surface. The latter concentration is approximated with the average concentration $x_{\underline{G}}^L$ of dissolved gas in the melt:

$$\frac{dn_{\underline{G}}}{dt} = -D_{\underline{G}} \times 4\pi r^2 \frac{x_{\underline{G}}^L - x_{\underline{G}}^{L/pore}}{V_m r} \tag{9.11}$$

where

$n_{\underline{G}}$ = kmol of gas atoms in the pore
V_m = molar volume (has to be added because the concentrations are measured in mole fractions and the mass in kilomoles)

$D_{\underline{G}}$ = diffusion constant of the dissolved gas atoms in the melt
$x_{\underline{G}}^L$ = average mole fraction (kmol) of the gas in the melt
$x_{\underline{G}}^{L/pore}$ = mole fraction (kmol) of dissolved gas atoms in the melt close to the pore surface.

We combine Equations (9.10) and (9.11) and solve dr/dt from the new equation. The result is:

$$\frac{dr}{dt} = \frac{3RD_{\underline{G}}T}{2V_m} \frac{x_{\underline{G}}^L - x_{\underline{G}}^{L/pore}}{3p_{total}r + 4\sigma} \tag{9.12}$$

This is the desired expression of the growth rate of spherical pores as a function of time. It is necessary to know the concentration of the gas in the melt and in the pore close to the pore surface if the equation is to be useful.

$x_{\underline{G}}^L$ is the original mole fraction of the gas in the melt. $x_{\underline{G}}^{L/pore}$ is strongly temperature dependent and can be calculated from the following expression [compare Equation (9.5)]:

$$x_{\underline{G}}^{L/pore} = \text{constant} \times \sqrt{\frac{p_{total} + \frac{2\sigma}{r}}{p_0}} \times \exp\left(-\frac{\text{constant}}{T}\right) \tag{9.13}$$

where p_0 is the standard pressure of the gas, normally 1 atm. The root expression originates from application of Sievert's (page 257) because the pressure is not 1 atm but has the value given in the numerator within the root expression.

Elongated Pores

Elongated pores are formed when the original gas concentration in the melt is lower than the solubility of the gas, i.e. lower than the eutectic concentration of the dissolved gas atoms (Figure 9.6). This type of pore is nucleated at the solidification front and grows successively during the solidification process.

As a concrete example of pore growth, we choose a steel melt, which can dissolve 8 ppm of hydrogen in the solid phase and 27 ppm at 1 atm (Figure 9.8) in the liquid phase.

Figure 9.8 (b) shows the concentration of $\underline{H}$ in the melt ($c_{\underline{H}}^L$) and in the solid phase ($c_{\underline{H}}^s$) as a function of the distance from the solidification front (100% solid phase at $y = y_{end}$). The initial concentration of $\underline{H}$ in the melt is 10 ppm [point A in Figure 9.8 (b)]. It then increases slightly in the area ahead of the tips of the dendrites. At point B (y_{front}), solid phase with a low concentration of $\underline{H}$ starts to precipitate ($c_{\underline{H}}^s = 3$ ppm). The more solid phase is precipitated, the higher will be the hydrogen concentration in the remaining melt.

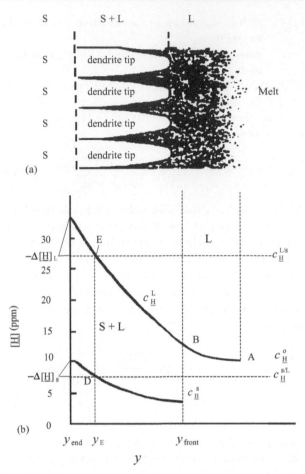

Figure 9.8 (a) Sketch of the melt and the two-phase region between the two solidification fronts. (b) Principal sketch of the solubility of $\underline{H}$ in the melt and in the solid phase within the two-phase region close to the solidification front. The concentration difference $\Delta[\underline{H}]$ between the supersaturated melt and the supersaturated solid phase and the $\underline{H}$ concentration in the melt and in the solid phase, close to the pores, are the driving forces of the diffusion of $\underline{H}$ atoms from the melt and the solid phase to the pores (Figure 9.9). The underlined $\underline{H}$ is the symbol for hydrogen atoms which are dissolved in the melt. S = solid phase; S + L = solid + liquid phase ; L = liquid phase.

The ratio of the concentrations $c_{\underline{H}}^s / c_{\underline{H}}^L$ is constant, which means that the hydrogen concentration in the solid phase also increases with decreasing distance to the solid front y_{end}. The lower dotted line corresponds to the saturation value of the hydrogen concentration in the solid phase.

At distances closer to the end of the two-phase region than y_E, the solid phase is supersaturated. At $y = y_{end}$, the hydrogen concentration is 10 ppm and equal to the original hydrogen concentration in the melt as no hydrogen has disappeared.

The upper dotted line corresponds to the saturation value of 27 ppm of $\underline{H}$ in the melt. At distances closer to $y = y_{end}$ than point E (y_E), the melt is supersaturated. At the end of

the solidification front ($y = y_{end}$), the hydrogen concentration in the last solidified melt has increased to 33 ppm.

It is evident from the analysis above that there are theoretical possibilities for precipitation of hydrogen in the steel melt. Often higher concentrations are required than the theoretical supersaturation limits for spontaneous pore formation. In the present case this condition is fulfilled. The nucleation of pores is very difficult and may occur by heterogeneous nucleation. When a pore is nucleated, $\underline{H}$ atoms from the supersaturated solid or liquid will diffuse into the pore. A common situation is that the pore grows at the same rate as the solidification front.

Figure 9.9 shows a sketch of pores, which grow at the same rate as the solidification front. If the pressure of the hydrogen pore corresponds to 1 atm outer pressure it will grow by influx of $\underline{H}$ from the supersaturated melt and likewise from the solid phase (Figure 9.9). The influx occurs by diffusion in both cases. At the interface there is equilibrium between the dissolved $\underline{H}$ atoms and gaseous diatomic hydrogen:

$$2\underline{H} \rightleftharpoons H_2$$

The growth rate, i.e. the velocity of the solidification front, and the hydrogen concentrations control the pore growth. If the solidification front moves slowly, the pore volume

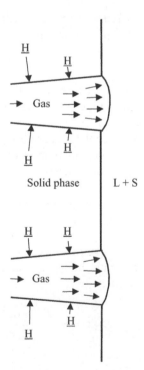

Figure 9.9 Principal sketch of the growth of hydrogen pores in the solid phase. The solidification front corresponds to y_{end} in Figure 9.8 (b). The diffusion of $\underline{H}$ atoms from the melt is not indicated in Figure 9.8 (a).

increases more rapidly than in the case when its velocity is large. The higher the hydrogen concentration, the more rapidly will the volume of the pore grow.

At low growth rate of the solidification front and/or at high supersaturations, the growth of the pore may be so fast that it loses contact with the solidification front and flows away into the melt ahead of the solidification front.

The sizes, shapes and distributions of the pores in various cases depend strongly on the conditions in each special case. Figure 9.10 gives some examples of different alternatives.

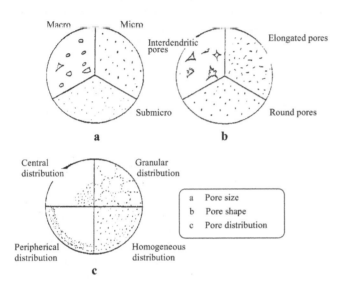

Figure 9.10 Examples of different types of pore structure. Reproduced with permission from Hodder Headline.

9.4 PRECIPITATION OF INCLUSIONS IN METAL MELTS

Slag inclusions are chemical compounds of a metal and non-metal components or only non-metal components. Examples are metal oxides, nitrides and sulfides, which are hard to dissolve in metal melts. Some metallurgical methods to remove them are filtration, flotation (keep in ladle or tundish for some time), vacuum degassing and/or gas flushing.

Before casting, the metals are molten in a furnace or a ladle. The metal may react with the refractory covering of the ladle, which results in a change of the composition of the melt. Refractory materials often consist of oxides. During the casting and solidification processes refractory components may precipitate.

Chemical reactions can be analysed in terms of thermodynamics, i.e. in terms of chemical potentials or with the aid of free energy functions. We will choose the latter method and briefly discuss metallurgical reactions in terms of free energy of the reacting system.

9.4.1 Thermodynamics of Chemical Reactions

All systems strive to achieve equilibrium, i.e. a state that corresponds to an *energy minimum*. The energy function, which describes a system, consisting of a number of reactants and reaction products, is the Gibbs' free energy:

$$G = H - TS \qquad (9.14)$$

where

$G =$ free energy of the system
$H =$ enthalpy of the system
$T =$ absolute temperature
$S =$ entropy of the system.

The condition for a spontaneous chemical reaction is that the total change in free energy of the system is negative, i.e. the energy of the system decreases. The change in free energy is defined as:

$$\Delta G = G_{\text{products}} - G_{\text{reactants}} \qquad (9.15)$$

The condition for a *spontaneous chemical reaction* can then be written as:

$$\Delta G = G_{\text{products}} - G_{\text{reactants}} < 0 \qquad (9.16)$$

The greater the change in the free energy of the system, the greater is the driving force of the reaction, i.e. the easier it will occur.

The condition for *equilibrium of the system* is:

$$\Delta G = G_{\text{products}} - G_{\text{reactants}} = 0 \qquad (9.17)$$

There is a relation between the Gibbs' free energy of an element and its concentration. For an ideal solution, this relation can be written as:

$$G = G^{\circ} + RT \ln x \qquad (9.18)$$

where

$G =$ free energy of the element
$G^{\circ} =$ standard free energy of the element
$T =$ absolute temperature
$x =$ concentration of the element (mole fraction).

9.4.2 Solubility Product

Guldberg–Waage's Law
Consider a chemical reaction

$$a\underline{A} + b\underline{B} \rightleftharpoons c\underline{C} + d\underline{D}$$

where A, B, C, and D are reactants and products, and a, b, c, and d are coefficients in the chemical reaction.

With the aid of the condition (9.17) and Equation (9.18), applied to all elements involved in the chemical reaction, we obtain the condition for equilibrium of the system:

$$G_{\text{products}} = G_{\text{reactants}}$$

or

$$c(G_{\underline{C}}^{\circ} + RT \ln x_{\underline{C}}) + d(G_{\underline{D}}^{\circ} + RT \ln x_{\underline{D}})$$
$$= a(G_{\underline{D}}^{\circ} + RT \ln x_{\underline{A}}) + b\left(G_{\underline{B}}^{\circ} + RT \ln x_{\underline{B}}\right)$$

which can be written as:

$$cRT \ln x_{\underline{C}} + dRT \ln x_{\underline{D}} - aRT \ln x_{\underline{A}} - bRT \ln x_{\underline{B}} = aG_{\underline{A}}^{\circ} + bG_{\underline{B}}^{\circ} - cG_{\underline{C}}^{\circ} - dG_{\underline{D}}^{\circ}$$

or

$$RT(\ln x_{\underline{C}}^{c} + \ln x_{\underline{D}}^{d} - \ln x_{\underline{A}}^{a} - \ln x_{\underline{B}}^{b}) = aG_{\underline{A}}^{\circ} + bG_{\underline{B}}^{\circ} - cG_{\underline{C}}^{\circ} - dG_{\underline{D}}^{\circ}$$

which can be transformed into:

$$\ln\left(\frac{x_{\underline{C}}^{c} x_{\underline{D}}^{d}}{x_{\underline{A}}^{a} x_{\underline{B}}^{b}}\right) = \frac{aG_{\underline{A}}^{\circ} + bG_{\underline{B}}^{\circ} - cG_{\underline{C}}^{\circ} - dG_{\underline{D}}^{\circ}}{RT}$$

or

$$\frac{x_{\underline{C}}^{c} x_{\underline{D}}^{d}}{x_{\underline{A}}^{a} x_{\underline{B}}^{b}} = \exp\left(\frac{aG_{\underline{A}}^{\circ} + bG_{\underline{B}}^{\circ} - cG_{\underline{C}}^{\circ} - dG_{\underline{D}}^{\circ}}{RT}\right) \qquad (9.19)$$

At a given temperature, Equation (9.19) can be written as:

$$\frac{x_{\underline{C}}^{c} x_{\underline{D}}^{d}}{x_{\underline{A}}^{a} x_{\underline{B}}^{b}} = \text{constant}$$

The value of the constant depends on the choice of concentration units. If we leave this choice open, Equation (9.19) can be written as:

$$\frac{[C]^{c}[D]^{d}}{[A]^{a}[B]^{b}} = \text{constant} \qquad (9.20)$$

This is Guldberg–Waage's law for the special case of two reactants and two products. In the general case it is valid at equilibrium for arbitrary numbers of reactants and products. The value of the constant depends on the temperature and the choice of concentration units.

Solubility Product

Consider two kinds of atoms dissolved in a melt, which react chemically with each other to form a reaction product with low solubility in the melt:

$$a\underline{A} + b\underline{B} \rightleftharpoons A_{\underline{a}}B_{\underline{b}}$$

We apply Guldberg–Waage's law to the system at equilibrium:

$$\frac{[A_{\underline{a}}B_{b}]}{[A]^{a}[b]^{b}} = \text{constant} \qquad (9.21)$$

As the reaction product $A_{a}B_{b}$ has low solubility in the melt, most of it precipitates and appears as an inclusion. The

factor $[A_{a}B_{b}]$ is constant and can be included in the constant in Equation (9.21) which therefore be written as:

$$[A]^{a}[b]^{b} = K_{A_{a}B_{b}} \qquad (9.22)$$

where $K_{A_{a}B_{b}}$ is called the solubility product.

The constants a and b are integers. In the case of metallurgical reactions one of the components is a metal and the other a nonmetallic compound, for example O, N, S, C or Si. Examples are oxides, nitrides, sulfides and carbides. The values of the integers depend on the composition of the inclusion. Examples are $Cr_{23}C_{6}$, $Cr_{7}C_{3}$, and $Fe_{3}C$, which were discussed in Chapter 7.

Chemical Reactions and Free Energy Diagrams

The standard free energies are known and listed in tables for most elements. The standard Gibbs' free energy change can be calculated for a reaction when the standard free energies of all the elements involved in the reaction are known.

The standard free energies at various temperatures of a great number of chemical reactions are known. If ΔG° is plotted as a function of temperature, conclusions concerning the reaction can be drawn. As an example we will discuss the formation of two metal oxides from oxygen and a metal A or B. The reactions can be written as:

$$2A + 2\underline{O} \rightleftharpoons 2AO$$
$$B + 2\underline{O} \rightleftharpoons BO_{2}$$

The standard free energy–temperature curves of these two reactions can be seen in Figure 9.11.

It can be seen from the diagram that $(-\Delta G^{\circ})$ increases for both reactions when the temperature increases. An increase in temperature promotes the decomposition of

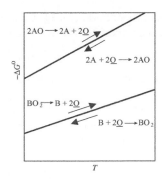

Figure 9.11 Standard free energy changes of two reactions as a function of temperature. Reproduced with permission from Hodder Headline.

the metal oxides in both cases. Oxides of refractory metals are used in tundishes and ladles for molten metals. At high temperatures the oxides become less stable and there is a risk of contamination.

The slope of the curve is steeper for metal A than for metal B. Metal B tends to oxidize easier than metal A. This property can be used to deoxidize metal A.

If metal B is added to a metal A melt containing $\underline{O}$ atoms, the B atoms react with most of the available oxygen atoms and form BO_2 molecules. If the oxide BO_2 is less dense than the liquid matrix at the prevailing temperature it can be separated mechanically from the melt.

Solubility Product as a Function of Temperature
With the aid of the free energy functions, it is possible to analyse the precipitation of inclusions, the composition changes of the melt and various types of faults which arise during the casting operation. Here we will analyse the reaction

$$B + 2\underline{O} \rightleftharpoons BO_2$$

where B and O are dissolved in a melt of element A.

The solubility product for a non-ideal solution can be written as:

$$x_{\underline{B}}x_{\underline{O}}^2 = \frac{1}{\gamma_{\underline{B}}\gamma_{\underline{O}}^2}\exp\left(-\frac{\Delta G^\circ}{RT}\right) \qquad (9.23)$$

where
$x_{\underline{B},\underline{O}}$ = concentration (mole fraction) of dissolved elements $\underline{B}$ and $\underline{O}$ in the melt
$\gamma_{\underline{B},\underline{O}}$ = activity coefficient of dissolved elements $\underline{B}$ and $\underline{O}$ in the melt
$-\Delta G^\circ$ = standard free energy change of the reaction.

If the temperature is constant, Equation (9.23) can be written with the aid of the definition of Gibbs' free

energy ($G = H - TS$):

$$x_{\underline{B}}x_{\underline{O}}^2 = \frac{1}{\gamma_{\underline{B}}\gamma_{\underline{O}}^2}\exp\left[-\left(\frac{\Delta S^\circ}{R} - \frac{\Delta H^\circ}{RT}\right)\right] \qquad (9.24)$$

where
ΔS° = standard entropy change of the reaction
$-\Delta H^\circ$ = standard enthalpy change of the reaction.

In many systems the activity coefficients are approximately constant and Equation (9.24) can be simplified to:

$$x_{\underline{B}}x_{\underline{O}}^2 = \text{constant} \times \exp\left(\frac{-\Delta H^\circ}{RT}\right) \qquad (9.25)$$

which describes the temperature dependence of the solubility product.

In Chapter 7, we found that phase diagrams are very useful tools for analysis of a solidification process. Equation (9.25) represents a surface in the three-phase system A–B–O.

In the following sections of this chapter we will give a survey of some very important metallurgical reactions for the most common metals and relate them to problems during casting of metals.

9.5 ALUMINIUM AND ALUMINIUM ALLOYS

9.5.1 Aluminium Oxide

An aluminium melt is easily oxidized at moderate temperatures and a thin film of Al_2O_3 is formed, which protects the surface of the melt from further oxidation. The oxidation rate increases with increasing temperature. A thin layer of oxide immediately covers the molten aluminium surface.

The purer the Al metal, the slower is the oxidation process with the oxygen in the air. In molten aluminium the oxide is also formed by a chemical reaction with the water vapour in the air:

$$2Al + 3H_2O \rightleftharpoons Al_2O_3 + 6H$$

During the reaction hydrogen is released, which diffuses into the metal melt. It is problematic that aluminium oxide and hydrogen in Al melts frequently occur together. Another Al_2O_3 source in aluminium melts is Al scrap with an oxidized surface if scrap is used in the production of aluminium. Turbulence in the melt caused by careless stirring in furnaces and ladles also contributes to an increased concentration of Al_2O_3.

Fine-grained oxide and other non-metallic impurities are present as a suspension in the aluminium melt. The density of the aluminium oxide is higher than that of the metal melt but no sedimentation or very slow sedimentation occurs. One reason for this is that the Al_2O_3 particles are porous and contain gas included in the pores. The average density of the particles is lower than that of nonporous Al_2O_3 and nearly equal to that of the Al melt.

Removal of Al_2O_3 and Other Impurities from Al melts

Aluminium oxide is a very stable chemical compound, which cannot be reduced to aluminium under normal conditions. To refine the melt from Al_2O_3 it is necessary to use mixtures of salt melts as so-called fusing agents, usually fluorides or chlorides, which attack the impurities either mechanically or chemically.

Often the surface of the melt is covered by a fusing agent, which protects the surface from the oxygen and water vapour in the air and prevents hydrogen and oxygen more or less effectively from diffusing into the melt. Some fusing agents react chemically with the aluminium oxide and refine the melt effectively.

Aluminium is produced by means of electrolysis. Removal of Al_2O_3 slag and other non-metallic impurities from Al melts with the aid of salt compounds, especially fluorides, plays an important role and has been studied carefully for this reason. In the presence of an excess of NaF the following reactions occur, for example:

$$4NaF + 2Al_2O_3 \rightleftharpoons 3NaAlO_2 + NaAlF_4$$

$$2NaF + Al_2O_3 \rightleftharpoons NaAlO_2 + NaAlOF_2$$

A simple mechanical method to remove Al_2O_3 slag inclusions and other impurities from Al melts, which is generally used in industry, is to filter the molten metal through some suitable material. The method has been used on a reduced scale in aluminium foundries for a long time.

Filters have been treated in Section 3.6. The melt can be filtered during transfer between containers or directly during casting. Sometimes the filter is integrated with the mould. Most filter systems contain porous refractory materials, which are able to remove very fine non-metal inclusions with a size down to $1 \mu m$.

9.5.2 Hydrogen in Aluminium and Aluminium Alloys

The gas that mainly is responsible for gas precipitation in aluminium is *hydrogen*. Hydrogen comes from water via the reaction:

$$3H_2O + 2Al \rightleftharpoons Al_2O_3 + 3H_2$$

The hydrogen dissociates at the surface of the metal melt:

$$M + H_2 \rightleftharpoons 2H + M$$

The equilibrium concentration of $\underline{H}$ in the melt is proportional to the square root of the hydrogen pressure above the melt:

$$[\underline{H}] = \text{constant} \times \sqrt{p_{H_2}} \quad \text{Sievert's law} \quad (9.26)$$

Figure 9.12 shows that the solubility of hydrogen in the melt is much higher than that in the solid phase. The solution of $\underline{H}$ is seldom saturated from the beginning. In spite of this, there is a risk of gas precipitation because the hydrogen concentration increases so much during the solidification process that the saturation value may be exceeded.

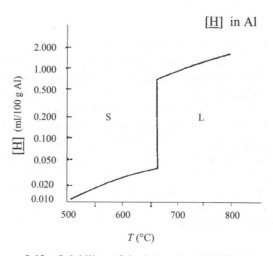

Figure 9.12 Solubility of hydrogen in aluminium at 1 atm. Reproduced with permission from Addison-Wesley Publishing Co. Inc., Pearson.

If the $\underline{H}$ concentration from the beginning is, for example, 0.4 ppm at 750 °C in an aluminium melt, calculations (the lever rule) show that it would increase to 3.6 ppm in the remaining melt when 90 % of the metal has solidified. However, gas precipitation will occur, as the solubility limit has been exceeded by a factor of 3 (Table 9.3), and the concentration $[\underline{H}] = 3.6$ ppm in the melt will never be reached.

It can be seen from Table 9.3 that the solubility of hydrogen decays when other elements are added.

TABLE 9.3 Solubility of hydrogen in melts of aluminium and aluminium alloys at 750 °C and 1 atm.

Alloy	ppm
Al 100 %	1.20
Al + 7 % Si + 0.3 % Mg	0.81
Al + 4.5 % Cu	0.88
Al + 6 % Si + 4.5 % Cu	0.67
Al + 4 % Mg + 2 % Si	1.15

Removal of Hydrogen in Aluminium and Aluminium Alloys

Hydrogen can be removed by flushing with argon or nitrogen. A large total bubble area favours the process, i.e. the bubbles should be *small*.

If carrier gas is bubbled through a melt of 10 tons at a rate of $0.01 \, m^3/s$, it is possible to decrease the hydrogen concentration from 0.9 to 0.2 ppm in 100 s. The action is most effective at the beginning. When the hydrogen concentration decreases it becomes more and more difficult to lower it further.

It has been found that addition of chlorine improves the result. Degassing with the aid of chlorine is more efficient than degassing by use of an inert gas, as can be seen from the Figure 9.13 (a) and (b), which show the density of the Al metal after solidification as a function of the degassing time in an open furnace. When the hydrogen concentration decreases, the density of the metal increases. The result is that the fraction of the light pores decreases.

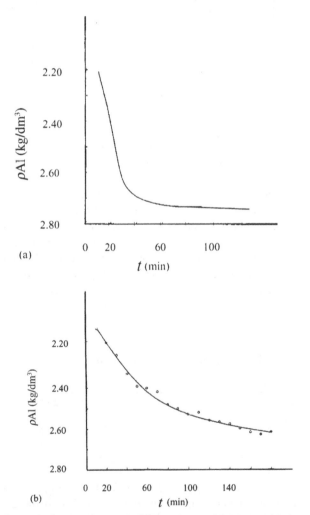

Figure 9.13 Degassing of a 200 kg Al melt with (a) pure chlorine gas and (b) pure nitrogen gas in an open oven. Temperature 750 °C, gas flow 140 l/min. Reproduced with permission from Elsevier Science.

When chlorine is added, hydrochloric acid is formed:

$$2H + Cl_2 \rightleftharpoons 2HCl$$

The carrier gas carries away hydrogen partly as H_2 and partly as HCl. In addition, the chlorine gas reacts with aluminium:

$$2Al\,(l) + 3Cl_2(g) \rightleftharpoons 2AlCl_3(g)$$

and forms gaseous aluminium chloride, which is introduced into the melt in the shape of very small bubbles. They also help to remove H_2 and HCl.

Often a mixture of chlorine and nitrogen is used. Such a mixture has the same effect as degassing with pure chlorine, but the processes require a longer time.

Example 9.3

In casting of Al-base alloys, the formation of hydrogen pores is a great problem. In order to decrease the pore formation during casting and solidification the melt is often degassed before casting.

How long a time will be required to degas a 400-kg aluminium melt with a gas flow of 200 l/min in the same furnace as was used for the degassing described in Figure 9.13 (a)? The gas consists of pure chlorine.

Solution:

The basis for degassing of a molten metal with the aid of flushing is that the concentration of the dissolved hydrogen gas [H] in the melt is not in equilibrium with the partial pressure of this gas in the bubbles. The partial pressure is lower than the pressure that corresponds to Sievert's law. The lack of equilibrium causes the dissolved hydrogen atoms to diffuse from the melt into the gas bubble and form hydrogen molecules. They mix with the bubble gas and are carried away with the bubbles out of the melt.

The factors which determine the efficiency of the refining are as follows:

1. the size of the total exposed bubble area, i.e. the gas/melt contact area;
2. the exposure time, i.e. the time which each bubble spend in contact with the melt.

The conclusion is that many small bubbles with a low rising rate give an optimal refining.

Figure 9.13 (a) shows the relation between the density of the Al metal and the length of the degassing time in an open-hearth oven for a melt. The density of the metal is a measure of its gas concentration: the lower it is, the higher is the density of the Al melt. The density becomes constant when practically all the hydrogen is gone.

In the present case we will calculate the degassing time for a melt of 400 kg and a gas flow of 200 l/min. A larger melt means a larger quantity of dissolved gas, i.e. the degassing time will be prolonged proportionally to the mass of the melt at a constant carrier gas flow.

An increased flow implies a greater number of bubbles per unit time. The degassing of a constant amount of melt will be more rapid when the volume flow is increased. This can be expressed mathematically:

$$t = \text{constant} \times \frac{m}{\text{volume flow}} \qquad (1')$$

where t is the degassing time and m the mass of the melt.

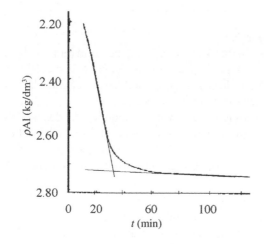

We construct a tangent to the curve in Figure 9.13 (a). The slope of this tangent represents the increase in density per unit time. We read the time when the tangent intersects the horizontal line, which represents the density of aluminium free from hydrogen:

$$t = 33 \, \text{min}$$

The desired time t_x for a large melt with mass M is:

$$t_x = \text{constant} \times \frac{M}{\text{Volume flow}} \qquad (2')$$

We divide Equation (2′) by Equation (1′), solve t_x and use the information given in Figure 9.13 (a), which gives:

$$t_x = t \times \frac{M}{m} \times \frac{\text{volume flow}}{\text{Volume flow}} = 33 \times \frac{400}{200} \times \frac{140}{200} = 46 \, \text{min}$$

Answer:
The time required to degas the aluminium melt is about 45 min.

9.5.3 Al-Fe Compounds

Iron is the dominant impurity in commercial aluminium and occurs also in raw aluminium, directly after electrolysis. Iron is often unintentionally added when steel and/or cast iron equipment is used at melting and casting of aluminium. Alternatively, iron or rust may happen to be added at remelting.

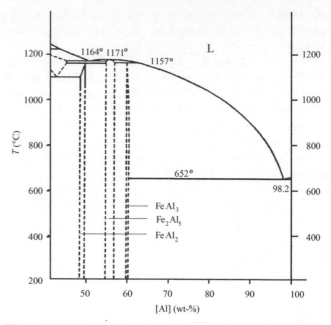

Figure 9.14 Part of the phase diagram of the system Al–Fe. Reproduced by permission of ASM International.

The phase diagram of the Al–Fe system is given in Figure 9.14. The phase which is in equilibrium with aluminium is normally written Al_3Fe. Nevertheless, crystals with a composition that approximately corresponds to Fe_2Al_7 have been found. The crystal structure indicates that the crystals may have an alternative composition between Fe_4Al_{13} and Fe_6Al_{19}. As can be seen from Table 9.4, all these compounds contain approximately the same weight percent of Fe.

TABLE 9.4 Compositions of some Al-Fe.

Formula	Fe (wt-%)
$FeAl_3$	40.7
Fe_2Al_7	37.3
Fe_4Al_{13}	38.9
Fe_6Al_{19}	39.5

The mechanical properties of cast aluminium are affected negatively by iron compounds. The iron is present as crystals of iron aluminates of the type discussed above or as Al–Fe–Si compounds. They increase the strength of the metal but its fatigue resistance decays.

Removal of Iron from Aluminium Melts

There are no simple chemical methods to separate iron from Fe–Al compounds. In spite of numerous patents on suggested methods there is none that works in a satisfactory way.

The only successful method to separate iron from aluminium melts is electrolysis. It is possible to refine aluminium with high concentrations of impurities, especially Fe, but also Si, Cu, Ni, Mg, and Zn up to a total of 15 wt-%, to give aluminium with an extremely high degree of refining, 99.990–99.999 wt-%, with this method. The refining process requires much energy. The same is true for the production of aluminium which also is done by electrolysis.

9.5.4 Grain Refinement of Aluminium Alloys

Grains in solidification structures of aluminium alloys can be seen with the naked eye when the surface of a sample is macroetched. The sizes of the grains depend strongly on the solidification rate and composition, but a diameter larger than 1 mm is not uncommon in sand mould castings.

The structure will be much finer than 1 mm if a so-called *grain refiner* is added. Elements such as Ti, B, Zr, V, and Co have been proved to refine the structure. The most effective way is to add Ti and B simultaneously. The addition of Ti and B can be done in several ways. One way is to add the potassium salts KBF_4 and K_2TiF_6 to the melt. A different and more popular way is to add so-called master alloys, normally containing 5 % Ti and 1 % B. The Ti and B atoms dissolve into the melt and precipitate as intermediate phases, e.g. AlB_2, TiB_2, and Al_3Ti.

Another potential refiner is Sc. It is believed that primary Al is nucleated on fine Al_3Sc particles. On the other hand, there are alloy systems, e.g. Al–Mg systems, which block the refinement effect by reacting with boron.

9.5.5 Modification of the Eutectic Structure of Al–Si Alloys

In Chapter 6 we discussed unmodified and modified eutectic Al–Si structures in Section 6.4.3. The unmodified eutectic structure in Al–Si alloys consists of Si flakes, which grow in a fan-shaped morphology in cooperation with Al (FCC). By adding Na or Sr, the morphology of the alloys can be changed to a much finer structure, where Si rods with a diameter of a few micrometres or less are growing with Al (FCC). The modification improves the mechanical properties of the alloys.

Na oxidizes rapidly and has a high vapour pressure. Consequently, it is volatile and has to be added in excess immediately before casting; 150–200 ppm is a recommended concentration for casting in a sand mould.

Sr gives a slightly coarser microstructure and somewhat less favourable mechanical properties than Na but has the advantage of disappearing from the melt more slowly. It is possible to remelt the material without loss of the modifying effect. Usually only alloys which contain 7–13 % Si are worth modifying.

Both Na and Si increase the risk of hydrogen absorption.

9.6 COPPER AND COPPER ALLOYS

It is fairly easy to melt copper and copper-base alloys but there is one complication. The melt will often react with the water vapour in the air, and the hydrogen which is formed in the reaction will dissolve into the copper melt. During the cooling and solidification processes hydrogen may precipitate as gas pores or react with oxygen and precipitate as pores of water vapour in the melt or casting.

Nitrogen does not cause gas precipitations in copper and copper-base alloys but if the copper melt contains sulfur, pores of sulfur dioxide may precipitate.

It is important to prevent these complications to obtain castings of acceptable quality.

9.6.1 Hydrogen in Copper and Copper Alloys

Even small amounts of hydrogen cause great problems. Even such a low concentration as 3 ppm of H causes, unless special steps are taken, hydrogen precipitation, which corresponds to approximately 45 % of the metal volume. Figure 9.15 shows the solubility of hydrogen in copper, copper alloys, and tin.

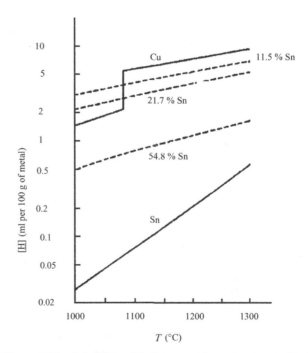

Figure 9.15 Solubility of hydrogen at 1 atm in copper, Cu–Sn alloys, and tin as a function of temperature. Reproduced with permission from Addison-Wesley Publishing Co. Inc., Pearson.

The solubility of hydrogen in the melt is higher than that in the solid phase. The mechanism of hydrogen precipitation is the same as that described in detail earlier in this chapter. Hydrogen concentrates in the melt until the solubility limit has been exceeded.

Since the concentrations of $\underline{H}$ and $\underline{O}$ are high at the end of the solidification process, the chemical reactions may result in the formation of H_2O, which is precipitated in the metal. This process occurs most often at a late stage of the precipitation. This will be discussed below.

It is also true for copper that addition of other substances in most cases reduces the solubility of hydrogen, which is illustrated in Figure 9.15. The solubility of hydrogen in pure copper at 1200 °C is 6.5 ppm. In an alloy which contains about 50 % Cu and 50 % Sn, the solubility of hydrogen is ∼1 ppm. However, addition of Ni increases the solubility of hydrogen. In a 90 % Cu + 10 % Ni alloy it is, for example, 10 ppm (not shown in the Figure).

Removal of Hydrogen in Copper and Copper Alloys

Two methods are used to prevent hydrogen precipitation: oxidation and flushing. The oxidation method will be discussed in the next section (page 273).

Flushing with nitrogen as carrier gas is used especially for copper alloys, which contain phosphorus and tin. In Cu–Zn alloys there are seldom any problems with hydrogen precipitation. Zn has a high vapour pressure and carries away the dissolved hydrogen in the melt by internal flushing.

Example 9.4

The deoxidation process frequently determines the hydrogen concentration in a copper melt during the melting. If the hydrogen concentration becomes too high, hydrogen pores are precipitated during the casting. In order to avoid hydrogen pores, a foundry has developed a casting process where the pressure is increased during casting.

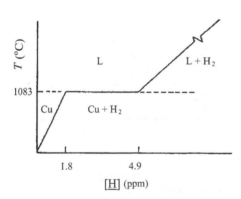

Simplified phase diagram of the system Cu–H at 1 atm.

Find the relation between the original hydrogen concentration in the melt and the minimum hydrogen pressure necessary for avoiding pore precipitation when the casting solidifies. Plot the function in a diagram. The phase diagram of Cu–H is given in the figure above.

Solution:
Hydrogen has a high diffusion rate. It is therefore reasonable to assume that the different phases are in equilibrium with each other. The instantaneous hydrogen concentration (ppm) in the melt at a solidification fraction f_c can, with this assumption, be written [the lever rule, Equation (7.15) on page 189] as:

$$c^L = \frac{c_0^L}{1 - f_c(1 - k_{partx})} \qquad (1')$$

where k_{partx} is the partition constant:

$$k_{partx} = \frac{c^s}{c^L} \qquad (2')$$

By applying Sievert's law, we obtain the hydrogen pressure (in atm) in equilibrium with the instantaneous hydrogen concentration in the melt:

$$c^L = K\sqrt{p} \qquad (3')$$

In the phase diagram we can read the saturation value of the hydrogen concentration in the melt at 1 atm to be 4.9 ppm. These values, inserted into Equation (3′), give:

$$4.9 = K\sqrt{1}$$

or $\qquad\qquad K = 4.9\,\text{ppm}/\sqrt{\text{atm.}}$

The hydrogen concentration is highest in the last part of the melt before complete solidification. We obtain the minimum pressure p by inserting the solidification fraction $= 1$ into Equation (1′) and combine it with Sievert's law:

$$c^L = \frac{c_0^L}{k_{partx}} = 4.9\sqrt{p} \qquad (4')$$

k_{partx} can be calculated with the aid of the figure in the text above:

$$k_{partx} = \frac{c^s}{c^L} = \frac{1.8}{4.9} \qquad (5')$$

and we obtain with the aid of Equation (5′):

$$c_0^L = k_{partx} \times 4.9\sqrt{p} = \frac{1.8}{4.9} \times 4.9\sqrt{p} = 1.8\sqrt{p}$$

Answer:

$$p = \left(\frac{c_0^L}{1.8}\right)^2 = 0.31\left(c_0^L\right)^2 \text{ atm}$$

$(c_0^L$ in ppm).

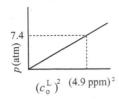

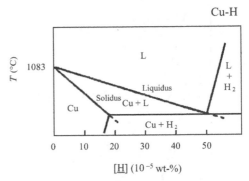

Figure 9.16 Simplified phase diagram of the system Cu–H at 1 atm.

9.6.2 Oxygen, Hydrogen, and Water Vapour in Copper and Copper Alloys

Water vapour–oxygen–hydrogen is a complex system, which generates much more severe problems than hydrogen alone.

H_2O, O_2, and $\underline{O}$ in the solid and liquid phases, and H_2 and $\underline{H}$ in the solid and liquid phases, have to be in equilibrium with each other simultaneously. With the aid of valid equilibrium constants (Table 9.5) and the phase diagrams of Cu–O and Cu–H (Figures 9.16 and 9.17), curves can be drawn which describe the concentrations of oxygen and hydrogen at various water vapour pressures. These curves, together with knowledge of the partial pressures of the gases in the furnace atmosphere, can be used to estimate the risk of precipitation of hydrogen and water vapour.

TABLE 9.5 Solubility of hydrogen in copper melts at two different temperatures according to Sievert's Law.

$H_2O \rightleftharpoons 2\,H + O$		
$[\underline{H}] = 3.0 \times 10^{-7} \times \sqrt{\dfrac{p_{H_2O}}{[\underline{O}]}}$		at 1083 °C
$[\underline{H}] = 8.0 \times 10^{-7} \times \sqrt{\dfrac{p_{H_2O}}{[\underline{O}]}}$		at 1250 °C
$2\,\underline{H} \rightleftharpoons H_2$		
$[\underline{H}]_{\text{melt}} = 2.5 \times 10^{-4} \times \sqrt{p_{H_2}}$		at 1083 °C
$[\underline{H}]_{\text{solid}} = 1.8 \times 10^{-4} \times \sqrt{p_{H_2}}$		at 1083 °C

The pressures are measured in atm and the concentrations in wt-%. The melting point of Cu = 1083 °C

Earlier we have shown solubility curves of the type where the solubility of a gas is plotted as a function of temperature. In this more complicated case we cannot show the $\underline{H}$ and $\underline{O}$ curves separately as they depend on each other. The solubility curves are therefore given as a relation between $\underline{O}$ and $\underline{H}$ with the water vapour pressure as a parameter, i.e. as a ternary diagram for a given temperature. The equilibrium varies with temperature, which gives different

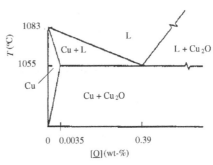

Figure 9.17 Simplified phase diagram of the system Cu–O at 1 atm.

values of the equilibrium constants at different temperatures (Table 9.5).

Figure 9.18 shows the ternary phase diagram of the water vapour–hydrogen–oxygen–copper system. It represents the projection of the ternary phase diagram on the bottom in analogy with the ternary alloy on page 212 in Chapter 7. Points in Figure 9.18 have different temperatures (the T axis is not seen in this representation). The phase diagrams of the binary systems Cu–H and Cu–O are included in the vertical planes through the coordinate axes.

The curves in Figure 9.18 illustrate the oxygen and hydrogen concentrations at some different water vapour pressures when the Cu melt is at equilibrium with a surrounding atmosphere of *pure water vapour*. The set of curves corresponds to the relation between $[\underline{H}]$ and $[\underline{O}]$ for various values of the pressure of the water vapour (marked in the Figure). Each curve is obtained with the aid of the solubility product of water at the given temperature. As the solubility product varies with temperature it is necessary to specify the temperature, in this case 1083 °C.

At a given water vapour pressure and a given solubility of one gas in the melt, the concentration of the other gas in the melt can immediately be read from the phase diagram.

The concentrations of $\underline{O}$ and $\underline{H}$ which together generate a water vapour pressure of 1 atm are surprisingly small. The

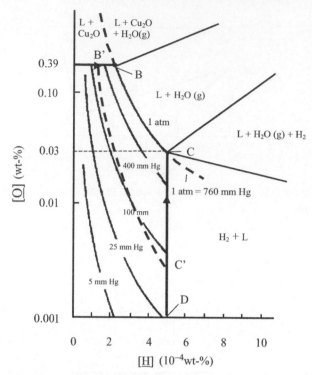

Figure 9.18 Ternary phase diagram of oxygen, hydrogen, and water vapour (parameter) in equilibrium with molten copper at different water vapour pressures. The temperature equals the melting point of Cu, which is 1083 °C. Reproduced with permission from Addison-Wesley Publishing Co. Inc., Pearson.

highest concentration of oxygen which can exist in the Cu melt is found in the binary phase diagram of Cu–O (Figure 9.17). It is found at the eutectic point and is 0.39 %. The line through point B in Figure 9.18 corresponds to this $\underline{O}$ concentration. The corresponding eutectic point in the binary phase diagram of Cu–H (Figure 9.16) is $\sim 5 \times 10^{-4}$ wt-% at point C. Hence the concentration of $\underline{H}$ is $\sim 5 \times 10^{-4}$ wt-% at point C. It represents the maximum $\underline{H}$ concentration in the Cu melt at 1 atm. The corresponding $\underline{O}$ concentration is 0.03 wt-%. It should be noted that the curve on the $[\underline{O}]$ axis is logarithmic.

We will analyse Figure 9.18 in more detail. The horizontal line in the upper part of the figure represents the oxygen concentration 0.39 wt-%, which corresponds to the eutectic point in the phase diagram of Cu–O (Figure 9.17). The oxygen concentration cannot exceed this value. Instead, Cu_2O is precipitated as slag:

$$2Cu(L) + \underline{O} \rightleftharpoons Cu_2O$$

A copper melt at equilibrium with water vapour of 1 atm can have an oxygen concentration between 0.03 and 0.39 wt-%. Analogously, the hydrogen concentration cannot exceed 5×10^{-4} wt-%. Instead, precipitation of H_2 will occur.

Precipitation of a Mixture of H_2 and Water Vapour

Figure 9.18 illustrates approximately the conditions for copper melts at the beginning of a casting process. Even at low water vapour pressure the concentrations of $\underline{O}$ and $\underline{H}$ in the melt will increase during the solidification because oxygen and hydrogen become enriched in the melt owing to microsegregation. The condition for gas precipitation is, as before, that the sum of partial pressures of the precipitated gases is equal to the external pressure. Even at small fractions of solidified phase, the equilibrium pressure of hydrogen and water vapour will reach 1 atm. When this pressure is reached, hydrogen + water vapour or both will precipitate as pores during the rest of the solidification process.

The condition for the formation of gas-filled pores with both hydrogen and water vapour inside the melt is that the gas pressure must be at least equal to the outer pressure, i.e. atmospheric pressure, plus the hydrostatic pressure of the melt. If we consider pores near the upper surface, we can neglect the hydrostatic pressure and obtain:

$$p_{H_2} + p_{H_2O} = p_{total} \qquad (9.27)$$

where p_{total} is the *total* pressure of the surrounding atmosphere, *not* the partial pressures of hydrogen + water vapour in the atmosphere.

Equation (9.27) is combined with the relations of the equilibrium constants given in Table 9.5. The result is a relationship [Equation (9.28)] between the $\underline{H}$ and $\underline{O}$ concentrations, which describes the risk of formation of hydrogen + water vapour pores in copper:

$$\frac{(c_{\underline{H}})^2}{(2.5 \times 10^{-4})^2} + \frac{(c_{\underline{H}})^2 c_{\underline{O}}}{(3.0 \times 10^{-7})^2} = 1 \qquad (9.28)$$

Equation (9.28) is plotted in Figure 9.18 and corresponds to the dashed curve B′C′. The sum of the water vapour pressure and the hydrogen pressure is equal to 1 atm. For other external pressures, other similar curves are valid. Curve BC in Figure 9.18 represents Equation (9.28) when $p_{H_2O} = 1$ atm and $p_{H_2} = 0$.

Removal of Hydrogen and Oxygen in Copper and Copper Alloys

In order to avoid the severe problems with water vapour precipitation when copper melts and molten copper-base alloys are cast and solidify, it is desirable to reduce both the hydrogen and oxygen concentrations in the melt before casting.

The hydrogen concentration is reduced by using an oxygen-rich copper melt as the basis for the casting operation. Then, immediately before casting, a deoxidation agent (page 282) is added, which reduces the oxygen concentration strongly. Usually phosphorus (addition 0.02–0.005 %) is used, but calcium boride, calcium carbide and lithium can also be used.

9.6.3 Carbon Monoxide in Copper and Copper Alloys

Carbon and oxygen dissolved in a copper melt may generate gas precipitation of carbon monoxide:

$$\underline{C} + \underline{O} \rightleftharpoons CO$$

If the oxygen concentration is less than 0.01 % the risk of CO precipitation is generally small. By deoxidation (page 282) precipitation of both H_2O and CO is prevented.

9.6.4 Sulfur Dioxide in Copper and Copper Alloys

The copper metal may either contain dissolved sulfur from the beginning or sulfur may dissolve during the melting or casting. Sulfur dioxide may be present in the furnace atmosphere or form in reactions between the melt and the mould.

The solubility of sulfur in copper is lower in the solid phase than in the melt, which gives a risk of precipitation of sulfur dioxide during solidification.

The equilibrium constants are given in Table 9.6.

TABLE 9.6 **Solubility products of sulfur and oxygen in copper melts at two different temperatures.**

$$SO_2 \rightleftharpoons \underline{S} + 2\,\underline{O}$$

$$p_{SO_2} = \frac{[\underline{S}] \times [\underline{O}]^2}{0.98 \times 10^{-5}} \quad \text{at } 1083\,°C$$

$$p_{SO_2} = \frac{[\underline{S}] \times [\underline{O}]^2}{3.3 \times 10^{-5}} \quad \text{at } 1250\,°C$$

The pressures are measured in atm and the concentrations in wt-%. The melting point of Cu = 1083 °C

It is possible to make the corresponding calculations of the system Cu–S–O–SO$_2$ as for the system Cu–H–O–H$_2$O (Section 9.6.2).

Figure 9.19 shows the phase diagram of the system Cu–S. Figure 9.20 shows the ternary diagram of the system Cu–O–S at 1083 °C. The horizontal line represents the oxygen concentration 0.39 %, which corresponds to the eutectic point in the Cu–O phase diagram (Figure 9.17). At this oxygen concentration Cu$_2$O starts to precipitate. The limiting curve AB corresponds to an SO$_2$ pressure of 1 atm. At point B Cu$_2$S starts to precipitate. The vertical line in Figure 9.20 corresponds to the sulfur concentration of the eutectic point of the binary phase diagram in Figure 9.19. In a solidification process, which ends at the vertical line on the right-hand side in Figure 9.20, Cu$_2$S will precipitate.

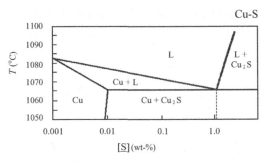

Figure 9.19 Simplified phase diagram of the system Cu–S.

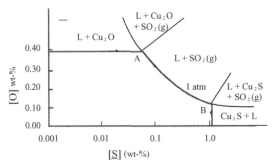

Figure 9.20 Ternary phase diagram of SO$_2$ at equilibrium with molten copper at 1083 °C. Reproduced with permission from Addison-Wesley Publishing Co. Inc., Pearson.

Removal of Sulfur Dioxide in Copper and Copper Alloys

The risk of precipitation of SO$_2$ is eliminated most effectively by reduction of the $\underline{O}$ concentration. This is done by deoxidation (see page 282), which reduces the amount of dissolved $\underline{O}$ in the melt.

9.7 STEEL AND IRON ALLOYS

The initial material for production of cast iron and steel is iron ores. They consist of various mixtures of iron oxides, which often contain small amounts of phosphorus, sulfur, and other undesired components. Figure 9.21 shows the manufacturing processes for steel and iron products which are the most common methods today. The dominant method is continuous casting (more than 90 % of the total world production) and the rest is cast as ingots.

The essential chemical processes are as follows:

1. reduction of the iron ores from oxide to steel with the aid of carbon and carbon monoxide in blast furnaces:

$$Fe_2O_3 + 3CO \rightarrow 2Fe + 3CO_2$$

2. reduction of the carbon content in the steel by oxidizing the carbon.

Production of Iron and Steel

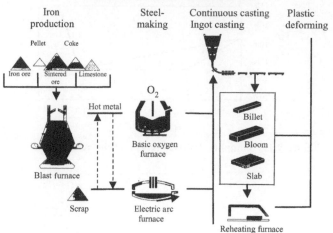

Figure 9.21 Survey of the manufacturing process for steel and iron. Reproduced by permission of JFE 21st Century Foundation.

The two production methods differ in the second step. The most common method is to use a basic oxygen furnace. Alternatively, scrap is added to an iron melt in an electric arc furnace. Oxygen is added either as oxygen gas or as iron ore. From the electric arc furnace the molten steel is subsequently cast in a continuous casting machine or cast as ingots.

Through the blast furnace–basic oxygen furnace process the molten steel achieves a desired composition and is then cast at a given temperature to ingots or in a continuous casting machine to give slabs, blooms, and billets.

During reduction in the blast furnace, carbon monoxide is generated by oxidation of coke. The molten iron contains about 4 % of carbon, which corresponds to the eutectic Fe–C composition (page 150 in Chapter 6), and is called *pig iron* or *raw iron*.

The carbon content in the steel is reduced in the basic oxygen furnace by oxidizing carbon with the aid of pure O_2 gas and iron oxide, Fe_2O_3. The remaining oxygen is removed with the aid of deoxidation agents (page 282) such as Si and Al or removed as CO_2 in a subsequent vacuum degassing process. The molten steel is then ready for casting.

During the casting operation, gases might precipitate as pores. *Hydrogen*, *nitrogen*, *oxygen*, and *carbon monoxide* are the gases which give a risk of gas precipitation in steel and iron alloys.

During the deoxidizing operations slag inclusions are formed and float around in the melt and follow it during the casting process. In Section 9.7.5 (page 285) we will discuss different types of slag inclusions during the casting and solidification processes. The slag particles consist of either sulfides or oxides.

9.7.1 Hydrogen in Steel and Iron Alloys

Hydrogen is generated by dissociation of water vapour and by a chemical reaction between water vapour in the mould and dissolved carbon in the steel melt:

$$2H_2O \rightleftharpoons 2H_2 + O_2$$

$$\underline{C} + H_2O \rightleftharpoons CO + H_2$$

At the surface of the melt dissociation of hydrogen occurs:

$$M + H_2 \rightleftharpoons M + 2\underline{H}$$

After the reaction, the monoatomic hydrogen diffuses into the steel melt and stays there as a solution. The hydrogen concentration can be calculated using Sievert's law:

$$[\underline{H}] = \text{constant} \times \sqrt{p_{H_2}} \qquad (9.29)$$

Sources of Hydrogen in Iron Melts

Liquid steel absorbs hydrogen very easily. The major source of hydrogen in steel is water vapour and the major source of water vapour is the water vapour in the atmosphere in contact with the steel bath. Other sources are water included in slag-making and lining materials.

The concentration of $\underline{H}$ in a steel melt depends on the external partial hydrogen pressure of the surrounding atmosphere, but also of the concentration of $\underline{O}$ in the melt. The solubility product limits the hydrogen concentration and relates it to the oxygen concentration:

$$[\underline{H}]^2 \times [\underline{O}] = \text{constant} \qquad (9.30)$$

The value of the constant varies with temperature.

Figure 9.22 illustrates the magnitude of the pressure of water vapour in various environments and under different circumstances and the resulting hydrogen concentrations in the steel melt.

At low temperatures, the partial pressure of water vapour is very low. The air is dry during the winter. The opposite is true during the summer near the sea, with the exception of deserts. Humid air has a much higher partial pressure than dry winter air. Burning oil furnaces give a higher water vapour pressure than the partial pressure during an average summer day as organic compounds often contain much hydrogen.

Figure 9.22 illustrates the strong influence of the oxygen concentration. It is important to keep the oxygen concentration as low as possible in steel (see Figure 9.1, page 256) and deoxidation agents are added to the melt (page 282). The decrease in the oxygen concentration leads to an increased hydrogen concentration as a consequence of the solubility product of water. The slopes of all the curves

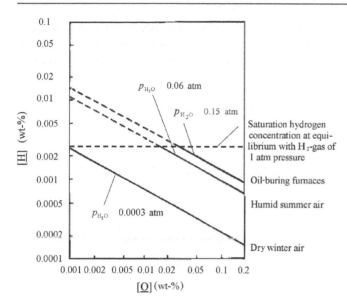

Figure 9.22 Hydrogen concentration in a steel melt as a function of the oxygen concentration in the melt with the partial pressure of water vapour of the surroundings as a parameter. Pressure of water vapour: in dry air, ~2 mm Hg; in moist air, ~32 mm Hg; in oil combustion gases, ~90 mm Hg; in natural gas, ~150 mm Hg. Reproduced by permission of Castings Technology International.

are negative. The curves are straight lines because the scales on both coordinate axes are logarithmic.

Hydrogen Solubility in Steel

Figure 9.23 shows the solubility of hydrogen in iron as a function of temperature. It can be seen that

(i) The solubility of hydrogen is much larger in the melt than in the solid phase.

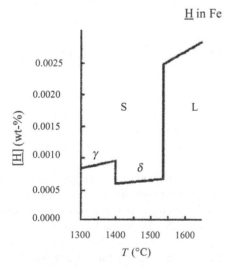

Figure 9.23 Solubility of hydrogen in iron at equilibrium with 1 atm H_2 as a function of temperature. Reproduced by permission of Castings Technology International.

(ii) The solubility of hydrogen depends on the type of solid iron phase. The solubility is larger for austenite (γ) than for ferrite (δ).

(iii) The solubility of hydrogen increases with temperature.

The melt normally contains less hydrogen than the solubility value. This means that a large amount of solid phase can form before hydrogen starts to precipitate. We can calculate the fraction of the melt which can solidify before gas precipitation starts, provided that the supersaturation phenomena can be neglected.

Consider a small volume that is representative of an alloy with dendritic solidification. We disregard all sorts of macrosegregation and assume that the average composition within the volume is constant and equal to the original hydrogen concentration of the alloy melt, the mole fraction x_0^L or the concentration in weight percent, c_0^L. We also assume that the considered volume has a uniform temperature and that each phase has a uniform composition. The last assumption would be unrealistic for alloying metals but is realistic for gases, which have high diffusion rates.

We call the instantaneous hydrogen concentration of the melt x^L or c^L and that of the solid phase x^s or c^s. At equilibrium between the two phases within the volume in question we have:

$$\frac{x^s}{x^L} = k_{\mathrm{part}\,x} \quad \mathrm{or} \quad \frac{c^s}{c^L} = k_{\mathrm{part}\,c} \tag{9.31}$$

where the partition constant $k_{\mathrm{part}\,x}$ can be derived from the phase diagram in Figure 9.24.

The assumption we have made allows us to use the lever rule [Equation (7.15) on page 189] to calculate the instantaneous hydrogen concentration in the melt. The fraction of solidification f, expressed as the fraction solid phase, will therefore be:

$$f = \frac{x^L - x_0^L}{x^L - x^s} = \frac{x^L - x_0^L}{x^L \left(1 - k_{\mathrm{part}\,x}\right)} \tag{9.32}$$

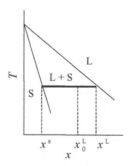

Figure 9.24 Phase diagram of Fe–H.

We solve for x^L in Equation (9.32) and obtain:

$$x^L = \frac{x_0^L}{1 - f(1 - k_{part\,x})} \qquad (9.33)$$

If the concentration is expressed in weight percent we have:

$$c^L = \frac{c_0^L}{1 - f(1 - k_{part\,c})} \qquad (9.34)$$

It is important to note that the definition of k_{part} is different, and depends on the unit of concentration, mole fraction or weight percent. c_0^L and $k_{part\,c}$ are known quantities. If the solubility limit of hydrogen in the melt is known, it is possible to calculate at what fraction of solid phase the limit will be exceeded and the risk of gas precipitation exists. It is convenient to make the calculation graphically.

Example 9.5
A specific sort of steel contains 12 ppm of $\underline{H}$ and is able to dissolve 24 ppm in the melt and 10 ppm in the solid phase at a pressure of 1 atm.

(a) Calculate the hydrogen concentration in the melt as a function of the solidification fraction during the solidification process and illustrate the function graphically.

(b) At what solidification fraction will the solubility limit be exceeded at a pressure of 1 atm?

Solution:

f	c^L (ppm)
0	12
0.2	13.6
0.3	14.5
0.4	15.6
0.5	16.9
0.6	18.5
0.7	20.3
0.8	22.5
0.9	25.3
1	28.8

(a) From the text we obtain $k_{part\,x} = 10/24$ and $c_0^L = 12$ ppm. These values are introduced into Equation (9.33):

$$c^L = \frac{c_0^L}{1 - f(1 - k_{part\,x})} = \frac{12}{1 - \dfrac{14f}{24}} \qquad (1')$$

Some f values are chosen arbitrarily and the corresponding c^L values are calculated. The curve is plotted as concentration of $\underline{H}$ in a steel melt versus the solidification fraction.

(b) From the diagram we can read that the solubility limit is reached at a solidification fraction 0.86.

Answer:

(a) See Equation (1') above and the corresponding diagram below.

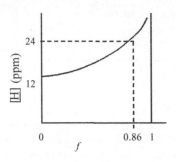

(b) The solidification fraction is 0.86 at the solubility limit of hydrogen in the melt.

Removal of Hydrogen in Steel and Iron Alloys
In order to reduce the risk of precipitation of hydrogen, the partial pressure of hydrogen in contact with the melt should be lowered as much as possible.

The melt must be carefully protected against contact with water vapour. Lining materials have to be preheated before contact with a steel melt. Hygroscopic materials, such as lime and other slag-forming materials, must be preheated and then protected from contact with water before they come into contact with a steel melt.

According to Sievert's law, i.e. the equilibrium between dissolved $\underline{H}$ atoms and H_2 gas in the surrounding atmosphere, [$\underline{H}$] decreases or disappears when the hydrogen pressure is reduced or becomes practically zero.

A high oxygen concentration reduces the dissociation of water vapour and consequently also the possibility of dissolving hydrogen. Deoxidation in a later stage of the production is favourable.

Another suitable method is to let some inert gas, for example Ar, bubble through the melt. $\underline{H}$ diffuses into the carrier gas rapidly enough to make the method useful in practice. In 20 min the $\underline{H}$ concentration can be lowered from 4 to 1.6 ppm in an Fe + 4.5 % C melt of 11 tons if it is treated with bubbling Ar at a rate of 0.005 m³/s.

The solubility of hydrogen in Fe–C alloys is also lowered by their carbon concentrations, as can be seen in Figure 9.25.

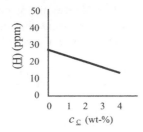

Figure 9.25 Solubility of hydrogen in Fe–C alloys at 1550 °C as a function of its $\underline{C}$ concentration. Reproduced with permission from the American Society for Metals (ASM).

9.7.2 Nitrogen in Steel and Iron Alloys

Nitrogen is in most cases an undesired gas in connection with casting. Precipitation of nitrides adversely affects the elastic properties of the casting. Nitrogen follows Sievert's law:

$$[\underline{N}] = \text{constant} \times \sqrt{p_{N_2}} \qquad (9.35)$$

and its partial pressure in air is large.

The solubility of nitrogen in iron at 1 atm at various temperatures and structures of the solid phase is shown in Figure 9.26, which indicates that the solubility of nitrogen in the *melt* at equilibrium with the atmosphere increases with increasing temperature. The solution process is endothermic in this case. This is in agreement with Figure 9.3 and the statements on page 259.

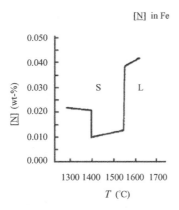

Figure 9.26 Solubility of nitrogen in iron at a pressure of 1 atm as a function of temperature. Reproduced with permission from the American Society for Metals (ASM).

For *austenite*, the solution process is exothermic and the solubility of nitrogen in austenite at equilibrium with the atmosphere decreases with increasing temperature (Figure 9.26). This is in agreement with Equation (9.6) and Figure 9.4 (page 259).

In analogy with hydrogen, the difference in solubility of nitrogen between the melt and the solid phase at the melting temperature is very large. The risk of nitrogen precipitation is reduced, as in the case of hydrogen, because the $\underline{C}$

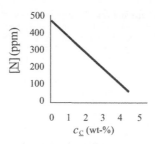

Figure 9.27 Solubility of nitrogen in Fe–C alloys at 1550 °C as a function of the $\underline{C}$ concentration. Reproduced with permission from the American Society for Metals (ASM).

concentration in the steel lowers the solubility of $\underline{N}$ considerably (Figure 9.27). Addition of Si reduces the solubility of nitrogen still further.

The $\underline{N}$ concentration decreases more in the melt than in austenite with increasing $\underline{C}$ concentration. At high $\underline{C}$ concentrations the solubility of $\underline{N}$ is higher in austenite than in the melt. This reduces the risk of nitrogen pore formation considerably in high-carbon steel alloys. It is not zero, however, owing to microsegregation in the melt at the end of the solidification process.

The solubilities of $\underline{H}$ and $\underline{N}$ in steel melts are not independent of each other. As soon as the added gas pressures equal 1 atm there is a risk of pore precipitation. It can be seen from Figure 9.28 that there is a risk of precipitation of N_2 and H_2 already at 80 ppm $\underline{N}$ + 4 ppm $\underline{H}$. In the area above the curve, pore precipitation is likely. In the area below the curve, the solubility limits are smaller than the maximum values and no pore precipitation occurs. However, it is not possible to be absolutely certain because of the effect of high $\underline{C}$ concentration described above.

There is a distinction between low-nitrogen steel with $<0.02\%$ N_2 and high-nitrogen steel with $>0.04\%$ N_2. In certain cases nitrogen is added on purpose to give some special quality to the alloy, for example to stabilize the austenite instead of the ferrite structure. Another example follows below.

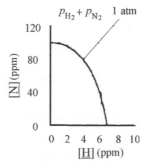

Figure 9.28 Relationship between the concentrations of $\underline{H}$ and $\underline{N}$ at the end of the solidification process of the alloy Fe + 3.8 % C at 1 atm. Reproduced with permission from The Metal Society, Institute of Materials, Minerals & Mining.

TABLE 9.7 Compositions of two stainless-steel alloys.

Alloy	C (%)	Cr (%)	Ni (%)	Si (%)	N (%)
1	0.15	25	20	2.0	0.0
2	0.15	22	13	1.4	0.2

The properties of the two alloys 1 and 2 described in Table 9.7 are fairly equal. The lower one is apparently cheaper in terms of raw material costs because the nickel concentration can be lowered at an increased nitrogen concentration.

Removal of Nitrogen in Steel and Iron Alloys

Flushing with argon cannot eliminate nitrogen to any striking extent in a reasonable time. Some other and better ways to reduce the risk of nitrogen precipitation are the following:

1. The alloy composition may be changed in such a way that austenite (γ-iron) instead of ferrite (δ-iron) is formed during the solidification process. In this case the difference in the solubility of nitrogen between melt and solid phase will be smaller.

2. The nitrogen solubility in the solid phase may be increased by addition of other metals.

3. Addition of surface-active substances such as O (0.03 %) and S (0.03 %), which cover most of the surface of the melt, may have the consequence that the dissociation of the nitrogen in the air becomes very slow.

4. The reaction rate may be lowered by addition of small amounts of P, Pb, Bi and Te (0.50 ppm of the last two metals give a decrease in [$\underline{N}$] of 80 % and 90 %, respectively).

5. Addition of preferably Ti or Zr, but also Al, may lead to precipitation of nitride instead of N_2 pores, which is easier to handle.

9.7.3 Oxygen and Carbon Monoxide in Steel and Iron Alloys

Both $\underline{C}$ and $\underline{O}$ are present in all steel and iron melts. When the steel solidifies both $\underline{C}$ and $\underline{O}$ concentrate in the melt and the saturation limit may be exceeded:

$$\underline{C} + \underline{O} \rightleftharpoons \underline{CO}$$

Guldberg–Waage's law is applied to the CO gas:

$$\frac{c_{\underline{C}} c_{\underline{O}}}{p_{\text{CO}}} = \text{constant} \qquad (9.36)$$

This equilibrium is of great importance at the production process of steel from iron ore.

The temperature dependence of the constant is rather weak and an average value, valid for the middle of the temperature interval 1500–1550 °C, can often be used:

$$c_{\underline{C}}^L c_{\underline{O}}^L = 0.0019 p_{\text{CO}} \quad \text{(wt-% and atm)} \qquad (9.37)$$

If the temperature dependence is considered the following empirical relation is often used:

$$c_{\underline{C}}^L \times c_{\underline{O}}^L = p_{\text{CO}} \times \exp\left(\frac{-2960}{T} - 4.75\right) \quad \text{(wt-%, atm and K)} \qquad (9.38)$$

We will examine the conditions for CO precipitation in an interdendritic region. The advance of the solidification process is represented by the fraction f of solid material.

The same reasoning as was applied to hydrogen in a steel melt (page 277) can be used here and the same equations are valid. The following values of the partition coefficients are valid:

$$k_{\text{part}_{\underline{O}}} = c_{\underline{O}}^s / c_{\underline{O}}^L = 0.054 \qquad (9.39)$$

$$k_{\text{part}_{\underline{C}}} = c_{\underline{C}}^s / c_{\underline{C}}^L = 0.20 \qquad (9.40)$$

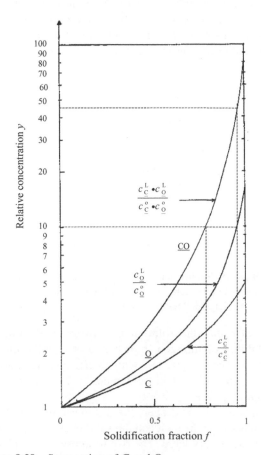

Figure 9.29 Segregation of $\underline{C}$ and $\underline{O}$.

With the aid of equation (9.34), we can calculate how the relative concentrations of $\underline{O}$ and $\underline{C}$ in the interdendritic melt increase during the solidification process as a function of the solidification fraction f:

$$y_{\underline{O}} = c_{\underline{O}}^{L}/c_{\underline{O}}^{\circ} = \frac{1}{1 - f(1 - 0.054)} \qquad (9.41)$$

$$y_{\underline{C}} = c_{\underline{C}}^{L}/c_{\underline{C}}^{\circ} = \frac{1}{1 - f(1 - 0.20)} \qquad (9.42)$$

In the same way as in Example 9.5 (page 278), arbitrary values of f can be chosen and the two curves can be drawn. In this case another curve has been drawn, which represents the product of the two curves, i.e.

$$y_{\underline{CO}} = y_{\underline{C}} y_{\underline{O}} = \frac{c_{\underline{C}}^{L} c_{\underline{O}}^{L}}{c_{\underline{C}}^{\circ} c_{\underline{O}}^{\circ}} \qquad (9.43)$$

The curves are plotted in Figure 9.29, where the upper curve represents $\underline{CO}$.

Example 9.6

Find the solidification fraction at which gas precipitation of CO can be expected to start in an interdendritic region for a steel melt with 0.050 wt-% $\underline{C}$ and 0.0040 wt-% $\underline{O}$. The effects of possible nucleation difficulties and the hydrostatic pressure can be neglected.

Solution:

The composition of the steel is given in the text.

$$c_{\underline{C}}^{\circ} c_{\underline{O}}^{\circ} = 0.050 \times 0.0040 = 0.000200 \, (\text{wt-\%})^2$$

Equation (9.37) gives the saturation value at $p_{co} = 1$ atm:

$$c_{\underline{C}}^{L} c_{\underline{O}}^{L} = 0.0019 \times 1 \, (\text{wt-\%})^2$$

The value of f which corresponds to:

$$y_{\underline{CO}} = \frac{c_{\underline{C}}^{L} c_{\underline{O}}^{L}}{c_{\underline{C}}^{\circ} c_{\underline{O}}^{\circ}} = \frac{0.0019}{0.00020} \approx 10$$

is read from Figure 9.29. We obtain $f = 0.78$.

Answer:

There is a risk of gas precipitation if $f \geq 0.78$.

To understand the interaction between $\underline{C}$ and $\underline{O}$ in the melt we have to analyse the ternary system Fe–O–C. The phase diagrams of the binary systems Fe–O and Fe–C are

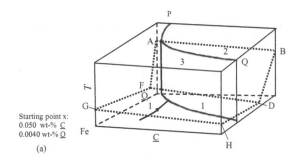

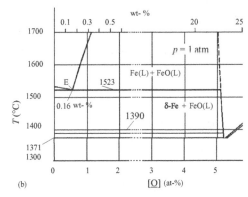

Figure 9.30 (a) Ternary phase diagram of the Fe–O–C system. 1, Liquidus surface 1 of ferrite (δ-iron): FGHD; 2, liquidus surface 2 of FeO (slag): ABDF; 3, liquidus surface 3 of CO gas: vertical curved surface including the upper curve PQ. Coordinate axes: Fe–O, Fe–C. **(b)** Phase diagram of Fe–O. Reproduced with permission from Jernkontorets Annaler (a) and the American Society for Metals (ASM) (b).

shown in Figure 9.30 (b) and Figure 7.21 on page 204, as respectively. These two-phase diagrams are included in two vertical planes of the Fe–O–C system in Figure 9.30.

The liquidus line in the Fe–O system is represented by a liquidus surface in the ternary Fe–O–C system, marked by 2 in Figure 9.30 (a), and describes the precipitation of FeO. The precipitation of CO is described by Equation (9.36). This equation, applied at a CO pressure of 1 atm, represents the surface marked by 3 in Figure 9.30 (a).

Figure 9.30 (a) can be used to discuss the possibilities of CO and FeO precipitation. The phase diagram shows that formation of FeO may occur if the $\underline{C}$ concentration is low enough and the melt reaches point E in Figure 9.30 (b). The condition for the FeO precipitation will be analysed in Example 9.7.

Example 9.7

What demand has to be made on the carbon concentration in a steel melt with 0.016 wt-% oxygen in order to avoid gas precipitation of CO in favour of the formation of FeO? FeO is formed at an oxygen concentration of 0.155 wt-% at 1500 °C.

Solution:
FeO is formed at the solidification fraction, which corresponds to the relative concentration:

$$y_{\underline{O}} = \frac{c_{\underline{O}}^{L}}{c_{\underline{O}}^{\circ}} = \frac{0.155}{0.016} \approx 10 \qquad (1')$$

Using the $\underline{O}$ curve in Figure 9.29, we find that $y_{\underline{O}} \approx 10$ is valid for $f \approx 0.95$. Simultaneously, the $y_{\underline{CO}}$ value must be lower than the value which corresponds to $\underline{CO}$ saturation. This value can be read from the $\underline{CO}$ curve in Figure 9.29 for $f \approx 0.95$:

$$y_{\underline{CO}} = \frac{c_{\underline{C}}^{L}c_{\underline{O}}^{L}}{c_{\underline{C}}^{\circ}c_{\underline{O}}^{\circ}} \approx 45 \qquad (2')$$

In addition, the equilibrium equation for CO [Equation (9.37)] is valid at $p_{CO} = 1$ atm:

$$c_{\underline{C}}^{L}c_{\underline{O}}^{L} = 0.0019 \times 1 \, (\text{wt-\%})^2 \qquad (3')$$

Combining Equations (1'), (2'), and (3') we obtain:

$$c_{\underline{C}}^{\circ} = \frac{c_{\underline{C}}^{L}c_{\underline{O}}^{L}}{y_{\underline{CO}}c_{\underline{O}}^{\circ}} \approx \frac{0.0019}{45 \times 0.016} \approx 0.0026 \, \text{wt-\%} \qquad (4')$$

Answer:
The carbon concentration must be <0.0026 wt-% or <0.002 wt-% for safety.

Removal of Oxygen in Steel and Iron Alloys

Lowering of the $\underline{O}$ concentration in steel is done by so-called *deoxidation*. Small amounts of a substance M_b are added to the molten steel M_a which contains $\underline{O}$. The metal M_b must fulfil the following conditions:

1. $M_b O$ must be more stable than $M_a O$.
2. $M_b O$ must be easy to separate from the melt.
3. M_a must not be influenced negatively by remaining M_b in the melt.
4. Remaining $\underline{O}$ must not deteriorate the properties of the cast metal alloy.

In steel melts Si, Mg, and Al are used as deoxidation agents.

9.7.4 Gas Precipitation in Rimming Iron Ingots

Killed steel is compact whereas rimming steel contains pores of various extents. The difference is due to gas preci-

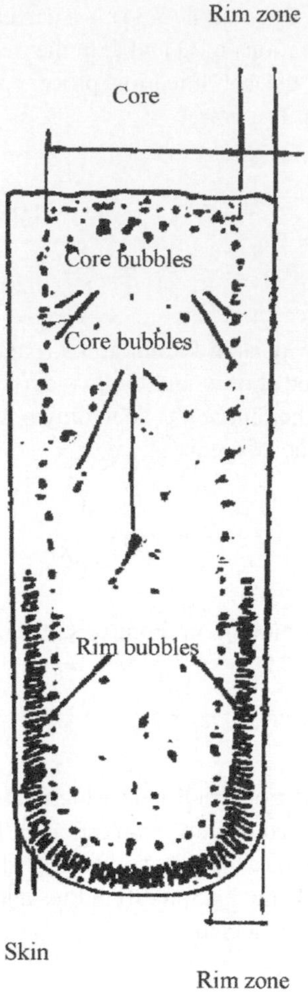

Figure 9.31 Cross-section of a rimming ingot. Reproduced with permission from the Scandinavian Journal of Metallurgy, Blackwell.

pitation during the solidification process. The main component of the gas is *carbon monoxide*. The gas precipitation that occurs when rimming steel solidifies thus depends mainly on its concentrations of carbon and oxygen.

An ingot of rimming steel is illustrated in Figure 9.31. It solidifies roughly in the following way:

Already during or immediately after casting there is a violent gas precipitation, which initiates movements in the steel bath. The motion occurs upwards at the periphery and downwards at the centre. The mould cools the steel melt relatively strongly and a certain nucleation occurs at the mould wall. These primarily formed nuclei are carried into the melt owing to the strong motion in the steel bath. There they float in the melt, grow, split up and form new nuclei. Owing to the great number of new crystals which are formed and grow in this way, the melt becomes viscous.

During this process, a crystal zone, the so-called rimmed zone, is developed at the surface of the ingot. The zone consists of relatively solid material and is formed as a result of the strong cooling from the mould.

The zone consists of many small crystals, which initially have been free and then have been added to what remains of a solidification front. The motion in the bath mainly determines the appearance of the rim zone.

At slow or low gas precipitation, *rim bubbles* are formed. Their length direction is perpendicular to the mould wall owing to the strong cooling from the mould and the solid material. With strong gas precipitation or violent bath movement the rim zone may be practically free from bubbles. At transition from the first to the last case the gas bubbles first disappear from the upper part of the rim zone. With increasing gas precipitation the bubbles also disappear in the lower part of the zone.

The reason why no bubbles are formed in the rim zone on violent gas precipitation is probably that floating crystals and gas bubbles efficiently wash the solidification front. The result is that the gas, which has been formed at the front, is carried away and gains no hold at the front. This obviously happens most easily in the upper part of the ingot, where more numerous and larger bubbles pass the solidification front.

During the formation of the rim zone, an edge is formed at the upper surface of the ingot, a so-called rim (Figure 9.32). This edge grows towards the centre. Finally, a crust of solid material is formed, which covers the whole upper surface of the ingot.

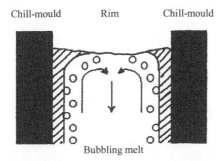

Chill-mould   Rim   Chill-mould

Bubbling melt

Figure 9.32 Formation of a rim. Reproduced with permission from the Scandanvian Journal of Metallurgy, Blackwell.

When the crust has been formed, the pressure in the centre of the melt increases and the gas precipitation stops. In some cases the pressure can be high enough to break the crust and the gas precipitation may start again. The result will be that a new violent movement in the bath starts at the centre of the ingot.

The more violent this motion is the fewer bubbles are formed inside the ingot. The positions of the so-called *core bubbles* can be seen in Figure 9.31. A strong bath motion causes a bubble-free zone between the rim and core bubbles at the moment when the top crust or the rim breaks. Core bubbles are formed when the bath motion or gas precipitation decreases once more. When the rim has been thick enough to resist the pressure from the inner

parts of the ingot, the rest of it solidifies. The inner parts of the ingot generally consist of a viscous mixture of melt and crystals. The melt may contain some gas pores, which may grow somewhat during the solidification process. These pores are the so-called nucleation bubbles.

With the aid of the phase diagram for Fe–C–O it is possible to explain what happens when the rims are formed and the pressure increases. Figure 9.33 shows a projection on the bottom plane of the liquidus surface, marked with 1 in Figure 9.30 (a).

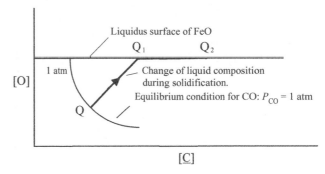

Figure 9.33 Solidification process in a rimming ingot when gas precipitation is prevented. The process is illustrated with the phase diagram of the system Fe–C–O. The figure is a projection of the ternary phase diagram in Figure 9.30. Reproduced with permission from Jernkontorets Annaler.

We assume that the pressure is high enough to prevent gas precipitation. In a steel melt with a composition corresponding to point Q, δ-iron is precipitated. If the pressure is 1 atm, precipitation of CO gas starts simultaneously. During the continuing solidification, the composition of the solid phase is changed owing to precipitation of δ-iron at the same time as the temperature decreases. The change in composition is described by some curve QQ_1. We know nothing of the shape of this curve and have drawn it as a straight line.

When the composition has been changed so much that the point Q_1 is reached, simultaneous precipitation of δ-iron, FeO, and CO occurs. The composition of the melt will continue to change along the monotechtic line Q_1Q_2. This explains why oxide precipitations are found in the centre of a rimming ingot.

If the steel releases very large quantities of gas at the beginning of the casting process, a so-called bootleg is formed [Figure 9.34 (a)]. When the mould is filled, the molten steel contains a great amount of gas bubbles. When they have disappeared, the upper surface of the melt in the mould sinks and the result is that the upper part of the ingot consists of a thin shell.

If the gas precipitation is somewhat weaker, ingots characteristic of rimming steel are obtained [Figure 9.34 (b) and (c)]. The amount of rim bubbles may be very small and occur, for example, as a few bubbles close to the bottom if the gas precipitation is extensive enough.

If the gas precipitation is even weaker, bubbles are formed in the majority of the rim zone. Owing to this bubble formation, the level of the steel rises in the mould. The steel is said to *ferment* [Figure 9.34 (d)].

In many cases an artificial rim is created a certain time after the casting by blowing gas on the surface to cool it or by covering the surface with a steel plate. The bubbles in the rim stop and a central pore is formed.

Fermentation after casting

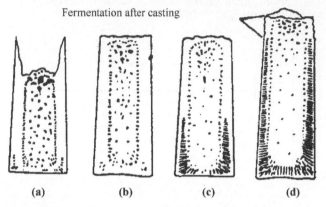

(a)   (b)   (c)   (d)

Figure 9.34 (a) Very strong gas precipitation. Boot leg ingot. Rim zone nearly free from bubbles. (b) Strong gas precipitation. Rim zone nearly free from bubbles. (c) Not particularly strong gas precipitation. Bubbles in the lower part of the ingot. (d) Rather weak gas precipitation. Bubbles in the majority of the rim zone. The steel ferments during the solidification. Reproduced with permission from the Scandanvian Journal of Metallurgy. Blackwell Publishing.

The gas precipitation rate influences not only the vertical position of the rim bubbles but also the thickness of the skin, i.e. the thickness of the part of the ingot situated between the rim zone and the mould (Figure 9.31). This part is of great importance for the properties of the ingot surface. Bubbles too close to the surface may form during the following heating and cause severe surface defects. It is generally very difficult to predict what oxygen concentration the steel should have to obtain the best possible structure. It is important to consider not only the carbon and oxygen concentrations but also its silicon and aluminium concentrations.

It is also important to adjust the casting rate correctly. If a rimming steel fills the mould rapidly, the gas precipitation would start at the top of the ingot where the hydrostatic pressure is zero. Rim bubbles would form close to the surface because there is no motion in the bath.

At a properly adjusted casting rate, the increasing ferrostatic pressure, caused by the increasing height of the iron column close to the bottom, will delay the gas precipitation so much that no gas bubbles are formed close to the bottom. When the casting is finished, the solidification in the lower part causes supersaturation of $\underline{CO}$ in the melt inside the shell, large enough to compensate the higher pressure. For this reason, gas precipitation occurs approximately

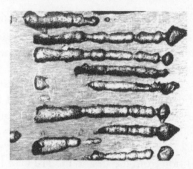

Figure 9.35 Example of rim pores with periodic contractions and expansions. This appearance is seen in *lower* parts of ingots. Reproduced with permission from the Scandanvian Journal of Metallurgy. Blackwell Publishing.

simultaneously from the bottom to the top of the ingot in the ideal case.

The shapes of the rim pores are different in the lower and upper parts of the ingot. In the *lower* part the pores have the shape of a rotation volume with periodic contractions and expansions formed by release of bubbles as is shown in Figures 9.35. In the *upper* part the rim pores are less well developed. They look like rows of pearls as is seen in Figure 9.36. Each pore is surrounded by a V-shaped segregation pattern.

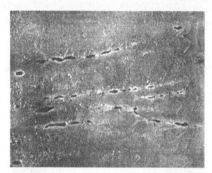

Figure 9.36 Rim pores of the appearance in the figure are common in the *upper* parts of ingots. The melt penetrates the pore walls when the rim pores are formed. Reproduced with permission from the Scandanvian Journal of Metallurgy. Blackwell Publishing

The formation of these two morphologies is assumed to be as follows. The growth mechanism of rim pores is the same as the growth process of hydrogen pores, described in Figure 9.9 (page 264). In this case it is $\underline{C}$ and $\underline{O}$, instead of $\underline{H}$, which diffuse to the pore surfaces. The process is assumed to occur in steps, as outlined below.

1. A pore is formed at the solidification front [Figure 9.37 (a)].
2. The solidification front advances and $\underline{C}$ and $\underline{O}$, concentrated in the interdendritic regions, diffuse to the pores. The pore grows faster than the solidification front [Figure 9.37 (b)].
3. When the pore has reached a certain critical size, which depends on the flow rate in the melt, some of the gas in

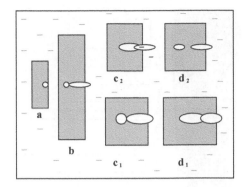

Figure 9.37 Formation of rim bubbles and rim channels. Upper part of an ingot: a–b–c_2–d_2. Lower part of an ingot: a–b–c_1–d_1. The solid phase is marked by shaded squares. The schematic drawing in (c_2) means that melt partly penetrates into the gas pore and reduces its size and unlaces it into two pores.

the pore is released and forms a rising bubble, which is unlaced [Figure 9.37 (b) and (c_1)]. If the flow in the melt is *slow* or zero, there will be enough gas left to form a new pore, smaller than the earlier one owing to loss of gas. The steel layer, which solidifies next, starts to unlace the pore [Figure 9.37 (c_1) and (d_1)].

4. A new expansion replaces the unlace process owing to continued gas precipitation [Figure 9.37 (d_1)]. Figure 9.35 shows an example of this type of rim bubble.

5. If the melt which passes a growing pore has a *high* flow rate, gas may be carried away and the melt may enter the oval pore wall [Figure 9.37 (c_2)]. When the next layer of steel solidifies at the oval wall, an indentation is formed or a total separation [Figure 9.37 (d_2)] occurs. The outer bubble grows. The next time gas is released from the pore the melt enters the pore again and the process is repeated. Figure 9.36 shows an example of this type of rim pore.

The process described in step 5 is common in the upper part of the ingot. The gas precipitation is faster there, owing to the lower ferrostatic pressure. Large pores from the lower part of the ingot pass through the melt and bring the small upper pores up to the ingot surface. The amount of gas becomes large enough to cause gas stirring. The stirring releases bubbles which otherwise would have been caught by the solidification front in the upper part of the ingot and caused bubbles close to the upper surface.

9.7.5 Macro- and Micro-slag Inclusions in Steel

In carbon elimination from raw iron, the carbon is removed by addition of oxygen. This is usually done in a converter or by remelting iron scrap together with raw iron in an electric steel furnace. At the same time, most of the sulfur and phosphorus are removed by addition of lime, $CaCO_3$.

When carbon is oxidized during the carbon elimination process, some silicon is simultaneously oxidized to silicon dioxide and iron to iron oxides. Some oxygen also dissolves in the melt. During the following casting, the dissolved oxygen reacts with the iron and carbon atoms. In the absence of special measures, inclusions of iron oxides and gas bubbles of carbon monoxide are formed.

In order to avoid the precipitation of iron oxide, inclusions, which react more easily with oxygen than iron, are added (page 282). Examples of such substances are Si, Mn, and Al. The steel is *deoxidized*. In this deoxidation process slag particles or slag droplets are formed in the melt. These slag inclusions follow the steel melt during casting and are included in the structure at solidification.

Slag inclusions in steel adversely affect the material and destroy its properties strongly. By machining of the material after casting, these inclusions are in many cases very annoying and result in a considerable reduction in the yield since much material is wasted and has to be rejected.

There is a distinction between macro- and microslag inclusions. Macroslag inclusions are defined as inclusions visible to the naked eye on a lathed or milled surface. The practical border between macro- and microinclusions is 0.1 mm:

$$\text{microslag inclusions} < \text{visibility limit } 0.1 \text{ mm}$$
$$\text{visibility limit } 0.1 \text{ mm} < \text{macroslag inclusions}$$

The macroslag inclusions in the steel melt are formed either during the casting, during the secondary treatment or during the cooling of the melt when the solubilities of the slag substances decrease. The macroslag inclusions are thus formed *before* the crystallization process and consist mainly of:

- reaction products from chemical reactions between the melt and the lining materials;
- deoxidation products.

The microslag inclusions are most frequently formed *during* the solidification process in the interdendritic regions as a consequence of microsegregation.

We will start by a discussion of macroslag formation in steel and then treat the formation of microslag inclusions in steel.

9.7.6 Macroslag Inclusions of Oxides in Steel

Slag Inclusions from Lining Materials
Furnaces, ladles and casting boxes are lined with a ceramic material, usually brick. The material must be able to resist the mechanical stress at high temperatures, be chemically resistant to the melt, and be a good thermal insulator.

Almost all types of lining materials consist of oxides. The most common ones are MgO, SiO_2, and Al_2O_3.

The lining materials are given a porous structure in order to achieve good thermal insulation. More compact materials have better mechanical properties than porous materials. In spite of the aim of obtaining good mechanical properties of the lining materials, erosion occurs and fragments are torn away during the casting. These fragments are carried away by the melt and become inclusions in the completed steel. They are also efficient heterogeneities for nucleation and precipitation of deoxidation products.

Many investigations show that more than 50 % of the macroinclusions consist of reaction products of lining materials.

Slag Inclusions from Deoxidation Products

Even before the addition of deoxidation agents, the melt contains particles of silicon dioxide and iron oxides. On addition of the deoxidation substance a large number of small oxide particles are normally formed. These would not adversely influence the quality of the steel if they could retain their small sizes. However, during the time before casting and during the casting and solidification processes the inclusions grow, because of collisions between the particles, sintering, or coalescence.

In the case of SiO_2 this process occurs very rapidly. For Al_2O_3 inclusions, which are crystalline, the sintering process occurs much more slowly than for SiO_2. The result is that long-chain, so-called *clusters* are formed. They contain hundreds or even more primary inclusions. Figure 9.38 shows an example of such clusters.

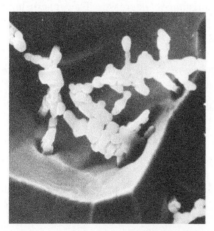

Figure 9.38 Electron microscope picture of Al_2O_3 clusters in steel. Reproduced with permission from the American Society for Metals.

Macroslag Distribution at Ingot Casting

The distribution of macroslag inclusions in a downhill-cast ingot is shown in Figure 9.39. The dark shaded areas in the picture contain high slag inclusion concentrations.

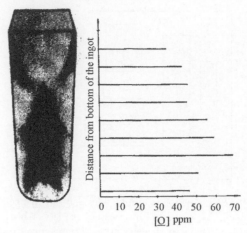

Figure 9.39 Distribution of large slag inclusions in a downhill-cast 5 ton ingot as a function of position. The right part of the figure shows the oxygen concentration along the central axis of the ingot. Reproduced with permission from Jernkontorets Annaler.

Figure 9.39 shows that the macroslag inclusions are concentrated partly in the lower part of the ingot – generally known as 'the slag inclusion drift' – and partly in the upper part of the cylindrical region of the ingot. The right part of the figure shows the oxygen concentration along the central axis of the ingot. The presence of slag inclusions is roughly proportional to the oxygen concentration in the ingot.

There may be several reasons why the slag inclusions are concentrated in the lower part of the ingot. Because the inclusions preferably exist in the so-called *sedimentation crystal cone*, it is reasonable to assume that the slag inclusions constitute nucleation agents for the free crystals in the melt. It is also possible that growing free crystals trap the slag inclusions before they settle at the bottom of the ingot.

Natural convection may also be very important. The large inclusions are carried away with the melt. When they are close enough to the solidification front it may incorporate them. This would explain the concentration of macroslag inclusions in the upper part of the cylindrical region of the ingot.

Heating of the upper surface of the ingot implies a reduction in the macroslag inclusion concentration. This strongly supports the theory that natural convection influences the distribution of macroslag inclusions in the ingot (Section 5.2). It is well known that natural convection decreases when the upper surface of the ingot is heated.

Macroslag Distribution in Continuous Casting

Macroslag inclusions are as frequently present in continuously cast billets as in ingots. The sources and formation mechanisms are also the same in both cases. On the other hand, the distributions of macroslag inclusions in ingot-cast and continuously cast billets are different. The sizes of the slag particles are also smaller in continuous casting than in ingot casting.

The macroslag inclusions in continuously cast billets occur partly as surface slag inclusions and partly inside the material. The formation of surface slag inclusions depends on:

1. the flow pattern in the mould when the melt from the ladle falls into the mould (Figure 9.5, page 261).
2. the fact that air bubbles are carried downwards into the steel bath where the oxygen of the air reacts with the steel melt.

In favourable precipitation cases, casting slag inclusions, which have been carried into the mould by the steel jet stream, may rise to the surface for two reasons. Their density is lower than that of the steel melt and the gas bubbles may carry the macroslag inclusions up to the surface on their way upwards. Part of the slag inclusions, which rise to the surface of the bath in the mould, may then be drawn downwards between the mould wall and the steel surface. This results in slag inclusions on the billet surfaces or transfer of slag inclusions to the casting powder in the cases when such powder is used.

The conditions for good separation decrease with decreasing billet size. When the cross-section decreases, the casting rate increases. The larger are the dimensions of the strand and the machine radius, the more the slag inclusions concentrate towards the centre of the strand. The centre slag inclusions have a different distribution in vertical machines than in bent machines, as is shown in Figure 9.40.

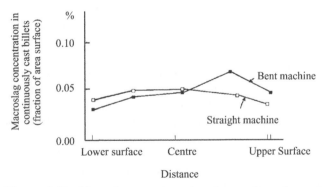

Figure 9.40 Macroslag concentration in continuously cast billets.

Methods to Avoid Macroslag Formation in Steel
Survey of Methods
Since slag inclusions have a strongly negative influence on the quality of steel, it is desirable to reduce the slag formation during the casting process as much as possible. The slag inclusions have considerably lower density than the melt and the tundish is designed in such a way that as

much as possible of the slag inclusions are separated before the casting (page 21).

By suitable design of the casting process, efforts are made to prevent the appearance of slag inclusions. The available methods are as follows:

1. choice of suitable material for lining of furnace, ladle, tundish, and mould;
2. shielding of the melt from the oxygen in air;
3. choice of optimal casting temperature;
4. use of a so-called 'hot top' to insulate the upper surface at ingot casting;
5. optimal design of the machine at continuous casting.

Point 1 has been discussed earlier in this chapter (page 285). The majority of the slag inclusions are oxides.

If *efficient shielding of the melt from contact with the oxygen in air* can be achieved, this will reduce the presence of macroslag inclusions. The method will be described below.

By *insulating the upper surface in ingot casting*, slag inclusion separation is facilitated and reduces the natural convection in the melt. Experience shows that a hot top made of some exothermic material, which generates much heat at the beginning of the solidification process, favours the separation. This is clearly seen in Figure 9.41.

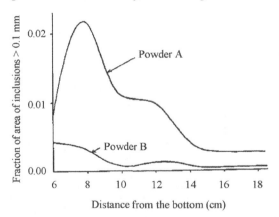

Figure 9.41 Influence of different hot top materials on the separation of macroslag inclusions > 0.1 mm. Reproduced with permission from the Scandanvian Journal of Metallurgy, Blackwell Publishing.

When the natural convection decreases, the possibility increases that slag particles are carried upwards in the melt and stick to the casting powder. This reduces the presence of macroslag in ingots.

Experience shows that the *casting temperature* influences the macroslag concentration. A *high* casting temperature often gives a *low* fraction of macroslag inclusions. The reasons for this are that the solubility of oxygen in the melt increases with temperature and that the equilibrium is changed owing to the temperature dependence of the equilibrium constant. The large inclusions dissolve and numerous small ones are then precipitated during cooling.

However, it is not possible to increase the casting temperature only with regard to the occurrence of slag inclusions. Other factors must also be considered. A high casting temperature is accompanied by disadvantages such as increased corrosion of the lining material and an increased tendency for crack formation.

Shielded Casting

So-called *shielded casting* in an N_2 or Ar atmosphere gives a strongly reduced slag concentration. Hence there has been great interest in shielded casting for a long time. It is, for example, very urgent to protect vacuum-treated steel from reabsorbing gases during the casting process.

Shielded casting assumes *both* that the air in the mould is replaced by an inert gas by careful 'lubrication' *and* that protecting gas surrounds the metal jet stream all the time. In continuous casting, the best method to shield the jet stream between tundish and mould is to use a ceramic tube, which ends under the bath level in the mould. A cover of casting powder protects the upper surface. In this way, two sources of gas absorption, at the metal jet stream and at the upper surface, are eliminated.

With mould sizes smaller than 14 cm, it is impossible to use a casting tube (Figure 9.42). For handling reasons no tubes can be produced for the smallest billet sizes, hence it is difficult to find suitable casting shields for small billets. Sometimes a tube 'curtain', within which N_2 or Ar is supplied, surrounds the jet stream.

Figure 9.42 Flow pattern in the melt when a casting tube is used.

However, the most common way is to add colza oil. The oil is decomposed and burns with the oxygen in the air and an inert atmosphere is formed. This is similar to the method used in downhill casting of ingots where tar is used to obtain an inert atmosphere.

Electroslag Refining

Whatever precautionary measures are taken, it is impossible to avoid macroslag inclusions completely in cast materials. In some cases slag inclusions at the surface may be removed by grinding the surface.

This method cannot be used if the inclusions are located in the interior of the material. In this case, electroslag

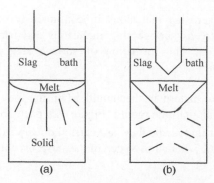

Figure 9.43 (**a**) and (**b**). Directions of temperature gradient and crystal directions at ESR at two electrode positions.

refining (ESR) can be used. The principle of ESR has been described in Chapter 2 (page 27) and the process is further discussed in Chapter 5 (pages 117–118).

During the remelting process, the crystals grow in the direction of the temperature gradient, i.e. perpendicular to the solidification front. It is possible to control the direction of the temperature gradient by changing the position of the electrode and thereby direct the crystals at arbitrary angles between the radial and axial positions. Examples of this are given in Figure 9.43 (a) and (b). The material acquires the best properties if the temperature gradient and thus the directions of the crystals are axial [Figure 9.43 (a)].

Slag Refining Using ESR Remelting

Slag refining by ESR occurs by chemical and physical refining at the electrode tip and in the slag bath. On the electrode tip, a thin film of molten metal forms, which increases the boundary surface between the metal and slag. The temperature in this region is high, which means that the conditions for separation of the slag inclusions at the interface are favourable.

When the droplets which have been formed at the electrode tip fall through the slag, there is an additional opportunity for good contact between the metal and slag. Finally, there is a third contact surface, between the metal bath and slag bath.

Hence there are three different areas where refining may occur. The first is considered to be the most important. Some workers have reported that up to 80 % of the refining occurs at the electrode tip and the rest during the passage of the droplet through the molten slag.

Slag refining by ESR is effective. In particular, the inclusions are reduced in size to a maximum of 5 mm and the fractions of silicates and sulfides decrease strongly compared with non-ESR remelted materials. Instead, there are small aluminate slag inclusions if Al_2O_3-bearing slag is used, otherwise small silicate slag inclusions.

This is illustrated in Figure 9.44, where the occurrence of slag in the electrode material is compared with the

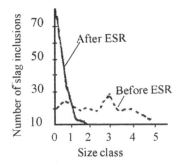

Figure 9.44 Frequency of slag inclusions of various sizes before and after ESR remelting.

ESR remelted material. In the first case, inclusions of all sizes are comparatively evenly distributed within five size classes. On the other hand, in the ESR material there are only inclusions which belong to the two smallest size classes.

9.7.7 Microslag Inclusions of Oxides in Steel

The majority of the slag inclusions in steel consists of oxides, mainly FeO, SiO_2, MnO and Al_2O_3. In Example 9.7 (page 281) we discussed the possibility of precipitating FeO instead of CO during a solidification process. Normally it is essential to avoid both FeO and CO precipitation.

In the steel melt, the concentrations of $\underline{C}$ and $\underline{O}$ are given. By addition of a suitable amount of a deoxidation agent, the effective concentration of $\underline{O}$ is decreased. The principle is simple.

The original concentrations of $\underline{C}$ and $\underline{O}$ in the steel melt are known and Equations (9.41) and (9.42) (page 280) describe how they are changed during the solidification process. A graphical representation of the segregation of $\underline{C}$ and $\underline{O}$ as functions of the solidification fraction is given in Figure 9.29 on page 280.

The deoxidation substance M reacts with $\underline{O}$ in the melt:

$$xM + y\underline{O} \rightleftharpoons M_xO_y$$

The solubility product of the oxide is known:

$$[\underline{M}]^x \times [\underline{O}]^y = K_{M_xO_y} \qquad (9.44)$$

The solubility product $K_{M_xO_y}$ depends on the temperature. This temperature dependence is given in tables for the substance in question.

All deoxidation agents must have $K_{M_xO_y}$ values such that slag precipitation occurs before the formation of both FeO and CO precipitation. Deoxidation slags are often problematic but many times better than precipitation of CO gas, which often has negative consequences for deoxidized steel (Section 9.7.4, page 282).

In this section we will treat the precipitation of SiO_2 and Al_2O_3 in a steel melt during the solidification process. The slag inclusions which are formed within the interdendritic regions are in most cases small and are thus characterized as microslag inclusions. The precipitation process of these inclusions will be discussed below. The section starts with a short recollection of the theory of homogeneous nucleation, which is applied to oxide inclusions in steel.

Homogeneous Nucleation of Oxide Inclusions

In Chapter 6 (pages 141–142) we discussed the theory of homogeneous nucleation. Here we will apply this theory to deoxidation processes (page 282) where deoxidation agents are added to remove oxygen from the melt. The $\underline{O}$ atoms become neutralized when they precipitate as inclusions of oxide. The surface tension strongly influences the nucleation process of inclusions.

According to Equations (6.2) and (6.4) (pages 140–14), the activation energy ($-\Delta G^*$) of homogeneous nucleation can be written as:

$$-\Delta G^* = 60\,kT^* = \frac{16\pi\sigma^3 V_m^2}{3(-\Delta G_m)^2} \qquad (9.45)$$

or

$$-\Delta G_m = \sqrt{\frac{16\pi}{3} \times \frac{\sigma^3 V_m^2}{60 k_B T^*}} \qquad (9.46)$$

where

$-\Delta G^*$ = activation energy of nucleation, i.e. the minimum energy required to form an oxide particle

$-\Delta G_m$ = driving force of nucleation or molar change of free energy (kmol) on nucleation

k_B = Boltzmann's constant

T^* = critical nucleation temperature

σ = surface tension between the inclusion oxide and the surrounding melt

V_m = molar volume of the oxide (m³/kmol).

In Chapter 6, we derived an expression [Equation (6.6) (page 142)] for the necessary supersaturation for nucleation of a primary phase with low solubility in the melt. The theory can be extended to be valid also for the precipitation of chemical compounds in the melt. Application of Equation (6.6) and the use of the thermodynamic relations on pages 265–267 give the value of $-\Delta G_m$, required for homogeneous nucleation of an inclusion M_xO_y:

$$-\Delta G_m = RT \ln\left[\frac{a_M^x a_O^y}{(a_{\underline{M}}^x a_{\underline{O}}^y)^{eq}}\right] \qquad (9.47)$$

where

$a_{\underline{M}}$ = activity of the $\underline{M}$ atoms (metal, nonmetal) in the melt

$a_{\underline{O}}$ = activity of $\underline{O}$ atoms in the melt

$(a_{\underline{M}})^{eq}$ = activity of the $\underline{M}$ atoms in the melt at equilibrium

$(a_{\underline{O}})^{eq}$ = activity of $\underline{O}$ atoms in the melt at equilibrium.

If we combine Equations (9.46) and (9.47), we obtain

$$-\Delta G_m = RT \ln\left[\frac{a_{\underline{M}}^x a_{\underline{O}}^y}{(a_{\underline{M}}^x a_{\underline{O}}^y)^{eq}}\right] = \sqrt{\frac{16\pi}{3} \times \frac{\sigma^3 V_m^2}{60 k_B T^*}} \quad (9.48)$$

where σ is the surface tension between the precipitated particles and the surrounding melt and T is the temperature of the melt.

In Table 9.8, the necessary driving force $(-\Delta G_m)$ for homogeneous nucleation is listed for an oxide inclusion an iron melt for various values of the surface tension.

TABLE 9.8 $-\Delta G_m$ **values for a solid oxide inclusion as functions of** σ **($V_m = 25 \times 10^{-3}$ m^3/kmole).**

σ (J/m^2)	$-\Delta G_m$ (10^{-3} J/kmole) Oxide inclusion
0.50	7.0
1.00	19.9
1.50	36.5
2.00	56.3
2.50	78.6

As a concrete example we will consider an Fe–50 % Ni melt at 1550 °C. The solubility products of some oxides at equilibrium with this melt, which contains oxygen, are given in Table 9.9.

TABLE 9.9 **Solubility products of oxides in an Fe–50 % Ni melt at 1550 °C.**

Oxide	K
FeO	9.12×10^{-2} (wt-%)^2
SiO$_2$	2.09×10^{-6} (wt-%)^3
Al$_2$O$_3$	1.72×10^{-16} (wt-%)^5

Figure 9.45 is based on the solubility product of SiO$_2$ in the melt:

$$[\underline{Si}] \times [\underline{O}]^2 = K_{SiO_2} \quad (9.49)$$

If we replace the activities with concentrations, Equations (9.48) and (9.49) can be written as:

$$\ln[\underline{Si}] + 2\ln[\underline{O}] = \text{constant} \times \sigma^{1.5} \quad (9.50)$$

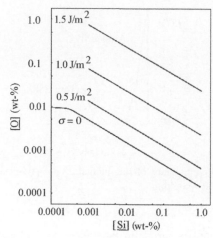

Figure 9.45 Effect of surface tension and melt composition on nucleation of SiO$_2$(s). Reproduced by permission of ASM International.

or

$$\ln[\underline{O}] = \frac{1}{2} \times \text{constant} \times \sigma^{1.5} - \frac{\ln[\underline{Si}]}{2} \quad (9.51)$$

Equation (9.51) is plotted in Figure 9.45. The concentration of $\underline{O}$ is at equilibrium with the $\underline{Si}$ concentration but also with the Fe concentration. $[\underline{O}]$ is read from the figure at $\sigma = 0$. Close to the $\underline{O}$ axis the straight line is bent and approaches the $\underline{O}$ concentration at equilibrium with FeO.

Figure 9.46 is based on the solubility product of Al$_2$O$_3$ in the melt. An analogous equation to Equation (9.50) can be

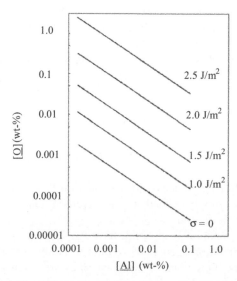

Figure 9.46 Effect of surface tension and melt composition on nucleation of Al$_2$O$_3$(s). Reproduced by permission of ASM International.

derived for Al_2O_3:

$$2\ln[\underline{Al}] + 3\ln[\underline{O}] = \text{constant} \times \sigma^{1.5} \qquad (9.52)$$

or

$$\ln[\underline{O}] = \frac{1}{3} \times \text{constant} \times \sigma^{1.5} - \frac{2\ln[\underline{Al}]}{3} \qquad (9.53)$$

[$\underline{O}$] is read from the figure at $\sigma = 0$. The upper curves in Figures 9.45 and 9.46 describe the necessary supersaturation of oxygen and Si and Al, respectively, for homogeneous nucleation of SiO_2 and Al_2O_3, respectively, for different values of the surface tension between the melt and the oxide particles. Often it is difficult to estimate the surface tensions. The curves can be used for this purpose, as is illustrated in Example 9.8.

___

Example 9.8

The steel melt presented in Figures 9.45 and 9.46 is deoxidized with the aid of Al, which is added at a concentration of up to 0.01 wt-%. The melt was in equilibrium with FeO before the deoxidation agent was added. Precipitation of many small oxide particles was observed when Al was added.

Assume that the oxide particles consist of Al_2O_3 and that they have been formed during a homogeneous nucleation process. Estimate the surface tension between the Al_2O_3 particles and the melt with the aid of Figures 9.45 and 9.46.

Solution:

Figure 9.45 shows that the initial concentration of oxygen in the melt was 0.01 wt-% ($\sigma = 0$). This value is also valid in Figure 9.46 because the melt is the same.

The Al concentration, which is given in the text, is 0.01 wt-%. These values define a point, which is marked in the figure above. Through this point a curve, parallel with the other lines, is drawn. The desired value of the surface tension is obtained by interpolation.

Answer:

The estimated value of the surface tension between the melt and the Al_2O_3 particles is about 1.8 J/m^2.

___

Conditions for Precipitation of Oxides

In most cases the slag oxides are precipitated *before* the solidification starts. However, oxides are also frequently observed in the interdendritic regions, which indicates that inclusions also have been precipitated *during* the solidification process. One theory which explains this fact is that the inclusions are pushed ahead of the solidification front. This possibility will be discussed on pages 292–293.

Another possibility is that new inclusions may be nucleated within the interdendritic regions. If this is the case, high supersaturation is a necessary condition for nucleation (page 289). Example 9.9 illustrates this.

___

Example 9.9

An iron-base alloy melt contains 0.4 wt-% $\underline{Si}$ and 0.004 wt-% $\underline{O}$ and it is in equilibrium with SiO_2.

(a) What combined supersaturation of $\underline{Si}$ and $\underline{O}$ is required for homogeneous nucleation of SiO_2 inclusions?

(b) At what fraction solidified material is this supersaturation reached by microsegregation?

The temperature of the melt is 1500 °C. Material constants are given below.

Material constant	Value
$\sigma_{\text{oxide/melt}}$	1.0 J/m^2
$V_m^{SiO_2}$	25×10^{-3} m^3/kmol
k_B	1.38×10^{-23} J/K
R	8.31 kJ/kmol K
$k_{\underline{O}}$	0.054
$k_{\underline{Si}}$	2/3

Solution:

(a) The critical step for the formation of SiO_2 inclusions is homogeneous nucleation. Equation (9.47) gives the activation energy per kilomol for nucleation. If $T^* \approx T$ it can be written as:

$$-\Delta G_m = RT\ln\left[\frac{a_{\underline{Si}}a_{\underline{O}}^2}{(a_{\underline{Si}}a_{\underline{O}}^2)^{eq}}\right] = \sqrt{\frac{16\pi}{3} \times \frac{\sigma^3(V_m^{SiO_2})^2}{60\,k_BT}} \qquad (1')$$

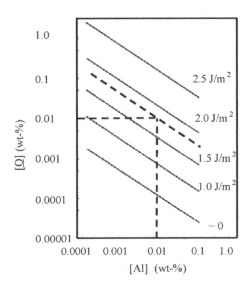

In this case, we have low concentrations and it is reasonable to replace the activities with the concentrations. If we do so and solve the combined supersaturation, i.e. the solubility product, we obtain:

$$c_{\underline{Si}} c_{\underline{O}}^2 = c_{Si}^\circ (c_O^\circ)^2 \times \exp\left[\frac{\sqrt{\dfrac{16\pi}{3}\dfrac{\sigma^3 (V_m^{SiO_2})^2}{60 k_B T}}}{RT}\right] \quad (2')$$

Inserting the numerical values of the material constants and other given quantities gives:

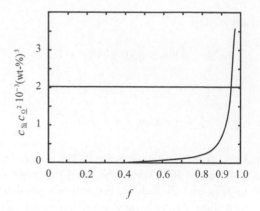

$$c_{\underline{Si}} c_{\underline{O}}^2 = 0.4 \times (0.004)^2 \times \exp\left[\frac{\sqrt{\dfrac{16\pi}{3} \times \dfrac{10^3 \times (25 \times 10^{-3})^2}{60 \times 1.38 \times 10^{-23}(1500 + 273)}}}{8.31 \times 10^3 (1500 + 273)}\right] = 0.0020 \ (wt\text{-}\%)^3$$

(b) Segregation of $\underline{O}$ is described by the lever rule, because $\underline{O}$ is a comparatively small atom, which is dissolved interstitially and diffuses fairly rapidly through the solid phase:

$$c_{\underline{O}} = \frac{c_{\underline{O}}^\circ}{1 - f(1 - k_{part\underline{O}})} \quad (3')$$

The $\underline{Si}$ atoms are larger than the $\underline{O}$ atoms but not as large as the Fe atoms. Neither Scheil's equation nor the lever rule is an ideal model to describe the Si segregation in the melt and in the solid phase. However, Scheil's equation is a somewhat better model than the lever rule. It will be used here instead of accurate numerical computer calculations. Hence we will neglect the diffusion in the solid phase and obtain:

$$c_{\underline{Si}} = c_{Si}^\circ (1 - f)^{-(1 - k_{part\,\underline{Si}})} \quad (4')$$

Next we set up the combined supersaturation:

$$c_{\underline{Si}}(c_{\underline{O}})^2 = c_{Si}^\circ (1-f)^{-(1-k_{part\,\underline{Si}})} \left[\frac{c_{\underline{O}}^\circ}{1 - f(1 - k_{part\,\underline{O}})}\right]^2$$

or, after rearrangement:

$$c_{\underline{Si}}(c_{\underline{O}})^2 = c_{Si}^\circ (c_{\underline{O}}^\circ)^2 \frac{(1-f)^{-(1-k_{part\,\underline{Si}})}}{[1 - f(1 - k_{part\,\underline{O}})]^2} \quad (5')$$

Inserting the known values gives:

$$c_{\underline{Si}} c_{\underline{O}}^2 = 0.4 \times 0.004^2 \times \frac{(1-f)^{-(1-2/3)}}{[1 - f(1 - 0.054)]^2} \ (wt\text{-}\%)^3 \quad (6')$$

This equation is solved graphically. We plot the function versus the fraction of solidified material.

The horizontal line in the diagram is the necessary combined supersaturation $0.0020 \ (wt\text{-}\%)^3$. It intersects the curve at $f \approx 0.95$.

Answer:

(a) The combined supersaturation is $0.0020 \ (wt\text{-}\%)^3$.
(b) SiO_2 inclusions are formed when about 95 % of the material has solidified.

Behaviour of Microslag Inclusions during the Solidification Process

Precipitation of slag inclusions depends on what happens to the small single slag particles during the solidification process after the casting operation. The question is whether the particles are incorporated by the solidification front and become included in the solid phase, or whether they are pushed ahead of the solidification front and gradually accumulate in the remaining melt.

The interaction during the solidification process between small slag particles and the melt and the solid phase has been carefully studied in transparent material under well-controlled conditions. The results of the experiments were as follows.

The slag particles remain at rest until they are reached by the solidification front. At moderate solidification rates the particles are pushed ahead of the solidification front and follow it. It was also found that at a certain critical solidification rate, the particles become trapped by the solidification front and incorporated into the solid phase.

The critical solidification rate depends on the particle radius. The surface tension is also very important. It can be shown that a condition for the particles to be pushed ahead of the solidification front and not to be absorbed by the solid phase is:

$$\sigma_{sp} > \sigma_{sL} + \sigma_{Lp} \qquad (9.54)$$

where

σ_{sp} = surface energy/m^2 between the solid phase and the particle

σ_{sL} = surface energy/m^2 between the solid phase and the melt

σ_{Lp} = surface energy/m^2 between the melt and the particle.

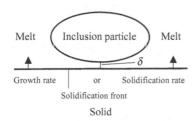

Figure 9.47 The distance δ is small and capillary forces act on the particle. A thin layer of melt is sucked into the region between the particle and the solidification front.

If the condition in Equation (9.54) is fulfilled, the surface tension forces cause a thin film of melt to be continuously supplied at the region between the particle and the solid phase. This process is in its turn controlled, not only by the surface tension condition and the solidification rate but also by the *viscosity of the melt*, the *self-diffusion constant*, and the *shape and smoothness of the particle*.

Similar investigations of slag precipitation have been performed during the solidification process in steel. It was found that molten iron oxide/sulfide precipitations immediately are incorporated into the solid phase, whereas silicates at low solidification rates are pushed ahead of the solidification front and accumulated in the very last solidified parts of the interdendritic regions.

The fact that the precipitation conditions and the shapes of the particles are so different implies that their distribution in the matrix becomes different.

Precipitated particles, which have been absorbed easily by the solid phase, will be much more disintegrated and more evenly distributed in the matrix than those which are accumulated in the last solidified parts. The latter particles are generally relatively coarse and may in principle increase in size during the precipitation process by coalescence (growth of the large particles at the expense of the small ones) or collision with each other.

9.7.8 Microslag Inclusions of Sulfides in Steel

In the preceding section we discussed the conditions for precipitation of oxide slag inclusions in steel. Correspond-

ing conditions can be set up for sulfides. However, the conditions are more complicated in this case since oxygen is always present and a mixture of oxide and sulfide slag inclusions is obtained. We restrict the discussion below to a qualitative description of precipitation of sulfide slag inclusions.

The most important slag inclusions which contain sulfur consist of MnS. As oxygen normally is available in the melt, a mixture of MnS slag inclusions and oxide inclusions, such as MnO–MnS, are also present.

Before we discuss precipitation of MnS, we describe the precipitation of FeS–FeO. It has been analysed by Flemings and is a good example of slag precipitation during solidification.

Precipitation of FeO–FeS Slag Inclusions

Figure 9.48 (a), (b), and (c) show a series of precipitated slag particles of various compositions and appearances. The dark phase represents FeO and the bright phase FeS. They have precipitated in the same steel sample but at different positions in the specimen.

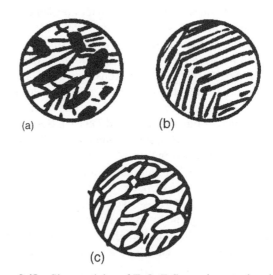

Figure 9.48 Slag particles of FeO–FeS type in a steel melt.

In order to understand the formation process of the three slag inclusion types in Figure 9.48 and to find out when the precipitations have occurred in relation to each other, we use the phase diagram of the ternary system as a basis.

Figure 9.49 shows a projection of the ternary phase diagram of the system Fe–FeO–FeS in analogy with the ternary diagram in Chapter 7 (page 212). The temperature axis is not seen in this representation. Different points in Figure 9.49 do not have the same temperature. The ternary eutectic point E has the lowest temperature. The binary phase diagrams of Fe–O and Fe–S are included in the planes which contain the lines from the Fe corner.

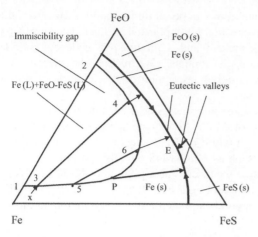

Figure 9.49 Projection of the ternary phase diagram of the system Fe–FeO–FeS on the 'bottom plane' [compare Figure 7.36 (a) and (b) on page 212 in chapter 7]. The point E corresponds to the ternary eutectic point where the three eutectic valleys meet. They emanate from the eutectic points of the three binary phase diagrams of the systems Fe–O, Fe–S, and FeO–FeS.

Figure 9.49 shows a schematic diagram of the liquidus surfaces in the ternary system Fe–FeO–FeS. We start with the binary phase diagram of Fe–O at a certain temperature. The region between the Fe corner and point 1 consists of a one-phase region (melt). The regions between point 2 and the FeO corner are also one-phase regions. The part of the line between points 1 and 2 represents a two-phase region of the two liquids Fe(L) and FeO(L). Such a region is called an *immiscibility gap*. Points 1 and 2 define the composition of the two phases.

The 'parabola part' of the triangle area represents a two-phase region with two melts of varying composition [Fe(L) + FeO–FeS(L)]. This two-phase region intersects the liquidus surface along the intersection curve 1–3–5–P–6–4–2 between the immiscibility gap and the liquidus surface of iron. From each of the eutectic points in the binary diagrams, which are included in the triangle sides, three eutectic valleys run downwards. They describe the changing composition of the melt when the temperature decreases. The valleys meet at the ternary eutectic point, marked E in the Figure, which describes the composition of the last solidified melt.

When an alloy melt with initial composition x cools, the following solidification process will happen. On cooling, primary precipitation of δ- or γ-phase in the shape of dendrites occurs. When the dendrites grow, oxygen and sulfur are rapidly enriched in the remaining melt between the dendrite arms and its composition changes. This process may be described in the diagram by the path from point x to point 3. When the composition of the melt corresponds to point 3, droplets of FeO–FeS(L) melt with a composition defined by point 4 are formed. These droplets are rapidly incorporated into the solid phase and become insulated from the remaining interdendritic melt.

The solidification will then continue by precipitation of solid Fe and FeO–FeS(L) by a monotectic reaction. The composition of the remaining matrix melt will then gradually change. This is described as a movement from point 3 to point 5 along the curve. For example, when the composition of the melt corresponds to point 5, droplets of FeO–FeS(L) melt with a composition defined by point 6 are formed. They become incorporated into the solid phase and insulated from the remaining melt, and so on. When the liquid has achieved the composition which corresponds to point P, it leaves the immiscibility gap and only solid Fe is formed.

Each droplet can be regarded as a separate melt. During the continued cooling process the composition of the inclusions will follow the paths indicated by the arrows in Figure 9.49, down to the eutectic valleys and further on to point E.

From analysis of the structure of the samples in Figure 9.48 and the discussion above, we can conclude that slag particles of the type in Figure 9.48 (a) will primarily solidify with a composition which corresponds to points 3 and 4. Those of the type in Figure 9.48 (b) will solidify with a composition which corresponds to points 5 and 6, and those of the type in Figure 9.48 (c) with a composition which corresponds to point P in Figure 9.49.

It is very important to avoid FeS inclusions as they are very harmful.

Precipitation of MnS Slag Inclusions

Manganese has a pronounced tendency to react with sulfur, hence MnS is a very common inclusion. The formation processes of MnS, and the conditions which affect them, have frequently been discussed in the literature. The English metallurgist Sims showed that the properties of cast steel are strongly affected by the morphology (shape) of the manganese sulfide inclusions.

Four types of MnS can be distinguished and are described in Figures 9.50 (a)–(d).

Types II and IV are much more harmful than types I and III. It is especially important to avoid precipitation of type II, which is *very* harmful. Type II has the maximum influence on the mechanical properties of the material and causes strong deterioration of the toughness and rupture limit. Steel with type II inclusions loses much of its impact strength in the transverse direction on rolling. Type II also strongly reduces the cutting ability for so-called automatic cutting steels.

Formation Mechanisms of MnS Slag Inclusions

The different types of MnS inclusions and the conditions which cause their precipitation can be described by the following short survey.

Type I	Type II	Type III	Type IV
Degenerated monotectic reaction	Normal monotectic reaction	Degenerated eutectic reaction	Normal eutectic reaction
	Forms at slow cooling rate or at low nucleation rate		
Forms at rapid cooling rate or at high nucleation rate			

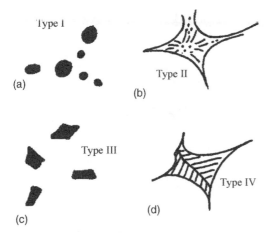

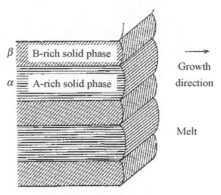

Figure 9.51 Normal eutectic growth.

Below we will discuss the precipitation processes of the four basic types of MnS and methods available to avoid the most harmful ones.

Precipitation of Type I and II Inclusions

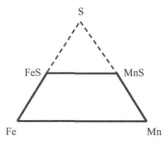

Figure 9.52 Phase diagram of the four-component system Fe–Mn–MnS–FeS.

Figure 9.50 (a) The inclusions are spherical and distributed at random in the matrix. Their sizes vary within very wide limits. (b) Small, apparently spherical inclusions arranged like a string of pearls in the matrix. The string may partly be prolonged and appear as rods. (c) Large edged massive facetted inclusions, which generally are distributed at random in the matrix. (d) Eutecticum of plates, which often form a certain angle with each other.

Figure 9.53 shows the Fe corner of a projection on the bottom plane of the ternary phase diagram of the system Fe–Mn–MnS–FeS (Figure 9.52), which can be regarded

A *eutectic reaction* is a solidification process in which *a liquid* L *solidifies to two solid phases* α *and* β:

$$L \rightarrow \alpha + \beta$$

A *monotectic reaction* is a process in which *a melt* L_1 *gives a solid phase* α *and another liquid phase* L_2:

$$L_1 \rightarrow \alpha + L_2$$

In normal eutectic or monotectic growth, the two phases grow at the same growth rate (Chapter 6, pages 146–147). The phases are said to cooperate (Figure 9.51). This is said to be a *normal eutectic or normal monotectic reaction*.

If the two solid phases do not cooperate on solidification they precipitate with different growth rates and grow independently of each other. This is called a *degenerated eutectic or monotectic reaction*.

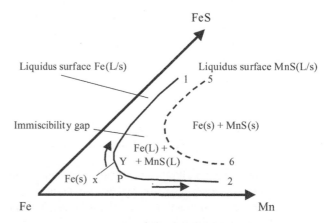

Figure 9.53 The Fe corner of the ternary phase diagram of the system Fe–Mn–MnS–FeS. The temperature axis of the phase diagram is not shown. The temperature decreases along the path x–Y owing to cooling. The primary δ-phase is gradually transformed into γ-phase. For this reason, the solid Fe phase is denoted Fe(s).

as part of the ternary system Fe–Mn–S (compare the ternary phase diagram on page 212 in Chapter 7).

The phase diagram in Figure 9.53 shows that the two-phase region [Fe(L) + MnS(L)] intersects the liquidus surface Fe(L/s) along the bent curve 1–P–2. The two-phase region consists of two liquid phases and is an immiscibility gap, in analogy with FeS (Figure 9.49).

In addition, the phase diagram shows the intersection between the two-phase region and the liquidus surface MnS(L/s). It is a metastable part of the system. The intersection curve is represented by the dashed line in Figure 9.53.

MnS inclusions of type I and II are formed with the aid of a monotectic reaction. This will occur in steels with *low* carbon concentration. In analogy with the discussion of the system Fe–FeO–FeS on pages 293–294, we will discuss the solidification process of an alloy with low S̲, M̲n̲, and C̲ concentrations by use of the ternary phase diagram in Figure 9.53.

Assume that the composition of such a melt is described by point x close to the Fe corner in the projection of the ternary phase diagram in Figure 9.53. When the melt cools, the solidification process starts by a primary precipitation of *ferrite* [Fe(δ)] and a dendrite network is formed. During the precipitation process, sulfur and manganese will be enriched in the remaining melt, owing to microsegregation. The composition of the remaining melt gradually changes along the line x–Y until point Y on the liquidus curve 1–P–2 is reached. At this moment, precipitation of liquid MnS starts in the melt by a *monotectic reaction*:

$$Fe(L) + \underline{Mn} + \underline{S} \rightarrow Fe(\delta) + MnS(L)$$

which occurs along the curve 1–P–2.

The point P corresponds to a temperature maximum along the curve 1–P–2. During the continued solidification process the interdendritic melt will change its composition to either lower or higher Mn content during the monotectic reaction, depending on the composition Y of the melt relative to point P.

If the melt has a composition which corresponds to point Y on the left-hand side of point P it will be enriched in S̲ and depleted in M̲n̲ during the reaction. The opposite is true if the composition Y of the melt is located on the right-hand side of point P. The former case is illustrated in Figure 9.53.

The formation of the two types of MnS inclusions, I and II, in low-carbon steel depends on the *nucleation rate*. Type I (Figure 9.54) is favoured by a *high* nucleation rate and type II by a *low* nucleation rate.

The MnS inclusions of type II are frequently observed as colonies. Figure 9.55 (a) shows such a case, where the colonies of type II have grown spherically from a central point and the MnS melt has been precipitated in the

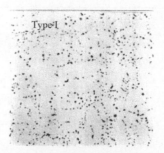

Figure 9.54 Degenerated monotectic precipitation of MnS type I. Reproduced with permission from the Scandanvian Journal of Metallurgy, Blackwell.

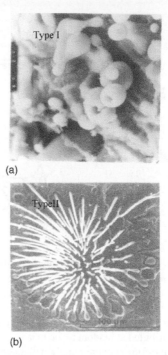

(a)

(b)

Figure 9.55 (a) MnS inclusions of type I. (b) MnS inclusions of type II seen through a scanning electron microscope. Reproduced with permission from the Scandanvian Journal of Metallurgy, Blackwell.

shape of *rods*. The reaction resembles a normal eutectic reaction but is a monotectic reaction.

Type II is the most harmful of all the MnS inclusions and should be avoided as much as possible (page 298).

Risk of FeS Inclusions
If the Mn concentration in the steel is low (point Y is situated on the left-hand side of point P), the composition of the melt becomes more and more enriched in S̲ and finally FeS will precipitate. This process can be prevented by addition of Mn. Then the point Y will be situated on the right-hand side of P (Figure 9.53) and MnS(L) will precipitate instead of FeS.

FeS precipitations often have a film-like appearance and have a very strong and adverse influence on the final properties of the material. They are even worse than MnS inclusions of type II and it is therefore *very* important to avoid FeS inclusions.

Precipitation of Types III and IV Inclusions

Above we noted that MnS inclusions of types I and II are caused by a monotectic reaction in low-carbon steel alloys. Types III and IV are formed in *high-carbon steel alloys* and *cast iron*. Another difference is that the primary precipitation consists of *austenite* instead of ferrite.

The mechanisms for precipitation of type III and IV MnS inclusions are the same as those described for types I and II but the high carbon content influences the ternary phase diagram of the system Fe–Mn–MnS–FeS. The liquidus surface of iron is displaced towards lower temperatures. This stabilizes the metastable eutectic reaction. A sketch of the changed phase diagram is shown in Figure 9.56, which can be assumed to be valid at high carbon contents.

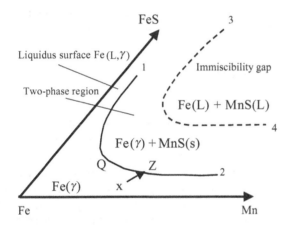

Figure 9.56 Sketch of the modified phase diagram of the system Fe–Mn–MnS–FeS. The curves appear in reverse order compared with Figure 9.53. The change in caused by the high carbon concentration of the alloy shown here.

When a melt of composition x cools, precipitation of γ-dendrites will occur. The $\underline{S}$ and $\underline{Mn}$ concentrations become enriched in the interdendritic melt. When the melt composition reaches the eutectic 1–Q–2 line at point Z, MnS(s) starts to precipitate.

If the nucleation rate is *high*, MnS inclusions normally grow separately by a *degenerated eutectic reaction*, freely floating in the melt. Under these circumstances, the degree of cooperation is small and the growths of the two phases occur independently of each other. The two phases have appearances which indicate that both are primary precipitations. The MnS inclusions will in this case have the shape of sharp-cornered equiaxed crystals of type III. Figure 9.57 (a) shows an example of such type III inclusions.

If the nucleation rate is *slow*, a *normal eutectic reaction will normally occur*. The two phases, MnS(s) and austenite,

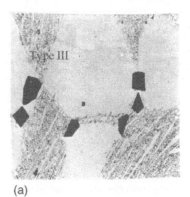

(a)

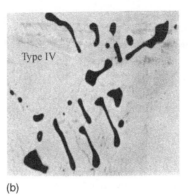

(b)

Figure 9.57 (a) MnS inclusions of type III. **(b)** MnS inclusions of type IV. Reproduced with permission from The Metal Society, Institute of Materials, Minerals & Mining.

cooperate and normal eutectic colonies of type IV are precipitated. Figure 9.57 (b) shows an example.

The precipitation of type III or IV inclusions is influenced by the $\underline{S}$ concentration and also by the cooling rate. This is the reason why melts, which solidify under nearly identical conditions, may give MnS precipitates of completely different appearances.

Transfer from One MnS Type to Another

The type of MnS inclusion depends on many variables. The main factors are:

- the cooling rate, i.e. solidification rate of the melt;
- the concentrations of $\underline{Mn}$ and $\underline{S}$ in the melt;
- the $\underline{C}$ concentration in the melt;
- the solubilities of the alloying elements in molten MnS + Fe;
- the deoxidation process of the melt.

The first three points have been discussed in connection with the precipitation processes on pages 295–297.

The solubilities of the alloying elements are closely related to transfer from one MnS type to another. The influence of the deoxidation process will be discussed below.

---

van't Hoff's rule:

$\Delta T_M = -\text{constant} \times c$

The melting point decrease of a solution (metal melt) is proportional to the concentration of the dissolved substance (alloying element).

As was discussed above for carbon, which is easily dissolved in molten iron but hardly at all in molten MnS, an alloying element, will lower the melting point more for iron than for MnS (follows from van't Hoff's rule above). The result is that the eutectic reaction will dominate over the monotectic reaction and precipitations of types III and IV are favoured.

On the other hand, with an alloying element, for example oxygen, which dissolves more easily in the MnS(L) melt than in the iron melt, the monotectic reaction will dominate over the eutectic reaction. The immiscibility gap will increase in size and the monotectic reaction will approach the Fe corner (Figure 9.53). The result is that MnS precipitations of types I and II are favoured.

Influence of the Deoxidation Process on the MnS Type
Deoxidation has been described earlier (page 282). Common deoxidants are Si and Al. Another method to reduce the $\underline{O}$ concentration is vacuum degassing. The deoxidation means also influence the MnS nucleation.

Inclusions of various sizes are observed, proving that *new inclusions are nucleated during the whole solidification process.*

On deoxidation with Si, spherical inclusions of $(MnSi)O_2$ and MnS of type I are formed. Thus Si has a favourable influence on the MnS precipitation.

Aluminium is a good deoxidant but it does not work effectively enough as a nucleating agent for MnS. On addition of Al as deoxidation agent a normal monotectic precipitation of MnS occurs, i.e. MnS inclusions of the bad type II are formed.

Oxide inclusions have a very decisive influence on the precipitation of MnS inclusions. The nucleation conditions for MnS are very sensitive to the available amount of oxide inclusions. When there are few oxide inclusions it is difficult to nucleate MnS inclusions. Degassing results in a very low $\underline{O}$ concentration of the melt. In the absence of sufficient amounts of a nucleating agent, MnS is precipitated as eutectic or monotectic colonies.

Methods to Avoid Undesirable Types of MnS Inclusions
The previous sections are the basis for suggesting measures to suppress the bad types of MnS inclusions. Suitable measures are:

- use of a high cooling rate;
- desulfurization;

- addition of elements which favour types I and III at the expense of types II and IV.

The use of a high cooling rate has to be a compromise. High cooling rates have disadvantages, for example risk of cracks. The second and third demands can be combined. Addition of substances which are even more inclined than Mn to form stable sulfides reduces the total amount of MnS inclusions, which is favourable. Such additional substances, which result in precipitation of type I and III MnS are chosen, as they cause minimum damage. Thermodynamic data show that such conceivable substances are Zr, Ti, Mg, Ca, and Ce.

Zirconium and titanium can *not* be used in steels with high carbon and nitrogen concentrations because of precipitation of nitrides and carbides, which make the steel brittle.

Cerium for control of the sulfide morphology is added as 'misch metal', which is a mixture of the rare earth metals with cerium as the main component. In addition, cerium works as a good deoxidation agent. The precipitated nonmetallic inclusions thus consist of oxysulfides.

9.8 CAST IRON

Cast iron is an important and special Fe–C alloy, which is widely used because of its good casting ability and other favourable properties in combination with its low price. For this reason, we will discuss it separately and illustrate the balance between different demands and the compromises, which is the everyday task in the casting industry.

Cast iron is a carbon-rich alloy (see page 150 in Chapter 6). It is produced from iron ores in the way described in Section 9.7 (pages 275–276). The subsections of Section 9.7 which treat steel and iron alloys in combination with gases and slag inclusions are also valid for cast iron. Here we will restrict the discussion to problems which concern cast iron specifically.

9.8.1 Melting Processes

The dominant equipment for melting all kinds of metals in the foundry industry is electric furnaces. They are used both for melting and as holder furnaces prior to casting. The most important types of electric furnaces are, in order of importance:

- induction furnaces;
- arc furnaces;
- resistance furnaces.

Induction furnaces are the most frequently used type of furnace and are used for cast iron, steel, and non-furnace metals.

Conventional electric arc furnaces are normally only used for melting and refinement of different steels. So-called vacuum arc furnaces are primarily used for melting of reactive metals, for example titanium.

Resistance furnaces are found in small- and medium-sized foundries. They are primarily used for uncomplicated melting of nonferrous metals.

Induction Furnaces

Induction furnaces have gradually become the most widely used furnaces for melting of iron and nonferrous metals. The advantages of this type of furnace are excellent metallurgical control and relatively pollution-free operation.

There are two kinds of induction furnaces, the coreless and the channel type. Although the two types use similar electrical principles, they differ radically in capacity and operation. The coreless furnace is widely used for melting. The channel-type furnace is better suited for holding and superheating than the coreless furnace.

Cupola Furnaces

The most frequently used type of furnace for the production of cast iron and particularly for melting of grey cast iron is the so-called cupola furnace. It consists basically of a cylindrical shaft furnace, which burns coke intensified by blowing air through tuyères. Alternating layers of metal and coke are charged into the top of the furnace. During its descent, the metal is melted by direct contact with a counter-current flow of hot gases from the coke combustion. The molten metal accumulates at the bottom of the well, from which it is discharged by intermittent tapping or continuous flow.

The cupola furnace can melt a wide range of bulk scrap, for example cast iron scrap, steel scrap, foundry returns, and pig iron (raw iron). It is more difficult to control cupola iron with respect to composition of alloying elements because it is a raw-melting process.

The problem with fluctuating composition can be overcome by combining the cupola furnace with an electric furnace. The latter, which acts as a holding furnace for the molten iron, offers time and opportunity for composition and temperature control and adjustment. Simultaneously, it is a buffer between the melting and casting units.

9.8.2 Production of Nodular Cast Iron

Structure and Distribution of Carbon in Cast Iron

Carbon appears as graphite and cementite (Fe_3C) in eutectic cast iron. In the old days cast iron was characterized by the colour of the fracture surface.

White iron has a reflecting surface and is characterized by precipitation of cementite.

Grey iron has a greyish appearance and its matrix contains precipitated graphite. The shape and appearance of graphite in grey iron vary widely. It is classified in five different types (Chapter 6, page 158). The main types are flake cast iron and nodular cast iron.

The mechanical properties (ductility) of cast iron improved considerably when the production of nodular cast iron started in the middle of the 20th century. New fields of application of cast iron were opened up. Nodular cast iron or spheroidal graphite (SG) consists of spherical nodules, surrounded by austenite.

Production of Nodular Cast Iron. Mg Treatment

Nodular cast iron has excellent mechanical properties, especially ductility. For this reason it is alternatively called *ductile cast iron*.

The production of nodular cast iron is performed under careful metallurgical and metallographic control. The basic material is pure cast iron, free from impurities and with a low phosphorus content. The average composition of nodular cast iron is given in Table 9.10. A vital part of the production process is addition of magnesium to the melt.

TABLE 9.10 Alloying elements in nodular cast iron.

Element	Content (wt-%)
C	3.3–3.8
Si	1.8–2.5
Mn	0.1–0.7
P	≤ 0.1
S	≤ 0.1

Magnesium treatment during the production process gives the nodular cast iron its characteristic spheroidal structure of the graphite inclusions and its good mechanical properties [Figure 9.58 (a)]. Magnesium boils at about 1100 °C and cannot be added to an iron melt ($T_M \approx$ 1200 °C) owing to the risk of explosion. For safe and effective addition of Mg to the cast iron melt, Mg is often alloyed with Ni or Si.

After the Mg treatment, the magnesium fades away as a result of oxidation and evaporation. It is necessary to cast the melt immediately and within a limited total casting time after the Mg treatment, otherwise the nodular structure partly disappears and the product will not be acceptable [Figure 9.58 (b)]. Figure 9.59 shows some decay curves for the determination of the maximum time interval between Mg treatment and casting as a function of various initial Mg concentrations.

Cerium can be used as an alternative to Mg to obtain nodular cast iron. This is done by addition of either NiCe or SiCe to the melt. The cheapest alternative is SiCe addition, but there are several practical problems, for example heat generation, rapid reaction, and flotation. The tolerance limits of Ce must be kept within a narrow interval because

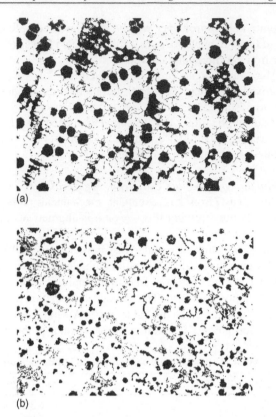

(a)

(b)

Figure 9.58 (a) Nodular cast iron with high nodular density – a very useful product. (b) Nodular cast iron with low nodular density – an unacceptable product.

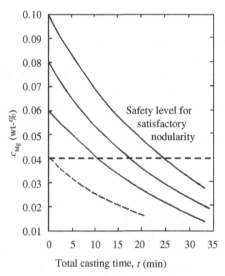

Total casting time, t (min)

Figure 9.59 Decay curves of Mg. Reproduced with permission from the Swedish Foundry Association.

otherwise Ce may react with carbon and form carbide. It also reacts easily with both oxygen and sulfur. NiCe is often used and is easier to handle than SiCe.

9.8.3 Removal of Gases and Sulfur. Metallurgical Treatments of Cast Iron

Removal of Gases
$\underline{H}$, $\underline{N}$, $\underline{O}$ and $\underline{CO}$ have to be removed from cast iron melts for the same reasons and by the same means as for steel (Sections 9.7.1–9.7.3).

Oxygen Control
Oxygen in cast iron follows the same relationships as in steel but the $\underline{C}$ and also $\underline{Si}$ contents are normally much larger in cast iron than in steel. Figure 9.60 shows the equilibrium relationship between $\underline{Si}$ and $\underline{O}$ in a cast iron melt at different temperatures and also the equilibrium condition for $\underline{C}$ and $\underline{O}$. The latter is influenced very little by temperature fluctuations. For this reason, only one curve is given.

Figure 9.60 shows that SiO_2 is the stable oxide at temperatures below 1400–1450 °C. At higher temperatures formation of CO will occur. However, owing to the very low $\underline{O}$ concentration in the melt no precipitation of CO bubbles will occur.

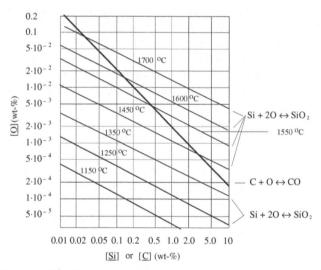

Figure 9.60 Equilibrium relationship between $[\underline{Si}]$ and $[\underline{O}]$ and between $[\underline{C}]$ and $[\underline{O}]$ respectively, in a cast iron melt. Reproduced with permission from the American Society for Metals (A.S.M.)

Al and Mg are stronger deoxidation agents than Si and C and will therefore determine the equilibrium concentration of $\underline{O}$ if they are added to the melt.

Reduction of Sulfur
The main measure is to decrease the sulfur content as much as possible. A low sulfur content must be achieved otherwise the shape of the graphite will be affected and no ductile iron or compacted graphite iron can be formed.

Treatment with Mg (MgNi) or Ce (SiCe), which are used to obtain a nodular structure of cast iron, is simultaneously an effective way to reduce the sulfur concentration in cast iron.

Sometimes more complex combinations of Mg, Ce, Ca, Ti, Si, and Al are used. Alternatively, CaO can be used for desulfurization. It reacts chemically with the sulfur in the melt:

$$CaO + \underline{S} \rightleftharpoons CaS + \underline{O}$$

The oxygen content is reduced in the melt by addition of Si, which gives SiO_2 (see above).

Inoculation Process
In order to decrease the risk of white cast iron formation, cast iron is normally inoculated. This is of the utmost importance for the production of a high-quality ductile cast iron. Various types of inoculants are supplied to the melt in an effective and controlled way. The addition of inoculants can be done in two ways, either early or as late as possible in the casting process.

Early Inoculation
In early inoculation, the inoculant is added into the ladle. Three alternative ways are used:

1. Granular material is poured into the melt in the ladle.
2. Air-blown small particles are injected into the melt in the ladle.
3. A wire of the inoculant is fed into the melt in the ladle.

The wire method is more effective than the air-assisted injection method as there is little loss of inoculant material. The early method has the disadvantage that the inoculant fades away and only a small fraction may remain when it reaches the mould.

Late Inoculation
Late inoculation is favourable and very efficient when fading is a problem, for example in alloys treated with magnesium. Three alternative methods are used:

- in-stream inoculation of the melt on its way into the mould;
- wire inoculation of the melt on its way into the mould;
- in-mould inoculation.

In-mould inoculation means that the inoculant is placed at some suitable point in the gating system, for example at the base of the down sprue or in the runner bar. Two examples are given in Figure 9.61. The metal stream passes around or over the small inoculant chamber. In this way the inoculant is dissolved gradually, which results in an even and continuous supply of the inoculant.

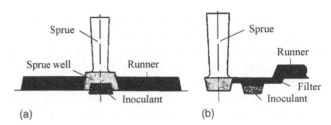

Figure 9.61 Two locations of the inoculation chamber during in-mould inoculation: (a) below the down sprue and (b) in the runner, between $\underline{Si}$ and $\underline{O}$ and between $\underline{C}$ and $\underline{O}$ in a cast iron melt. Reproduced with permission from Elsevier Science.

9.9 NICKEL AND NICKEL-BASE ALLOYS

Ni-base alloys constitute a group of special alloys. In contrast to cast iron, which is cheap and very common, the nickel-base alloys are expensive and used only for special purposes. These alloys are widely used in highly corrosive environments and for high-temperature applications, for example aircraft engines and gas turbine systems.

9.9.1 Demands on Nickel-base Alloys

The demands on engine materials which will be exposed to high temperatures and corrosive environments are extremely high. The demands are also very high on the production procedure of such materials, especially concerning the consistency and reproducibility of the process. The demands on Ni-base alloys are as follows:

- low content of gases;
- low content of slag inclusions;
- resistance to corrosion;
- resistance to high temperatures.

The last two demands are fulfilled by the composition of the alloy. The first two demands depend on the production process and will be discussed below.

9.9.2 Removal of Gases from Nickel-base Alloys

The high demands on Ni-base alloys require a different technique to that normally used in metal production and casting.

In most cases the metal melts react with the gases in the air and with water vapour and the lining of the ladle and other containers. This type of chemical reaction is prevented simply by working in vacuum or in an atmosphere of inert gases.

The only undesired gases which are present are the gases dissolved in the metal. These are removed by keeping the melt in vacuum. The dissolved gas atoms diffuse to the surface as the partial pressure of the gas at the surface of the melt is almost zero and therefore the concentration of the gas in the melt, close to the surface, will also be almost zero according to Sievert's law. The gas atoms diffuse through the melt to the surface, leave the melt, and are pumped away continuously. In this way, the concentration of the dissolved gas in the melt is strongly reduced and can be kept very low.

9.9.3 Removal of Slag Inclusions from Nickel-base Alloys

It is important to avoid slag inclusions as much as possible. For this reason, scrap is often used together with virgin material as raw material. The virgin material consists of chemically pure metals, which are produced either electrolytically or by special chemical separation processes.

The control of the scrap is rigorous. Scrap which contains traces of elements such as P, Pb, Sn, Sb, and As is not accepted. Sulfur can be removed by slag treatment and therefore it is not necessary to abandon sulfur-bearing materials.

If the alloy contains more than 0.2 % of the metals Al, Ti, and Zr, it is not suitable for melting and casting in air. At higher alloying concentrations, these elements readily oxidize, which results in large inclusions, oxide lumps, and poor composition control. If an inert atmosphere or vacuum is used, the metals are shielded from oxygen and no slag inclusions can be formed.

9.9.4 Melting and Refinement of Nickel-base Alloys

Processing

Ni-base alloys are affected at *all* stages of the manufacturing, casting, and refining processes. It is therefore necessary to use vacuum or an inert atmosphere during all the processes. Many types of electrical arc processes have been developed and various techniques are used. Vacuum-induction melting is of major commercial importance. Vacuum refining removes volatile metallic impurities such as Bi and Pb.

The utmost care is necessary to avoid contamination of the Ni-base alloy at casting. Cleaning of the charge, melting, and mould chambers is vital. Removal of oxides from the bottom and walls of the crucible is also important. The equipment has to be carefully examined before use in order to avoid opening the furnace during operation. Great care must be taken in charging, especially handling large pieces of scrap in order to avoid cracking or damage to the crucible refractory material.

Pure magnesium oxide is favoured for most Ni-base alloy melting, but aluminium oxide is also used. Magnesia runners and sprouts have to be preheated in order to avoid thermal shock during casting.

Refining

Deoxidation during vacuum melting does not achieve complete elimination of dissolved gases such as $\underline{H}$, $\underline{O}$, and $\underline{N}$. Hydrogen diffuses easily through metal melts and low hydrogen contents of the order of 5 ppm can be achieved. It is more difficult to remove oxygen and nitrogen. The oxygen levels can be reduced to and kept at 20–30 ppm.

Nitrogen is of critical importance in Ni-base alloy melting because the alloys invariably contain Ti and V, which form nitrides and are a disaster for good mechanical properties of Ni-base alloys. It is difficult to remove nitrogen. One is fortunate if nitrogen residuals as low as 40–50 ppm can be achieved. There is only one radical solution to the nitrogen problem, namely to maintain the lowest possible nitrogen content in the charge material as is economically possible.

Removal of sulfur during vacuum is a problem. Sulfur is another substance which can best be handled by keeping it to a minimum in the original charge. High-sulfur charge materials can be lowered in sulfur content by prerefining the material with lime in a separate electrical furnace.

Methods for refining are vacuum-arc remelting and ESR refining. Process advantages of vacuum-arc remelting are low gas content, no melt-crucible reaction, very low inclusion content and rigid control of solidification structure with low macro- and microsegregations. The disadvantages are expensive high-quality electrodes, relatively low productivity and very high capital costs for the complex equipment.

During the ESR process, droplets of molten metal from the electrode fall through a slag bath and are refined. The slag bath protects the refined molten metal at the bottom from contamination with the surrounding air and provides a cleaning and refining action on the droplets. This process is considered to be more effective than any other process of desulfurisation. Ni-base alloys with high Ti and Al contents present a special problem in electroslag melting since both Ti and Al tend to be lost into the slag during the process.

SUMMARY

■ Solubility of Gases in Metals

See Figure 9.3.

Gas is in most cases more soluble in a metal melt than in the solid phase. When a metal melt solidifies, this can result in gas precipitation in the interdendritic areas.

If heat has to be added in order to dissolve the gas in the metal melt (*endothermic reaction*), the solubility curves have a *positive* slope and gas precipitation may occur. This is true for most of the common metals, for example Fe, Al, Cu, and Mg.

If the solubility process is *exothermic*, the solubility curves have *negative* slopes and gas precipitation does not occur.

Pores

Pores reduce the quality of castings and ingots. Efforts are made by various methods to prevent the formation of gas pores.

Gas Concentrations

The gas concentration in the melt and the *partial* pressure of the gas in the surrounding atmosphere are in equilibrium with each other:

$$[G] = \text{constant} \times \sqrt{p_{G_2}} \qquad \text{Sievert's law}$$

Even if the gas concentration in the melt initially is low and far from the solubility limit, gas precipitation may occur when the melt has reached a certain solidification fraction.

The relation between gas concentration in the melt, the solidification fraction and the partition coefficient is given by:

$$x^L = \frac{x_0^L}{1 - f(1 - k_{\text{part} \, x})} \qquad c^L = \frac{c_0^L}{1 - f(1 - k_{\text{part} \, c})}$$

■ Choice of Concentration Units in Fick's First Law

Case I: Weight Percent or ppm

Dimensional analysis of Fick's law in the usual SI units:

$$\frac{dm}{dt} = -DA\frac{dc}{dy}$$

$$\left[\frac{\text{kg}}{s}\right] = \left[\frac{\text{m}^2}{s} \times \text{m}^2\right] \times \frac{[dc]}{[\text{m}]} \quad \rightarrow \quad [dc] = \left[\frac{\text{kg}}{\text{m}^3}\right]$$

The concentration dc in Fick's law has the dimensions mass/unit volume.

One *must not, without anything further, insert values of dc expressed in units of weight percent or ppm, which both have the dimension zero.*

With the aid of a special factor, the given concentrations in weight percent or ppm must be transformed into a unit, which suits Fick's law

In case I the following relation is valid:

$$c = w\rho$$

where
w = weight fraction
ρ = density of the metal (kg/m^3).

The weight fraction w can be calculated if the concentration in wt-% or ppm is known:

$$w = (w \times 100)\text{wt-\%} = (w \times 10^6)\,\text{ppm}$$
$$\Rightarrow 1\text{ppm} = 10^{-4}\text{wt-\%}$$

Case II: Mole Fraction or Atomic Percent

If the mass is given in kmol, a dimensional analysis of Fick's law in the same way as in case I gives:

$$\frac{dm}{dt} = -DA\frac{dx_A}{dy}$$

$$\left[\frac{\text{kmol}}{s}\right] = \left[\frac{\text{m}^2}{s} \times \text{m}^2\right] \times \frac{[dx_A]}{[\text{m}]} \qquad [dx_A] = \left[\frac{\text{kmol}}{\text{m}^3}\right]$$

The concentration dx_A in Fick's law above has the dimensions kmol/unit volume.

One *must not, without anything further, insert values of dx_A expressed in units of mole fraction or atomic percent, which both have the dimension zero.*

In case II the following relation is valid:

$$c = \frac{x_A}{V_m}$$

where
V_m = the molar volume, i.e. the volume in m^3 of 1 kmol of the metal.

Here c can be calculated if the mole fraction x_A or the atomic percent is known:

$$x_A = \text{mole fraction} = \frac{\text{at-\%}}{100}$$

■ Sources of Gas Precipitation

The most important sources of gas precipitation are ranked in three groups, 1, 2, 3, in order of difficulty: group 1 offers the worst problems. The table shows that oxygen is involved in all gases except $\underline{H}$ and $\underline{N}$.

Gas	$\underline{H}$	$\underline{O}$	$\underline{N}$	$\underline{H_2O}$	$\underline{CO}$	$\underline{SO_2}$
Metal/alloy:						
Endothermic reaction:						
Aluminium	1					
Copper	1			1	3	2
Iron and low-$\underline{C}$ steel	2				1	
Steel alloys	1		2		1	
Cast iron	1		2			
Nickel alloys	1	1	1			
Exothermic reaction:						
Titanium	No gas precipitation					
Zirconium	No gas precipitation					

■ Slag Inclusions

Slag inclusions are chemical compounds of a metal and nonmetal components or only nonmetal components. The inclusions, for example metal oxides, nitrides, and sulfides, are difficult to dissolve in the metal melt. Precipitation occurs when the low solubility product in the metal melt is exceeded:

$$[\underline{A}]^a \times [\underline{B}]^b = K_{A_aB_b}$$

which corresponds to the reaction $\quad a\underline{A} + b\underline{B} \rightleftharpoons A_aB_b$

Macroslag Inclusions

Macroslag inclusions: visibility limit > 0.1 mm.

Macroslag inclusions are formed in the melt before the crystallization process. They consist of reaction products with lining material and deoxidation products. Small particles are formed, which sinter together into long-chained *clusters*. The process is rapid for SiO_2 and slow for Al_2O_3.

Microslag Inclusions

Microslag inclusions: visibility limit < 0.1 mm.

Microslag inclusions are often formed during the solidification process in the interdendritic regions owing to microsegregation.

■ Methods to Remove or Reduce Macroslag Inclusions

Slag refining using ESR occurs mainly at the electrode tip but also during the passage of the droplet through the molten slag layer.

ESR remelting reduces the sizes of the slag inclusions and results in strongly improved mechanical properties of the material.

Filtering.

■ Methods to Avoid Macroslag Formation

- Suitable material for lining of furnace, ladle, tundish, and mould.
- Shielding of the melt from the oxygen in air by means of shielded casting in an N_2 or Ar atmosphere.
- Choice of optimal casting temperature. Higher temperature gives less slag inclusions.
- Use of a so-called hot top to insulate the upper surface during ingot casting.
- Optimal design of the casting machine during continuous casting.

■ Inclusions in Metals and Alloys

The sources of inclusions in alloys are gases and solid slag, precipitated during the solidification process. Solid slag consists of compounds of a metal and one or several nonmetal components or only nonmetal elements.

They are not ranked according to difficulty because the inclusion problems vary with the composition of the alloy and the phase diagrams of the metal systems involved.

Depending on the composition of the alloy, precipitations may appear as either gas pores or solid slag inclusions. Hence there is no reason to keep a rigid distinction between gas and slag precipitations.

Inclusion	Oxide	Nitride	Sulfide	Others
Metal/alloy:				
Aluminium	Al_2O_3			Al_3Fe
Copper	Cu_2O		Cu_2S	
Steel alloys	Al_2O_3, SiO_2		MnS	
Cast iron			MnS	Fe_3C, Al_3Fe

Aluminium and Aluminium-base Alloys

Metal or Alloy	Reacting agent	Precipitated gas or slag inclusions	Methods to prevent or remove the inclusions
Al and Al-base alloys	H_2O	H_2	Flushing with Ar or N_2 (eventually with Cl_2)
	O_2	Al_2O_3	Filtering or refining with the aid of NaF or NaCl melts
	Fe	Fe-Al compounds	Electrolysis

Copper and Copper-base Alloys

Metal or alloy	Reacting agent	Precipitated gas or slag inclusions	Methods to prevent or remove the inclusions
Cu and Cu-base alloys	H_2O	H_2, H_2O	Flushing with N_2 followed by deoxidation in the same way as below
	O_2	Cu_2O	Deoxidation with C, P, Li, or Ca
	S	SO_2, Cu_2S	Deoxidation in the same way as above

Iron and Iron-base Alloys

Metal or alloy	Reacting agent	Precipitated gas or slag inclusions	Methods to prevent or remove the inclusions
Iron and steel alloys	H_2	H_2, H_2O	CO 'carbon boil'
	O_2	CO, CO_2, complex Fe oxides	Deoxidation with Si, Mn, Al, Ti, Zr, Ca, Mg or combinations of them
	Lining material	MgO, SiO_2 Al_2O_3	See text below
	N_2	N_2, complex nitrides	CO 'carbon boil' Neutralization with e.g Al or Ti
	S	FeS, complex Fe–O-S compounds	Addition of Mn
		MnS, complex Mn–O–S compounds	Addition of alloying elements which form stable sulfides more easily than Mn. Such elements are Zr and Ti (only if [C] < 0.2 %), Mg, Ca, and Ce (as misch metal). The effect is precipitation of MnS inclusions of types I and III, which cause less damage than types II and IV
Cr-bearing steels		Cr carbides	Limitation of the C̲r and C̲ concentrations
Cast iron	H_2, N_2, O_2	H_2, N_2, O_2, CO_2, nitrides, carbides	The same methods as for steel
	S	SO_2, sulfides	Addition of NiMg or SiCe or a complex mixture of Mg, Ce, Ca, Ti, Si, and Al

■ Macroslag Distribution in Cast House Processes

Macroslag Distribution in Ingot Casting

The main part of the macroslag inclusions consists of oxides. The macroslag inclusions are concentrated partly to the lower part of the ingot and partly to the upper part of the cylindrical region of the ingot.

The oxygen concentration in the ingot follows approximately the presence of macroslag inclusions. Natural convection is of great importance for the distribution of the macroslag inclusions of the ingot.

Gas Precipitation in Ingots

Rimming ingots are good examples of gas precipitation during casting with no protection towards the surrounding atmosphere. Characteristic gas bubbles of mainly CO are formed.

Macroslag Distribution in Continuous Casting

Macroslag inclusions in continuous casting are similar to those in ingot casting. Macroslag inclusions in continuously cast billets occur as both surface slag inclusions and inside material. The distribution of the macroslag inclusions depends on the machine.

Slag inclusions are formed when the melt from the ladle falls into the mould and air bubbles are carried away with the melt. The oxygen in the air reacts with the melt.

■ Precipitation of Microslag Inclusions during the Solidification Process

Below a certain critical solidification rate, which depends on the particle radius, the slag particles are pushed ahead of the solidification front.

Above the critical solidification rate, the particles are trapped by the solidification front and become incorporated into the solid phase.

The surface tension also affects the slag distribution in the material. The condition for the particles to be pushed ahead of the solidification front and not be absorbed by the solid phase is:

$$\sigma_{sp} > \sigma_{sL} + \sigma_{Lp}$$

Microslag Inclusions of Oxides in Steel

The main part of slag inclusions in steel consists of oxides, especially FeO, SiO_2, MnO, and Al_2O_3.

Addition of a means of deoxidation M (a metal or nonmetal) to a steel melt may prevent the precipitation of both

CO gas and FeO. Instead, precipitation of M_xO_y occurs when its solubility product is exceeded:

$$[\underline{M}]^x \times [\underline{O}]^y = K_{M_xO_y}$$

$K_{M_xO_y}$ is temperature dependent.

Microslag Inclusions of Sulfides in Steel

Oxygen is always present in steel melts. Mn is the usual deoxidation agent. In the presence of sulfur it is necessary to consider both the system Fe–O–S and the system Mn–O–S. The most important slag inclusions which contain sulfur consist of MnS.

FeO–FeS Slag Inclusions

Fe–FeO-FeS slag inclusions are formed when the steel melt, free from $\underline{Mn}$, cools and solidifies. The formation process depends on the energy phase diagram. The compositions of the slag particles change gradually at decreasing temperature.

FeS slag inclusions are very harmful and should be avoided. This can be achieved by adding Mn to the melt.

Mn Slag Inclusions

The MnS inclusions are classified as four different types, I, II, III, and IV. Type II has the most negative influence on the mechanical properties of the material. It is especially desirable to avoid types II and IV.

The formation process of MnS inclusions can be described with the aid of the phase diagram for Mn–O–S. It is discussed briefly in this chapter.

Types I and II are common in low-carbon steel alloys. They are caused by precipitation of MnS by a *monotectic* reaction:

$$Fe(L) + \underline{Mn} + \underline{S} \rightarrow Fe(\delta) + MnS(L)$$

Fe(L) solidifies as ferrite, Fe(δ), which later is transformed into austenite, Fe(γ).

Types III and IV are usually formed in high-carbon steel alloys and in cast iron which contains manganese and sulfur. The MnS inclusions are caused by precipitation of MnS by a *eutectic* reaction:

$$Fe(L) + \underline{Mn} + \underline{S} \rightarrow Fe(\gamma) + MnS(s)$$

Fe(L) solidifies as austenite, Fe(γ).

■ Cast Iron

When Mg alloyed with Ni is added to a cast iron melt, nodular or spheroidal cast iron with very good mechanical properties is obtained.

Desulfurization is achieved by addition of MgNi or CaO.

Cast iron contains high concentrations of carbon and silicon, which act as deoxidation agents. SiO_2 is the stable oxide below 1400 °C.

Cast iron melts are inoculated to avoid white solidification.

■ Nickel and Nickel-base Alloys

Ni-base alloys are a special group of alloys, designed for use at high temperatures and/or in strongly corrosive environments.

It is necessary to remove all sorts of impurities from the raw materials and the casting equipment, for example slag inclusions and gases, particularly oxygen. The raw material consists of virgin material and scrap, which have to be as free as possible from impurity metals.

Methods of Refining

Gases are removed with the aid of vacuum degassing.

Methods of slag refining Ni-base alloys are vacuum-arc remelting and ESR refining.

EXERCISES

9.1 In cast metal alloys one can often find three kinds of pores:

1. Interdendritic pores, often called shrinkage pores.
2. Spherical pores, which are either randomly distributed in the material or more often concentrated close to the upper surface.
3. Elongated pores in the columnar crystal zone.

Explain the mechanism of formation of the three types and mention likely gas concentrations in the melt, required for formation of the three types (see the diagram below).

Hint B20

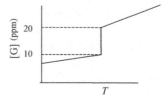

9.2 (a) When castings are produced, bubbles of the type shown in the left figure appear. One explanation for the formation of the bubbles is that hydrogen atoms at the surface of the casting diffuse through the solid/melt interface and are precipitated on the surface of the bubbles.

Which parameters other than the diffusion rate will influence the pore growth? In what way?

Hint B61

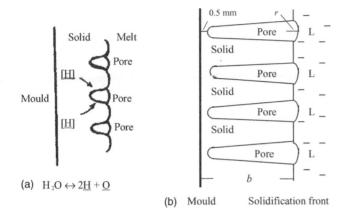

(a) $H_2O \leftrightarrow 2\underline{H} + \underline{O}$

(b) Mould   Solidification front

(b) During casting of steel, it is often observed that the castings contain bubbles of the type shown in the right figure. The explanation for the formation of the bubbles, indicated above and discussed in the literature, is that the hydrogen atoms from the surface of the casting diffuse through the solid surface layer and are precipitated at the surface of the bubbles.

Check the possibility that this is a reasonable explanation by calculation of the growth rate of a pore and give the condition for pore growth.

Hint B102

Assume in your calculations that the hydrogen concentration at the solid surface of the casting is given by the equilibrium concentration at a pressure of 2 atm and that the hydrogen pressure inside the bubbles is 1 atm.

The mole fraction of $\underline{H}$ in the solid phase at equilibrium with hydrogen gas at 1450 °C is described by the relation

$$x_{\underline{H}} = 4 \times 10^{-4} \sqrt{p} \quad (p \text{ in atm})$$

The diffusion constant of $\underline{H}$-atoms in the solid phase is $1 \times 10^{-6}\ \mathrm{m^2/s}$. The molar volume of $\underline{H}$ is $7.5 \times 10^{-3}\ \mathrm{m^3/kmol}$.

9.3 In the production of copper alloys, hydrogen and oxygen are dissolved in the melt. This often results in pore formation when the metal is cast and allowed to solidify. The pores are formed by precipitation of water vapour.

Calculate the maximum hydrogen concentration in a Cu melt with 0.01 wt-% $\underline{O}$, which can be present from the beginning, if formation of water vapour during the solidification process is to be avoided. Cu_2O will precipitate instead of water vapour.

Hint B158

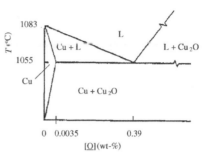

Simplified phase diagram of the system Cu–O at 1 atm O_2.

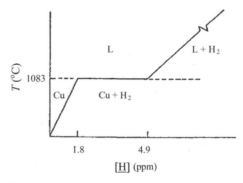

Simplified phase diagram of the system Cu–H at 1 atm H_2.

The solubility product of the reaction $2\underline{H} + \underline{O} \rightleftharpoons H_2O$ is:

$$c_{\underline{H}}^2 c_{\underline{O}} = 2 \times 10^{-11} p_{H_2O}$$

where p_{H_2O} is measured in atm and $c_{\underline{H}}$ and $c_{\underline{O}}$ are expressed in wt-%.

The binary phase diagrams of the systems Cu–O and Cu–H are given above.

9.4 When rimming ingots solidify, the CO gas precipitation ceases when the upper surface freezes. Subsequently the pressure inside the ingot increases during the continued solidification process. As a consequence of the pressure increase, the CO precipitation is replaced by FeO precipitation.

(a) Calculate the CO pressure inside the ingot and the carbon concentration when the FeO precipitation starts in a steel with 0.050 wt-% C and 0.050 wt-% O.

Hint B130

(b). Calculate the pressure (maximum pressure) in the interior of the ingot when the whole ingot has solidified.

Hint B296

The equilibrium $C + O \rightleftharpoons CO$ is described by the relation

$$c_{\underline{C}}^{L} c_{\underline{O}}^{L} = \exp\left(\frac{-2960}{T} - 4.75\right) p_{CO}$$

where the concentrations are expressed in wt-% and the pressure in atm.

The partition constants for carbon and oxygen are 0.20 and 0.054, respectively (page 280). The phase diagrams of systems Fe–C and Fe–C–O–CO are given below.

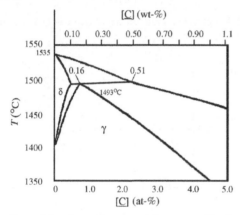

Part of the phase diagram of the system Fe–C.

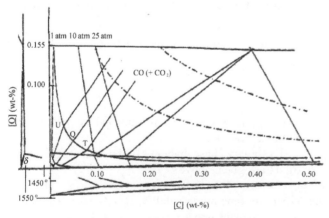

Phase diagram of the system Fe–C–O–CO.

9.5 At a steelworks, steel with 0.050 % C is cast. The gas precipitation during the solidification process of rimming steel is determined by the sum of the partial pressures of the dissolved gases.

Describe graphically how the oxygen and hydrogen concentrations have to vary to ensure that the gas precipitation starts immediately after the casting.

Hint B314

The equilibrium between the concentrations of O and C in the melt and the pressure of CO gas outside the melt is assumed to be described by the relation on page 280:

$$c_{\underline{C}}^{L} c_{\underline{O}}^{L} = 0.0019 p_{CO} \quad \text{(wt-\% and atm).}$$

The solubility of hydrogen in iron at equilibrium with 1 atm hydrogen as a function of temperature is illustrated in Figure 9.23 on page 277.

9.6 A steel melt contains 0.040 wt-% O and 0.030 wt-% C. Manganese is added to the melt in order to prevent CO precipitation. The temperature of the melt is 1500 °C.

Calculate the minimum Mn concentration required to prevent formation of gaseous CO.

Hint B128

The partition coefficients of O and C are 0.054 and 0.20, respectively. The partition coefficient of Mn is 0.67. The solubility product of MnO in the melt can be written as:

$$K_{MnO} = c_{\underline{Mn}}^{L} c_{\underline{O}}^{L} = \exp\left[-\left(\frac{12\,760}{T} - 5.68\right)\right] \text{ (wt-\%)}^2$$

9.7 Sulfur often occurs as an impurity in steel and adversely affects the properties of the steel. In order to reduce this negative effect, substances are added that bond the sulfur as sulfides. The morphologies of these sulfides decide the properties of the material. In most types of steel, manganese is used for formation of manganese sulfides.

There are four different types of manganese sulfides. It is known that the carbon concentration has an influence on the type of MnS morphology which is obtained.

(a) Give an account of the influence of the carbon concentration on the precipitation process.

Hint B303

(b) In some types of steel, small amounts of titanium are added in order to form titanium sulfides instead of manganese sulfides. Titanium can only be used in low-carbon steels. Explain why titanium cannot be used in high-carbon steels.

Hint B255

9.8 Sulfur in steel is often harmful and adversely affects the mechanical properties of the steel. In order to reduce the negative effect of the sulfur, Mn is added and MnS is precipitated during the solidification process.

At what solidification fraction during the solidification process does MnS begin to precipitate in a steel which contains 0.50 wt-% <u>Mn</u> and 0.0020 wt-% <u>S</u>?

Hint B81

The solubility product of MnS is:

$$K_{MnS} = c_{\underline{Mn}}^L c_{\underline{S}}^L = \exp\left[-\left(\frac{34\,200}{T} - 15.4\right)\right](\text{wt-\%})^2$$

The partition coefficient $k_{\text{part}\,\underline{S}}$ of sulfur is 0.010. For manganese $k_{\text{part}\,\underline{Mn}} = 0.67$. The temperature of the melt is 1500 °C.

9.9 Cerium is often used instead of Mg for the production of nodular cast iron. It has been claimed that Ce_2O_3 acts as an inoculation agent for graphite.

Check by calculations if Ce_2O_3 can form by homogeneous nucleation if 0.05 wt-% Ce is added to a cast iron melt where the oxygen concentration is in equilibrium with the silicon concentration.

Hint B277

The <u>Si</u> concentration of the melt is 2.0 wt-%. The solubility product of Ce_2O_3 at an absolute temperature T in the iron melt can be written as:

$$[\underline{Ce}]^2 \times [\underline{O}]^3 = 10^{-\frac{341\,810 + 86T}{4.575T}}(\text{wt-\%})^5$$

The melting point of cast iron is 1150 °C. The equilibrium relationship between the [<u>Si</u>] and [<u>O</u>] concentrations is described in the diagram on page 300. The molar volume of Ce_2O_3 is estimated as 25×10^{-3} m^3/kmol. The surface tension between the melt and the particles is 1.5 J/m^2.

9.10 Stokes' law is very useful when the possibility of catching slag inclusions by flotation during casting operations is discussed.

Stokes' law:

$$F = 6\pi \eta r v$$

is valid for spherical particles with radius r, which move in a viscous medium with viscosity coefficient η under the influence of a force F. The velocity of the particles relative to the melt is v.

(a) Apply Stokes' law to SiO_2 slag particles with radius r in the melt and set up the particle velocity as a function of the radius r. What direction has the particle velocity?

Hint B2

(b) Calculate the velocity of SiO_2 particles with radii 1 and 10 μm, respectively.

Hint B247

The density of the melt is 7.8×10^3 kg/m^3. The density of SiO_2 is 2.2 kg/m^3. The viscosity coefficient of the steel can be written as:

$$\ln \eta = \frac{13\,368}{RT} - 2.08$$

where T is the temperature of the melt (1550 °C) measured in kelvin. The viscosity coefficient η is measured in Pa (N s/m^2).

(c) Draw two curves in a diagram which show the distances L through which the particles in (b) move as functions of time t.

Hint B336

(d) Consider an SiO_2 particle close to the lower surface of the melt at the moment when the melt starts to solidify. What is the condition that the slag particle must fulfil to escape, i.e. to avoid being trapped in the solid?

Hint B150

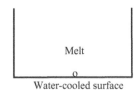

The heat transfer coefficient between the melt and the water-cooled surface is 500 W/m^2 K. The heat of fusion of the steel is 272×10^3 J/kg. Room temperature is 25 °C.

10 Solidification and Cooling Shrinkage of Metals and Alloys

10.1   INTRODUCTION

When a metal melt or a molten alloy solidifies and cools, its volume decreases in most cases. The shrinkage of castings and ingots, owing to this volume decrease, is relatively large and may cause severe problems. The most important of them are:

- deformation of the casting by collapse of the surface shell at the beginning of the solidification process;
- formation of cavities in the interior of the casting;
- pipe formation;
- appearance of casting strain, shrinkage, cracks, and defects in the castings.

The volume decrease can in some cases cause such severe difficulties that it is necessary to give up the use of an alloy even if it has excellent properties in other respects.

In this chapter we discuss the causes and the effects of solidification and cooling shrinkage of metals and alloys and the available methods to eliminate or reduce their negative effects during casting.

10.2   SOLIDIFICATION AND COOLING SHRINKAGE

10.2.1   Origin of Solidification and Cooling Shrinkage

All matter is in a state of incessant motion. The atoms/ions in a crystal or solid phase are exposed to forces from their neighbours and vibrate back and forth around their equilibrium positions. The higher the temperature, the larger will

Materials Processing during Casting   H. Fredriksson and U. Åkerlind
Copyright © 2006 John Wiley & Sons, Ltd.

be the amplitudes of the oscillations and the more space is required for each atom/ion. Hence a piece of metal expands when its temperature increases.

When a metal melts, the strong bonds between the atoms/ions break. Even if strong forces still act between nearest neighbour atoms/ions and a certain short-distance order remains, they can move fairly freely relative to each other. The average distances between the atoms/ions are larger in the melt than in the solid phase. In most cases, melting of the metal thus causes an increase of its volume.

If the temperature of the melt is increased, the kinetic motion of the atoms/ions increases and causes further volume strain. During solidification and cooling, the opposite processes to those described above occur.

10.2.2 Solidification and Cooling Shrinkage of Pure Metals and Alloys

The solidification and cooling of a metal melt after casting occurs in three stages:

- cooling shrinkage of the melt from the casting temperature to the temperature when the solidification starts;
- solidification shrinkage at the transition from melt to solid phase;
- cooling shrinkage at the cooling of the solid phase down to room temperature.

The cooling shrinkage of the melt causes no major problems in most cases. Addition of more melt can compensate for the shrinkage.

Figures 10.1 and 10.2 show how the specific volume, i.e. the volume of 1 kg of the metal, changes as a function of temperature during the cooling and solidification processes of a pure metal and an alloy, respectively.

A *pure* metal has a well-defined melting and solidification temperature, as can be seen in Figure 10.1. On the other hand, an *alloy* solidifies within a *temperature interval* defined by the liquidus and solidus temperatures, as illustrated in Figure 10.2.

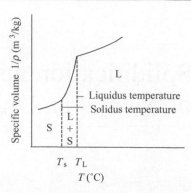

Figure 10.2 The solidification and cooling processes of an alloy.

The difference between these two cases of *solidification shrinkage* is very important and results in completely different solidification processes. These are discussed in Sections 10.2.3 and 10.2.4.

Cooling shrinkage causes undesired volume and shape changes of the casting and mechanical strain in the material. The shrinkage process is discussed in Sections 10.6 and 10.7.

Solidification and cooling shrinkage during casting is extensively discussed in Sections 10.4–10.6. In Section 10.7, thermal stress and crack formation are briefly discussed.

10.2.3 Solidification Shrinkage of a Pure Metal Pipe Formation in an Ingot

A pure metal solidifies along a well-defined solidification front at a rate that is controlled by the heat transport to the surroundings. When a melt is teemed into a mould, it starts to solidify by heat transport away from the melt and out through the mould. The solid phase is nucleated on the cold mould surface and forms a layer, which grows inwards to the melt, perpendicular to the surface.

Solidification shrinkage is in most cases inevitable during ingot casting. One distinguishes between internal cavities or *shrinkage pores* and more open types of solidification shrinkage cavities, so-called *pipes*. It is of great practical importance *where* the cavities are situated in an ingot. There are in principle three different positions (Figure 10.3).

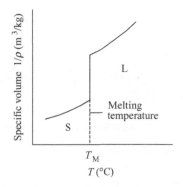

Figure 10.1 The solidification and cooling processes of a pure metal.

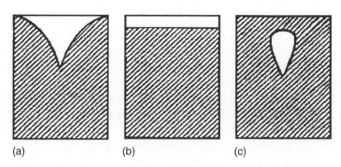

Figure 10.3 Three alternative positions of shrinkage cavities, so-called *pipes,* in an ingot. Reproduced with permission from John Wiley & Sons, Inc.

In reality, combinations between these distinguished cases occur. The positions and shapes of the shrinkage cavities or pipes depend on the composition of the melt and the mode of solidification, especially on the mechanism of heat conduction away from the ingot.

In *case a* the ingot has been cooled from the bottom and the sides or periphery. This case represents a typical ingot solidification. This type of cavity is called a pipe. *Case b* shows an example of ingot solidification from the bottom only. In *case c* the upper surface, the sides and the bottom have been cooled simultanously. In this case we obtain a shrinkage pore or an internal cavity.

The final products must *not* contain internal cavities or any part of a pipe. During the production of castings it is desirable to control the solidification process in such a way that the pipe ends up in an extra container with molten metal, conveniently situated outside the mould. Such a container is called a *feeder* or a *casting head*. It is important to let the melt in the feeder cool more slowly than the last melt in the mould. The method is discussed in detail in Section 10.5.1.

10.2.4 Solidification Shrinkage of an Alloy

An alloy solidifies within a temperature interval defined by the composition of the alloy, i.e. its phase diagram.

Characteristic for the solidification process of alloys, which has been described in Chapter 6, is that there is no well-defined solidification front. Instead, solid phase precipitates anywhere in the melt. Dendrite crystals grow and form a network surrounded by melt in the shape of more or less curved channels between the dendrites.

Initially the dendrites are thin and the channels wide enough to allow transport of melt. The solidification shrinkage can be compensated by melt, which is sucked through channels and fills the pores. The solidification process is generally the same as in a pure metal and may result in pipe formation. However, bridges of dendrites are formed across the pipe by crystal formation. Examples are seen on page 327.

When the solidification process proceeds, the dendrites become gradually thicker and the channels gradually thinner. It becomes more and more difficult to supply melt to the first solidified parts, close to the mould, through the channels. A lack of metal arises, which results in an *internal pore*. The shrinkage in the last solidified parts may result in *shrinkage cavities* in the shape of pores everywhere in the casting or the ingot (Figure 10.4). At the beginning of the solidification process, when the strength of the surface shell is not especially large, the shrinkage may result in a collapse of the whole casting shell, a so-called *external shrinkage cavity*. External and internal pores are called macropores with a common name. The volume decrease of the precipitated crystals may result in microscopic pores between the crystals, *micropores* or intercrystalline pores.

A shrinkage cavity is characterized by walls of a rough, crystal-line appearance. In steel the 'Norway spruce' structure,

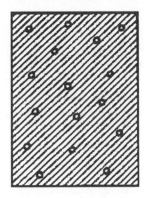

Figure 10.4 Appearance of shrinkage pores in an alloy. Reproduced with permission from John Wiley & Sons, Inc.

which is characteristic for this metal, can be seen. It is easy to distinguish these pores from gas pores, which have smooth surfaces.

Shrinkage cavities and micropores reduce the mechanical strength of a casting and make it weak. The extension of the problem is difficult to judge because the defect in the material cannot be seen on the surface and is impossible to estimate with reference to the linear shrinkage of the metal. X-ray and gamma-radiation examinations are expensive examination methods, which are used only in cases where a very high quality of the casting is required.

Efforts are made to prevent the formation of shrinkage cavities as much as possible, among other things by:

- design with uniform thickness of the casting dimensions;
- use of moulds and cooling bodies;
- control of the casting temperature;
- proper location and design of sprues and feeders;
- optimal composition of the alloy;
- optimal cooling conditions.

The last two measures are the most important ones. The choice of alloy is of great importance. The problems can be mastered if the solidification interval of the alloy is comparably small, which is the case with brass and low-carbon steel. Cast iron also works well because it expands rather than shrinks. Corresponding circumstances are present in the case of 'red metal' (85% Cu, 5% Zn, 5% Sn, 5% Pb) by precipitation of lead at the end of the solidification.

The upper curve in Figure 10.5 shows the total volume of the shrinkage cavities and the pipe volume, which arises when a binary alloy solidifies. Inside area 1, i.e. under the bent curve, the shrinkage appears in the shape of microcavities or pores distributed within the whole casting and the ingot. The remaining volume forms a pipe. Because the composition of the alloy changes during the solidification process (Figure 10.6), both a pipe and distributed shrinkage cavities are obtained.

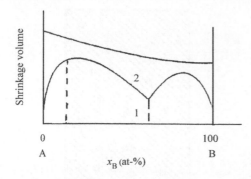

Figure 10.5 The shrinkage volume as a function of the composition of the alloy. Area 1, shrinkage cavities; area 2, pipe. Reproduced with permission from Edward Arnold Hodder Headline.

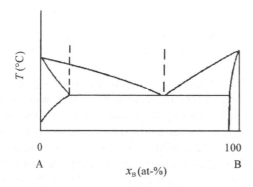

Figure 10.6 Phase diagram of the binary system related to the alloy in Figure 10.5. Reproduced with permission from Edward Arnold Hodder Headline.

10.3 CONCEPTS AND LAWS. METHODS OF MEASUREMENT

10.3.1 Concepts and Laws

To be able to design a feeder or a hot top properly, it is necessary to set up models for the solidification process and perform theoretical calculations of, for example, the volume of the feeder, the depth of the pipe and the height of the hot top. These calculations require the concepts and laws in the box.

10.3.2 Methods of Measurement

Measurement of Small Volume Changes
Determination of densities, volume strain coefficients, and solidification shrinkage of metals presumes careful measurements of small volume changes. It is also necessary to measure the temperature of the sample. Temperature measurements are normally performed with a thermocouple.

The volume determinations are often prevented by the high melting points of the metals. Many direct and indirect methods are used, among others the ones outlined below.

- *Specific volume* = the volume of a mass unit (m^3/kg)
- *Length strain*

$$l = l_0[1 + \alpha_l(T - T_0)]$$

where
$\alpha_l = \Delta l / l_0 \Delta T$ = the length strain coefficient
T = temperature

- *Volume strain*

$$V = V_0[1 + \alpha_V(T - T_0)]$$

where
$\alpha_V = \Delta V / V_0 \Delta T$ = the volume strain coefficient

Approximately we have $\alpha_V = 3\alpha_l$

- *Solidification shrinkage* is defined as:

$$\beta = \frac{\rho_s - \rho_L}{\rho_s} = \frac{V_L - V_s}{V_L}$$

where
ρ_s = density of the solid phase
ρ_L = density of the melt
V_s = volume of the solid phase
V_L = volume of the melt

Chvorinov's rule (Chapter 4, pages 80–81):

$$t = C\left(\frac{V}{A}\right)^2$$

where
A = contact area between mould and melt
V = solidified volume at time t
The constant C can be calculated from the relation:

$$C = \frac{\pi}{4} \frac{\rho_{\text{metal}}^2 (-\Delta H)^2}{(T_i - T_0)^2 k_{\text{mould}} \rho_{\text{mould}} c_p^{\text{mould}}}$$

where
ρ_{metal} = density of the metal
$-\Delta H$ = heat of fusion of the metal
T_1 = temperature of the metal surface close to the sand mould $\approx T_L$
T_0 = room temperature
k_{mould} = thermal conductivity of the mould material
ρ_{mould} = density of the mould material
c_p^{mould} = thermal capacitivity of the mould material.

Determination of Density and Volume Strain Coefficient of the Solid Phase

- Measurements of the lattice constant (distance between adjacent atoms in a crystal) in a metal sample at different temperatures are performed with the aid of X-ray diffraction.

The mass and volume of the sample are also measured. From the measured data, the density of the metal at various temperatures and the volume expansion coefficient can be calculated.

- Direct measurement of length expansion of a metal sample as a function of temperature.

   From the measured data the volume expansion coefficient can be calculated.

Determination of Densities of Melts

- Accurate measurements of melt densities can be performed with the aid of the apparatus shown in Figure 10.7. The basis of the method is the equilibrium between the argon pressure p in the tube of the crucible and the pressure in the melt at the tip of an argon bubble when it just can be released.

   The pressure p is measured for two different h values, which eliminates the surface tension of the melt. The density of the melt at the given temperature can be calculated from the measured quantities.

   The method gives very accurate values of the densities of melts. The errors are estimated to be about 0.05 %. In addition, if the density of the melt is known as a function of temperature, its volume strain coefficient can also be calculated.

- Weighing of a known volume of melt enclosed in an pycnometer of aluminium oxide. The pycnometer is submerged in the melt, filled there, removed and allowed to cool before weighing.

- Weighing of a plumb body, both in air and submerged in the melt.

- Measurement of the pressure required to release a gas bubble from a tube emerging at a certain depth under the free surface of the melt. From measurements at two or several different depths, the density of the melt can be calculated.

Determination of Solidification Shrinkage

- Indirect calculation from the densities of the melt and the solid phase at the melting temperature.

- Measurement of the pressure in a gas with constant external volume in contact with a metal sample before and after solidification close to the melting temperature. When the sample solidifies, its volume decreases and the internal volume of the gas increases and the pressure decreases (Figure 10.8).

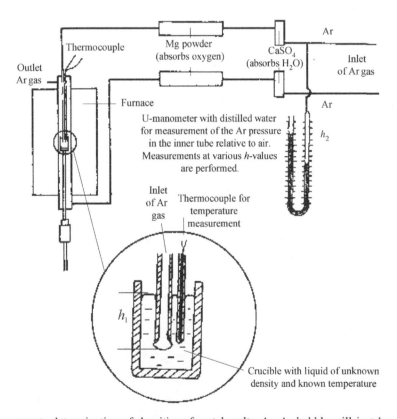

Figure 10.7 Apparatus for accurate determination of densities of metal melts. An Ar bubble will just be released when the condition $p_1 = p_0 + \rho_L g h_1 + 2\sigma/r$ is fulfilled, where p_0 is the pressure at the surface and r is the radius of the tube in the crucible. For the height h_2 we have $p_2 = p_0 + \rho_L g h_2 + 2\sigma/r$. The difference between the equations is $p_2 - p_1 = \rho g(h_2 - h_1)$. The four measured quantities are known and ρ_L can be calculated.

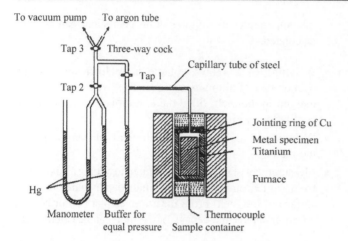

Figure 10.8 Apparatus for determination of solidification shrinkage. A gas is enclosed in the sample container, which is connected to a manometer via a thin steel tube. Tap 2 is opened and the pressure in the two middle manometer legs is adjusted with the aid of tap 3 until the heights of the legs in the buffer manometer become equal (see the figure). The pressure in the right leg of the left manometer is then equal to the pressure of the enclosed gas. It can easily be measured by reading the left manometer.

From the pressure measurements, it is possible to calculate the internal volume increase of the gas, i.e. the volume decrease of the sample, and the solidification shrinkage of the metal. The ranges of application, sources of defects, and accuracies of the various methods cannot be discussed here for space reasons.

Some Results of Measurements

The most examined and best known metals are iron and its alloys and extensive data have been published in the literature. Crystalline transformations in steel, for example from ferrite to austenite, also cause volume changes.

In Table 10.1, examples of data for some different metals are given.

10.4 SOLIDIFICATION AND COOLING SHRINKAGE DURING CASTING

When a casting solidifies, a central cavity is in most cases formed in the casting owing to solidification and cooling shrinkage. It is not desirable that any part of the pipe be situated within the casting. An extra container, a so-called *feeder* or *casting head*, filled with molten metal during the casting can be placed on top of the mould to avoid the situation in Figure 10.9 (b).

10.4.1 Feeder System

If the method with a feeder will work, it is necessary to control the solidification process during casting in such a

TABLE 10.1 **Volume strain and solidification shrinkage of some different metals.**

Metal Alloy	Melting point/melting interval (°C)	Density of the solid phase at 20 °C (kg/m^3)	Density of the solid phase at the melting point (kg/m^3)	Density of the melt at the melting point at (kg/m^3)	Length strain coefficient in the solid phase at 20 °C (K^{-1})	Length strain coefficient in the solid phase at high temperature (K^{-1})	Solidification shrinkage β close to the melting point (%) indicates expansion
Pure iron	1535	7878	7276	7036	12.2×10^{-6}	14.6×10^{-6} (800 °C)	3.3
Cast iron:							
Grey iron	1090–1310	7000–7500			10.6×10^{-6}	14.3×10^{-6} (500 °C)	−1.9*
White iron		7700					4.0–5.5
Low-C steel	1430–1500	7878–7866			11.8×10^{-6}	14.2×10^{-6} (600 °C)	2.5–3.0
High-C steel	1180–1460	7800–7860			12.5×10^{-6}	13.6×10^{-6} (600 °C)	4.0
18-8 stainless steel		7500			15.9×10^{-6}	28.1×10^{-6} (1000 °C)	4
Al	660	2699		2365	23.5×10^{-6}	26.5×10^{-6} (400 °C)	6.6
Al-bronze: 90 % Cu + 10 % Al	1070	7500			18.0×10^{-6} (20–300 °C)		~5
Cu	1083	8940		7930	17.0×10^{-6}	20.3×10^{-6} (1000 °C)	3.8
Brass: 63 % Cu + 37 % Zn	915	8400			20.5×10^{-6} (20–300 °C)		~5

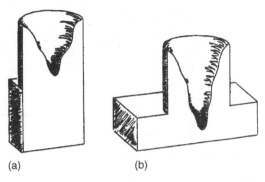

(a)   (b)

Figure 10.9 Cylindrical feeder placed on two different castings, namely (a) a cube and (b) a plate. The plate has the same solidification time as the cube but a larger volume. The volume, shape, and solidification time of the feeder are the same in both cases. Reproduced with permission from John Wiley & Sons, Inc.

way that all the shrinkage pores, formed on solidification, end up in the feeder. Efforts are made to direct the solidification by suitable cooling in such a way that the melt in the feeder solidifies more slowly than the last parts of the casting.

It is necessary to determine the volume of the feeder. It can possibly be easily found by applying Chvorinov's rule (Chapter 4, page 314). By choosing the solidification time of the casting equal to that of the feeder, this equation may allow the calculation of the feeder volume when the volume of the casting is known.

As can be concluded from Figure 10.9, it is not sufficient with such a coarse calculation. In Figure 10.9 (a) and (b), the castings have different shapes and volumes but their solidification times are equal. Chvorinov's rule gives the same value of the feeder in both cases.

Obviously the feeder volume is not large enough in case (b) to compensate for the total solidification shrinkage of the casting. It is larger in case (b) than in case (a) since the volume of the plate is greater than that of the cube.

The volume of the feeder has to be adapted in such a way that two conditions are fulfilled:

- The solidification time of the feeder must be equal to or longer than that of the casting. Then the melt in the feeder compensates for the solidification shrinkage of the casting.
- The volume of the feeder must have a size which at least is large enough to enclose the whole solidification shrinkage inside the feeder.

Hence we have a time condition and a volume condition, both of which must be fulfilled. Both will be treated theoretically below.

Time Condition of the Feeder
At the beginning of the 1950s, the American metallurgist Caine was the first to point out that equal solidification time is not a sufficient condition for efficient functioning

of a feeder. He also pointed out that the pipe at equal solidification time often reaches the casting. He claimed that if the feeder solidifies at exactly the same time as the casting, the feeder volume has to be very much larger than that of the casting. On the other hand, if the feeder solidifies more slowly than the casting, the feeder volume can be chosen only slightly larger than the volume decrease of the casting during the solidification process.

Adam and Taylor fulfilled Caine's ideas theoretically. They derived a relation between the solidification shrinkage and the volumes of the feeder and the casting as a function of their corresponding contact areas with the melt.

Below we will use the following designations:

A_c = contact area between the casting and the mould
A_f = contact area between the feeder and the mould
V_c = volume of the casting
V_f = volume of the feeder
V_{sm} = volume of solidified metal in the feeder
β = solidification shrinkage = $\dfrac{\rho_s - \rho_L}{\rho_s}$.

The amount of solidified metal in the feeder is equal to the feeder volume minus the *total* solidification shrinkage in the feeder *and* the casting:

$$V_{sm} = V_f - \beta(V_c + V_f) \qquad (10.1)$$

Suppose that Chvorinov's rule (page 314) is valid for casting and feeder. The two solidification times are then: $\quad$ Q1

$$t_c = C_c \left(\frac{V_c}{A_c}\right)^2 \qquad (10.2)$$

and

$$t_f = C_f \left(\frac{V_{sm}}{A_f}\right)^2 \qquad (10.3)$$

where C_c and C_f are proportionality constants. The minimum solidification time of the feeder equals the solidification time of the casting. This condition gives the relation:

$$C_c \left(\frac{V_c}{A_c}\right)^2 = C_f \left(\frac{V_{sm}}{A_f}\right)^2 \qquad (10.4)$$

V_{sm}/V_c is solved from Equation (10.4):

$$\frac{V_{sm}}{V_c} = \left(\frac{C_c}{C_f}\right)^{\frac{1}{2}} \frac{A_f}{A_c} \qquad (10.5)$$

Equation (10.1) is divided by V_c, which gives:

$$\frac{V_{sm}}{V_c} = \frac{V_f}{V_c} - \beta\left(1 + \frac{V_f}{V_c}\right) \qquad (10.6)$$

After replacement of terms, Equation (10.6) can be written as:

$$(1 - \beta)\frac{V_f}{V_c} = \frac{V_{sm}}{V_c} + \beta \qquad (10.7)$$

The value of V_{sm}/V_c [Equation (10.5)] is inserted into Equation (10.7):

$$(1 - \beta)\frac{V_f}{V_c} = \left(\frac{C_c}{C_f}\right)^{\frac{1}{2}}\frac{A_f}{A_c} + \beta \qquad (10.8)$$

or

$$V_f = V_c \frac{\left(\frac{C_c}{C_f}\right)^{\frac{1}{2}}\frac{A_f}{A_c} + \beta}{(1 - \beta)} \qquad (10.9)$$

Equation (10.9) is the *general time condition for the feeder* volume. If both the feeder and casting are made of sand, C_c is equal to C_f and $C_c/C_f = 1$, and we obtain a special case of Equation (10.9):

$$V_f = V_c \frac{\frac{A_f}{A_c} + \beta}{(1 - \beta)} \qquad \text{Sand mould} \qquad (10.10)$$

It is possible to calculate the volume of the feeder from Equation (10.9) or (10.10) when all other quantities are known.

Volume Condition of the Feeder

The *efficiency of the feeder* is defined as the ratio ε of the volume of melt in the feeder which has been added to the casting and the volume of the feeder:

$$V_{add} = V_f - V_{sm} \qquad (10.11)$$

and we obtain

$$\varepsilon = \frac{V_f - V_{sm}}{V_f} \qquad (10.12)$$

The melt needed from the feeder is equal to the sum of the solidification shrinkage in the casting *and* the feeder:

$$V_f - V_{sm} = \beta(V_f + V_c) \qquad (10.13)$$

From Equations (10.12) and (10.13), we obtain the volume condition of the feeder:

$$V_f = \frac{\beta V_c}{\varepsilon - \beta} \qquad (10.14)$$

A common cylindrical feeder with a height 1.5 times its diameter (Figure 10.10) has an efficiency of 14 %. If such a

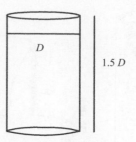

Figure 10.10   Common type of feeder.

feeder is used during aluminium casting, which has a solidification shrinkage of ~7 %, Equation (10.14) gives:

$$V_f = V_c$$

which means that the volume of the feeder must be the same as that of the casting.

For steel with a solidification shrinkage of 3 % and the same shape of feeder and efficiency as above, the value of the feeder volume becomes far more favourable:

$$V_f = 0.27\, V_c$$

The efficiency of the feeder is obviously very important. It is desirable to have the smallest possible feeder volume, especially when casting large products. Equation (10.14) shows that the feeder volume decreases when the efficiency increases.

It is not an efficient approach to raise the temperature of the melt in the feeder in order to reduce its size. By insulation of the feeder, it is possible to prolong its solidification time and increase its efficiency, which is apparent from Figure 10.11.

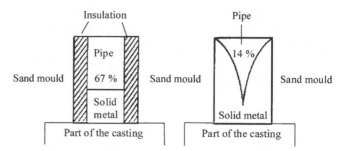

Figure 10.11   Feeder with and without insulation of the envelope surface. The efficiencies ε (%) are given. Reproduced with permission from Elsevier Science.

A feeder will work in a satisfactory way only if both the time and volume conditions are fulfilled.

For *compact castings*, the *time condition* gives a feeder volume that is large enough for the pipe to be completely enclosed within the feeder. For *thin flat castings* (Figure 10.12), the feeder volume is determined by the *volume condition*. Such castings solidify rapidly and much earlier than the melt in the feeder. The time condition is therefore fulfilled

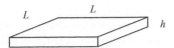

Figure 10.12 'Rule of thumb': if $h > L/7$ and the feeder volume is large enough, the plate solidifies without any shrinkage pores; if $h < L/7$ shrinkage pores are formed inside the plate even if the feeder volume is large enough.

simultaneously in this case. For thicker castings, the time condition has to be considered.

Example 10.1

A cube with side 20 cm will be cast in carbon steel, which solidifies as austenite. To obtain a compact casting, a cubic feeder is used. Owing to heat losses, the temperature of the melt in the casting is lowered 200 °C below the temperature of the feeder, when the casting starts to solidify. How large must the side of the feeder at least be in order to give a compact casting? The solidification shrinkage is 0.050.

Solution:

We assume that the side of the cubic feeder $= x$, and use the following designation:

T_L Liquidus temperature and solidification temperature
T_0 Temperature of the surroundings
ρ Density of the steel melt
$-\Delta H$ Heat of fusion of steel
c_p Thermal capacitivity of steel
T_f Temperature of the feeder
k_{mould} Thermal conductivity of the mould
ρ_{mould} Density of the mould
c_p^{mould} Thermal capacity of the mould.

To solve the problem, we will apply Equation (10.8). For this purpose, we set up expressions for the constants C_f and C_c in Chvorinov's rule for the feeder and the casting, respectively, by use of the equation given in the box on page 314. The cooling heat before solidification has to be included in the calculations.

Feeder constant:

$$C_f = \frac{\pi}{4} \left[\frac{\rho(-\Delta H) + c_p \rho (T_f - T_L)}{T_L - T_0} \right]^2 \frac{1}{k_{mould}\, \rho_{mould}\, c_p^{mould}}$$
(1')

Casting constant:

$$C_c = \frac{\pi}{4} \left[\frac{\rho(-\Delta H)}{T_L - T_0} \right]^2 \frac{1}{k_{mould}\, \rho_{mould}\, c_p^{mould}}$$
(2')

If we form the ratio C_c/C_f the quantities ρ, k_{mould}, ρ_{mould}, c_p^{mould} and $(T_L - T_0)$ disappear on the division and we obtain:

$$\frac{C_c}{C_f} = \left[\frac{-\Delta H}{(-\Delta H) + c_p(T_f - T_L)} \right]^2$$
(3')

This expression is inserted into Equation (10.8) and we have:

$$(1 - \beta)\frac{V_f}{V_c} = \frac{A_f}{A_c}\left(\frac{C_c}{C_f}\right)^{\frac{1}{2}} + \beta = \frac{-\Delta H}{(-\Delta H) + c_p(T_f - T_L)}\frac{A_f}{A_c} + \beta$$
(4')

If we insert given values and table values ($c_p = c_p^{Fe} = 0.45$ J/kg K) into Equation (10.8), we obtain:

$$0.95 \times \frac{x^3}{0.20^3} = \frac{276}{276 + 0.45 \times 200} \cdot \frac{5x^2}{5 \times 0.20^2} + 0.050 \quad (5')$$

One root of the equation is $x \approx 0.173$ m. The side of the cubic feeder ought to be chosen with a good margin.

Answer:

The feeder should be as large as the casting. i.e. the feeder cube should have side 20 cm.

Example 10.2

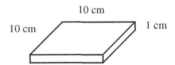

A square plate with side 10 cm and height 1.0 cm is to be cast. What volume should the feeder at least have to ensure that the casting will be compact?

The efficiency of the feeder is 14 % and the solidification shrinkage of steel is 4 %.

Solution:

The height of the casting is small compared with its side. It is reasonable to use the volume condition for calculation of the feeder volume. Equation (10.14) will therefore be used:

$$V_f = \frac{\beta V_c}{\varepsilon - \beta} = \frac{0.04 \times 0.01 \times 0.10^2}{0.14 - 0.04} = 40 \times 10^{-6} \text{ m}^3$$

Answer:

The feeder volume must be at least 40 cm³, which corresponds to 40 % of the volume of the casting.

10.4.2 Influence of Alloying Elements and Mould Material on Pipe Formation – Centreline Feeding Resistance

Influence of Alloying Elements on Pipe Formation

The shape and extension of the pipe are strongly influenced by the concentration of alloying elements and by the mould material. There is a great difference in the appearance of the pipe of a pure metal (Figure 10.13) and that of an alloy with a wide solidification interval (Figure 10.14).

- For a pure metal, the pipe is concentrated to the upper, central parts of the casting.
- For an alloy, the solidification shrinkage is distributed as pores over the whole volume.

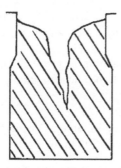

Figure 10.13 Pipe in a pure metal casting. Reproduced with permission from Merton C. Flemings.

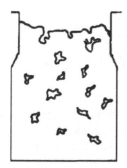

Figure 10.14 Pipe in a casting of an alloy with a wide solidification interval. Reproduced with permission from Merton C. Flemings.

The reason for this basic difference is, as has been mentioned before, that the pure metal solidifies along an even front at a well-defined temperature whereas the alloy solidifies over a more or less wide solidification temperature interval. In the latter case there are a dendrite network and melt simultaneously over the whole volume. As will be seen from Examples 10.3 and 10.4 below, it is more difficult to supply melt as compensation for the solidification shrinkage to regions which are solidifying through such a network of dendrites than in a homogeneous melt without a solid phase.

Fall of Pressure in Feeding Channels

In each more or less viscous liquid, a driving force must exist that overcomes the internal friction in the liquid and forces the liquid through, for example, a tube or a channel. The driving force may be the gravitation force or be caused by an applied external pressure difference between the two ends of the tube. The pressure in the streaming liquid – or, in our case, in the melt – decreases in the direction of its motion. The fall of pressure can be calculated with the aid of the laws of hydromechanics.

Example 10.3

Calculate the fall of pressure in the melt as a function of time t during the solidification and distance x along the casting in the figure shown. The length L, the radius r_0, and the viscosity η of the metal melt are known.

Darcy's law:

$$\frac{dP}{dx} = -\frac{8\eta}{r^2}v$$

describes the fall of pressure P in channels with streaming fluids.

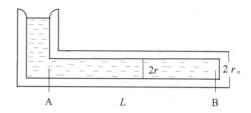

Solution:

When a pure metal melt solidifies in a mould of the shape shown in the figure, solidification and cooling shrinkage occur. The shrinkage cavities are filled with melt, which continuously flows through the open central channel available for the melt. Owing to this flow, a fall of pressure in the melt arises in the direction of the flow.

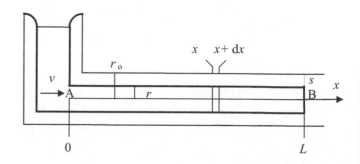

The pressure at point B is lower than that at point A because the melt flows in the channel to compensate for the solidification and cooling shrinkages.

We introduce the following designations:

η = viscosity of the metal melt
r = radius of the channel
v = rate of flow of the melt
s = thickness of the solidifying shell = $r_0 - r$.

The pressure fall P in channels with a streaming fluid can be written as:

$$\frac{dP}{dx} = -\frac{8\eta}{r^2}v \qquad (1')$$

During the time Δt a layer with volume:

$$\Delta V_s = \Delta s \times 2\pi rx \qquad (2')$$

will solidify. Before this layer solidified, it had the liquid volume:

$$\Delta V_L = \frac{\Delta s \times 2\pi rx}{1 - \beta} \qquad (3')$$

Both sides of Equation (3') are divided by Δt and then $\Delta t \to 0$:

$$\frac{dV_L}{dt} = \frac{ds}{dt}\frac{2\pi rx}{1-\beta} \qquad (4')$$

In the same way, we obtain from Equation (2'):

$$\frac{dV_s}{dt} = \frac{ds}{dt}2\pi rx \qquad (5')$$

We introduce the solidification shrinkage $V = V_L - V_s$ and, by subtracting Equation (5') from (4'), we obtain:

$$\frac{dV}{dt} = \frac{dV_L}{dt} - \frac{dV_s}{dt} = \frac{ds}{dt}2\pi rx\left(\frac{1}{1-\beta}-1\right) = \frac{ds}{dt}2\pi rx\frac{\beta}{1-\beta} \qquad (6')$$

The solidification shrinkage per unit time is replaced by new melt, which has the velocity v:

$$\frac{dV}{dt} = v\pi r^2 \qquad (7')$$

Dividing Equations (6') and (7') by πr^2 gives the equality:

$$v = \frac{\beta}{1-\beta}\frac{ds}{dt}\frac{2\pi rx}{\pi r^2} \qquad (8')$$

We apply Chvorinov's rule (page 314):

$$t = C\left(\frac{V}{A}\right)^2 = C\left(\frac{\pi r_0^2 x - \pi r^2 x}{2\pi\frac{r_0 + r}{2}x}\right)^2 = C(r_0 - r)^2 = Cs^2 \qquad (9')$$

where we have chosen an average value of the area A during the time interval 0 to t. C is the constant in Chvorinov's rule.

$$s = \sqrt{\frac{t}{C}} \qquad (10')$$

We differentiate Equation (10') with respect to t:

$$\frac{ds}{dt} = -\frac{dr}{dt} = \frac{1}{2\sqrt{Ct}} \qquad (11')$$

We combine Equations (1'), (8'), and (11') to give:

$$dP = -\frac{8\eta}{r^2}\frac{\beta}{1-\beta}\frac{1}{2\sqrt{Ct}}\frac{2\pi r}{\pi r^2}xdx \qquad (12')$$

If we assume that r is independent of x, we can integrate both sides of Equation (12'):

$$\int_{P_0}^{P_x} dP = P_x - P_0 = -8\eta\frac{\beta}{1-\beta}\frac{1}{r^3\sqrt{Ct}}\int_0^x xdx \qquad (13')$$

Inserting:

$$r = r_0 - s = r_0 - \sqrt{\frac{t}{C}}$$

gives:

$$P_x - P_0 = -4\eta\frac{\beta}{1-\beta}\frac{x^2}{\left(r_0 - \sqrt{\frac{t}{C}}\right)^3\sqrt{Ct}} \qquad (14')$$

Answer:

$$P_0 - P_x = 4\eta\frac{\beta}{1-\beta}\frac{x^2}{\left(r_0 - \sqrt{\frac{t}{C}}\right)^3\sqrt{Ct}}$$

The answer in Example 10.3 is valid for a pure metal. In Example 10.4, the conditions which are present when an alloy with a wide solidification interval solidifies will be illustrated.

Example 10.4
Calculate the fall of pressure as a function of time t during the solidification and the distance x along the casting in the first figure shown here. The length L, radius r_0, and viscosity of the melt are known. The second figure illustrates the solidification process of the alloy.

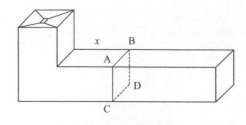

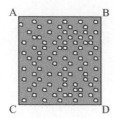

Dark areas are solidified metal. Bright areas represent the cross-sections of the narrow channels, which contain melt.

Solution:

In each single small and narrow channel, a solidification shrinkage per unit time equal to:

$$\frac{dV}{dt} = \frac{-dr}{dt} 2\pi rx \frac{\beta}{1-\beta} \qquad (1')$$

is obtained [compare Equation (6′) in Example 10.3 and use the relation $ds = -dr$], where r is the radius of the channel. The solidification shrinkage will be compensated for by new melt, which has the velocity v and flows through the channel with radius r. In analogy with Example 10.3, we obtain:

$$\frac{dV}{dt} = v\pi r^2 \qquad (2')$$

If we equate the two expressions in Equations (1′) and (2′), we have:

$$v = \frac{\beta}{1-\beta} \frac{2x}{r} \left(\frac{-dr}{dt} \right) \qquad (3')$$

The further reasoning is identical with that in Example 10.3. The desired pressure difference will be the same as in Example 10.3 if r_0 is replaced by $\lambda_{den}/2$.

Answer:

$$P_0 - P_x = 4\eta \frac{\beta}{1-\beta} \frac{x^2}{\left(\frac{\lambda_{den}}{2} - \sqrt{\frac{t}{C}} \right)^3 \sqrt{Ct}}$$

where λ_{den} is the dendrite arm distance and equal to the average distance between the channels in the alloy.

The reason why r_0 equals $\lambda_{den}/2$ is as follows. In Example 10.3, the radius of the channel is equal to r_0 before the solidification has started. The corresponding state in Example 10.4 is when there is no solid phase and the many channels touch each other. Hence their initial radii equal half the average distance between the channels (upper part of Figure 10.15).

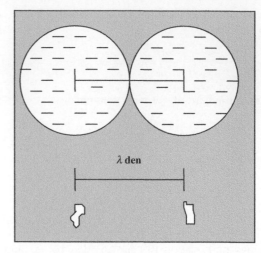

Figure 10.15 Top: appearance at the beginning of the solidification process; bottom: appearance at the end of the solidification process.

When there are many narrow channels instead of a wide one, the fall of pressure will be much larger and $\lambda_{den}/2 \ll r_0$. For this reason, it is much more difficult to transport melt through materials with a wide solidification interval than through a pure metal. The pressure drop will be much larger than for pure metals and the formation of shrinkage pores will thus increase.

Effective Feeding Distance

It can be concluded from Examples 10.3 and 10.4 that the pressure in the channels which supply melt to compensate for the solidification shrinkage decreases with the length of the channel.

The location of the feeder is therefore very important. The feeder can only provide melt to a casting within a certain distance, the so-called *effective feeding distance*. If only one feeder is used and the casting has larger dimensions than the effective feeding distance, a thin pipe appears at the centreline and the product will be wasted.

The shape of each casting can be described approximately as a composition of cubes, plates, and rods. If the effective feeding distances are known for these basic units, it is possible to estimate the effective feeding distance for the product in question.

Cubes offer minor feeding problems but rods must not be too long if the product is to be free from defects. The solidification processes in rods of different lengths and square cross-sections have been examined by careful and extensive experiments. Corresponding examinations have been performed for square plates of various thickness and cross-sections. The results of these examinations are described below.

The general main rule for the location of feeders is:

- It is necessary to control the solidification process in such a way that the melt solidifies later at the position of the feeder than in all other regions.

As an example, we choose the square rod in Figure 10.16. In Chapter 4 we discussed the influence of the geometry of the casting and found that the cooling and solidification rates are more rapid at corners than along plane surfaces. This fact affects the cooling at the end surface in this case. The solidification is therefore more rapid in these regions than in other parts of the casting. Melt is easily transported from the warmer parts to the colder regions to compensate for the solidification shrinkage there.

In Chapter 4, we also found that the cooling and solidification rates are much slower in transition regions than elsewhere. The cooling in such regions, for example that between a feeder and a casting, is therefore much slower than that at the end surface, because heat is conducted into the mould partly from the feeder and partly from the casting. Consequently, the melt flows easily from the feeder to the end regions of the casting and fills the shrinkage pores as long as the channels are wide enough.

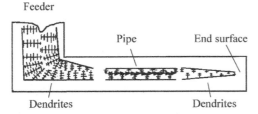

Figure 10.16 Solidifying square rod with feeder. The pipe and parts of the crystal structure have been drawn. Reproduced with permission from Elsevier Science.

It is apparent from the example in Figure 10.16 that it is not difficult to get a compact casting close to the end surfaces and close to the feeder. To obtain a casting free from defects, it is necessary to avoid pipe formation in the intermediate region. A useful concept is the *effective feeding distance*, defined as

D_{max} = the longest total length, measured from the edge of the feeder, that a pipe-free casting can have if pipe formation is to be avoided.

Effective Feeding Distance of a Square Rod
If the rod is longer than the effective feeding distance, a narrow central pipe is obtained. Owing to boundary effects, caused by high temperature gradients at the feeder edge and the end surface of the rod, part of the central line becomes free from defects (Figure 10.16).

For a square rod, cast in a low-carbon steel alloy, it has been found by experiments that the effective feeding distance D_{max}

is between two and four times the side d of the cross-section square:

$$2d \leq D_{max} \leq 4d \qquad (10.15)$$

This condition is illustrated in Figure 10.17.

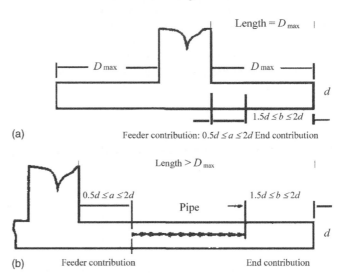

Figure 10.17 Effective feeding distance of a rod with a square cross-section in the case of a single feeder. (a) Rod length = effective feeding distance; (b) rod length > effective feeding distance. Reproduced with permission from Elsevier Science.

To avoid a result such as that in Figure 10.17 (b), two feeders are used if one is not sufficient. Then there is no end surface and the cooling conditions will be different. Each feeder provides in this case a maximum distance of $2d$ with melt. The conditions are illustrated in Figure 10.18.

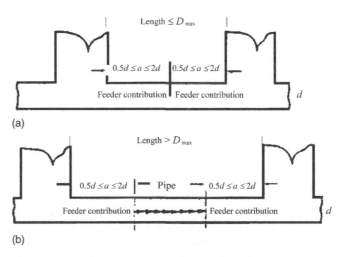

Figure 10.18 Effective feeding distance of a rod with a square cross-section in the case of double feeders. (a) Distance between the feeders = effective feeding distance; (b) distance between the feeders > effective feeding distance. Reproduced with permission from Elsevier Science.

Effective Feeding Distance of a Plate
For a square plate of thickness d, cast in low-carbon steel, it has been found that the temperature gradient at the feeder is responsible for $5d$ and the temperature gradient at the corner edge of the plate for more than $6d$. Hence *one* feeder is sufficient for a plate with a diagonal equal to the width of the feeder plus $(5d + 6d) \times 2 = 22d$. If the feeder is not sufficient, a square-like pipe in the centre of the plate is obtained. With the aid of *four* feeders, conveniently located, plates with a diagonal equal to the widths of two feeders plus $5d \times 2 + (5d + 6d) \times 2 = 32d$ can be cast without trouble (Figure 10.19).

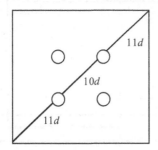

Figure 10.19 Square plate of thickness d with four feeders.

Another way to increase the effective feeding distance is to use special cooling bodies in the way illustrated in Figure 10.20.

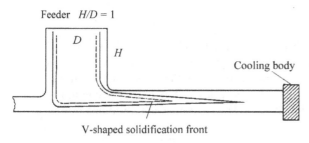

Figure 10.20 Directed cooling is used to increase the effective feeding distances. From K. Strauss, Applied Science in the Casting of Materials, reproduced with kind permission of Foseco International Limited.

Influence of Solidification Shrinkage on the Solubility Limit and Nuclei Formation During Gas Precipitation
In Examples 10.3 and 10.4, we have calculated the pressure fall which arises in the melt during the solidification process due to solidification shrinkage. As a consequence of the reduced pressure in the interdendritic region, the solubility limits of gases will decrease in these regions. The more material that solidifies, the more the pressure and also the solubilities of the gases in the melt will decrease.

As an example to illustrate this effect, we choose the segregation of hydrogen in iron-base alloys. This has been treated earlier in the shape of the solved Example 9.5 on page 278 in Chapter 9 (Figure 10.21).

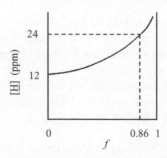

Figure 10.21 Concentration distribution of hydrogen in a steel melt as a function of the degree of solidification without consideration of solidification shrinkage.

The pressure fall in the melt can be calculated from the answer in Example 10.4. This equation is combined with Sievert's law (Chapter 9, page 257) for calculation of the hydrogen concentration in the melt as a function of the degree of solidification and the pressure in the melt. The degree of solidification is obtained by a heat balance equation. The result is described graphically in Figure 10.22.

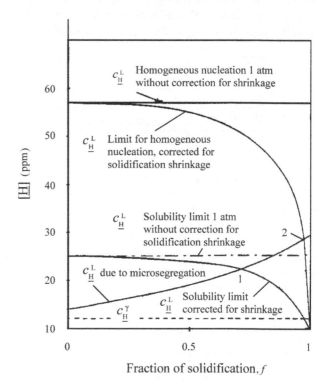

Figure 10.22 The concentration distribution of hydrogen in the same steel melt as in Figure 10.21, but with consideration of the influence of solidification shrinkage on the solubility limit of hydrogen precipitation. The figure shows the influence of the pressure due to the solidification shrinkage on the solubility limit of hydrogen. The possibility of nucleating a pore increases drastically at the end of the solidification process when the critical pressure is strongly reduced. The figure is valid for a given particular solidification rate. Reproduced by permission of ASM International.

If the solidification shrinkage is disregarded (see Figure 10.22), the solubility limit at the pressure 1 atm is constant during the whole solidification process and equal to the equilibrium concentration of 24 ppm (Chapter 9, page 278). As can be seen in Figure 10.22, the solubility limit of hydrogen decreases in reality strongly at the end of the solidification process, owing to solidification shrinkage. The risk of pore formation and precipitation of hydrogen is great.

It is often difficult to nucleate a pore. The supersaturation necessary for homogeneous nucleation of a pore can be calculated with the aid of thermodynamic relations. For hydrogen the concentration for homogeneous nucleation was found to be 57 ppm at 1 atm (upper horizontal line in Figure 10.22). This value decreases with decreasing pressure, which is an effect of the solidification shrinkage.

At point 1 the melt becomes supersaturated with $\underline{H}$ due to hydrogen segregation and pressure drop in the interdendritic regions. It will be difficult, however, to nucleate pores at this low supersaturation. Point 2 illustrates the possibility of nucleating a pore homogeneously at a fraction of solidification $f = 0.96$. As can be seen from Figure 10.22, the nucleation is strongly promoted by the solidification shrinkage.

Influence of Mould Material on the Solidification Process

The *mould material* also influences the solidification process strongly. In a sand mould, a mixture of solid and liquid phases exists in the centre of the casting during a very long period. In a cast iron chill-mould, which conducts heat more rapidly, this period is considerably shorter. These differences influence the appearance of the pipe in the same way as the differences in solidification interval of alloys do.

In a strongly cooled chill-mould [Figure 10.23 (a)], the pipe has the appearance illustrated in Figure 10.13, i.e. the same as in a pure metal. In a sand mould [Figure 10.23 (b)], the pipe has a similar appearance to the pipe in an alloy with a wide solidification interval (Figure 10.14).

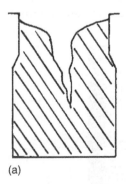 

(a)   (b)

Figure 10.23 (a) Appearance of a pipe on strong cooling of a metal mould during the solidification process. (b) Appearance of a pipe on casting in a sand mould. Reproduced with permission from McGraw-Hill.

Feeding of Alloys with Solidification Intervals – Feeding Resistance

Feeding of melt during casting of alloys with solidification intervals is more difficult than feeding castings of pure metals or eutectic alloys. A quantitative measure, which describes the type of feeding, is required. It is desirable that this measure be independent of the total solidification time. If so, it can be used to compare different mould materials, different mould sizes, and different alloys.

The longer there is a solid phase at the centre of the casting, the more difficult will be the feeding. As a quantitative measure of feeding, the concept of *centreline feeding resistance* (CFR) is often used. It is defined as follows:

$$\text{CFR} = \frac{t_{\text{total}} - t_{\text{initial}}}{t_{\text{total}}} \times 100 \qquad (10.16)$$

where

CFR = measure of the feeding difficulty (%)
t_{total} = total solidification time
t_{initial} = time when the solidification starts at the centreline.

It should be noted that *the factor of solidification at the centreline* in reality is a *measure of the width of the solidification interval*. The essential fact is that CFR represents a *fraction* or a percentage part of the total solidification time and is *not* an absolute time interval.

The centreline is most difficult to feed, hence knowledge of the width of the solidification interval in this part of the casting is especially important. However, we cannot conclude that the presence of small amounts of solid phase at the centreline prevents feeding seriously. The start of the solidification process is just an experimentally suitable reference point on the time scale. It would be even better to use the time at which the melt cannot penetrate the crystal network, but this is more difficult to determine experimentally and in addition is dependent of the gas concentration of the metal.

If the solid phase of a pure metal is precipitated at the centreline for the first time at the end of the solidification time, the time of beginning solidification is approximately equal to the total solidification time, i.e. CFR ≈ 0. The wider the solidification interval of an alloy, the higher will be its CFR.

TABLE 10.2 Central feeding resistance (CFR) of some common alloys, solidified in a sand mould.

Alloy	CFR (%)
18-8 steel (0.2 % C)	35
Cr steel (12 % C)	38
Cu	0
Brass	26
Steel (0.6 % C)	54
Monel	64
Al-bronze	95
92 % Al + 8 % Mg	91
95 % Al + 4.5 % Cu	96

An alloy has a CFR value equal to 100 % if the solidification at the centre starts at $t = 0$. Table 10.2 gives examples of CFR values for some alloys, solidified in a sand mould.

Feeder Volume as a Function of CFR

It is possible to use CFR to calculate the size of a feeder. Instead of comparing the total solidification time t_c of the casting and the total solidification time of the feeder t_f, we will compare t_c with the time of the beginning of solidification at the centre of the feeder t_{bscf}.

We apply the definition of CFR to the feeder:

$$\text{CFR} = \frac{t_f - t_{bscf}}{t_f} \times 100 = \left(1 - \frac{t_{bscf}}{t_f}\right) \times 100 \quad (10.17)$$

or

$$t_{bscf} = t_f \left(1 - \frac{\text{CFR}}{100}\right) \quad (10.18)$$

The value of t_f is taken from Equation (10.3) on page 317 and inserted into Equation (10.18):

$$t_{bscf} = C_f \left(\frac{V_{sm}}{A_f}\right)^2 \left(1 - \frac{\text{CFR}}{100}\right) \quad (10.19)$$

If we introduce the condition that the solidification time of the casting is equal to the time of the beginning of solidification at the centre of the feeder:

$$t_c = t_{bscf} \quad (10.20)$$

we obtain with the aid of Chvorinov's rule:

$$C_c \left(\frac{V_c}{A_c}\right)^2 = C_f \left(\frac{V_{sm}}{A_f}\right)^2 \left(1 - \frac{\text{CFR}}{100}\right) \quad (10.21)$$

If C_f is replaced by $C_f(1 - \text{CFR}/100)$ in Equation (10.4) in page 317, it will be identical with Equation (10.21). We can modify Equation (10.8) in the same way. If we assume that the condition $t_c = t_{bscf}$ is valid, C_f in Equation (10.8) is replaced with $(C_f - \text{CFR}/100)$ and we obtain the relation:

$$(1 - \beta)\frac{V_f}{V_c} = \left[\frac{C_c}{C_f \left(1 - \frac{\text{CFR}}{100}\right)}\right]^{\frac{1}{2}} \frac{A_f}{A_c} + \beta \quad (10.22)$$

This is analogous to Equation (10.21) and can be used for the calculation of the feeder size. Apparently a different result is obtained when we use the condition $t_c = t_{bscf}$ than results when the relation $t_c = t_f$ is applied. The higher CFR is, the larger will be the feeder volume.

CFR increases with decreasing cooling rate and increasing solidification temperature interval. The use of CFR is a great help in understanding the effects of mould materials and alloy composition during casting. CFR is normally determined experimentally with the aid of cooling curves (Chapter 6).

10.5 SOLIDIFICATION SHRINKAGE DURING INGOT CASTING

10.5.1 Pipe Formation in Ingots

When a melt is teemed into a chill-mould, it starts to solidify because heat is transported away from the melt and outwards through the chill-mould. The solid phase nucleates on the cold mould surface and forms a layer, which grows inwards, perpendicular to the surface.

A thin film of solid phase is formed at the bottom and walls of the cold chill-mould, which fixes the external shape of the ingot [Figure 10.24 (a)]. The cooling shrinkage results in a lowering of the free surface of the melt [Figure 10.24 (b)].

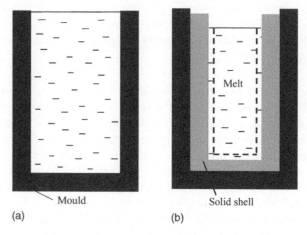

Melt

Mould

Solid shell

(a)   (b)

Figure 10.24 Solidification process on solidification shrinkage. (a) No solidification (b) steel shell + melt.

The continued solidification is illustrated in Figure 10.25. As a result of the proceeding solidification shrinkage, the free surface of the remaining melt in the centre is gradually lowered. When

Figure 10.25 Typical pipe of an unshielded ingot.

the entire ingot has solidified, a funnel-shaped cavity has been formed at the top of the ingot. Such a cavity is called a *pipe*.

The walls in a pipe oxidize and do *not* seal on rolling. The entire part of an ingot that contains a pipe is therefore wasted. In order to achieve a maximum yield, it is desirable to make the pipe volume as small as possible. To reduce the height of the pipe, a so-called *hot top* is used to insulate the upper part of the ingot. Then this part cools more slowly than the rest of the ingot. This method is successful when properly applied, which can be seen in Figure 10.26. The better the hot top is, the smaller will be the pipe volume.

The method is treated in detail on page 331.

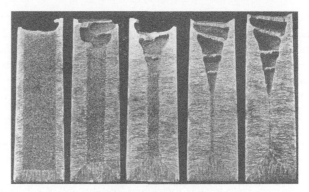

Figure 10.27 Five different stages in the formation process of a pipe.

Figure 10.26 Solidified ingot with a hot top. From K. Strauss, *Applied Science in the Casting of Materials*, reproduced with kind permission of Foseco International Limited.

Pipe Formation in Ingots

At the beginning of the 20th century, the English metallurgist Brearley used a stearine melt in his simulation experiments to study the formation of a pipe at solidification of an ingot.

Molten stearine was cast in an ordinary small chill-mould and was allowed to solidify under specified times of various length. Then the ingots were rapidly decanted (emptied of molten stearine). The stearine ingots were cut lengthways and a series of instant pictures of the formation and development of the pipe was obtained.

Figure 10.27 shows sketches from Brearley's original experiments. The solidification process proceeds by growth of the stearine crystals perpendicularly from the sides towards the centre of the ingot. Simultaneously a shrinkage pore, shaped as an upside-down cone, is formed at the top of the ingot. Later experiments on metal ingots have shown good agreement with Brearley's observations.

It should be noted that the bridges of solid phase appear across the pipe for metals also. When a closed pore is formed between the solid crust and the metal melt, the heat emission from the upper surface is drastically reduced. The upper surface is equivalent to an insulating cover. This model will be used in the theoretical treatment in Sections 10.5.2 and 10.5.3.

Brearley also illustrated the difference in pipe extension in ingots with only equiaxed crystals (Figure 10.29) and ingots

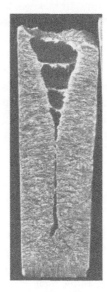

Figure 10.28 The extension of the pipe in an ingot with columnar crystal structure.

Figure 10.29 The appearance of the pipe in an ingot with equiaxed crystal structure.

with a columnar structure (Figure 10.28) with the aid of stearine ingots. A comparison between Figures 10.28 and 10.29 shows that *ingots with an equiaxed crystal structure have smaller pipes than ingots with a columnar crystal structure*. The reason is probably that ingots with equiaxed crystals in addition to the pipe also contain a large number of micropores.

In ingots with a columnar crystal structure, the transport of melt through the dendrite network occurs more and more slowly the thinner the free channels between the dendrite arms are.

The pipe has a similar appearance in alloys with a wide solidification interval. In these cases the solidification front is very irregular. For this reason, one may obtain a deep pipe, which partly consists of dendrites, from which the rest of the melt has been sucked away.

We mentioned earlier that the volume of the pipe can be reduced by insulation of the upper surface of the ingot during the solidification process with a so-called hot top. In this way, the yield of useful material can be increased.

In the following sections, theoretical treatments of the pipe formation with and without a hot top are presented. The advantage of using a hot top is illustrated by comparison of two solved examples.

10.5.2 Theory of Pipe Formation in Ingots with no Hot Top

Brearley's simulation experiments with stearine melts initiated a theoretical model of pipe formation.

The solidification front moves from the outer parts of the ingot towards the centre of the melt. For each layer which solidifies, the upper surface of the melt is lowered by a distance that corresponds to the solidification shrinkage of the solidified layer (Figure 10.30). This model forms the basis of the theoretical calculation of the shape of the pipe.

We start with a chill-mould with the data given below and make the following assumptions:

1. internal width of the chill-mould $= x_0$;

2. internal length of the chill-mould $= y_0$;

3. internal height of the chill-mould $= z_0$.

We neglect the cooling shrinkage and assume further that

4. the chill-mould is filled with melt when the solidification starts;

5. the chill-mould cools the melt uniformly;

6. the melt has a very narrow solidification interval.

We will try to describe the shape of the pipe mathematically. A good method is to regard the solidification as a discontinuous process, where thin layers gradually solidify and their

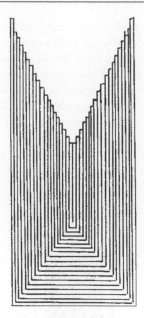

Figure 10.30 The shell model of pipe formation.

solidification shrinkage causes stepwise lowering of the free surface of the melt.

We look at a chill-mould (Figure 10.31) from above in Figure 10.32. When the solidification has proceeded for some time, layers of solid metal with thicknesses Δx and Δy

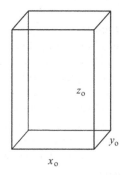

Figure 10.31 Chill-mould with sides x_0, y_0, and z_0.

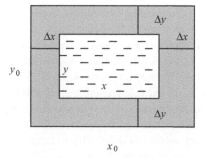

Figure 10.32 Horizontal cross-section of the chill-mould seen from above. Reproduced with permission from the Scandinavian Journal of Metallurgy, Blackwell Publishing.

have been formed in the chill-mould. The region between the outer rectangle x_0y_0 and the inner rectangle xy is filled with solid metal. Inside the rectangle xy there is melt.

Figure 10.33 shows two vertical cross-sections of the chill-mould. The solidification shrinkage of the layers of solidified metal around the inner walls of the mould with thicknesses Δx, Δy, and Δz at the bottom has resulted in a lowering of the upper surface by an amount dZ. The thin layers with thicknesses dx and dz, which will solidify in the next step, are drawn in Figure 10.33(b).

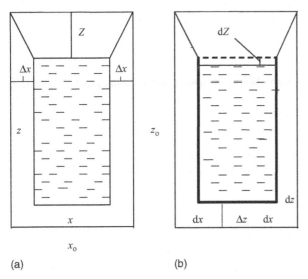

(a)   (b)

Figure 10.33 (a) Vertical cross-section of the chill-mould. Consider the melt in the volume xyz before the thin layer in (b) has solidified. (b) The thin layer has thickness dx and dy on the side walls and dz on the bottom surface. The solidification shrinkage of the layer has caused a lowering of the upper surface of the melt by an amount dZ.

With the aid of a material balance, we can obtain a relationship between the quantities given above and the density of the solid metal ρ_s and the density ρ_L of the melt.

The volume of the melt before the thin layer has solidified is xyz. When the thin layer has solidified, the remaining melt has a volume $(x - 2dx)(y - 2dy)(z - dz - dZ)$. The number 2 in the first two factors originates from *two* layers with thickness dx and dy on the four vertical sides which enclose the melt.

The thin layer consists of five planar plates with a total volume $2yzdx + 2xzdy + xydz$. The mass is not changed by the fact that part of the melt solidifies:

$$\rho_L xyz = \rho_L(x - 2dx)(y - 2dy)(z - dz - dZ) + \rho_s(2yzdx + 2xzdy + xydz)$$

All terms of second order (of the type $dxdy$) can be neglected. After reduction, the relation can be written as

$$\rho_L xydZ = (\rho_s - \rho_L)\,2yzdx + (\rho_s - \rho_L)\,2xzdy + (\rho_s - \rho_L)xydz$$

or

$$\frac{dZ}{dx}xy = \frac{\rho_s - \rho_L}{\rho_L}\left[\left(2yz + 2xz\frac{dy}{dx}\right) + xy\frac{dz}{dx}\right] \quad (10.23)$$

Equation (10.23) is combined with the definition of solidification shrinkage:

$$\beta = \frac{\rho_s - \rho_L}{\rho_s} \quad (10.24)$$

which can be written as:

$$\beta = 1 - \frac{\rho_L}{\rho_s} \Rightarrow \frac{\rho_s}{\rho_L} = \frac{1}{1 - \beta} \quad (10.25)$$

Equation (10.25) can be used to calculate the expression:

$$\frac{\rho_s - \rho_L}{\rho_L} = \frac{\rho_s}{\rho_L} - 1 = \frac{1}{1 - \beta} - 1 = \frac{\beta}{1 - \beta} \quad (10.26)$$

Equation (10.26) is inserted into Equation (10.23) to give:

$$\frac{1 - \beta}{\beta}\frac{dZ}{dx} = \frac{2yz + 2xz\frac{dy}{dx}}{xy} + \frac{dz}{dx} \quad (10.27)$$

This is the differential equation that has to be solved to obtain Z as a function of x, y, and z. It is rather complicated and in most cases has to be solved numerically.

If the solidification rates are equal along the vertical sides, we have $dx = dy$. In most cases we also have $dz/dx = C$, where C is a constant. The differential Equation (10.27) is then simplified to:

$$\frac{1 - \beta}{\beta}\frac{dZ}{dx} = \frac{2(x + y)z}{xy} + C \quad (10.28)$$

This equation cannot be solved exactly, other than in some special cases. To proceed further, we replace x, y, and z with:

$$x = x_0 - 2\Delta x$$
$$y = y_0 - 2\Delta y = y_0 - 2\Delta x$$
$$z = z_0 - C\Delta x - Z$$

Inserting these expressions into Equation (10.28), we obtain:

$$\frac{1-\beta}{\beta}\frac{dZ}{dx} = \frac{2(x_0+y_0-4\Delta x)(z_0-C\Delta x-Z)}{(x_0-2\Delta x)(y_0-2\Delta x)} + C \quad (10.29)$$

In this version, the differential equation can be solved numerically and the result is shown in Figure 10.34. It can be seen that the pipe is very deep in the ingot. To achieve a better yield it is obviously advisable to use a hot top.

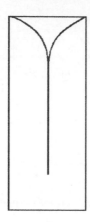

Figure 10.34 Calculated ingot pipe.

Analytical solutions of Equation (10.29) exist in some simple special cases, for example when the chill-mould has a square or circular cross-section and if the solidification rates from the sides and the bottom are equal. In these cases, Equation (10.29) is transformed into a linear differential equation, which can be solved by standard methods.

If the cross-section of the chill-mould is a square and the solidification rates at the sides and at the bottom are equal, the following relationships are valid:

$$x_0 = y_0 \qquad C = 1$$
$$x = y$$

and

$$dx = dy = dz \qquad \Delta x = \Delta y = \Delta z$$

If these expressions are introduced into Equation (10.29) and the differential equation is solved, the solution will be in this special case:

$$Z = z_0 - \frac{x_0}{2} - \frac{3\beta}{1-5\beta}(x_0 - 2\Delta x) + \left(\frac{3\beta x_0}{1-5\beta} - z_0 + \frac{x_0}{2}\right)\left(\frac{x_0 - 2\Delta x}{x_0}\right)^{\frac{4\beta}{1-\beta}} \quad (10.30)$$

Example 10.5
A 2.6 ton ingot is to be cast in a chill-mould with the dimensions given in the figure shown. The chill-mould has a variable square cross-section. The solidification shrinkage is 0.050.

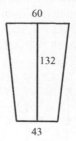

(a) Calculate the concave profile of the pipe as a function of the distance to the chill-mould wall and plot the result in a figure.
(b) Calculate the maximum depth of the pipe.

It is advisable to make reasonable approximations of the shape of the chill-mould to simplify the calculations.

Solution:

(a) Since the chill-mould has a square cross-section, it is reasonable to approximate the real chill-mould with a straight chill-mould with a constant cross-sectional area. We choose its side as 60 cm and adapt the height in such a way that the volume becomes equal in the two cases.

$$x_0 = 0.60 \text{ m}$$
$$z_0 = 0.98 \text{ m}$$

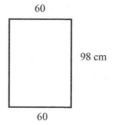

These values are inserted into Equation (10.30), which is valid in this case:

$$Z = 0.98 - \frac{0.60}{2} - \frac{3 \times 0.050}{1-5 \times 0.050}(0.60 - 2\Delta x) + \left(\frac{3 \times 0.050}{1-5 \times 0.050} - 0.98 + \frac{0.60}{2}\right)\left(\frac{0.60 - 2\Delta x}{0.60}\right)^{\frac{4 \times 0.050}{1-0.050}} \quad (1')$$

which can be reduced to:

$$Z = 0.68 - 0.200(0.60 - 2\Delta x) - 0.48\left(\frac{0.60 - 2\Delta x}{0.60}\right)^{0.210}$$

$$(2')$$

where $0 < \Delta x < x_0/2 = 0.30$.

Equation $(2')$ is the desired function. In order to plot it, we choose some suitable values of Δx and calculate the corresponding values of Z.

Δx	Z
0.00	0.00
0.05	0.04
0.10	0.09
0.15	0.14
0.20	0.20
0.25	0.28
0.27	0.32
0.29	0.40
0.295	0.44
0.30	0.68

(b) The maximum depth of the pipe corresponds to $\Delta x = x_0/2$. The value equals $z_0 - x_0/2 = 0.98 - 0.60/2 = 0.68\,\mathrm{m}$, since we have assumed that the solidification rate is the same from all sides in the whole ingot.

Answer:

(a) See Equation $(1')$ and the figure above.

(b) The depth of the pipe is $\sim$70 cm.

10.5.3 Theory of Pipe Formation in Ingots with a Hot Top

In ingots, which will be warm-rolled later, insulated pores will be closed. Provided that air has no admission to the walls of the pores and does not oxidize them, such pores have no destructive effect on the quality of the ingot.

On the other hand, a pipe is always exposed to contact with air and oxidation cannot be avoided. When an ingot with a pipe is warm-rolled, a so-called 'fish-tail' appears (Figure 10.35). Remaining oxide on the surface of the pipe causes this defect. The oxide prevents the pipe from sealing. The damage will be more pronounced the larger the pipe is and may have the consequence that a large fraction of the rolled ingot has to be discarded.

Figure 10.35 'Fish-tail'. From K. Strauss, Applied Science in the Casting of Materials, reproduced with kind permission of Foseco International Limited.

To avoid this, efforts are made to insulate or heat the upper surface of the ingot to make it solidify later than all other regions. The best method is to place a so-called hot top on the upper surface of the ingot,

A hot top can be described as an insulated container. The insulation causes the ingot to solidify more slowly than the ingot as a whole. This is just the effect which is desired. The slow solidification in the centre prevents the appearance of a deep pipe and the yield becomes essentially much better with than without a hot top (Figure 10.36).

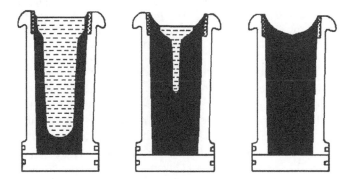

Figure 10.36 Solidification process in ingots with a hot top. Reproduced with permission from Elsevier Science.

The volume of a hot top equals 10–15 % of the total volume of the ingot. The better the hot top is, the smaller will be its volume. A good hot top is made of a material with low thermal capacity and low thermal conduction. Unfortunately, highly insulating and porous materials often have low mechanical strength and cannot be used. The top must not be deformed by the melt.

Incorrect dimensioning of the top, i.e. too small a height, may have the consequence that the pipe creeps into the ingot below the hot top, which has to be avoided. If this were to occur before the centre of the ingot has solidified, the pipe might force its way very far down into the ingot.

Theoretical Model for a Pipe with a Hot Top

By using a theoretical model of a hot top, it is possible to make a rough estimation and calculate a suitable size of the top.

Figure 10.37 shows calculated solidus isotherms close to the hot top in an ingot. It can be seen that the isotherms are perturbed only within a relatively narrow region close to the boundary between the ingot and the hot top. The conclusion is that the ingot and the top can be treated as two separate bodies in a simple analysis and that the solidification time of each of them can be calculated separately.

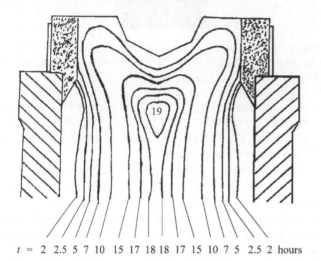

$t = 2 \quad 2.5 \quad 5 \quad 7 \quad 10 \quad 15 \quad 17 \quad 18 \quad 18 \quad 17 \quad 15 \quad 10 \quad 7 \quad 5 \quad 2.5 \quad 2$ hours

Figure 10.37 The positions of the solidification front in an ingot with hot top at various times. The time is given in hours after the start of the solidification. Reproduced with permission from the Scandanavian Journal of Metallurgy, Blackwell.

With the aid of mathematical models for the heat flow in a solidifying melt, it is possible to calculate the temperature field and the position of the solidification front in the melt with the aid of a computer. These methods do not consider the shrinkage in the hot top during the solidification process. For such calculations, it is convenient to use the same simple method as above for the solidification of the ingot and concentrate on the solidification shrinkage.

If we assume that the upper surface of the top does not solidify, we can set up a simple material balance which gives the shrinkage in the top as a function of the amount of solidified metal in the top and ingot. When a thin layer solidifies in the top and the ingot, the solidification shrinkage is compensated for by melt, which fills this volume difference. The upper surface of the remaining melt sinks in proportion to the withdrawn melt.

The mass of the sucked melt is equal to the contribution of mass to the ingot (Figure 10.38):

$$\rho_L A_{top} \, dZ_{top} = (\rho_s - \rho_L)(dV_{top} + dV_{ingot}) \quad (10.31)$$

where

A_{top} = cross-sectional area of the hot top

dZ_{top} = lowered distance of the upper surface of the melt due to the total solidification shrinkage of the thin layer in the ingot and hot top

dV_{top} = volume of the thin layer in the top which has solidified

dV_{ingot} = volume of the thin layer in the ingot which has solidified

ρ_s = density of the solid metal

ρ_L = density of the melt.

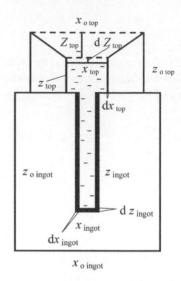

Figure 10.38 Ingot with a hot top. Definition of designations.

To derive an expression for the solidified volumes, we need to know the relative solidification rates in the hot top, the ingot, and the ingot bottom. Thus we introduce special designations for the relative solidification rates in the top and the ingot bottom with the solidification rate in the x-direction as a basis:

$$n = \frac{dx_{top}}{dx_{ingot}} \quad (10.32)$$

$$m = \frac{dz_{ingot}}{dx_{ingot}} \quad (10.33)$$

We disregard the fact that the chill-mould and the top are conical. This means that A_{top} is independent of the height at a given time. The solidified volumes are also simple to calculate. For a rectangular ingot, we have:

$$A_{top} = x_{top} \, y_{top} \quad (10.34)$$

The solidified layer in the top has four vertical sides. If the solidification rates in the x- and y-directions are equal, we obtain:

$$dV_{top} = 2(x_{top} + y_{top}) \, z_{top} \, dx_{top} \quad (10.35)$$

The solidified layer in the ingot consists of four vertical sides plus bottom:

$$dV_{ingot} = 2(x_{ingot} + y_{ingot}) \, z_{ingot} \, dx_{ingot} + x_{ingot} \, y_{ingot} \, dz_{ingot} \quad (10.36)$$

Equations (10.32)–(10.36) are inserted into Equation (10.31):

$$\rho_L x_{top} y_{top} dZ_{top} = (\rho_s - \rho_L)[2(x_{top} + y_{top})z_{top} dx_{top}$$
$$+ 2(x_{ingot} + y_{ingot})z_{ingot} dx_{ingot} + x_{ingot} y_{ingot} dz_{ingot}] \quad (10.37)$$

With the aid of the definition of solidification shrinkage $\beta = (\rho_s - \rho_L)/\rho_s$ we can express the ratio $(\rho_s - \rho_L)/\rho_L$ as a function of β as in Equation (10.26) on page 329. Equation (10.37) can be transformed into:

$$dZ_{top} = \frac{2\beta}{1-\beta} \cdot \frac{dx_{ingot}}{x_{top}\, y_{top}} \left[nz_{top}(x_{top} + y_{top}) + z_{ingot}(x_{ingot} + y_{ingot}) + \frac{mx_{ingot}\, y_{ingot}}{2} \right] \quad (10.38)$$

To facilitate a numerical solution of Equation (10.38), we also introduce:

$$x_{ingot} = x_{0\,ingot} - 2\Delta x \quad (10.39)$$
$$y_{ingot} = y_{0\,ingot} - 2\Delta x \quad (10.40)$$
$$z_{ingot} = z_{0\,ingot} - m\Delta x \quad (10.41)$$

We may replace x_{ingot}, y_{ingot}, and z_{ingot} with Equations (10.39), (10.40), and (10.41), respectively, in Equation (10.38). The equation we obtain is combined with a description of the solidification rate of the ingot, often in the shape of a computer program.

The method is illustrated in a simplified form in Example 10.6.

Example 10.6

A square ingot of steel with the dimensions $30 \times 30 \times 150$ cm³ is to be cast. The ingot will be equipped with a hot top in the shape of a layer of quartz sand, which will cover its surface. A suitable value of the height of the top has to be determined.

Assume an initial value of the height of the top.

(a) Determine the total solidification time of the ingot.

(b) Calculate the shape of the pipe and draw a figure which shows the ingot and the top when the last melt in the mould just has solidified and determine the depth of the pipe.

(c) Perform a calculation of the amount of melt which remains in the box when the whole ingot has solidified.

(d) What practical measures are to be recommended on basis of the calculations?

Reasonable values of the constants required for the calculations are taken from standard reference tables. The soli-

dification shrinkage of the steel is 0.050. Experimental data show that the constant in the parabolic growth law can be given the value $S_1 = 3.2 \times 10^{-3}$ m/s$^{0.5}$. For a casting of iron in quartz sand, the value of the constant is 8×10^{-4} m/s$^{0.5}$.

Preparatory Discussion. Choice of a Suitable
Value of the Height of the Top
In Chapters 4 and 5, we discussed various models for calculation of solidification rates and temperature distribution on solidification. The shell thickness in a solidifying melt can, to a good approximation, be described by a relationship of the type:

$$x(t) = S_1 \sqrt{t} \quad (1')$$

where
x = thickness of the metal shell
t = time
S_1 = a constant.

The growth rate of the shell is obtained by derivatization of Equation (1') with respect to time:

$$\frac{dx(t)}{dt} = \frac{S_1}{2\sqrt{t}} \quad (2')$$

Equation (2') is valid only with the assumption that the width of the ingot is larger than its thickness (one-dimensional solidification). In spite of this condition, we will use this simple relation and disregard the fact that the solidification rate increases at the end of the solidification for square cross-sections.

The hot top is most often made of sand or other insulating material. Sand has a lower density and lower melting point than the metal melt and has the advantage of not being changed chemically during the solidification process. The shell thickness in the top during the solidification can also be described by a parabolic growth law but with a *different* constant:

$$x(t) = S_2 \sqrt{t} \quad (3')$$

We make a rough calculation in order to find a suitable value of the height of the hot top. Its cross-section can suitably be chosen equal to the cross-section of the ingot.

The volume of the ingot before solidification is equal to $0.30 \times 0.30 \times 1.50 = 0.135$ m³. $\beta = 0.050$, which gives $1 - \beta = 0.95$. The volume of the ingot after solidification $\approx 0.95 \times 0.135 = 0.128$ m³. The volume of the ingot after solidification $= z_{ingot} \times 0.30 \times 0.30$

With the designations in the figure shown, we obtain:

$$z_{ingot} = \frac{0.128}{0.30 \times 0.30} = 1.42 \text{ m}$$

Hence the hot top must at least be:

$$z_{0\,top} = 1.50 - 1.42 = 0.08 \text{ m} = 8 \text{ cm}$$

In order to eliminate the risk of the pipe penetrating the hot top, we choose $z_{0\,top} = 15$ cm for safety, i.e. nearly the double roughly calculated value.

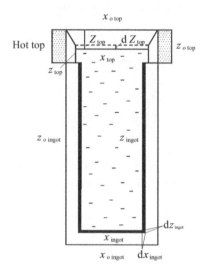

Hot top

Vertical cross-section of a solidifying ingot with a hot top.

Solution:
We apply Equation (10.31) on page 332, which gives the lowered distance in the hot top at time t as a function of solidified amount of melt in top and ingot:

$$\rho_L A_{top} dZ_{top} = (\rho_s - \rho_L)(dV_{top} + dV_{ingot}) \qquad (4')$$

The volume of the top is small compared with the volume of the ingot, and can therefore be neglected. $(\rho_s - \rho_L)/\rho_L$ can be transformed to $\beta/(1 - \beta)$ [Equation (10.26)].

A_{top} is the area of the melt in the top and equal to the square of the whole side minus two solidified layers:

$$A_{top} = x_{top}^2 = (x_{0\,top} - 2S_2\sqrt{t})^2 \qquad (5')$$

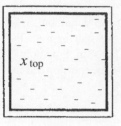

Solidification process in the hot top seen from above.

where the constant $S_2 = 8 \times 10^{-4}$ m/s$^{0.5}$ according to the text.

We assume that the solidification in the ingot occurs at the same rate in the x-, y-, and z-directions. The thickness of the solidified layer $= S_1\sqrt{t}$, where $S_1 = 3.2 \times 10^{-3} = $ m/s$_{0.5}$ according to the text, and we obtain:

$$x_{ingot} = (x_{0\,ingot} - 2S_1\sqrt{t}) \qquad (6')$$

$$y_{ingot} = (x_{0\,ingot} - 2S_1\sqrt{t}) \qquad (7')$$

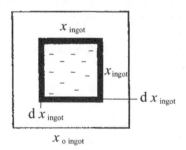

Solidification process in the ingot seen from above.

because the cross-section of the mould is square and:

$$z_{ingot} = (z_{0\,ingot} - S_1\sqrt{t}) \qquad (8')$$

The number of surfaces is five, i.e. four vertical side surfaces and one bottom surface. The total area of the solidified layer has to be multiplied by dx_{ingot} to give dV_{ingot}. We obtain dx_{ingot} by applying Equation (2'):

$$dx_{ingot} = \frac{S_1\,dt}{2\sqrt{t}} \qquad (9')$$

All these quantities are inserted into Equation (10.31), which gives

$$(x_{0\,top} - 2S_2\sqrt{t})^2 dZ_{top} = \frac{\beta}{1 - \beta}\left[4(x_{0\,ingot} - 2S_1 \cdot \sqrt{t})(z_{0\,ingot} - S_1\sqrt{t}) + (x_{0\,ingot} - 2S_1\sqrt{t})^2\right]\frac{S_1\,dt}{2\sqrt{t}} \qquad (10')$$

The volume of the top is approximately 10 % (15 cm/150 cm) of the ingot volume. Hence it is reasonable to neglect the solidification shrinkage in the top compared with that in the ingot.

(a) Initially we calculate the total solidification time of the ingot. The whole ingot has solidified when the thickness of the solidified metal equals half the square side of the mould:

$$S_1 \sqrt{t_{ingot}} = \frac{x_{0\,ingot}}{2}$$

or

$$t_{ingot} = \left(\frac{x_{0\,ingot}}{2S_1}\right)^2 = \frac{0.30^2}{(2 \times 3.2 \times 10^{-3})^2}$$

$$= 2197\,s \approx 2200\,s = 37\,min$$

t	dt	x_{top}	x_{ingot}	z_{ingot}	dZ_{top}	Z_{top}
0	—	0.30	0.30	1.50	0.000	0.000
220	220	0.28	0.21	1.45	0.020	0.020
440	220	0.27	0.17	1.43	0.012	0.032
660	220	0.26	0.14	1.42	0.008	0.041
880	220	0.25	0.11	1.41	0.006	0.047
1100	220	0.25	0.09	1.39	0.005	0.052
1320	220	0.24	0.07	1.38	0.003	0.055
1540	220	0.24	0.05	1.37	0.002	0.057
1760	220	0.23	0.03	1.37	0.001	0.059
1980	220	0.23	0.02	1.36	0.001	0.059
2200	220	0.22	0.00	1.35	0.000	0.059

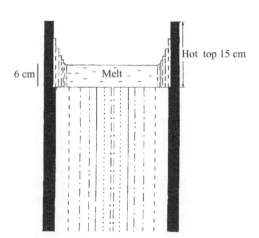

Melt + ingot which has started to solidify. A number of instant pictures of the position of the solidification front in the ingot and the top at various times during the solidification process have been drawn.

(b) We divide the time into ten equal intervals and insert these values into Equation (10′) and calculate successively x_{top}, x_{ingot}, z_{ingot}, dZ_{top}, and $Z_{top} = \Sigma dZ_{top}$ with the aid of known values of S_1 and S_2.

(c) From the table shown, it can be read that $x_{top} = 22\,cm$ and $Z_{top} = 6\,cm$ when the whole ingot just has solidified.

If we disregard the volume of solidified metal in the top, the volume of the remaining melt in the hot top, when the whole ingot has solidified, is equal to:

$$22 \times 22 \times (15 - 6)\,cm = 22 \times 22 \times 9\,cm$$

Since iron has the density $7.8 \times 10^3\,kg/m^3$, the mass of the remaining melt will be:

$$m = 0.22 \times 0.22 \times 0.09 \times 7.8 \times 10^3 = 34\,kg$$

Answer:

(a) The whole ingot solidifies in ~37 min.
(b) See the figure shown.
(c) The remaining melt in the top when the ingot just has solidified weighs ~34 kg.
(d) A hot top height of 15 cm seems to be unnecessarily large. Since we have neglected the cooling shrinkage of the top, we have to add ~10 % of the lowered distance 6 cm to take this into consideration. It ought to be sufficient with a layer of 80 mm of quartz sand, which is in perfect agreement with our rough preparatory calculation; 10 cm is definitely an upper limit.

A comparison between Examples 10.5 and 10.6 shows convincingly the benefit of hot tops.

10.6 SOLIDIFICATION AND COOLING SHRINKAGE DURING CONTINUOUS CASTING

During continuous casting, the melt must be strongly cooled since no metal melt must be left inside the strand when it leaves the machine (Figure 10.39).

If the walls and bottom of a chill-mould are strongly cooled, a large temperature gradient is formed in the melt. The dendrites grow mainly in the direction of the temperature gradient and the shrinkage cavities appear in the shape of a narrow pipe in the centre (Figure 10.40). This phenomenon has been treated for ingots in Section 10.5.

Corresponding cooling conditions with strong temperature gradients are also valid during continuous casting, when

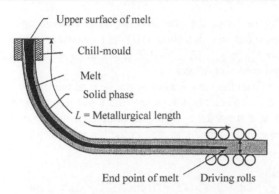

Figure 10.39 labels: Upper surface of melt; Chill-mould; Melt; Solid phase; L = Metallurgical length; End point of melt; Driving rolls

Figure 10.39 Bent continuous casting machine with a curvature below the chill-mould. Above the chill-mould there is a tundish, which is not shown. The bending is a practical measure to achieve a lower building with reduced mechanical stress on the workshop floor. Reproduced by permission of ASM International.

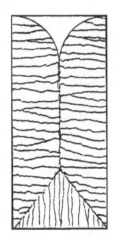

Figure 10.40 Dendrite growth during ingot solidification with a large temperature gradient.

the walls of the strand are cooled by water spraying in the chilled zone below the chill-mould. Pipe formation appears at the end of the solidification of the strand when the supply of new melt from the tundish has ceased.

Pipe formation at continuous casting is discussed below and a theoretical model, developed by Fredriksson and Raihle, is introduced.

10.6.1 Pipe Formation during Continuous Casting

Pipe formation is normally connected with ingot casting. During ingot casting the pipe is mainly caused by solidification shrinkage.

In continuous casting, melt is continuously added to the mould and the volume decrease due to solidification and cooling shrinkage is thus compensated. The situation is changed at the end of the casting process when the supply

of melt from the tundish above the mould ceases. Then the conditions become comparable to those which are valid in ingot casting and a pipe is formed at the end of the casting process.

The pipe volume can be reduced by different practical methods and is in most cases not a major problem in continuous casting processes. However, the mechanism of pipe formation in a continuously cast strand is the same that of centreline segregation, which is a severe problem. This will be discussed in Chapter 11.

The pipe is well developed in billets but less marked in casting of slabs. The reason for this difference is found in the design of the casting machine and the casting conditions in the two cases. Here we will mainly discuss billet casting. However, in some cases calculations valid for a one-dimensional model similar to a slab strand are included, in order to illustrate the general calculation process.

Experimental examinations show that the pipe volume during continuous casting is so large that it cannot be explained entirely by solidification shrinkage.

The central cavities in continuous casting are caused by combined solidification and cooling shrinkage.

10.6.2 Theory of Pipe Formation during Continuous Casting

Solidification shrinkage in combination with cooling shrinkage is the origin of pipe formation in the strand in the final stage of the casting process. Figure 10.41 shows a schematic diagram of the strand during the solidification process. We assume that the cross-sectional area of the strand is a square with the side x_0.

We assume that the supply of melt is stopped at $t = 0$. At that time, the upper surface of the melt had the initial position ABCD. At time t it had the position A'B'C'D' and at the time $t + dt$ the surface has sunk to the dotted position. During the time interval dt, the volume dV_s has solidified. This causes a solidification shrinkage $dV_{sol} = dV_s\beta$, where β is the solidification shrinkage (page 329), and a cooling shrinkage $= dV_{cool}$.

We introduce a coordinate system with the z-axis in the direction of the strand motion and the xy-plane perpendicular to the z-axis. In a bent machine the direction of the z-axis is initially vertical (Figure 10.41) and horizontal when all the melt has solidified (Figure 10.39).

The surface sinks by an amount dZ in the direction of the positive z-axis during the time dt, owing to solidification and cooling shrinkage. We also assume that the densities of the solid and the melt are independent of Z. Then a volume balance can replace the material balance. At a distance Z from point O in Figure 10.41, the volume element dV_{pipe} can be written as:

$$dV_{pipe} = AdZ = dV_{sol} + dV_{cool} = dV_s\beta + dV_{cool} \quad (10.42)$$

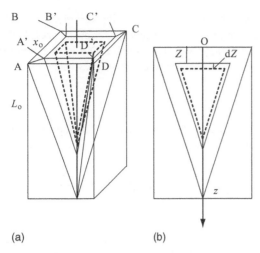

(a)   (b)

Figure 10.41 Schematic diagram of a solidifying strand. (a) Position of the solidification front at three different times: three pyramids inside of each other. The central one is dotted. (b) Cross-section of the strand in a vertical plane through the central axis and parallel with two of the vertical sides. In reality the strand is much longer than the figure shows and, in addition, bent. Reproduced by permission of ASM International.

where

V_{pipe} = volume of the pipe

dZ = change of distance Z during the time interval dt

A = cross-sectional area of the melt at distance z from O along the strand

dV_s = change in solidified volume during time interval dt

dV_{sol} = change in solidified volume during time interval dt due to solidification shrinkage

β = solidification shrinkage = $(\rho_s - \rho_L)/\rho_s$ (page 329)

dV_{cool} = change in solidified volume during time interval dt due to cooling shrinkage.

The left-hand side of Equation (10.42) represents the change in the pipe volume during the time dt. The first term on the right-hand side corresponds to the solidification shrinkage during the time dt and the second term to the cooling shrinkage during the time dt. The total pipe volume is obtained by integration of Equation (10.42) with respect to time.

Figure 10.42 shows the pipe in its horizontal position.

Determination of the Solidification Shrinkage

Solidification Shrinkage Volume

Calculation of the solidification shrinkage during continuous casting is performed mainly in the same way as was used for calculation of the pipe formed in an ingot without a hot top (page 328).

Consider the volume element in Figure 10.43 with a square cross-section and height dZ. The outer circumference is $4x_0$ and the thickness of the solidified layer is x. If the thickness increases

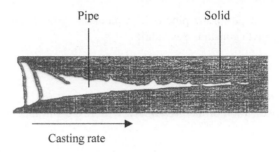

Figure 10.42 Vertical cross-section of a pipe in the central region of a strand during continuous casting. The asymmetry of the pipe is due to the fact that the casting machine is bent and the pipe is horizontal with the planar side downwards before the melt has solidified completely in the central region. Reproduced with permission from ASM International.

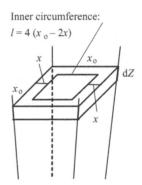

Figure 10.43 Volume element for calculation of the pipe in a strand during continuous casting.

by an amount dx during the time dt, the solidified volume increases by an amount dV_s during the time dt and we obtain:

$$\frac{dV_s}{dt} = 4(x_0 - 2x)\frac{dx}{dt}dZ \qquad (10.43)$$

where

x_0 = the side of the square strand

x = thickness of the solidified shell

Z = coordinate along the strand ($z = 0$ at O in Figure 10.41)

V_s = solidified volume.

Equation (10.43) is valid all the way along the strand. The thickness x and its time derivative dx/dt are not constant but vary with time and position. The total solidification volume can be calculated if the function dx/dt is known. A reasonable example with a parabolic growth law is given below. The shape of the pipe will be discussed later.

Example 10.7

Derive an expression for the volume of the solidification shrinkage of a square strand with side x_0 during continuous casting depth of the pipe is L. A parabolic growth law with the growth constant S is valid:

$$x = S\sqrt{t}$$

Solution:
Derivatization of the growth law with respect to time gives:

$$\frac{dx}{dt} = \frac{S}{2\sqrt{t}} \tag{1'}$$

If the values of x and dx/dt are inserted into Equation (10.43), we obtain with the aid of Figure 10.43:

$$V_s = \int_0^{V_s} dV_s = \int_0^Z dZ \int_0^t 4(x_0 - 2S\sqrt{t})\frac{Sdt}{2\sqrt{t}} \tag{2'}$$

where z is distance of the volume element below the upper level of the chill-mould.

The time t and the distance z are related to each other by the simple relation:

$$t = \frac{Z}{v_{cast}} \tag{3'}$$

if the casting rate v_{cast} is assumed to be constant. Equation (3') for t is introduced into Equation (2') and we obtain the total solidification shrinkage:

$$V_s = \int_0^L dZ \int_0^{\frac{Z}{v_{cast}}} 4(x_0 - 2S\sqrt{t})\frac{Sdt}{2\sqrt{t}} \tag{4'}$$

where L is the depth of the pipe. The double integral is solved in two steps. Integration with respect to t gives:

$$\int_0^{V_s} dV_s = \int_0^L dZ \int_0^{\frac{Z}{v_{cast}}} \left(2x_0 S\frac{dt}{\sqrt{t}} - 4S^2 dt\right)$$

$$= \int_0^L \left(4x_0 S\sqrt{\frac{Z}{v_{cast}}} - 4S^2 \frac{Z}{v_{cast}}\right)dZ$$

and integration with respect to z gives:

$$V_s = \int_0^{V_s} dV_s = \left[\frac{8x_0 S}{3\sqrt{v_{cast}}}Z^{\frac{3}{2}} - \frac{4S^2}{v_{cast}}\frac{Z^2}{2}\right]_0^L$$

or

$$V_s = \frac{8x_0 S}{3\sqrt{v_{cast}}}L^{\frac{3}{2}} - \frac{2S^2}{v_{cast}}L^2 \tag{5'}$$

We now have an expression that describes the solidified volume along the strand when the supply of melt has ceased and a parabolic growth law is valid. During the solidification the upper surface of the melt is successively lowered and a pipe is formed.

Answer:
The total solidification shrinkage contributes to the pipe volume with an amount:

$$\Delta V_{sol} = \beta V_s = \beta\left(\frac{8x_0 S}{3\sqrt{v_{cast}}}L^{\frac{3}{2}} - \frac{2S^2}{v_{cast}}L^2\right)$$

Determination of the Cooling Shrinkage
Alloys solidify over a temperature range equal to $T_L - T_s$. Within this temperature interval, solidification heat is gradually released and the central temperature of the strand is constant or changes very little (Figure 10.44), in contrast to the conditions valid after the solidification fronts have met. When all the melt has solidified, no further heat is released and the temperature at the centre then decreases rapidly (Figure 10.44).

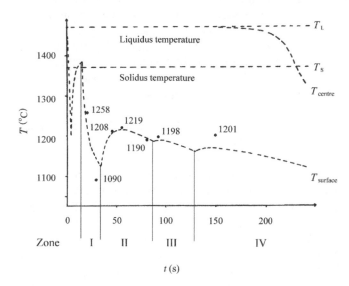

Figure 10.44 Temperature at the centre and at the middle of a side of a square strand during continuous casting as a function of time. Solidification interval = $T_L - T_s$.

The cooling shrinkage can be described as the sum of two parts and depends on the temperature conditions at the centre and at the surface of the strand.

Before the solidification fronts meet, the *first* part is the shrinkage of the solidified shell *before* the solidification fronts meet at the centre of the strand.

Before the solidification fronts meet, the temperature at the solidification fronts in the interior of the strand is approximately constant and equal to the liquidus temperature. Hence one may neglect the internal shrinkage caused by the temperature change at the solidification fronts. Its contribution to the pipe formation is negligible.

The water cooling is a tool for controlling the fluctuating surface temperature (to avoid cracks) and keeping it as constant as possible (better than in Figure 10.44). Hence the net effect on the surface shrinkage is relatively small and its contribution to the cooling shrinkage will be neglected here. After the solidification fronts have met, the *second* part is the shrinkage or strain *after* the union of the solidification fronts at the centre of the strand. The situation becomes completely different when the solidification fronts have met at the centre of the strand.

The sudden and rapid temperature decrease at the centre when the solidification fronts have met (Figure 10.44) causes a cooling shrinkage in the whole strand, which contributes to the total cooling shrinkage.

The major part of the cooling shrinkage arrives when the solidification fronts have met. The aim of the water cooling is now to lower the temperature of the strand. The temperature decreases *both* at the centre *and* at the surface affecting the cooling shrinkage in the whole strand. Both of these effects contribute to the total cooling shrinkage of the strand.

$$\Delta V_{cool} = \Delta V_{surface} + \Delta V_{centre} \qquad (10.44)$$

It is difficult to derive analytical expressions for these volume changes and we make some simplifications in our further treatment.

Cooling Shrinkage Volume

The temperature change as a function of time is much smaller in the length direction (along the z-axis) than in the perpendicular directions. Consequently, we can neglect the cooling shrinkage in the length direction and only take the changes of the cross-sectional *area* instead of the *volume* changes into consideration. We can then write:

$$\Delta A_{cool} = \Delta A_{surface} + \Delta A_{centre} \qquad (10.45)$$

or

$$\Delta A_{centre} = \Delta A_{cool} - \Delta A_{surface} \qquad (10.46)$$

The formation of a pipe, which is a cavity in the centre, shows that the cooling shrinkage at the centre is directed *outwards*. $\Delta A_{surface}$ and ΔA_{centre} have opposite signs, as shown in Figure 10.45.

Figure 10.45 can be interpreted in the following way:

1. If ΔA_{centre} is *larger* than $\Delta A_{surface}$, the central part of the strand shrinks more than the outer frame, i.e. the surface, allows. This means that *a cavity is formed* in the central

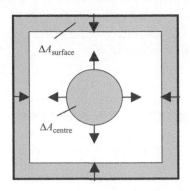

Figure 10.45 Area shrinkage at the surface and the centre of a strand, due to thermal contraction. If the central shrinkage exceeds the surface contraction, a pipe is formed at the top of the strand.

region. This cavity becomes filled with melt from above and hence it contributes to the pipe at the top of the strand.

2. If ΔA_{centre} is *smaller* than $\Delta A_{surface}$, the central part shrinks less than the outer frame, i.e. the surface. This means that the central region becomes compressed and *no cavity is formed*. No pipe is formed at the top of the strand.

The first alternative will be treated below. The simplest way to find the total cooling shrinkage ΔA_{cool} is to calculate $\Delta A_{surface}$ and ΔA_{centre} separately and then add them.

Calculation of $\Delta A_{surface}$

The surface is cooled from outside, which results in a cooling rate $dT_{surface}/dt$ at the surface of the strand. The temperature in the centre of the strand is constant (Figure 10.44).

We assume that the temperature decreases *linearly* from the centre and outwards to the surface. If the temperature decreases by an amount $dT_{surface}$ at the surface, it sinks at a distance x from the surface by an amount [see Figure 10.46 (b)]:

$$dT_{surface}(x) = \frac{\frac{x_0}{2} - x}{\frac{x_0}{2}} dT_{surface} \qquad (10.47)$$

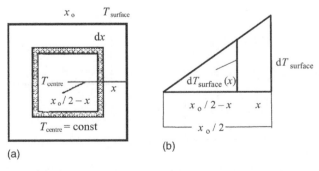

Figure 10.46 Temperature distribution when $T_{surface}$ varies linearly in the cross-sectional area.

We will calculate the contribution of the surface shrinkage to the total cooling shrinkage of the square cross-section. The basic equation of surface strain is:

$$\Delta A = \alpha_A A \Delta T \qquad (10.48)$$

We apply it to the dark area element in Figure 10.46 (a) and integrate over the whole cross-sectional area.

With the aid of Equations (10.47) and (10.48) and the relation $\alpha_A = 2\alpha_l$, we obtain the shrinkage of four area elements:

$$dA_{\text{surface}} = 2\alpha_l \times 4(x_0 - 2x)dx \times \frac{\dfrac{x_0}{2} - x}{\dfrac{x_0}{2}} dT_{\text{surface}}$$

$$\alpha_A \qquad \text{four area elements} \quad dT_{\text{surface}}(x)$$

The cooling shinkage of the whole cross-section, when T_{surface} decreases by an amount $\Delta T_{\text{surface}}$, is then:

$$\Delta A_{\text{surface}} = \int_0^{\frac{x_0}{2}} 2\alpha_l \times 4(x_0 - 2x)dx \frac{\dfrac{x_0}{2} - x}{\dfrac{x_0}{2}} \Delta T_{\text{surface}}$$

$$= \frac{4}{3}\alpha_l x_0^2 \Delta T_{\text{surface}} \qquad (10.49)$$

Calculation of ΔA_{centre}

When the solidification is complete, no more solidification heat is generated. The temperature T_{centre} decreases rapidly whereas T_{surface} can be regarded as constant.

As above, we assume that the temperature decreases linearly, in this case from the centre outwards to the surface of the strand. If the temperature decrease in the centre is

dT_{centre}, then the temperature fall in the dark area element [Figure 10.47 (f)] is

$$dT_{\text{centre}}(x) = \frac{x}{\dfrac{x_0}{2}} dT_{\text{centre}} \qquad (10.50)$$

In the same way as above, we obtain [Figure 10.47(a)] the shrinkage of the area element:

$$dA_{\text{centre}} = 2\alpha_l \times 4(x_0 - 2x)dx \times \frac{x}{\dfrac{x_0}{2}} dT_{\text{centre}}$$

$$\alpha_A \qquad \text{four area elements} \quad dT_{\text{surface}}(x)$$

The central cooling shrinkage of the whole cross-sectional area when T_{centre} decreases by an amount ΔT_{centre} equals:

$$\Delta A_{\text{centre}} = \int_0^{\frac{x_0}{2}} 2\alpha_l \times 4(x_0 - 2x)dx \frac{x}{\dfrac{x_0}{2}} \Delta T_{\text{centre}} = \frac{2}{3}\alpha_l x_0^2 \Delta T_{\text{centre}}$$

$$(10.51)$$

Calculation of the Total Cooling Shrinkage Area ΔA_{cool}

Equation (10.45) gives the net cooling shrinkage area when a pipe is formed:

$$\Delta A_{\text{cool}} = \Delta A_{\text{centre}} + \Delta A_{\text{surface}}$$
$$= \frac{2}{3}\alpha_l x_0^2 \Delta T_{\text{centre}} - \frac{4}{3}\alpha_l x_0^2 \Delta T_{\text{surface}} \qquad (10.52)$$

where

ΔA_{cool} = total shrinkage of the cross-sectional area = average area of the pipe, formed in the strand

$\Delta A_{\text{surface}}$ = fraction of ΔA_{cool} which emanates from the surface shrinkage

ΔA_{centre} = fraction of ΔA_{cool} which emanates from the centre shrinkage.

As mentioned on page 339, ΔA_{cool} has a physical significance only if it is positive. The minus sign appears in Equation (10.52) because the movement of the outer surface is opposite to the movement of the central region (see text and figure on page 399).

Condition for Pipe Formation in Terms of Cooling Rates

From the derivations on pages 399–340, we can conclude that

$$G = \frac{\Delta T_{\text{surface}}}{\Delta T_{\text{centre}}} = \frac{dT_{\text{surface}}}{dT_{\text{centre}}} = \frac{\dfrac{dT_{\text{surface}}}{dt}}{\dfrac{dT_{\text{centre}}}{dt}} \qquad (10.53)$$

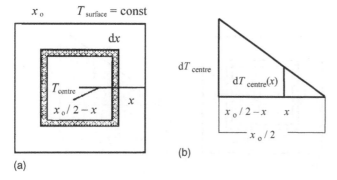

(a)

(b)

Figure 10.47 Temperature distribution when T_{centre} varies linearly in the cross-sectional area.

The G factor defined by Equation (10.53) can be used to express ΔA_{cool} in terms of ΔT_{centre} and the ratio of the cooling rates at the centre and at the strand surface. As $\Delta T_{surface} = G \Delta T_{centre}$, Equation (10.52) can be written as:

$$\Delta A_{cool} = \frac{2}{3} \alpha_l x_0^2 (1 - 2G) \Delta T_{centre} \qquad (10.54)$$

where

$$\alpha_l = \text{length strain coefficient}$$
$$x_0 = \text{cross-sectional side of the strand}$$
$$dT_{surface}/dt = \text{cooling rate at the surface of the strand}$$
$$dT_{centre}/dt = \text{cooling rate at the centre of the strand}$$
$$G = G\text{-factor, the ratio of the cooling rates,}$$
$$(dT_{surface}/dt)/(dT_{centre}/dt)$$
$$\Delta T_{centre} = \text{change of temperature at the centre.}$$

It can be concluded from Equation (10.54) that it is the G-factor, the ratio of the cooling rate at the surface to the cooling rate at the centre, that decides whether a pore is formed or not.

- If $G < 0.5 \Rightarrow \Delta A_{cool} > 0 \Rightarrow$ a pipe is formed.

  The cooling rate at the surface *is less* than half the cooling rate at the centre. A pore is formed and melt is sucked downwards to fill it. As the melt cannot advance owing to narrow channels, a pipe is formed in the strand.

- If $G > 0.5 \Rightarrow \Delta A_{cool} < 0 \Rightarrow$ no pipe.

  The cooling rate at the surface *exceeds* half the cooling rate at the centre. No pore is formed. The centre is exposed to thermal stress, which may cause central cracks. This matter is treated in Section 10.7.8.

Determination of the Volume and Shape of the Pipe

Volume of the Pipe

Equations (10.53) and (10.54) show that the cooling rate at the surface of the strand has a great influence on the pore volume.

The cooling rate is affected by the casting rate and the efficiency of the water cooling in the cooling zones of the casting machine. The cooling rates can be calculated by solving the heat equations given in Chapter 5.

ΔA_{cool} is calculated from Equation (10.54). The lower integration limit of the time integral is the time when the solidification fronts meet. The upper limit is the time when the temperature in the centre has reached the solidus temperature.

The area calculated in this way is the total area shrinkage due to the thermal contraction. In order to find the contribution to the volume of the pipe, one has to multiply the area by the length of the strand where this shrinkage occurs. ΔA_{cool} can be regarded as an average cross-section of the pipe. In this case, it is reasonable to use the total depth of the pipe, *i.e.* the metallurgical length L as the length of the strand.

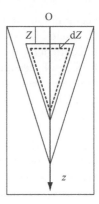

Figure 10.48 Two-phase region at the centre of the strand.

The volume of the pipe is the sum of the solidification shrinkage and the cooling shrinkage. According to Equations (10.42) and (10.52), the general expression for the pipe volume of a square strand (Figure 10.48) can be written as:

$$V_{pipe} = \beta V_s + \left(\frac{2}{3} \alpha_l x_0^2 \Delta T_{centre} - \frac{4}{3} \alpha_l x_0^2 \Delta T_{surface} \right) L \quad (10.55)$$

$$\underbrace{\hphantom{\beta V_s}}_{\Delta V_{sol}} \qquad \underbrace{\hphantom{\Delta A_{cool} L}}_{\Delta V_{cool} = \Delta A_{cool} L}$$

or, with the aid of Equation (10.54):

$$V_{pipe} = \beta V_s + \frac{2}{3} \alpha_l x_0^2 (1 - 2G) \Delta T_{centre} L \qquad (10.56)$$

The method for calculating the pipe volume is illustrated below.

Example 10.8

Set up an expression for the pipe volume for a square strand with a cross-sectional side x_0 during continuous casting when the temperature in the centre is equal to the solidus temperature of the alloy, T_s.

Assume that the thickness of the solidified layer of the strand follows the growth law $x = S\sqrt{t}$.

Solution and Answer:

The volume of the pipe is obtained by introducing the expressions in the answer in Example 10.7 (page 338) and the relation $\Delta T_{centre} = T_L - T_s$ into Equation (10.56):

$$V_{pipe} = \beta \left(\frac{8 x_0 S}{3 \sqrt{v_{cast}}} L^{\frac{3}{2}} - \frac{2 S^2}{v_{cast}} L^2 \right) + \frac{2}{3} \alpha_l x_0^2 (1 - 2G)(T_L - T_s) L$$

where

$$\beta = \text{solidification shrinkage} = (\rho_s - \rho_L)/\rho_s$$
$$S = \text{growth constant}$$
$$L = \text{metallurgical length (Figure 10.39)}$$
$$v_{cast} = \text{casting rate}$$

G = ratio of the cooling rates at the surface and at the
centre of the strand

α_l = length strain coefficient

x_0 = length of the side of the square strand

$T_{L,s}$ = liquidus and solidus temperatures of the alloy.

Shape of The Pipe

The shape of the pipe is described if we calculate the depth z as a function the thickness x of the solidified shell (Figure 10.49).

During the pipe formation, the upper level of the melt sinks when the supply of melt from the tundish has ceased. The decrease Z of the liquid surface can in principle be calculated in a similar way as in the case of ingot solidification, which was presented in Section 10.5.2 (page 328).

For a continuous square strand (see Figure 10.43), the following relation is valid:

$$AdZ = (x_0 - 2x)^2 dZ = \beta dV_s + L dA_{cool} \qquad (10.57)$$

However, the calculations in the case of a pipe during continuous casting are much more complicated than those in the case of an ingot pipe. In the latter case, only the solidification shrinkage is involved as the temperature can be regarded as constant.

During continuous casting, the temperature conditions and the cooling rates at the surface and at the centre of the strand and also the growth of the solid shell as a function of the geometric coordinates influence the shape of the pipe. When all functions and quantities are introduced into the differential Equation (10.57), it can be solved but the solution will be omitted here. In most cases only numerical solutions are possible.

Comparison Between Theory and Experiments

The theoretical calculations of the solidification and cooling shrinkages fit very well with the observed values of the pipe depth, as can be seen in Figure 10.49. The assumptions we

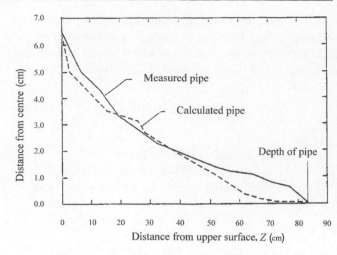

Figure 10.49 Pipe formed at the end of continuous casting (casting rate 1.9 m/min) compared with the shape of the calculated pipe. Reproduced by permission of ASM International.

have made are obviously realistic. The model of pipe formation during continuous casting given above is acceptable.

Figure 10.42 illustrates the appearance of a typical pipe during continuous casting.

10.7 THERMAL STRESS AND CRACK FORMATION DURING SOLIDIFICATION AND COOLING PROCESSES

10.7.1 Basic Concepts and Relationships

A bimetallic thermometer consists of two ribs made of different metals and firmly attached to each other. The double rib is straight at a certain temperature. If its temperature is increased or decreased, it bends in one direction or in the opposite direction. Its motion can be transferred to a pointer and the thermometer can be calibrated and graded (Figure 10.50).

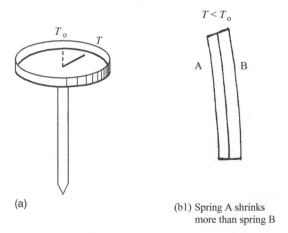

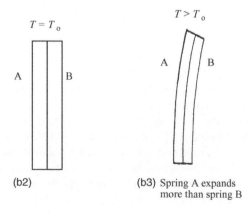

Figure 10.50 (a) Bimetallic thermometer. (b1–b3) Principle of a bimetallic thermometer.

Some Concepts and Definitions

In the example with the bimetallic thermometer, both ribs have the same length at the initial temperature. When the latter is changed, the ribs increase or decrease their lengths by different amounts. This causes mechanical forces in the materials, which result in bending of the ribs, which creates new equilibrium positions as a function of the temperature change.

On a temperature change, the material 'tries' to change its state of extension due to the thermal strain. If the thermal strain is allowed to develop freely, no stress will arise. If the thermal strain is prevented, stress will appear in the material.

Forces in solid materials, caused by temperature changes, are called *thermoforces*. To be able to treat them quantitatively we will repeat some concepts and laws of solid mechanics.

By *stress* we mean *force per unit area*. It the force F is perpendicular to the surface with area A, the stress is called *normal stress*, which is designated by σ:

$$\sigma = \frac{F}{A} \qquad (10.58)$$

If the force lies in the surface plane, the stress is called *shear stress* and is designated by τ:

$$\tau = \frac{F}{A} \qquad (10.59)$$

By *strain* we mean the *relative length change*, i.e. the *ratio of the length change to the length*:

$$\varepsilon = \frac{\Delta l}{l} \qquad (10.60)$$

where ε is a dimensionless quantity.

Hooke's Laws of Tensile and Shear Stresses

Bodies that are exposed to stresses will be deformed. The deformations may have two alternative sources:

- mechanical forces;
- temperature differences in the material.

Deformation on Tensile Stress

If the relative length change or the strain ε, which arises in a rod owing to the *tensile stress* σ, is small, then the tensile stress is proportional to the strain:

$$\sigma = E\varepsilon \qquad \text{Hooke's law} \qquad (10.61)$$

where the constant E is a material constant, which is called the *modulus of elasticity*.

The stress σ is *positive* for a *tensile* force and *negative* for a *compressive* force. Both are normal stresses. The strain ε is *positive* in the case of *elongation* and *negative* on *contraction*.

Lateral Contraction

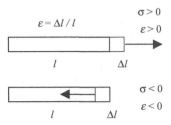

Figure 10.51 Deformation of a bar exposed to a tensile force (top) and a compressive force (bottom).

When the rod in Figure 10.51 is elongated owing to the tensile stress σ, its cross-section will shrink. At small tensile stresses the *relative transverse contraction* $\varepsilon_{\text{trans}}$ is proportional to the strain in the length direction:

$$\varepsilon_{\text{trans}} = -\nu\varepsilon \qquad (10.62)$$

where the proportionality constant ν is called *Poisson's number*. For metals this number usually has a value of ~ 0.3.

Deformation on Shear Stress

A shear stress (the force in the surface plane) gives rise to a deformation of a body or an element. The deformation caused by a shear stress is called *shear* or *shearing*. The shearing γ is the change of an originally right-angle (see Figure 10.52). At low loads, Hooke's law of shear is valid:

$$\tau = G\gamma \qquad (10.63)$$

where G is the *modulus of shearing*.

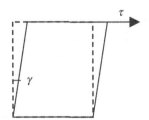

Figure 10.52 Shear stress on a rectangular plate.

Relation between Modulus of Elasticity and Modulus of Shearing

Many materials, especially metals, are isotropic, i.e. have the same properties in all directions. For such materials it is possible to derive a relation between the modulus of elasticity E, the modulus of shearing G, and Poisson's number ν:

$$G = \frac{E}{2(1+\nu)} \qquad (10.64)$$

Stresses and Strains in Materials caused by Temperature Differences

A rod of length l is exposed to a temperature change ΔT and no external forces act on it (Figure 10.53). Then we have:

$$l + \Delta l = l(1 + \alpha \Delta T) \tag{10.65}$$

where

l = length of rod at temperature T
Δl = change of length on temperature change ΔT
α = length strain coefficient.

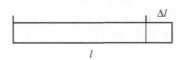

Figure 10.53 Thermal strain of a rod.

The temperature change ΔT gives rise to the length change Δl:

$$\Delta l = l\alpha\Delta T \tag{10.66}$$

$\Delta l/l$ is equal to the strain ε and we obtain:

$$\varepsilon = \alpha\Delta T \tag{10.67}$$

If the rod is exposed to *both* a temperature change ΔT *and* a stress σ simultaneously, we can calculate the strains which appear independently of each other and then add them (super-position principle). Hence the total strain can be written as:

$$\varepsilon = \frac{\sigma}{E} + \alpha\Delta T \tag{10.68}$$

Equation (10.68) can be used to calculate the thermal stress in a rod as follows. The rod has the initial position that is illustrated in Figure 10.54 (a). If the rod were free to expand in the x-direction [Figure 10.54 (b)], the strain would be $\alpha\Delta T$. On the other hand, if the rod were fixed between two rigid walls, it would be prevented from extension [Figure 10.54 (c)].

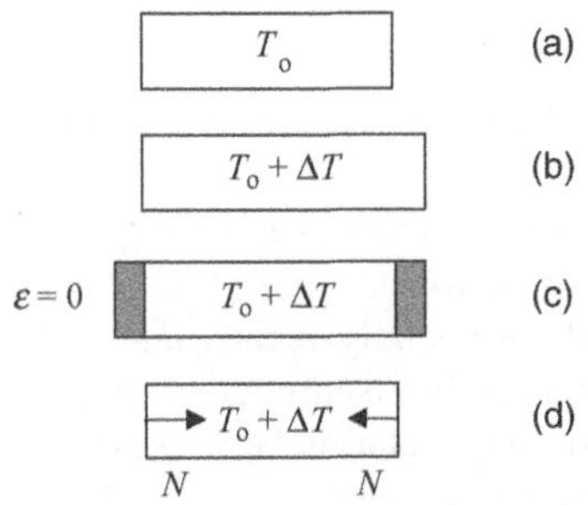

$$
\begin{array}{ll}
\boxed{\quad T_0 \quad} & \text{(a)} \\[4pt]
\boxed{\quad T_0 + \Delta T \quad} & \text{(b)} \\[4pt]
\varepsilon = 0 \quad \boxed{\quad T_0 + \Delta T \quad} & \text{(c)} \\[4pt]
\boxed{\rightarrow T_0 + \Delta T \leftarrow} & \text{(d)} \\
N \qquad\qquad N
\end{array}
$$

Figure 10.54 (a) Initial position; (b) free strain; (c) prevented strain; (d) prevented strain. Normal forces replace the rigid walls.

The condition $\Delta l = 0$ is valid and consequently $\varepsilon = 0$ and thermal stress arises in the material.

The thermal stress σ which acts on the cross-section of the fixed rod if the temperature increases by an amount ΔT can be calculated if we insert the condition $\varepsilon = 0$ into Equation (10.68):

$$0 = \frac{\sigma}{E} + \alpha\Delta T \tag{10.69}$$

i.e. the thermal stress can be written as:

$$\sigma = -E\alpha\Delta T \tag{10.70}$$

The negative sign in Equation (10.70) indicates that the stress is directed inwards and the forces are compressive.

Hooke's Generalized Law

Consider a small volume element of an isotropic material which is exposed to a tensile stress σ and a temperature increase ΔT (Figure 10.55). We want to calculate the strain which they cause.

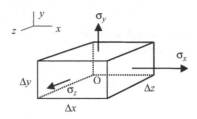

Figure 10.55 The components of the tensile stress in the x-, y-, and z-directions have only been drawn on three of the six surfaces for the sake of clarity. On the other surfaces there are normal stresses of corresponding sizes and opposite directions $-\sigma_x$, $-\sigma_y$, and $-\sigma_z$.

In the general case, we start with Equation (10.68) and calculate the components of the strain in the x-, y-, and z-directions. In the calculation the lateral contraction has to be considered. With the aid of Equations (10.62) and (10.68), we obtain Table 10.3, which gives Hooke's generalized law in components:

$$\varepsilon_x = \frac{1}{E}[\sigma_x - \nu(\sigma_y + \sigma_z)] + \alpha\Delta T \tag{10.71a}$$

$$\varepsilon_y = \frac{1}{E}[\sigma_y - \nu(\sigma_x + \sigma_z)] + \alpha\Delta T \tag{10.71b}$$

$$\varepsilon_z = \frac{1}{E}[\sigma_z - \nu(\sigma_x + \sigma_y)] + \alpha\Delta T \tag{10.71c}$$

TABLE 10.3 Hooke's generalized law in coefficients of the components.

	σ_x	σ_y	σ_z	ΔT
ε_x	$\dfrac{1}{E}$	$-\dfrac{\nu}{E}$	$-\dfrac{\nu}{E}$	α
ε_y	$-\dfrac{\nu}{E}$	$\dfrac{1}{E}$	$-\dfrac{\nu}{E}$	α
ε_z	$-\dfrac{\nu}{E}$	$-\dfrac{\nu}{E}$	$\dfrac{1}{E}$	α

Below we will restrict the treatment to one-dimensional calculations of thermal stress with known temperature distributions as a basis. Initially we will discuss some simple cases to show the principle of the calculations.

Thermal Stresses in Materials

Thermal stresses may exist in materials that have been produced from melts, which solidify or become heated in connection with design or use. Panes of glass and art glass may, for example, contain enclosed thermal stresses, which may result in a breakdown without any obvious external reason. Other examples are turbines, jet engines, and nuclear reactors, where thermal stresses have to be considered during construction and use.

The following types of equations are set up to solve thermoelastic problems:

1. Equilibrium equations, which involve the forces and stresses, which act on the material.
2. Geometric relationships, which the strains have to fulfil for compatibility reasons.
3. Equations which describe the materials (e.g. Hooke's law) and involve stresses, strains and temperature differences.

The above equations occur in all problems where thermal stresses and thermal strains are to be calculated. In each special problem there are, in addition:

4. Specific boundary conditions.

Many of the listed equations are linear partial differential equations. The solutions are *additive*, i.e. a combined problem can be divided into several separate sub-problems, which can be solved separately and then be added to give the final solution (the superposition principle).

In cases where one-dimensional strain is considered, the calculations will often be considerably simpler if a fictive so-called *fixed normal force*:

$$dN(\Delta T) = -E\alpha\Delta T dA$$

is introduced. It acts on each small element to prevent strain when the temperature increases by an amount ΔT (Figure 10.56). ΔT is a function of one or several position-determining coordinates. The fixed normal force does not exist in reality and it is therefore necessary to introduce an

additional force $F(\Delta T)$ of equal size and opposite direction to the fixed normal force $N(\Delta T)$:

$$F(\Delta T) = \iint\limits_{A} E\alpha\Delta T dA \qquad (10.72)$$

From Equation (10.72), the *average stress*, which acts on a cross-section A, can be calculated:

$$\sigma_{\text{ave}} = \frac{F}{A} \qquad (10.73)$$

The reason for using the *average stress* in the calculations is that it does not matter in what way all the sub-forces are distributed at positions which deviate slightly from the end surface. This statement is called *Saint-Venant's principle*. It is therefore possible to use F in the calculations instead of all the small real sub-forces and the average stress instead of the real stresses, which act on the surface elements.

The thermal stress, which acts on an internal area element, is equal to the sum of the average stress, caused by the force $F(\Delta T)$ and the negative fixed normal stress:

$$\sigma^{\text{T}} = \frac{\iint\limits_{A} E\alpha\Delta T dA}{A} - E\alpha\Delta T \qquad (10.74)$$

An example will illustrate the principle of calculation.

Example 10.9

The temperature in a rectangular metal plate with constant thickness $2c$ is a linear function of y and independent of x and z (see the figure). The plate can expand freely only in the y-direction. Calculate the thermal stresses in the x- and z-directions if:

$$\Delta T = \frac{y}{c}\Delta T_0$$

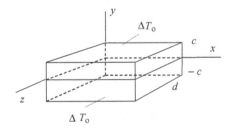

Temperature distribution in a thin rectangular metal plate with the sides $2l$ and $2c$.

Solution:

Since the plate can expand freely in the y-direction, no stress will appear in this direction, i.e.

$$\sigma_y = 0 \qquad (1')$$

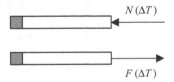

Figure 10.56 The fictive fixed normal force N and the opposite force F acting on a rod.

We also know that no expansion can occur in the *x*- and *z*-directions, i.e.

$$\varepsilon_x = \varepsilon_z = 0 \qquad (2')$$

Inserting Equations (1′) and (2′), into Equation (10.71a) gives:

$$0 = \varepsilon_x = \frac{1}{E}(\sigma_x - \nu\sigma_z) + \alpha\Delta T \qquad (3')$$

Inserting Equations (1′), (2′), and (3′) into Equation (10.71c) gives:

$$0 = \varepsilon_z = \frac{1}{E}(\sigma_z - \nu\sigma_x) + \alpha\Delta T \qquad (4')$$

By combining Equations (3′) and (4′) we obtain the fixed normal stresses:

$$\sigma_x = \sigma_z = \frac{-E\alpha\Delta T}{1 - \nu} \qquad (5')$$

We insert the relationship $\Delta T = (y/c)\Delta T_0$ into Equation (5′) to give:

$$\sigma_x = \sigma_z = \frac{-E\alpha\Delta T_0}{1 - \nu}\frac{y}{c} \qquad (6')$$

The average stresses will be

$$\overline{\sigma_x} = \overline{\sigma_z} = \frac{F_x}{2cd} = \frac{1}{2cd}\int_{-c}^{+c} E\alpha\,\Delta T\,d\,\mathrm{d}y = \frac{E\alpha\,\Delta T_0\,d}{2cd}\int_{-c}^{+c}\frac{y}{c}\,\mathrm{d}y = 0 \qquad (7')$$

Equations (6′), (7′), and the given relationship are inserted into Equation (10.74), which gives:

$$\sigma_x = \sigma_z = 0 - \frac{E\alpha\,\Delta T_0}{1 - \nu}\frac{y}{c}$$

The total thermal stress will therefore be equal to the expressions in Equation (6′).

Answer:
The stresses are $\sigma_x = \sigma_z = \dfrac{-E\alpha\,\Delta T_0}{1 - \nu}\dfrac{y}{c}$

The maximum thermal stress in Example 10.9 will be $(y = c)$:

$$\sigma_x = \sigma_z = \frac{-E\alpha\,\Delta T_0}{1 - \nu} \qquad (10.75)$$

It is true that the thickness of the plate is not included in Equation (7′), but a thicker plate generally has a greater temperature difference between its two parallel surfaces than a thin plate. The risk that a plate of brittle material breaks owing to thermal stresses is therefore greater for a thick than for a thin plate.

In many applications, one side of the plate stays in contact with hot gases with varying temperatures. The result is a varying heat flow through the plate, caused by the temperature variations and alternating positive and negative strains of the material. This gives a risk of cracks in the material after prolonged use.

10.7.2 Thermal Stress and Strain during Solidification and Casting Processes

During ingot casting and continuous casting, a solidifying shell is formed as soon as the melt comes into contact with the chill-mould. During the solidification and cooling process, the temperature varies strongly as a function of position and time. Strong thermal stresses appear as a consequence of the temperature gradients in the solidifying shell.

If the thermal stresses are so large that the elasticity limit is exceeded, surface cracks, intermediate cracks, or centre cracks appear in the material, and cause serious deterioration of the final product.

The sizes of the thermal stresses also depend on the rate of the temperature variations in the material. If they are very rapid, the effect will be larger than with slower temperature variations and is described as a *thermal shock*.

If a material is exposed to repeated cyclic temperature variations, the effect can be described as *thermal fatigue* of the material. The phenomenon must be taken into consideration during construction and production. In addition, it is necessary to watch thermal fatigue carefully because it leads to a risk of crack formation, which under special circumstances may have disastrous consequences, for example for jet engines in aircraft and tubes in nuclear power plants.

Below we will discuss different types of thermal stresses, mathematical models for calculation of thermal stresses and the corresponding strains, and also thermal stress and strain during continuous casting.

10.7.3 Mathematical Models of Thermal Stress during Unidirectional Solidification

The condition for the calculation of thermal stresses is complete knowledge of the temperature in the material as a function of position and time.

Two mathematical models are required for thermal stress calculations, one for calculation of the temperature variations and another for calculation of the mechanical stress strain process on the basis of the known temperature distribution in position and time.

This is a difficult and time-consuming task. In the general case there are seldom exact, analytical solutions. Many different calculation methods have been developed, which give more or less accurate solutions. The most common numerical method is based on the *finite element method* (FEM). Another method is based on bar theory.

In some cases, the calculations can be simplified by calculations in one or two instead of three dimensions. One may, for example, assume that a stress or a strain is planar.

One example of a planar stress is a thin plate with a temperature $T(x, y)$, which only varies in the plane of the plate and is constant in the z-direction, i.e. in the direction of thickness. The stresses in the z-direction can be neglected, i.e. stresses occur only in the xy-plane.

It is reasonable to assume a planar strain when one dimension dominates and the temperature is a function of this dimension. An example is a long cylinder with fixed ends. There is no axial motion, which means that $\varepsilon_z = 0$.

The simplest analysis of thermal stresses involves only elastic stresses (normal stresses and shear stresses). If the material is very ductile and the thermal stresses are strong enough, the strain will no longer depend linearly on the stress but a non-linear plastic term (plastic stress) has to be added, which complicates the solution considerably.

Thermal Stress During Unidirectional Solidification

During ingot casting, and especially continuous casting, different types of cracks and other defects appear in the solidified material. These cracks arise during the cooling process when the steel has a temperature slightly below the solidus temperature. Hence it is more important to study the strains than the stresses themselves.

Calculations of thermal stresses and strains during unidirectional solidification are much more complicated than in Example 10.9 (pages 345–346), where we had a stationary temperature field. On unidirectional solidification the temperature is a function of both position and time.

For a careful calculation of the temperature distribution during unidirectional solidification, it is necessary to consider the temperature field both in the solidifying shell and in the two-phase region.

Calculation of the Temperature Distribution in the Solid Phase

The temperature distribution during unidirectional solidification is in principle as shown in Figure 10.57.

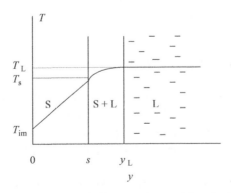

Figure 10.57 Temperature distribution in unidirectional solidification with no excess temperature.

The heat flow occurs in the y-direction and the problem can therefore be treated as one-dimensional. T_L is constant whereas the surface temperature T_{im} of the metallic shell and the thickness s and the position of the solidification front y_L are functions of time.

The calculation of the temperature distribution and the thermal stresses during the solidification and cooling process can be performed by three different methods:

1. We can assume $T(y, t)$ to be a function of second degree of y and determine the constants with the aid of known boundary conditions.
2. We assume $T(y, t)$ to be a function of third degree of y and determine the constants with the aid of known boundary conditions.
3. We use FEM.

FEM is the most accurate method and can be regarded as the best answer. A comparison between FEM calculations and calculations based on method 2 is given in Figure 10.58.

Calculation of Thermal Stresses and Strains

Calculation of thermal stresses and strains during unidirectional solidification involves the temperature T as a function of both the position y and the time t, as indicated above.

On the basis of the alternative temperature fields derived with the aid of FEM and with third-degree function of y, the strains and thermal stresses can be calculated. This task is much more difficult than the calculations we performed in Example 10.9. In that case we had a stationary temperature field, i.e. T was a function of position but not of time.

It can be shown that the total strain ε^u as a function of y and t during unidirectional solidification can be written as:

$$\varepsilon^u(y, t) = \int_{t*}^{t} \left[\frac{1}{s} \int_{0}^{s} \frac{\partial \varepsilon^T}{\partial t} dy \right] dt + \varepsilon^T(T_{rigid}) - \varepsilon^T(T) \quad (10.76)$$

where

ε^u = total strain on unidirectional solidification
ε^T = thermal strain caused by the sum of the average thermal stress and the negative fixed normal stress (page 345)
s = position of the solidification front at time t
t^* = the time when the solidification front reaches position y
T = temperature at position y and time t
T_{rigid} = temperature at which the material is rigid enough to resist compressive and tensile stresses, chosen between T_L and T_s and often at the temperature when the fraction of solid is 0.60.

The coordinate system has been chosen perpendicular to the solidification front with $y = 0$ at the surface of the shell; s is the position of the solidification front at time t and y is situated within the solid phase, i.e. within the limits $0 \leq y \leq s$.

At the first sight, it is difficult to see that the right-hand side of Equation (10.76) is a function of y. However, the lower limit t^* is a function of y as t^* is the time when the solidification front reaches position y. The relationship can be written as:

$$s(t^*) = y \qquad (10.77)$$

The result of the calculations [Equation (10.76) in combination with the known temperature distribution], which are performed by a computer, is shown in Figure 10.58. It can be seen that a temperature field of the third degree gives good agreement with 'the best answer', FEM, especially for low y-values.

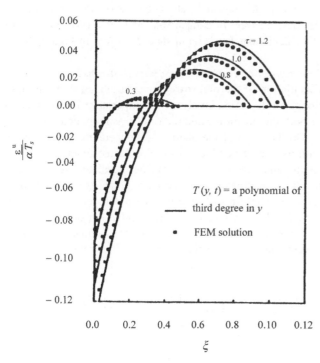

Figure 10.58 Calculated strains ε^u on unidirectional solidification as a function of the distance y from the surface of the solidified shell at various times. In the figure dimensionless quantities are used: $\xi = (Q_0^*/kT_s)y$, where $Q_0^* =$ thermal flow at $t = 0$ and $y = 0$; $\tau = (Q*/T_s)^2 (t/\rho ck)$, where $Q* =$ thermal flow at $t = 0$ and position y.

Thermal strain is a very important quantity because it is related to thermal stress. The basic relation is Equation (10.70). Stress causes strong forces in materials that may lead to crack formation. The stress must not exceed a certain maximum value to eliminate the risk of cracks. The maximum stress corresponds to a maximum strain, which must not be exceeded if cracks are to be avoided.

An old 'rule of thumb' is to keep the strain lower than 0.2 %. Today, 'sharper' but more complicated and less 'transparent' criteria for the allowed strain are used. They are based on the type of calculations which have been sketched above as the basis for Figure 10.58. A simplified version is given in Example 10.10 below.

Simplified Method for Calculation of the Strain During Unidirectional Solidification

The calculations behind the curves in Figure 10.58 are extensive and time consuming even if computers are used. It may be worth while to try to find simplified methods for the calculations even if simplified assumptions are necessary. Obviously, these simplifications will lead to a worse result, compared with reality, than the methods described above. The advantage is that comparatively rapid and fairly good information is obtained, which can immediately be applied and tested in industry. Further, a better understanding of the course of the calculations is achieved.

We make the following assumptions:

1. No consideration is taken to the solidification interval of the alloy. We start with a known temperature distribution, which is valid for a pure metal. The assumption implies that $T_s = T_L$.
2. Temperature distribution and functions for the surface temperature and thickness of the solidifying shell are taken from Chapters 4 and 5.

The calculation method is illustrated concretely as a solved example. In analogy with the more accurate calculations, we will use Equations (10.71) for calculation of the strains.

Example 10.10

In a cast house one worries about crack formation during continuous casting and wants to design the casting process in such a way that the risk of cracks and other defects, caused by thermal stresses, becomes a minimum. A first step is to map the risks of crack formation. The task to do this is entrusted to you.

Derive approximate expressions of:

(a) the thickness s of the solidifying shell as a function of time;
(b) the surface temperature T_{im} of the solidifying shell as a function of the thickness s of the shell;
(c) the strain ε^u as a function of position y;
(d) Plot ε^u as a function of position y in a diagram.

Perform calculations (a)–(c) and give the strain as a function of y in the shape of a table and a diagram at two alternative shell thicknesses, $s = 0.01$ and 0.02 m.

Material costants	
k	$30\,\text{W/m K}$
h	$750\,\text{W/m}^2\,\text{K}$
$-\Delta H$	$2.17 \times 10^{10}\,\text{J/m}^3$
T_s	$1500\,°\text{C}$
α	$10^{-5}\,\text{K}^{-1}$

For the planned casting process, the data given in the table are valid. The temperature of the cooling water $T_0 \le 100\,°\text{C}$.

In the calculations the known functions for the surface temperature T_im [Equation (4.45) on page 74] and the time required to achieve a given thickness y [Equation (4.48) on page 74] can be used. They are given below as Equations (1') and (2'). The relationship between T and y (Figure 4.17 on page 73 is also given below.)

$$T_\text{im} = \frac{T_\text{L} - T_0}{1 + \frac{h}{k} y_\text{L}(t)} + T_0 \tag{1'}$$

and

$$t = \frac{\rho(-\Delta H)}{T_\text{L} - T_0} \frac{y_\text{L}}{h} \left(1 + \frac{h}{2k} y_\text{L}\right) \tag{2'}$$

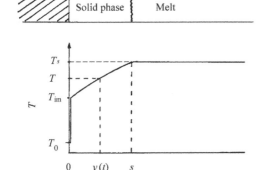

Solution:

(a) In order to obtain the thickness of the solidifying shell as a function of time, we start with Equation (2') with the new designations:

$$t = \frac{\rho(-\Delta H)}{h(T_\text{s} - T_0)} s \left(1 + \frac{h}{2k} s\right) \tag{3'}$$

and solve s as a function of t from this second-order equation:

$$s^2 + \frac{2k}{h} s = \frac{2k(T_\text{s} - T_0)}{\rho(-\Delta H)} t \tag{4'}$$

which has the positive solution:

$$s = -\frac{k}{h} + \sqrt{\left(\frac{k}{h}\right)^2 + \frac{2k(T_\text{s} - T_0)}{\rho(-\Delta H)} t} \tag{5'}$$

It is the desired solution.

(b) Equations (1') and (2') have to be adapted to the designations used in Equation (10.76). The designation y_L is replaced by s and T_L by T_s.

T_im is the temperature of the solidifying shell, close to the mould. Hence we obtain directly:

$$T_\text{im} = \frac{T_\text{s} - T_0}{1 + \frac{h}{k} s} + T_0 \tag{6'}$$

(c) To find the desired function, we start with Equation (10.76), which is valid for unidirectional solidification:

$$\varepsilon^\text{u}(y, t) = \int\limits_{t^*}^{t} \left[\frac{1}{s} \int\limits_0^s \frac{\partial \varepsilon^\text{T}}{\partial t} dy\right] dt + \varepsilon^\text{T}(T_\text{s}) - \varepsilon^\text{T}(T) \tag{7'}$$

where t^* is the time when the solidification front passes position y:

$$s(t^*) = y \tag{8'}$$

We know from Example 10.9 (page 345) that $\varepsilon^\text{T} = \alpha\Delta T$ and assume that α is a constant, independent of time and temperature. We have:

$$\alpha = \frac{d\varepsilon^\text{T}}{dT} = \text{constant}$$

and we obtain:

$$\varepsilon^\text{T}(T_\text{s}) - \varepsilon^\text{T}(T) = \alpha(T_\text{s} - T) \tag{9'}$$

We will also need the function $\partial \varepsilon^\text{T}/\partial t$, which is obtained by partial derivatization of Equation (9') with respect to time:

$$\frac{\partial \varepsilon^\text{T}}{\partial t} = \alpha \frac{\partial T}{\partial t} \tag{10'}$$

In order to obtain a useful expression for $\partial T/\partial t$, we will set up the equation of the inclined line in the figure on this page:

$$T(y, t) = \frac{T_\text{s} - T_\text{im}}{s} y + T_\text{im} \tag{11'}$$

Partial derivatization of Equation (11') with respect to time gives:

$$\frac{\partial T}{\partial t} = \frac{s \left(-\frac{\partial T_\text{im}}{\partial t}\right) - \frac{\partial s}{\partial t}(T_\text{s} - T_\text{im})}{s^2} y + \frac{\partial T_\text{im}}{\partial t} \tag{12'}$$

Q1

Equation (12′) introduces another unknown derivative, $\partial T_{im}/\partial t$. This can be obtained with the aid of Equation (6′), which can be transformed into:

$$T_{im} = \frac{T_s - T_0}{1 + \frac{h}{k}s} + T_0 = \frac{T_s + T_0 \frac{h}{k}s}{1 + \frac{h}{k}s} \qquad (13')$$

Equation (13′) is partially derivatized with respect to time to give:

$$\frac{\partial T_{im}}{\partial t} = \frac{T_0 \frac{h}{k}\frac{\partial s}{\partial t}\left(1 + \frac{h}{k}s\right) - \frac{h}{k}\frac{\partial s}{\partial t}\left(T_s + T_0 \frac{h}{k}s\right)}{\left(1 + \frac{h}{k}s\right)^2} \qquad (14')$$

We must also find an expression for $\partial s/\partial t$. It is obtained by partial derivatization of Equation (5′) with respect to time:

$$\frac{\partial s}{\partial t} = \frac{\frac{k(T_s - T_0)}{\rho(-\Delta H)}}{\sqrt{\left(\frac{k}{h}\right)^2 + \frac{2k(T_s - T_0)}{\rho(-\Delta H)}t}} \qquad (15')$$

Inserting Equations (9′), (10′), (12′), and (14′) into Equation (7′) and integration gives:

$$\varepsilon^u(y, t) = \alpha \int_{t^*}^{t} \left[\frac{1}{s} \int_0^s \frac{s\left(-\frac{\partial T_{im}}{\partial t}\right) - \frac{\partial s}{\partial t}(T_s - T_{im})}{s^2} y + \frac{\partial T_{im}}{\partial t} \right] dy\, dt + (T_s - T)$$

$$= \alpha \int_{t^*}^{t} \left[\frac{1}{s} \int_0^s \frac{s\left(-\frac{\partial T_{im}}{\partial t}\right) - \frac{\partial s}{\partial t}(T_s - T_{im})}{s^2} \frac{s^2}{2} + \frac{\partial T_{im}}{\partial t} s \right] dt + (T_s - T)$$

or

$$\varepsilon^u(y, t) = \alpha \int_{t^*}^{t} \left[\int_0^s \frac{\frac{\partial T_{im}}{\partial t}}{2} - \frac{\frac{\partial s}{\partial t}(T_s - T_{im})}{2s} \right] dt + (T_s - T)$$

$$(16')$$

which is a function of y as t^* is a function of y.
(d) The expressions for $\partial T_{im}/\partial t$, $\partial s/\partial t$, T_{im}, s, and T are taken from Equations (14′), (15′), (5′), (6′), and (11′), respectively, for calculation of ε^u. t^* is calculated from Equation (8′).

The calculations are performed with the aid of a computer. The result is given in the table shown.

y_t (m)	t^* (s)	$\varepsilon^u(y, t)$ for $s = 0.010$ m	y (m)	t^* (s)	$\varepsilon^u(y, t)$ for $s = 0.020$ m
0.0001	0.18	−0.156	0.0001	0.18	−0.49
0.0010	1.85	−0.099	0.0020	3.74	−0.30
0.0019	3.55	−0.052	0.0039	7.45	−0.14
0.0028	5.24	−0.015	0.0058	11.36	−0.02
0.0037	7.07	0.013	0.0077	15.41	0.06
0.0046	8.88	0.033	0.0096	19.63	0.11
0.0055	10.73	0.045	0.0115	24.01	0.14
0.0064	12.61	0.050	0.0134	28.56	0.15
0.0073	14.54	0.047	0.0153	33.27	0.13
0.0082	16.51	0.038	0.0172	38.15	0.09
0.0091	18.50	0.022	0.0191	43.20	0.03
0.0100	20.54	0.000	0.0200	45.64	0.00

Answer:

(a) $s = -\frac{k}{h} + \sqrt{\left(\frac{k}{h}\right)^2 + \frac{2k(T_s - T_0)}{\rho(-\Delta H)}t}$

(b) $T_{im} = \frac{T_s - T_0}{1 + \frac{h}{k}s} + T_0$

(c) See Equation (16′).

(d) See table above and diagram below.

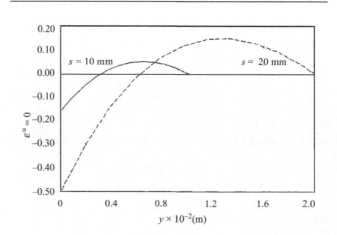

This type of strain calculation is used together with knowledge of the mechanical properties of solid alloys discussed in Section 10.7.6. This is illustrated later in Example 10.11. Such calculations result in concrete criteria for risk of cracks. The aim is to control the casting operations and solidification processes of metals and alloys to avoid different types of cracks. These topics are discussed in Sections 10.7.7–10.7.9.

10.7.4 Air Gap Formation due to Thermal Contraction

In Sections 5.5.1 and 5.5.2 (pages 105–110) we discussed the formation of an air gap between a mould and a solidifying shell. The basis for the discussion was the heat transport through the shell across the air gap. The formation of an air gap was briefly described as a contraction of the solidified shell.

Here we will complete the discussion on air gaps, analyse the phenomenon from the aspect of thermal stress, and report some general experimental results.

Principal Description of Air Gap Formation
Consider a cylindrical chill-mould wall (Figure 10.59) under a constant hydrostatic pressure. Immediately after the casting, which is assumed to occur instantaneously, a thin shell is formed close to the chill-mould wall. The thickness of the shell grows, its average temperature decreases and the tem-

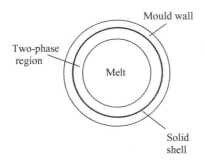

Figure 10.59 Cylindrical chill-mould with a solidifying melt. An air gap has been formed between the mould wall and the solid shell.

perature gradient becomes steeper. As the cooling shell gradually grows, the hydrostatic pressure from the melt will press the shell towards the mould but the temperature decrease dominates over this effect after a certain time and the cooling shell contracts in all directions.

If the cast metal is an alloy with a wide solidification interval, the whole shell is not compact at the beginning but contains melt between the crystals and between the dendrites of each crystal with a higher concentration of alloying elements than the solid phase. This is also the situation for the inner part of the casting during the continued solidification process.

The surface wall of the chill-mould warms up owing to the heat transport from the solid shell. Its temperature increases gradually and the surface wall *expands* slightly and moves

inwards at the beginning. However, the contraction of the shell is larger than the expansion of the mould wall. At a certain thickness the shell loses contact with the chill-mould and an air gap, which grows continuously, is formed.

Correlation between Air Gap Width and Heat Transfer Coefficient
Intense efforts have been made to determine the overall heat transfer coefficient and its time dependence. All such experiments show uniformly the same result. A typical example is given in Figure 10.60, which shows the heat transfer coefficient at permanent mould casting as a function of time.

The heat transfer coefficient reaches a maximum after a

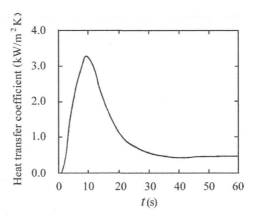

Figure 10.60 Heat transfer coefficient of an Al–Si alloy during permanent mould casting as a function of time. Reproduced by permission of ASM International.

rapid increase, followed by a gradual decrease to a constant value. The rapid increase during the first seconds can be explained as a gradual build-up of the metallostatic pressure during mould filling.

A large number of parameters, among them the metallostatic pressure from the melt and the roughness of the mould surface, influence the maximum value of h. After reaching the maximum value, h starts to decrease, owing to air gap formation. Measurements on the air gap formation process reveal a complicated problem. Nevertheless, a large number of successful results have been reported. In most cases transducers have been used. An example of such measurements will be presented here.

The relative displacements of the mould and the metal, due to expansion and contraction caused by temperature variations, are measured. The result is illustrated in Figure 10.61. Movements towards the centre of the cylinder are considered positive. Negative values correspond to movements towards the periphery.

The upper curve in Figure 10.61 shows the contraction of the metal. The lower curve represents the expansion of the

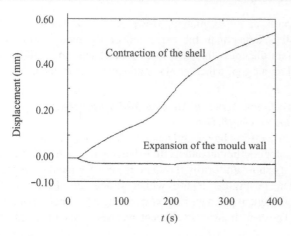

Figure 10.61 Average displacement of the metal (upper line) and the mould (lower line) as a function of time. Reproduced by permission of Castings Technology International.

mould when the solidified shell is strong enough to withstand the metallographic pressure.

The measurements have been analysed theoretically. The air gap was calculated by use of thermal contraction due to temperature gradients. During the solidification process, a certain fraction of lattice defects is formed. The defects become annealed and the material shrinks during this process.

The calculated air gap and the heat transfer coefficient, derived from the air gap, vary with time, as shown in Figures 10.62 and 10.63. Time is chosen as zero at the start of the solidification. In Figure 10.62, the calculated thermal air gap and the air gap due to vacancy condensation are shown together with the sum of the two effects.

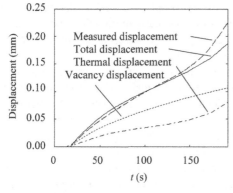

Figure 10.62 Displacement of solidifying shell due to thermal shrinkage, vacancy condensation, and the sum of the two effects. The calculated total displacement is compared with experimental values. Reproduced by permission of Castings Technology International.

The calculations show that it is not possible to fit the calculated thermal air gap, due to thermal shrinkage only, with the measured air gap. Addition of the effect from vacancy condensation to the thermal shrinkage gives

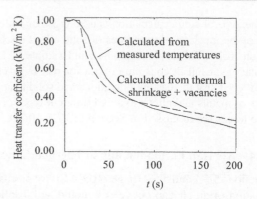

Figure 10.63 Heat transfer coefficient from temperature measurements and calculated from air gap due to thermal shrinkage and vacancy condensation. Reproduced by permission of Castings Technology International.

results which fit with the measured data for the contraction of the casting during the solidification.

Even the small quantities of entrapped vacancies, found in these experiments, have a large influence on the gap formation.

The effect of the air gap is expressed macroscopically as a heat transfer coefficient of the gap with the aid of the well-known relation:

$$h = \frac{k}{\delta} \tag{10.78}$$

where

h = heat transfer coefficient of air
k = thermal conductivity of air
δ = width of the air gap.

The heat transfer coefficients that correspond to the calculated air gaps can be derived with the aid of Equation (10.78). If these h values are plotted as a function of time, the dashed curve in Figure 10.63 is obtained.

The heat transfer coefficients were also determined from experimental temperature measurements. If these values are plotted as a function of time, the continuous curve in Figure 10.63 is obtained. The two curves show satisfactory agreement.

10.7.5 Air Gap Formation in Ingots

When a melt is teemed into a chill-mould, it starts to solidify because heat is transported away from the melt and outwards through the chill-mould. The solid phase nucleates on the cold mould surface and forms a layer, which grows inwards, perpendicular to the surface.

At the beginning of the solidification process, the solidification rate is mainly controlled by the heat transport from the solidifying shell to the chill-mould. A thin film of solidified steel is formed at the bottom and walls of the cold chill-mould, which fixes the external shape of the ingot. The thin shell cannot resist the pressure from the internal liquid steel mass but is pressed towards the chill-mould wall and becomes deformed in such a way that the outer dimensions of the shell initially remain constant.

When the thickness of the steel shell grows, its mechanical strength becomes large enough to resist the pressure from the liquid steel. The shell cools and shrinks and the ingot loses contact with the chill-mould wall, first in the upper part of the chill-mould where the pressure of the steel melt is lowest [Figure 10.64 (a)].

At the moment when the shell loses contact with the mould wall, the heat transport conditions becomes radically changed.

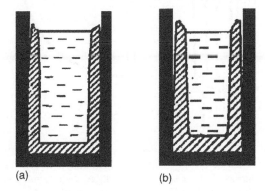

(a)   (b)

Figure 10.64 (a) The steel shell starts to lose contact with the upper part of the chill-mould. (b) The steel shell has partly lost contact with the chill-mould.

The heat transfer is strongly reduced owing to the air insulation between the ingot and the chill-mould. The temperature becomes lower in the lower part of the ingot because the heat transport is more efficient there. This results in an ingot shell with increasing thickness downwards. Later the ingot loses contact with the chill-mould wall entirely [Figure 10.64 (b)]. The continued heat transport through the steel shell controls the lateral solidification rate.

Figure 10.65 shows the width of the air gap as a function of the relative height of the ingot. The contact with the walls is not

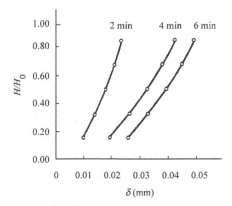

Figure 10.65 Air-gap thickness δ as a function of the relative height H/H_0 of the ingot and the solidification time (parameter). Reprinted with permission from Elsevier Science.

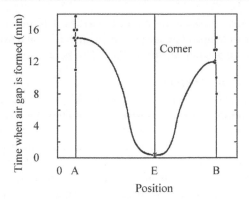

Figure 10.66 Time of air-gap formation as a function of position. $t = 0$ corresponds to the start of the casting process. Points A and B are explained in Figure 10.67. Reproduced by permission of Verlag Stahleisen GmbH, Düsseldorf, Germany © 1972.

lost uniformly. This is evident from Figure 10.66, which shows the time for air gap formation as a function of position. The air gap is initially formed at the corners and later on at the sides of the chill-mould. A cross-section of an ingot is shown in Figure 10.67.

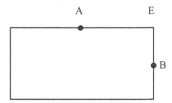

Figure 10.67 Cross-section of the ingot, seen from above.

In most cases, the shrinkage of the ingot shell causes a deformation of the ingot section because the shrinkage stress is inhomogeneously distributed in the corners and at the sides of the ingot, owing to cooling of various strengths.

10.7.6 Mechanical Properties of Steel at High Temperatures. Ductility. Hot Cracks

During solidification and/or hot working of metallic materials, so-called hot cracks will appear in the material at temperatures immediately below the solidus temperature. Bishop, Ackerlind, and Pellini have shown that thin films of melt occur in the interdendritic region and at the crystal boundaries. Such films lead to deterioration of the plastic properties and reduce the ductility of the material considerably. Hot cracks are caused by thermal strain in the material and propagate along the grain boundaries. Condensation of lattice defects also makes metals brittle. These matters are discussed on pages 356–357.

External forces and/or inconvenient design of the mould contribute to the hot cracks. By elimination of these sources of cracks as much as possible, the risk of hot cracks will be reduced. It is also important to know the plastic properties of the materials as a function of temperature for judgement of the size and extension of the thermal strain and hence the risk of hot cracks.

Experimental Method

Experimental determination of the ductility of steel as a function of temperature offers several difficulties. One reason is that the technique of measurement is difficult at the high temperatures in question. Another reason is that the microstructure of the material influences the result. The same ductility and thermal strain are *not* obtained at a certain temperature of measurement if the measurements are performed on directly solidified material (melt which has solidified and cooled to the temperature of measurement) or on reheated material (from room temperature to the temperature of measurement).

Fredriksson and Rogberg have developed equipment for measurement of ductile stresses at high temperatures (Figures 10.68 and 10.69). The heat source consists of three ellipsoid-shaped reflectors and three 800 W halogen lamps. The heating filament of each lamp is situated in one of the foci of the ellipsoid and the sample in the other common focus of all three ellipsoids. In this way, a sample can be heated to the liquidus temperature in 3–4 min. The molten zone adopts the shape of a pear and is kept in position by the surface tension. The temperature is measured with thermocouples.

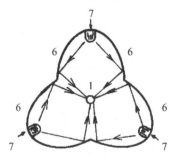

Figure 10.68 1, Sample material; 6, ellipsoid-shaped mirrors; 7, Heat sources. See also Figure 10.69. Reproduced with permission from the Scandinavian Journal of Metallurgy, Blackwell.

With the aid of thermocontrol equipment, the molten zone can be cooled at a predetermined cooling rate and in such a way a casting process can be simulated. When the temperature has reached the desired value, the sample is subjected to ductile stress. Ductile force and strain are measured simultaneously and are registered directly by a recorder.

Ductility of Steel at High Temperatures

As a measure of ductility, the relative area reduction ψ (%) is used:

$$\psi = \frac{A_{\text{before}} - A_{\text{after}}}{A_{\text{before}}} \times 100 \qquad (10.79)$$

where A_{before} and A_{after} are the cross-sectional area before and after the ductile test, respectively.

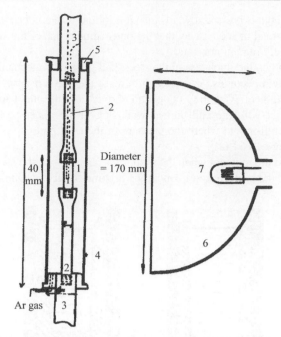

Figure 10.69 Sketch of a ductile test equipment. 1, Sample; 2, holder ; 3, tensile rods; 4, silica tube; 5, sealing joints; 6, ellipsoid-shaped mirrors; 7, heat source. Reproduced with permission from the Scandanvian Journal of Metallurgy, Blackwell.

Alloying elements or impurities with low solubility and low diffusion rate in the solid phase normally reduce the ductility and increase the tendency for crack formation strikingly. One example is phosphorus, which increases the crack formation in 25% Cr–10% Ni austenite steel. On the other hand, phosphorus does not change the tendency of crack formation in steel, which solidifies as ferrite, because its solubility and diffusion rate in ferrite are much higher than in austenite.

Figure 10.70 shows the area reduction ψ as a function of temperature for three different types of carbon steel, one

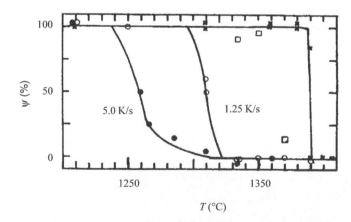

Figure 10.70 Ductility of steel as a function of temperature. × Reheated material; ○ directly solidified material, cooling rate 1.25 K/s; •, directly solidified material, cooling rate 5.0 K/s; □, heat-treated material. Reproduced with permission from the Scandinavian Journal of Metallurgy, Blackwell.

directly solidified, one reheated and one heat-treated. It can be seen that:

- The transition from ductile to brittle material (low ductility) occurs in a reheated material within a narrow temperature interval, which is called the *transition temperature*, T_{tr}. In the other cases the transition interval is wider.
- Reheated material has a considerably higher transition temperature than directly solidified material, 70–100 °C higher depending on the degree of homogenization.
- The higher the cooling rate, the lower will be the transition temperature.

The transition from brittle to ductile material can also be observed in other ways. One can measure the mechanical rupture limit σ_{rl}, which is required to pull apart the sample, as a function of temperature.

In Figure 10.71, σ_{rl} as a function of temperature within the temperature interval 1210–1410 °C is illustrated for the same carbon steel as in Figure 10.70. For reheated material, the transition from brittle to ductile material is as abrupt as in Figure 10.70. The transition temperature is 1390 °C.

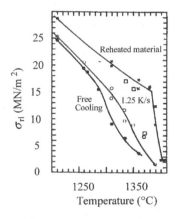

Figure 10.71 Ductile stress curves of carbon steel as a function of temperature. Symbols as in Figure 10.70. Reproduced with permission from the Scandinavian Journal of Metallurgy, Blackwell.

For directly solidified materials, the transition is not equally evident. At 1350 °C the samples with free cooling (5 K/s) still have a relatively high rupture limit (ductile strength) in spite of low ductility.

Ductility of Stainless Steel at High Temperatures
Stainless steels generally contain relatively high percentages of Cr and Ni, for example 18-8 steel, but also small amounts of other elements, for example C, Si, Mn, P, and N.

Fredriksson and Rogberg have performed ductility tests on about 20 different stainless-steel alloys, among them some with low Ni concentration, which solidify as ferrite (δ-solidification), and some with high Ni concentration, which solidify primarily as austenite (γ-solidification). This allowed a comparison between alloys with different structures.

Figure 10.72 (a) and (b) show the area reduction on δ- and γ-solidification, respectively, as a function of temperature. It is obvious that the transition temperatures of the γ-alloy are *lower* than those of the δ-alloy. Another difference between the γ- and δ-alloys is the influence of the cooling rate on the transition temperatures. These aspects are discussed on page 350.

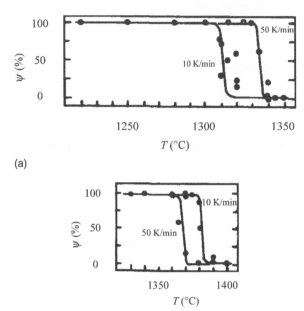

(a)

(b)

Figure 10.72 (a) Ductility as a function of temperature for an Fe(γ)–Cr–Ni alloy at two different cooling rates. The transition temperature T_{tr} increases with increasing cooling rate for the γ-alloy. (b) Ductility as a function of temperature for an Fe(δ)–Cr–Ni alloy at two different cooling rates. T_{tr} decreases with increasing cooling rate for the δ-alloy. Reproduced with permission from the Scandinavian Journal of Metallurgy, Blackwell.

The mechanical rupture limit σ_{ri} as a function of temperature for the δ- and γ-alloys is illustrated in Figure 10.73. Both alloys show similarities with the directly solidified steel

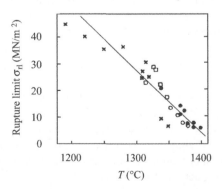

Figure 10.73 Maximum ductile stress, i.e. rupture limit, as a function of temperature at two different cooling rates for the γ-alloy and the δ-alloy. Cooling rates (K/min): γ-alloy, $\times$ 10 and $\square$ 50; δ-alloy, $\bullet$ 10 and $\circ$ 50. Reproduced with permission from the Scandinavian Journal of Metallurgy, Blackwell.

alloy (Figure 10.71). No sharp transition temperature of the rupture limit, like that measured for the ductility, can be distiguished.

The stainless steels also have considerable strength at temperatures above the transition temperature T_{tr}. Their values of σ_{rl} are of the same magnitude as that of carbon steel.

Theory of Hot Crack Formation

All alloys show a transition from brittle to ductile structure at a temperature close to the solidus temperature or just below this temperature. The brittle region causes crack formation during many casting and solidification processes. These cracks are called *hot cracks* or *solidification cracks.*

Solidification cracking is often described as occurring in the presence of liquid films when ductile stresses are applied, for example thermal stresses. These liquid films, which are found in the interdendritic areas or at grain boundaries, reduce the risk of hot cracks.

However, this is not always in agreement with the observations. Figure 10.72 (a) and (b) show that the transition temperature occurs far below the solidus temperature in Fe–Cr–Ni-based alloys when they solidify as austenite.

If the liquid film theory were correct, the transition temperature should decrease with increasing cooling rate, since microsegregation increases with increasing cooling rate and therefore the solidification temperature should decrease. This is true for Fe–Cr–Ni alloys that solidify with δ- structure. However, the experiments do not verify this prediction for Fe–Cr–Ni alloys that solidify with γ-structure. For the γ-alloys the transition temperature increases with increasing cooling rate (page 355).

In the 1990s, Fredriksson presented a new solidification model, which includes the effects of vacancies, formed during solidification. The vacancies formed during the solidification process will then condense at dislocations or grain boundaries. The condensation causes strain and stresses during this process and results in contraction of the solidified material. If the vacancies condense, for instance at grain boundaries, the result will be formation of hot cracks. A ductile stress will also generate new vacancies and the condensation of these contributes to the formation of hot cracks.

The theory claims that supersaturation of vacancies will nucleate cracks and favour the growth of cracks. The nucleation of a crack is assumed to occur by heterogeneous nucleation, which depends on the supersaturation of vacancies in the solid. Figure 10.74 shows the relation between the thermal stress, which has to be exceeded in order to form a crack, and the supersaturation of vacancies, required to nucleate a crack. The supersaturation is described as a factor times the equilibrium fraction of vacancies.

This thermal stress can easily be achieved in iron-base alloys, especially during addition of sulfur, phosphorus, and oxygen. This may be the explanation why sulfur and phosphorus are notorious elements for causing hot cracks during solidification, even if the alloy has such low concentrations of these elements

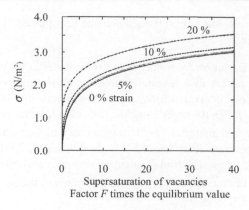

Figure 10.74 Thermal stress as a function of the supersaturation of vacancies for nucleation of a crack for different strain values on a sample of γ-Fe at 1500 °C.

that liquid films cannot possibly form. The dotted curve shows the case when the crack is nucleated in a strain field.

According to the model, the diffusion of vacancies from the bulk solid to the crack tip controls the crack growth. The driving force for crack propagation is the difference between the vacancy concentration at the supersaturated solid and the crack tip. It will therefore be proportional to the diffusion rate of vacancies from the supersaturated solid to the crack tip or simply a function of the supersaturation of vacancies.

The crack growth rate as a function of the supersaturation of vacancies is shown in Figure 10.75. With no strain at all the growth rate will be of the order 100 μm/s. The dotted curve in Figure 10.75 shows that a strain ≤ 5 % has little effect on the growth rate of the crack.

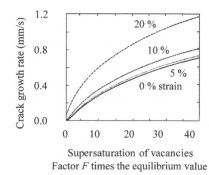

Figure 10.75 Growth rate of a crack as a function of the supersaturation of vacancies at three different strain values and at $\Delta\sigma = 1$ J/m^2. Sample of γ-Fe at $T = 1500$ °C.

The growth rate of cracks is fairly fast and the model therefore implies that the transition temperature of ductility is related to the nucleation of a crack. It is important for the nucleation of a crack that the supersaturation of vacancies, formed in the solid during the solidification process, can be maintained. This is not desirable and it is important to find an annealing process which reduces the supersaturation of vacancies, for example temperature decrease.

The transition temperature from brittle to ductile structure are influenced by the:

- supersaturation of vacancies;
- diffusion rate of vacancies;
- distance between the vacancy sinks, i.e. the structure coarseness;
- time of diffusion of vacancies, which is related to the cooling rate.

10.7.7 Hot Crack Formation during Component Casting

Plastic and elastic deformations of castings have been discussed in Chapter 8 (pages 242 and 244). Residual stresses often remain in cast components, which lead to risks of distortion in the components. The stresses can be released by heat treatment (Chapter 8).

In the case of component casting, the heat treatment temperature to release residual stresses is close to room temperature. The thermal contraction will also become critical at high temperatures when the alloy is in the brittle region. Hot cracking will thus occur.

A cast component never contracts completely freely during the casting process. The mould restrains the contraction, as illustrated in the Figure 10.76 (a) and (b).

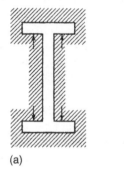
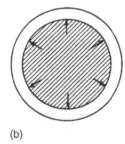

(a)

(b)

Figure 10.76 (a) Mould restraint, a typical design feature, which causes contraction stress. (b) Core restraint, a typical design feature, which causes contraction stress. Reproduced with permission from Elsevier Science.

As soon as the casting has reached a temperature below the transition temperature from brittle to ductile structure, there is very little risk of crack formation.

The risk of crack formation is therefore concentrated in the temperature interval between the transition temperature and the temperature in the solid–liquid two-phase region where the fraction of solid is large enough for the material to resist

stress. The larger this temperature interval, the greater is the risk of cracks in the material.

Example 10.11

A casting of the shape illustrated in Figure 10.76(a) is to be cast in a steel alloy with the properties described below. It is necessary to analyse the risk of crack formation. Experimental values of the stress as a function of temperature for the steel are known and illustrated in the figure shown.

The thermal expansion coefficient of the steel is $2.0 \times 10^{-5}\,\mathrm{K}^{-1}$. The elasticity modulus of the steel can be written as a function of temperature:

$$E - [1.1 + 0.10(T_{\mathrm{cr}} - T)] \times 10^3\,\mathrm{N/mm}^2$$

where T_{cr} is the critical temperature for elastic deformation.

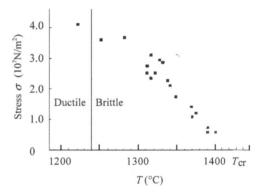

(a) Calculate the thermal stress in the cast material as a function of temperature if the material cannot contract freely. Hooke's law is assumed to be valid. Plot the function on the figure shown. The function represents the rupture limit of the material as a function of the test temperature. The transition temperature between brittle and ductile structure is shown in the figure.

(b) Discuss the risk of hot crack formation in the material on the basis of the experimental values in the figure.

Solution:

(a) Hooke's law [Equation (10.61)] can be written as

$$\sigma = \varepsilon E \qquad (1')$$

According to Equation (10.67), the strain ε can be written as:

$$\varepsilon = \alpha \Delta T = \alpha (T_{\mathrm{cr}} - T) \qquad (2')$$

Equations (1′) and (2′) are combined with the expression for E, given in the text, transformed into SI units:

$$\sigma = \alpha(T_{cr} - T)E = 2.0 \times 10^{-5}(T_{cr} - T)$$
$$[1.1 + 0.10(T_{cr} - T)] \times 10^3 \times 10^6 \qquad (3')$$

The function is plotted in the figure in the Answer below.

An average straight line is drawn in the figure in the text. It represents the rupture stress as a function of the temperature and is plotted in both figures in the Answer below.

The temperature at which the material breaks is obtained when the thermal stress is equal to or larger than the rupture limit, i.e. at the intersection point between the thermal stress curve and the straight line. If the intersection point is located in the brittle zone, the material breaks. If the point is situated in the ductile zone, the material expands and the rupture limit is never reached.

Answer:

(a) The desired function, which represents the thermal stress as a function of temperature, is:

$$\sigma = 2.0 \times 10^4(T_{cr} - T)[1.1 + 0.10(T_{cr} - T)]$$

It is plotted in the lower figure illustrated below.

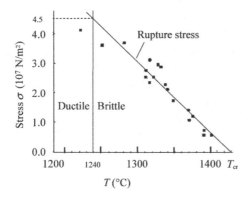

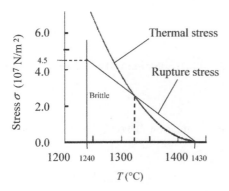

(b) The straight line in the left figure corresponds to the rupture stress in the material as a function of temperature. It is also drawn in the right figure with the aid of the two points (1240 °C, 4.5×10^7 N/m²) and (1430 °C, 0). The intersection point corresponds to $T = 1320$ °C. Hence there is a large risk of hot crack formation at temperatures ≤ 1320 °C where the thermal stress exceeds the rupture limit.

10.7.8 Thermal Stress and Crack Formation during Continuous Casting

During continuous casting, both the primary and secondary cooling are strong and the cooling rate is high. When the melt is poured into the chill-mould and meets its cold walls, a shell is formed very rapidly. The solidifying shell initially cools rapidly because the contact between shell and chill-mould is good. When the shell cools it shrinks and an air gap is formed between the chill-mould and the shell as soon as the latter has become stable enough to resist the ferrostatic pressure. The thermal conduction decreases with increasing air gap and the surface temperature of the shell increases.

A complex interaction between the thickness and the temperature of the solidifying shell results in a varying growth rate along the shell, both in its length direction and along its periphery. Every tiny variation of the growth rate of the shell automatically changes its surface temperature. These temperature variation cause thermal stresses and result in a risk of surface cracks.

Knowledge of the growth rate of the solidifying shell is therefore of utmost importance for the calculation of the sizes of the thermal stresses and for the possibility of designing the chill-mould and the secondary cooling in an optimal way.

When the shell has reached the centre of the cross-section of the strand at the end of the metallurgical length, the temperature in the centre, which initially has been fairly constant, will fall rapidly. This process causes thermal stresses within the interior of the strand and results in a risk of centre cracks.

The risk of crack formation is a severe problem in continuous casting. Below we will discuss the air gap more in detail and then return to thermal stresses and crack formation.

Air Gap Between Shell and Chill-Mould
Experimental investigations show that the thickness of the shell inside the chill-mould varies (Figure 10.77). These variations cause changes in the surface temperature and growth rate of the shell during the solidification process.

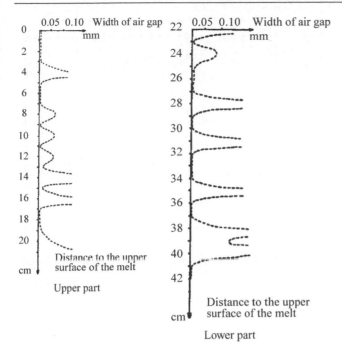

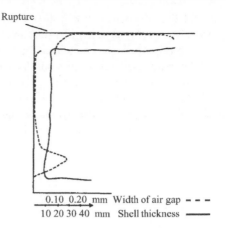

Figure 10.78 Shell thickness (continuous line) and air gap width (dashed line) in a steel billet 27 cm below the upper surface of the melt. Reproduced with permission from the Scandinavian Journal of Metallurgy, Blackwell.

Figure 10.77 Width of air gap between shell and chill-mould as a function of the depth under the upper surface of the melt for a steel billet $(14 \times 14\,cm)$. Reproduced with permission from the Scandinavian Journal of Metallurgy, Blackwell.

Figures 10.78 and 10.79 illustrate the variation of the air gap around the periphery of a strand. This type of variations also causes variations of the surface temperature of the shell around the periphery and in the length direction.

The air gap is primarily formed at the corners. Figure 10.80 (a) shows how the air gap is formed only a few seconds after the casting starts and how it increases with time.

Figure 10.80 (b) shows that the temperature near the corner, where the air gap is very large, is much higher than at the middle of the side. In order to counteract the reduced heat conduction, the chill-mould is made conical to decrease the air gap. The best result is obtained if the conicity of the chill-mould is larger at the upper part than in the lower part.

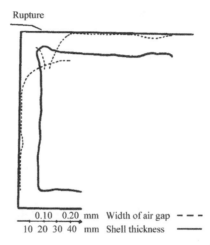

Figure 10.79 Shell thickness (continuous line) and air gap width (dashed line) in a steel billet 37 cm below the upper surface of the melt. Reproduced with permission from the Scandinavian Journal of Metallurgy, Blackwell.

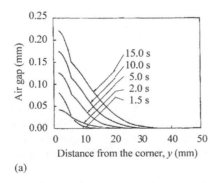

(a)

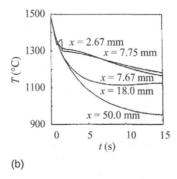

(b)

Figure 10.80 (a) Calculated air gap of a billet $10 \times 10\,cm$ at a casting rate of 2.8 m/s and chill-mould height of 70 cm. (b) Calculated surface temperatures along the strand at various distances from the corner for the same billet as in (a). Reproduced with permission from the Scandinavian Journal of Metallurgy, Blackwell.

There is a relation between the risk of surface cracks and the degree of use of the chill-mould. In order to examine this relation, a casting process has been simulated both for a chill-mould which has been used many times and for a new one.

Curves a and b in Figure 10.81 show the model chill-moulds and their real prototypes [see Figure 5.22 (page 109)]. For the sake of simplicity, the side of the square section of the model chill-mould has been chosen as 100 mm.

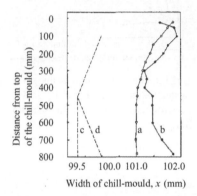

Figure 10.81 Shapes of billet profiles. (a) New industry chill-mould; (b) industry chill-mould, used 350 times; (c) simulated new chill-mould, (d) simulated used chill-mould. Reproduced with permission from the Scandinavian Journal of Metallurgy, Blackwell.

The results of the simulation calculations are shown in the Figures 10.82 and 10.83.

The results of the simulation calculations are summarized in Figure 10.84, which shows the calculated surface temperature as a function of time for four different distances from the corner in the two cases.

Figure 10.84 shows that the larger air gap in the old chill-mould causes a considerable temperature increase in its lower

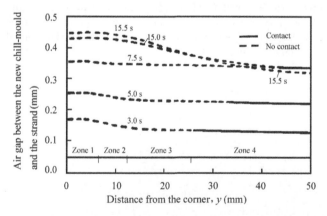

Figure 10.82 Calculated distance between the new chill-mould and the surface of the strand as a function of the distance from the corner at five different times. Continuous lines indicate contact between shell and chill-mould. Dashed lines indicate that the strand surface is surrounded by air. Reproduced with permission from the Scandinavian Journal of Metallurgy, Blackwell.

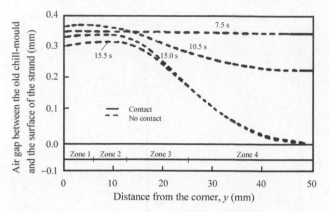

Figure 10.83 Calculated distance between the old chill-mould and the surface of the strand as a function of the distance from the corner at five different times. Continuous lines indicate contact between shell and chill-mould. Dashed lines indicate that the strand surface is surrounded by air. Reproduced with permission from the Scandinavian Journal of Metallurgy, Blackwell.

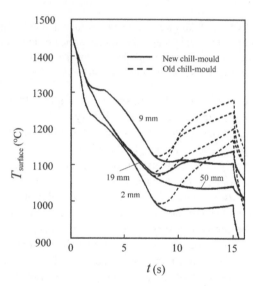

Figure 10.84 Calculated surface temperatures as a function of time at four different distances from the corner (parameters of the curves) for the new and old chill-moulds. The two cases are identical during the first 7.5 s but differ strongly later. An uneven distribution of the thermal stresses in the chill-moulds causes an irregular order of the curves corresponding to different parameter distances. Reproduced with permission from the Scandinavian Journal of Metallurgy, Blackwell.

part. It can also be seen that the temperatures at the corners in both the old and the new chill-moulds are lower than elsewhere. This is an effect of more effective heat transport at the corners than at other places. This was discussed in Chapter 4 (page 80).

Types of Hot Cracks During Continuous Casting

The risk of formation of hot cracks during continuous casting is very high. Figure 10.85 and Table 10.4 give a survey of the

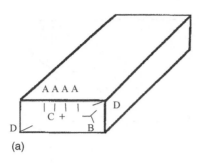

(a)

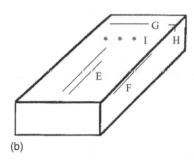

(b)

Figure 10.85 (a) Strand section with some different types of internal cracks. A, midway cracks; B, triple point cracks; C, centreline cracks; D, diagonal cracks. (b) Strand section with some different types of surface cracks. E, longitudinal midface cracks; F, longitudinal corner cracks; G, transverse midface cracks; H, transverse corner cracks; I, star cracks. Reproduced with permission from the Scandinavian Journal of Metallurgy, Blackwell.

TABLE 10.4 Survey of observed cracks in continuously cast steel strands.

Internal cracks	Surface cracks
A: Midway cracks	E: Longitudinal midface cracks
B: Triple point cracks	F: Longitudinal corner cracks
C: Centreline cracks	G: Transverse midface cracks
D: Diagonal cracks	H: Transverse corner cracks
	I: Star cracks

many different kinds of cracks which have been observed in continuously cast strands.

The cracks can be divided into two main groups:

- internal cracks;
- surface cracks.

A review of the most important types is given in Table 10.4.

Surface Cracks
Cracks Inside the Chill-Mould
Thermal stresses are caused by differences in cooling rate between different parts of the solidifying shell. As long as the strand is situated inside the chill-mould, the cooling rate in the interior parts of the shell will be larger than that at the surface. The result is thermal stresses near the solidification front and compression of the surface.

In order to give a concrete picture of the thermal stresses during continuous casting, we will continue the simulation calculations for the case with the old chill-mould [Figure 10.81 (d)]. The thermal stresses have been calculated at different points in the *xy*-plane (the length direction of the strand is the *z*-axis) at time $t = 15$ s and are plotted in a diagram as vectors. It takes 15.0 s for the strand to pass the chill-mould and Figure 10.86 thus shows the stress pattern just before the exit from the chill-mould.

In Figure 10.86, the solidus curve T_s and the calculated transition temperature curve $T_{tr} = T_s - 100\,^\circ C$ have been inserted. The material is assumed to be brittle within the region between these two curves.

In addition, the mechanical strains ε_x^m in the *x*-direction are plotted as figures. It can be seen that this strain has its maximum within the zone where the material is brittle. Hence there is a great risk of internal longitudinal cracks, perpendicular to the surface.

Outside the Chill-Mould
At the exit from the chill-mould and at the entrance into the secondary cooling zone, the sudden thermal conduction

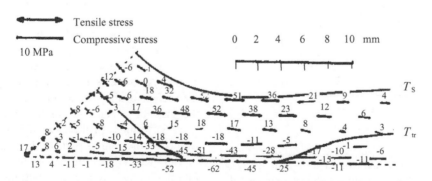

Figure 10.86 Calculated directions and sizes of the thermal stresses in the *xy*-plane (arrows) and calculated strains (figures times 10^{-4}) at the exit of the chill-mould ($t = 15.0$ s). Reproduced with permission from the Scandinavian Journal of Metallurgy, Blackwell.

away from the whole surface is caused by water cooling. The distribution of the thermal stresses and the mechanical stresses is totally changed and the strains appear in the surface region instead of the region inside the interior of the strand. Figure 10.87 illustrates the new stress conditions.

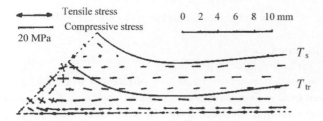

Figure 10.87 Stresses in the *xy*-plane after the exit out of the chill-mould ($t = 15.5$ s). The length direction of the strand is the *z*-axis. Reproduced with permission from the Scandinavian Journal of Metallurgy, Blackwell.

The stresses after casting in the old chill-mould have been calculated at $t = 15.5$ s, i.e. just below the chill-mould, and have been plotted into Figure 10.87 together with the isotherms T_s and T_{tr}. These thermal stresses decrease rapidly. The simulation calculations show that they are replaced by torsion stresses, caused by the ferrostatic pressure, after some seconds.

Simulation calculations for the new chill-mould show that the process in principle is the same as that described above, but there are also important differences, which we will now discuss (Figures 10.88 and 10.89).

Risk Regions of Crack Formation
In order to give a concrete example of mapping risk regions of crack formation, the simulation calculations and the comparison we made on page 360 between an old and frequently used chill-mould and a new one have been used as a basis for Figures 10.88 and 10.89.

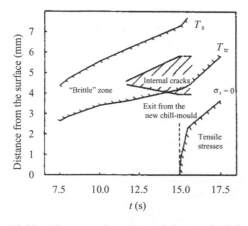

Figure 10.88 Distance *y* from the surface as a function of time for the new chill-mould. Position and extension of some regions of special interest at the width $x = 19$ mm are shown. Reproduced with permission from the Scandinavian Journal of Metallurgy, Blackwell.

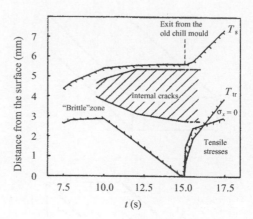

Figure 10.89 Distance *y* from the surface as a function of time for the old chill-mould. Position and extension of some regions of special interest at the width $x = 19$ mm are shown. Reproduced with permission from the Scandinavian Journal of Metallurgy, Blackwell.

Figures 10.88 and 10.89 give a comparison between the new and old chill-moulds with respect to risk regions of crack formation. They show calculated distances from the strand surface as a function of time for various lines/regions in a plane, perpendicular to the surface at a certain distance ($x = 19$ mm) from the corner. This distance corresponds to the distance from the corner to the centre of the hot surface region in both cases (see Figure 10.84).

The difference in width of the risk zones of internal crack formation ($\varepsilon_x^m > 0.2$ %) and brittleness illustrates the influence of a new versus an old chill-mould on the cooling process and on the risk of internal cracks.

After passage of the chill-mould ($t > 15$ s), the region of thermal stresses increases and advances, i.e. the distance from the surface increases. In the case of the old chill-mould (Figure 10.89), the tensile stresses even reach the brittle zone in less than 1 s, immediately after the exit from the chill-mould. This results, of course, in a severe risk of crack formation.

Internal Cracks
The major reason for centre and 'half-way' cracks in strands is the *thermal contraction of the solidified shell*. They are caused by differences in cooling rates between the surface and the centre at the end of the metallurgical length. When the centre starts to solidify there is a rapid drop in temperature along the central axis, as we found in Section 10.6.2. As there is no corresponding temperature drop at the surface of the strand, thermal contraction will occur. In the case of a slab the strand bulges and a longitudinal crack is formed. In the case of billets, star-like cracks are formed in the central parts of the strand.

Half-way cracks are one of the most common types of internal cracks found in continuously cast billets and slabs. The mechanism of formation can be postulated. The prediction is based on the results of both industrial measurements and thermomechanical strain analysis under different boundary conditions.

Half-way cracks form parallel to the primary dendrite arms in the columnar zone. The cracks run perpendicularly to the thermal strains that develop in the material. The dominant source of stresses, which causes crack formation, is reheating of the slab surface. The stresses which cause half-way cracks depend mainly on

- the magnitude of the reheating of the surface;
- the thickness of the solid shell;
- the width of the mushy zone.

The magnitude of the surface reheating is related to the design of the secondary cooling cycles and the level of the surface temperature.

10.7.9 Methods of Reduction of the Negative Effects of Thermal Stresses

General Precautions during Casting
It is impossible to avoid strong temperature changes in the material during casting and cooling. On the other hand, the production of castings may be designed in such a way that the risk of crack formation is minimized. Theoretical and practical knowledge of the kind discussed earlier in this section on thermal stresses constitute the basis of judgement of the best method.

The following general principles are valid:

- Moulds and chill-moulds should be designed to be as free from sharp edges and corners as possible.
- The compositions of the alloys to be used ought to be chosen in such a way that the materials have maximum ductility within the limits of other demands upon the products.
- The cooling should be designed in such a way that the temperature differences between different parts of the casting become as small as possible during the solidification and cooling process.
- The mechanical action on the casting during the sensitive temperature interval between the solidus temperature and the transition temperature from ductile to brittle material should be kept as small as possible, because then the risk of hot cracks is considerable.
- Strong cooling often results in more thermal stress than soft cooling (see Figure 11.34 on page 393).

Additional Precautions during Continuous Casting
In Section 10.7.8, we found that the risk of crack formation is often high during continuous casting, owing to the large temperature gradients and rapid changes of the surface temperature.

- The risk of *surface cracks* can be reduced by keeping the surface temperature as constant as possible during cooling. It is therefore important that the chill-mould is made conical, in order to keep the air gap as constant as possible during the solidification process. The best result is achieved if the conicity of the chill-mould is made larger in the upper than in the lower part. In this way it is possible to avoid the heat transport being strongly reduced, which would cause large changes in the temperature conditions (see Figure 10.86).
- The risk of *centre cracks* can be reduced by increasing the cooling of the surface strongly at the end of the metallurgical length (thermal reduction). This reduces the large difference between the cooling rates at the surface and at the centre, which normally arises when the centre has solidified. The same effect can be achieved by mechanical reduction, i.e. when the strand is exposed to a strong mechanical pressure at the end of the metallurgical length.

SUMMARY

■ Solidification and Cooling Shrinkage

When a pure metal melt or an alloy solidifies and cools, its volume is reduced in most cases. The shrinkage that appears in castings owing to this volume reduction is comparably large and can cause severe problems unless measures are taken. The most important problems are:

- deformation of the casting by collapse of the surface shell at the beginning of the solidification process;
- appearance of shrinkage cavities inside the casting;
- pipe formation;
- appearance of casting stresses, shrinkage cracks, and distorsions inside the casting.

■ Pipe Formation

Pure Metals
A pure metal has a well-defined melting point and a well-defined solidification front. A pipe is formed when the metal solidifies and cools. Strong cooling of the mould also favours the appearance of a pipe.

Alloys
An alloy solidifies over a temperature interval. On cooling, solid metal is formed all over the melt in the shape of dendrite arms. The solidification shrinkage cannot be compensated for because it is difficult to transport melt through the narrow channels between the dendrite arms.

Shrinkage cavities, so-called shrinkage pores, occur everywhere in the material.

Casting in a sand mould (low cooling rate) also causes a similar distribution of the shrinkage pores.

■ Methods to Minimize or Eliminate Pipes in Ingot Castings

Feeder
During the production of castings, efforts are made to control the casting process in such a way that the pipe as a whole is located in a feeder and the casting becomes compact.

The size and position of the feeder are of great importance. Two conditions have to be fulfilled:

- The solidification time of the feeder must be equal to or longer than that of the casting. Then the melt in the feeder compensates for the solidification shrinkage of the casting.
- The volume of the feeder must have a size that is at least large enough to enclose the whole solidification shrinkage inside the feeder.

There are several methods to calculate a suitable feeder volume.

As a measure of the feeding difficulty, the concept of centreline feeding resistance is used:

$$CFR = \frac{t_{total} - t_{initial}}{t_{total}}$$

Hot Top in Ingot Casting
During *ingot casting*, it is possible to prevent pipe formation successfully with the aid of a hot top. Without a hot top, a deep pipe is obtained, which destroys a large part of the ingot. The pipe cannot be welded during rolling.

■ Pipe Formation during Continuous Casting
During *continuous casting*, a pipe is obtained at the end of the casting process. The risk of a pipe can be prevented by careful design of the casting rate and the cooling rate at the end of the casting process.

$$\Delta V_{pipe} = \Delta V_{sol} + \Delta V_{cool}$$

Solidification Shrinkage
Solidification shrinkage = solidified volume × β:

$$\Delta V_{sol} = \Delta V_s \beta$$

Cooling Shrinkage
Cooling shrinkage = sum of centre cooling shrinkage and surface cooling shrinkage:

$$\Delta V_{cool} = \Delta V_{centre} + \Delta V_{surface}$$

or

$$\Delta A_{cool} = \Delta A_{centre} + \Delta A_{surface}$$

ΔA_{centre} and $\Delta A_{surface}$ have opposite signs.

- If ΔA_{centre} is *larger* than $\Delta A_{surface}$, the central part of the strand shrinks more than the outer frame, i.e. the surface, allows. This means that *a cavity is formed* in the central region. At the top of the strand the melt fills the cavity from above. Later, when the melt can no longer pass through the narrow channels, the cavity contributes to the pipe.
- If ΔA_{centre} is *smaller* than $\Delta A_{surface}$, the central part shrinks less than the outer frame, i.e. the surface. This means that the central region becomes compressed and *no cavity is formed*. No contribution to the pipe is obtained.

The total volume of the pipe due to thermal contraction equals ΔA_{cool} times the metallurgical length:

$$\Delta V_{cool} = \Delta A_{cool} L$$

Calculations are performed on the *special case* of a *square strand* with side x_0.

After the solidification fronts have met at the centre of the strand the temperature decreases both at the centre and at the surface of the strand. Both contribute to the total cooling shrinkage.

$$\Delta A_{surface} = \frac{4}{3} \alpha_l x_0^2 \Delta T_{surface}$$

$$\Delta A_{centre} = \frac{2}{3} \alpha_l x_0^2 \Delta T_{centre}$$

The total shrinkage area formed owing to thermal contraction is:

$$\Delta A_{cool} = \frac{2}{3} \alpha_l x_0^2 (1 - 2G) \Delta T_{centre}$$

The ratio G of the cooling rate at the surface and the cooling rate at the centre influences the pipe formation:

- If $G < 0.5 \Rightarrow \Delta A_{cool} > 0 \Rightarrow$ a pipe is formed
- If $G > 0.5 \Rightarrow \Delta A_{cool} < 0 \Rightarrow$ no pipe.

■ Thermal Stress and Crack Formation during Solidification and Cooling Processes

Thermal forces in solid materials are caused by temperature differences and cause thermal stresses and strains.

Basic Laws

The stress σ and the temperature difference ΔT cause a strain ε in the material.

Poisson's number: $v = -\frac{\varepsilon_{\text{trans}}}{\varepsilon}$

Basic equation in one dimension: $\varepsilon = \frac{\sigma}{E} + \alpha \Delta T$

Hooke's generalized law in three dimensions:

$$\varepsilon_x = \frac{1}{E}[\sigma_x - v(\sigma_y + \sigma_z)] + \alpha \Delta T$$

$$\varepsilon_y = \frac{1}{E}[\sigma_y - v(\sigma_x + \sigma_z)] + \alpha \Delta T$$

$$\varepsilon_z = \frac{1}{E}[\sigma_z - v(\sigma_x + \sigma_y)] + \alpha \Delta T$$

■ Thermal Stresses during Unidirectional Solidification

In order to calculate the strain ε, it is necessary to know the temperature as a function of position x and time t:

$$\varepsilon^{\text{u}}(y,t) = \int\limits_{t^*}^{t} \left[\frac{1}{s} \int\limits_0^s \frac{\partial \varepsilon^{\text{T}}}{\partial t} \mathrm{d}y \right] \mathrm{d}t + \varepsilon^{\text{T}}(T_{\text{rigid}}) - \varepsilon^{\text{T}}(T)$$

where t^* is the time when the solidification front reaches the position y.

■ Plastic Properties of Materials

Crack formation is closely related to the plastic properties of materials.

As a measure of the ductility of a material the relative area reduction at a ductility test is used:

$$\psi = \frac{A_{\text{before}} - A_{\text{after}}}{A_{\text{before}}} \times 100$$

where A_{before} and A_{after} are the cross-sectional area before and after the ductility test, respectively.

For *low* values of ψ the material is *brittle*. For high values of ψ the material is *ductility*. The smaller ψ is the higher is the risk of crack formation.

The transition from brittle to ductile material occurs within a narrow temperature interval, the so-called *transition temperature*. A material is brittle within the temperature interval between the solidus temperature and the transition temperature.

Alloying elements and impurities with low solubilities and low diffusion rates in the solid phase generally lower the ductility and increase the risk of crack formation considerably.

The cooling rate and the structure of a material influence the transition rate from brittle to ductile material.

The higher the cooling rate is, the larger will be the risk of cracks.

■ Air Gap Formation

Air gap formation is closely related to thermal stress. The mould wall expands slightly and the casting shrinks:

$$h = \frac{k}{\delta}$$

When an air gap has been formed, the heat transport, i.e. the cooling, is drastically reduced.

■ Cracks

Types of Cracks

Surface cracks and internal cracks.

Crack Formation during Continuous Casting

The problems with crack formation are especially severe during continuous casting as the temperature difference between the strand surface and the centre is large during the solidification and cooling process.

The size of the air gap in the mould is of a great importance. The shell thickness causes a varying surface temperature and growth rate of the upper part of the shell.

Used chill-moulds have larger air gaps than new ones, which results in higher surface temperatures in old than in new chill-moulds. This increases the risk of cracks.

In the temperature interval between the solidus temperature and the transition temperature between brittle and ductile material there is a pronounced risk of internal cracks and surface cracks.

■ Methods of Reduction of the Negative Effects of Thermal Stresses during Continuous Casting

- The chill-mould is made conical to create an air gap of constant size.
- During casting the surface temperature is kept as even as possible in order to avoid surface cracks.
- The cooling of the surface is increased strongly at the end of the metallurgical length (*soft thermal reduction*).

In this way, the large difference in cooling rate at the surface and at the centre which normally appears when the centre has solidified is reduced. The same effect can be achieved by *mechanical reduction*, i.e. when the strand is exposed to a strong mechanical pressure at the end of the metallurgical length.

EXERCISES

10.1 (a) A steel cube with a side of 20 cm is to be cast. A cylindrical feeder with a diameter of 20 cm is located on top of the cubic mould. Calculate the minimum height of the feeder if the feeder and the mould are made of sand.

Hint B36

(b) Alternatively, the cube can be cast in the sand mould with a cylindrical feeder, which is made of a more insulating material and has a diameter of 10 cm. In this case the solidification time is prolonged by a factor of four compared with that in (a). Calculate the minimum height of the new feeder. Which alternative is the best one?

Hint B101

Use Table 10.1 on page 316 to find a reasonable value of the solidification shrinkage.

10.2 A hollow cylinder with the dimensions external diameter 40 cm, internal diameter 20 cm and height 30 cm is to be cast in a copper alloy. The cylinder can be cast in either brass or Al-bronze.

Material constants	
Cylinder sand:	
k_c	14.5×10^{-4} J/m s K
ρ_c	1.5×10^3 kg/m^3
c_p^c	0.27 kJ/kg K
Feeder material	
k_f	4.1×10^{-4} J/m s K
ρ_f	0.90×10^3 kg/m^3
c_p^f	0.20 kJ/kg K

A demand on the cylinder is that it has to be completely compact. For this reason, two cylindrical feeders, each with a diameter of 10 cm, are chosen. The material, size and design of the feeders determine the choice of alloy. The size of the feeder will be chosen to be as small as possible.

The cylinder mould is made of sand and the feeders are made of a highly insulating ceramic material. Material constants of the mould and feeder materials are given in the table here. Tables 10.1 (page 316) and 10.2 (page 326) can be used to find the required material constants of the alloys.

Calculate the height of the feeders in the two alternatives and suggest a choice of material.

Hint B180

10.3 Figure 10.36 (page 331) shows the solidification process of an ingot with a hot top; the figure on the right shows the finally solidified ingot.

Show by three sketches the appearance the upper surface if we had chosen:

(a) a lower height of the hot top than the optimal one;

(b) a higher hot top than the optimal one;

(c) the optimal height of the hot top.

Explain the physical background of the appearance of the upper surface in the three cases.

Hint B244

10.4 An ingot with dimensions 0.20×1.00 m and a height of 1.30 m is to be cast in steel in a thick iron mould. The ingot is equipped with a hot top made of sand with the same cross-sectional area as the ingot. Immediately after casting, the upper surface of the ingot is covered with a layer of 30 mm of silica sand as heat insulation.

However, it is doubtful whether the height of the hot top has been chosen properly for satisfactory use. For this reason, it is important to calculate the minimum value of the height of the hot top.

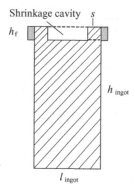

(a) As a first rough estimation, the following simple but rather unrealistic method can be used.

The total shrinkage volume is assumed to be equal to the hollow volume of the hot top. Perform an approximate calculation of the height h_f of the hot top if the thickness s of the solidified shell in the hot top is guessed to be 15 mm at the time of total solidification of the ingot.

Calculate the height of the hot top with the aid of this method.

Hint B348

(b) The method in (a) gives a too low value of the hot top as it must be large enough to keep some melt even after the solidification of the ingot.

Perform a more accurate calculation of the height of the hot top. The information that the thickness of the shell is 15 mm is no longer valid. The thickness of the solidified shell of the hot top at the time of total solidification of the ingot has to be calculated.

Hint B222

Material constants:

Steel:	*Hot top made of sand:*
$\rho = 7.5 \times 10^3 \, \text{kg/m}^3$	$\rho = 1.6 \times 10^3 \, \text{kg/m}^3$
$k = 30 \, \text{W/m K}$	$k - 0.63 \, \text{W/m K}$
$-\Delta H = 272 \times 10^3 \, \text{J/kg}$	$c_\text{p} = 1.05 \times 10^3 \, \text{J/kg K}$
$T_\text{L} = 1550 \, ^\circ\text{C}$	

(c) Do you find the method in (a) useful?

Hint B208

10.5 A metal works casts copper ingots with the dimensions $0.200 \times 0.800 \times 1.200 \, \text{m}^3$ (width × thickness × height) in sand moulds for the production of copper plates.

Material constants

k_Cu	398 W/m K
ρ_Cu	$8.94 \times 10^3 \, \text{kg/m}^3$
$T_\text{L}(\text{Cu})$	1083 °C
k_sand	0.63 W/m K
ρ_sand	$1.6 \times 10^3 \, \text{kg/m}^3$
c_p^sand	$1.05 \times 10^3 \, \text{J/kg K}$

The ingot is cast in a water-cooled chill-mould with an insulating hot top located in the upper part of the chill-mould. The hot top consists of sand. The chill-mould–ingot heat transfer coefficient is 400 W/m^2 K. Calculate the minimum height of the hot top for a satisfactory use.

Hint B164

10.6 Calculate the thermal stress in the x-direction of a thin rectangular metal plate of even thickness as a function of y. The temperature change of the plate which causes the thermal stress is a square function of y and independent of x and z:

$$\Delta T = \Delta T_0 \left(1 - \frac{y^2}{c^2} \right)$$

Hint B268

The right-hand figure shown illustrates the temperature distribution in the thin plate with the sides $2l$ and $2c$. The plate can expand freely in the y- and z-directions but not in the x-direction.

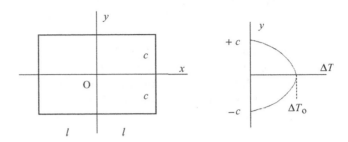

10.7 The figure shown illustrates a common temperature distribution along the surface during continuous casting of square billets. The reason for the temperature increase at a certain distance from the meniscus (upper liquid surface in the chill-mould) is that the water cooling of the surface of the strand is stopped for a short time between the cooling regions.

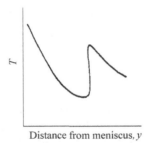

Distance from meniscus, y

Considering the two diagrams shown, discuss the possible defects which may form in billets at this temperature distribution in the following two cases:

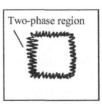

Two-phase region

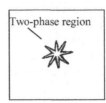

Two-phase region

(a) when two-thirds of the cross-sectional area has solidified and the water-cooling is stopped;

Hint B224

(b) when the solidification fronts have met at the centre and 50 % melt remains at the centre of the billet.

Hint B337

10.8 The air gap between the chill-mould and the strand is very important in continuous casting. Consider a chill-mould with the length 70 cm. The casting rate of an alloy with the liquidus temperature $T_L = 1450\,°C$ and the solidus temperature $T_s = 1350\,°C$, is 1.0 m/min. $T_0 = 100\,°C$.

Material costants	
k	29 W/m K
$-\Delta H$	276×10^3 kJ/kg
k_{air}	0.24 W/m K
ρ	7.3×10^3 kg/m^3

Calculate the shell thickness and the surface temperature of the strand, at the exit from the chill-mould, as functions of the width of the air gap. The air gap δ is assumed to be approximately constant along the chill-mould. List the values of the shell thickness and the surface temperature for $\delta = 10^{-5}$, 10^{-4}, and 10^{-3} m.

Hint B68

10.9 During the continuous casting of slabs, there is a risk of fracture at the centre of the strand due to the thermal stresses which arise when the solidification fronts meet.

Illustrate the risk of fracture by calculation of the linear dilatation and the thermal stress in the central region of a slab at the end of the solidification process and discuss the possible consequences of the stress in the material.

Hint B124

The slab material is low-carbon steel with the following material constants. The liquidus temperature of the alloy is 1480 °C and the solidus temperature is 1350 °C. The linear strain coefficient of the alloy is $2 \times 10^{-5}\,K^{-1}$. The modulus of elasticity of the alloy is $20 \times 10^{10}\,N/m^2$.

10.10 It has been proposed that the origin of half-way cracks during continuous casting is reheating of the surface when an air gap is formed during the casting process (Figure 10.84 on page 360). In order to confirm or reject this statement, we will analyse the thermal stresses at a continuous casting operation

illustrated in the figure shown. It shows the temperature at a point of the strand surface as a function of time.

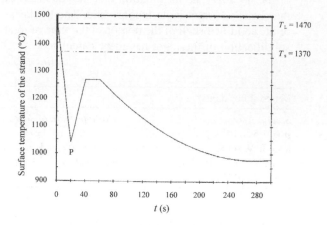

Material constants of the shell	
T_L	1470 °C
T_s	1370 °C
α	$5.0 \times 10^{-5}\,K^{-1}$
E	$100 \times 10^9\,N/m^2$
Transition temperature between the ductile and brittle zones:	
T_{tr}	1330 °C

Calculate:

(a) the temperature increase $\Delta T\,(y)$

Hint B15

(b) the strain $\varepsilon\,(y)$ and the stress $\sigma\,(y)$

Hint B56

across the solid shell after the heating period of the surface as functions of the distance y from the surface. Plot them as functions of y in two diagrams.

(c) Plot the stress σ as a function of the temperature and discuss the risk of crack formation.

Hint B190

At the time for reheating of the surface the shell thickness of the strand is 3.0 cm. In order to simplify your calculations, you may assume that the solidification rate is zero during the heating period, that the temperature distribution in the solidified shell is linear and that the shell is free from stresses at point P. You may also assume that Hooke's law is valid. Material constants of the steel alloy are given in the table.

11 Macrosegregation in Alloys

11.1 INTRODUCTION

In Chapter 7 we discussed microsegregation in solidifying alloys and introduced a simple mathematical model for microsegregaion, Scheil's equation [Equation (7.10) on page 186]. We assumed that the melt and the solid phase had the same molar volume, which meant that the melt and the solid phase had the same density.

This assumption does not agree with reality. When a metal solidifies, its density will increase, i.e. its molar volume will decrease. Owing to solidification and cooling shrinkage pores may form inside the metal, which was discussed in Chapter 10. The pressure inside the melt is very low, practically zero and, as a consequence of the low pressure, melt is sucked through the dendrite network. During this process, local deviations in the composition of the alloy, compared with its average composition, will appear. This phenomenon, which is called *macrosegregation*, will be treated in this chapter.

Macrosegregation, caused by solidification and cooling shrinkage, is especially important in continuous casting but appears also during casting with a feeder.

As an example, during continuous casting of a 10×10 cm billet with an average carbon concentration of 0.8 wt-%, a carbon concentration of more than 1.0 wt-% is obtained in the centre. Figure 11.1 shows the variation of the carbon concentration from the middle of a lateral surface to the centre of such a billet.

Chemical analysis of specimens at various distances from the lateral surface shows that the carbon concentration varies strongly. The carbon concentration is higher at the surface than further on in the interior. Close to the centre, the carbon concentration initially decreases and later increases to a sharp maximum.

The described carbon distribution is similar in practically all steel strands. The structure of the metal does not

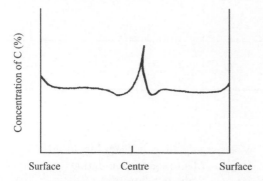

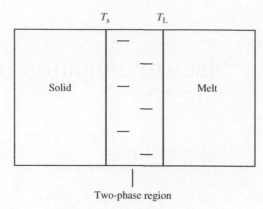

Figure 11.1 Distribution of the carbon concentration in a continuously cast billet as a function of the distance from a lateral surface. The C-concentrations have been measured along an axis through the midpoint of a lateral surface and the centre.

Figure 11.2 Definition of the two-phase region.

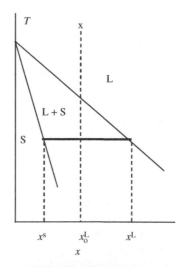

Figure 11.3 The zig-zag pattern is a symbol for the two-phase region.

influence the carbon distribution very much. On the other hand, the cooling conditions are of decisive importance. They control completely the macrosegregation pattern, which will be confirmed by the basic theory of macrosegregation.

In the following sections we discuss macrosegregation in the region close to the lateral surface (inverse segregation) and in the centre (centre segregation) separately. Later we also discuss other types of macrosegregation, among them that caused by natural convection and macrosegregation caused by crystal sedimentation.

11.2 MACROSEGREGATION DUE TO SOLIDIFICATION SHRINKAGE

To get an idea of the appearance of macrosegregation as a consequence of solidification shrinkage, we will start with the solidification process of an alloy.

In Section 4.3.1 and in Chapter 5, the solidification process of alloys was described and discussed extensively. An alloy solidifies within a solidification interval. The solidification starts at a solidification front, where the temperature equals the liquidus temperature T_L.

Behind the solidification front there is a two-phase region with both melt and solid phase in the shape of a dendritic network. The two-phase region is followed by a second solidification front, which indicates that 100 % of the melt has solidified. The temperature at this front equals the solidus temperature T_s (Figure 11.2).

The two-phase region is illustrated in Figure 11.3 by a zigzag pattern with tongues of melt and solid phase, which symbolizes the dendritic network.

The composition of the melt in the two-phase region is determined by the decreasing temperature and the phase diagram of the alloy. It is apparent from Figure 11.4 that the *lower* the temperature is, the *larger* is the difference

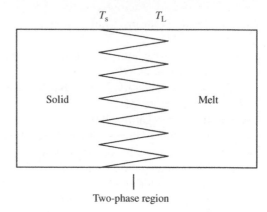

Figure 11.4 Phase diagram of a binary alloy.

in composition, i.e. concentration of the alloying elements, between the solid phase and the remaining melt.

One of the most common cases of macrosegregation is solidification and cooling shrinkage of the solid phase. Melt is sucked into the dendritic network owing to the depression in the interdendritic melt, caused by the shrinkage. During continuous casting the melt comes from above. During casting with a feeder it is the feeder that supplies the melt. It is sucked into the shrinkage spaces and fills them.

11.2.1 Simple Mathematical Model of Macrosegregation during Casting with a Feeder

In order to derive an expression for the concentration of the alloying element as a function of the fraction of solidification in a volume element, we will use a figure similar to that used to derive Scheil's equation for microsegregation (Chapter 7, page 186). The difference is that in this case we will consider the fact that the solid phase has a higher

density and a smaller molar volume than the melt and that melt is sucked into the volume element to compensate for the solidification shrinkage. The diffusion in the solid phase is assumed to be zero.

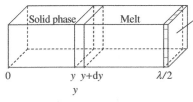

Figure 11.5 Volume element illustrating solidification with shrinkage.

Consider a volume element with cross-sectional area A and length $\lambda/2$ in Figure 11.5 and introduce the following designations:

$V_s =$ volume of the solid phase
$V_m^s =$ molar volume of the solid phase
$V_L =$ volume of the melt
$V_m^L =$ molar volume of the melt.

When the volume dV_L solidifies, a volume dV_s of solid phase is formed. The number of kilomoles which solidify, dV_L/V_m^L, is equal to the number of kilomols of newly formed solid phase, dV_s/V_m^s:

$$\frac{dV_L}{V_m^L} = \frac{dV_s}{V_m^s} \tag{11.1}$$

or

$$dV_L = V_m^L \frac{dV_s}{V_m^s} \tag{11.2}$$

When the thickness of the solid phase increases by an amount dy, a volume of melt equal to:

$$dV_L = Ady \frac{V_m^L}{V_m^s}$$

is consumed. If no pore is to be formed a volume $Ady(V_m^L/V_m^s - 1)$ of melt has to be supplied from outside. A material balance for the alloying element B within the volume $A(\lambda/2 - y)$ where a solid volume dy is formed can be written as:

$$\frac{\left(\frac{\lambda}{2} - y\right)Ax^L}{V_m^L} + Ady\left(\frac{V_m^L}{V_m^s} - 1\right)\frac{x^{L'}}{V_m^L} = Ady\frac{x^s}{V_m^s} + \frac{A\left[\frac{\lambda}{2} - (y + dy)\right]}{V_m^L}(x^L + dx^L) \tag{11.3}$$

where

$x^L =$ concentration of element B in the volume element
$x^{L'} =$ concentration of element B in the melt added from outside to the volume element
$x^s =$ concentration of element B in the solid phase in the volume element.

In analogy with the derivation of Scheil's equation in Chapter 7 [Equation (7.10), page 186], we have neglected the diffusion in the solid phase in Equation (11.3).

The left-hand side of Equation (11.3) describes how many kilomoles of substance B the volume element $A(\lambda/2 - y)$ initially contains plus the number of kilomoles of substance B supplied from outside as a consequence of the solidification shrinkage.

The right-hand side of Equation (11.3) represents the number of kilomoles of the alloying element B present in the volume element after solidification of the volume Ady. The first term equals the number of kilomoles in the solid phase and the second term how many kilomoles the remaining melt within the volume element contains. The composition of the sucked melt has been slightly changed.

Equation (11.3) is reduced and the term $dydx^L$ is neglected, which gives:

$$\frac{\left(\frac{\lambda}{2} - y\right)dx^L}{V_m^L} = \frac{x^L dy}{V_m^L} - \frac{x^s dy}{V_m^s} + \frac{x^{L'} dy}{V_m^L}\left(\frac{V_m^L}{V_m^s} - 1\right) \tag{11.4}$$

We multiply this equation by V_m^L and use the relation $k = x^s/x^L$ and obtain:

$$\left(\frac{\lambda}{2} - y\right)dx^L = \left[x^L\left(1 - k_{part}\frac{V_m^L}{V_m^s}\right) + x^{L'}\left(\frac{V_m^L}{V_m^s} - 1\right)\right]dy \tag{11.5}$$

This is the differential equation that has to be solved to give x^L as a function of y.

It is necessary to know $x^{L'}$ as a function of x^L to solve Equation (11.5). The relation between $x^{L'}$ and x^L depends on the temperature distribution during the solidification process. We will discuss four special cases below.

Case Ia
Assumptions:

1. The diffusion of the alloying element in the solid phase is low and can be neglected.
2. The added melt has the same composition as the melt inside the volume element, i.e. $x^L = x^{L'}$

Case Ia can be applied to casting with a feeder, which solidifies at the same rate as the casting (Figure 11.6). In this case Equation (11.5) can be written as:

$$\left(\frac{\lambda}{2} - y\right)dx^L = x^L(1 - k)\frac{V_m^L}{V_m^s}dy \qquad (11.6)$$

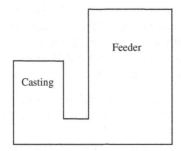

Figure 11.6 Case Ia.

Equation (11.6) is separable and can easily be integrated.

For comparison Scheil's equation for microsegregation:

$$x^L = \frac{x^s}{k_{part}} = x_0^L(1 - f)^{-(1-k)}$$

$$\int_0^y \frac{dy}{\frac{\lambda}{2} - y} = \int_{x_0^L}^{x^L} \frac{dx^L}{(1 - k_{part})\frac{V_m^L}{V_m^s}x^L} \qquad (11.7)$$

The solution can be written as:

$$x^L = \frac{x^s}{k_{part}} = x_0^L\left(1 - \frac{2y}{\lambda}\right)^{-(1-k)\frac{V_m^L}{V_m^s}} \qquad (11.8a)$$

or

$$x^L = \frac{x^s}{k_{part}} = x_0^L(1 - f)^{-(1-k)\frac{V_m^L}{V_m^s}} \qquad (11.8b)$$

where

x_0^L = the initial concentration of the alloying element in the melt

$2y/\lambda$ = fraction solid phase in the volume element

f = solidification fraction.

Equation (11.8b) is valid for case Ia when the diffusion rate of the alloying element in the solid phase is low and the concentration $x^{L'} = x^L$. It describes the concentration of the alloying element within the volume element during the solidification as a function of the solidification fraction f.

Equation (11.8b) is reminiscent of Scheil's equation. As the solidification shrinkage is considered, the ratio of the molar volumes appears as a factor in the exponent in Scheil's equation. In Figure 11.7, the concentration x^s is plotted as a function of the solidification fraction f with and without consideration of the solidification shrinkage. In the latter case the ratio of the molar volumes in the exponent is equal to 1.

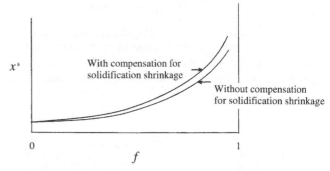

Figure 11.7 Concentration of the alloying element in the solid as a function of the solidification fraction with and without consideration of the solidification shrinkage.

It can be seen from Figure 11.7 that correction for solidification shrinkage increases the concentration of the alloying element.

The area under the curve within the region $f = 0$–1 is a measure of the amount of the alloying element within the volume element. It can be calculated by integration of Equation (11.8b). This topic will be treated later in connection with the calculation of the size of the macrosegregation.

Case Ib
Assumptions:

1. The diffusion of the alloying element in the solid phase is high.
2. The added melt has the same composition as the melt inside the volume element, i.e. $x^L = x^{L'}$.

Equation (11.8b) is not valid in this case since it assumes that the diffusion in the solid phase can be neglected. The diffusion rate is high for interstitially dissolved substances, e.g. carbon. A new derivation has to be performed for this case.

We use the same volume element as before (Figure 11.5) and set up a material balance. The number of kilomoles of the alloying element must be the same before and after the solidification if melt supplied from outside is included in the calculation.

The material balance can be written as:

$$\frac{x^s y A}{V_m^s} + \frac{x^L\left(\frac{\lambda}{2} - y\right)A}{V_m^L} + x^{L'}\left(\frac{1}{V_m^s} - \frac{1}{V_m^L}\right)\mathrm{d}yA = \frac{(x^s + \mathrm{d}x^s)(y + \mathrm{d}y)A}{V_m^s} + \frac{(x^L + \mathrm{d}x^L)\left[\frac{\lambda}{2} - (y + \mathrm{d}y)\right]A}{V_m^L} \quad (11.9)$$

No. of kmol in the solid phase   No. of kmol in the melt   No. of added kmol due to solidification shrinkage   No. of kmol in the solid phase   No. of kmol in the melt

After inserting $x^L = x^{L'}$ and reduction we obtain, if all terms of the type $\mathrm{d}x\mathrm{d}y$ are neglected,

$$\left(\frac{V_m^L}{V_m^s} - 1\right)\frac{x^L}{V_m^L}\mathrm{d}y + \left(\frac{x^L}{V_m^L} - \frac{x^s}{V_m^s}\right)\mathrm{d}y = \frac{y\mathrm{d}x^s}{V_m^s} + \frac{\left(\frac{\lambda}{2} - y\right)\mathrm{d}x^L}{V_m^L} \quad (11.10)$$

Inserting $x^s = k_{part} x^L$ and multiplication with V_m^L gives:

$$\left(yk_{part}\frac{V_m^L}{V_m^s} + \frac{\lambda}{2} - y\right)\mathrm{d}x^L = x^L\frac{V_m^L}{V_m^s}(1 - k_{part})\mathrm{d}y \quad (11.11)$$

Equation (11.11) is integrated between the limits x_0^L to x^L and 0 to y:

$$\int_{x_0^L}^{x^L} \frac{\mathrm{d}x^L}{x^L \frac{V_m^L}{V_m^s}(1 - k_{part})} = \int_0^y \frac{\mathrm{d}y}{y\left(k_{part}\frac{V_m^L}{V_m^s} - 1\right) + \frac{\lambda}{2}} \quad (11.12)$$

and we have:

$$\frac{1}{\frac{V_m^L}{V_m^s}(1 - k_{part})}\ln\left(\frac{x^L}{x_0^L}\right) = \frac{1}{k_{part}\frac{V_m^L}{V_m^s} - 1}\ln\left[\frac{y\left(k_{part}\frac{V_m^L}{V_m^s} - 1\right) + \frac{\lambda}{2}}{\frac{\lambda}{2}}\right] \quad (11.13)$$

Equation (11.13) can be transformed into:

$$x^L = \frac{x^s}{k_{part}} = x_0^L\left[1 - \frac{2y}{\lambda}\left(1 - k_{part}\frac{V_m^L}{V_m^s}\right)\right]^{-\frac{(1-k)\frac{V_m^L}{V_m^s}}{1 - k\frac{V_m^L}{V_m^s}}} \quad (11.14a)$$

or

$$x^L = \frac{x^s}{k_{part}} = x_0^L\left[1 - f\left(1 - k_{part}\frac{V_m^L}{V_m^s}\right)\right]^{-\frac{(1-k)\frac{V_m^L}{V_m^s}}{1 - k\frac{V_m^L}{V_m^s}}} \quad (11.14b)$$

Equation (11.14b) is valid in case Ib, i.e. when the diffusion rate in the solid phase of alloying element is

high and the concentration $x^{L'} = x^L$. It describes the concentration of the alloying element within the volume element during solidification as a function of the solidification fraction f.

Case IIa

Assumptions:

1. The diffusion of the alloying element in the solid phase is low and can be neglected.
2. The added melt has a constant composition $x^{L'} = x_0^L$.

Case II a can be applied to casting with a feeder, which is very well insulated and solidifies much more slowly than the casting (Figure 11.8). Equation (11.5) can in this case be written as:

$$\left(\frac{\lambda}{2} - y\right)\mathrm{d}x^L = \left[x^L\left(1 - k_{part}\frac{V_m^L}{V_m^s}\right) + x_0^L\left(\frac{V_m^L}{V_m^s} - 1\right)\right]\mathrm{d}y \quad (11.15)$$

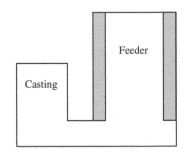

Figure 11.8   Case IIa.

Equation (11.15) is separable and can easily be integrated:

$$\int_{x_0^L}^{x^L} \frac{\mathrm{d}x^L}{x^L\left(1 - k_{part}\frac{V_m^L}{V_m^s}\right) + x_0^L\left(\frac{V_m^L}{V_m^s} - 1\right)} = \int_0^y \frac{\mathrm{d}y}{\frac{\lambda}{2} - y} \quad (11.16)$$

Integration gives:

$$\frac{1}{1 - k_{part}\frac{V_m^L}{V_m^s}}\ln\left[\frac{x^L\left(1 - k_{part}\frac{V_m^L}{V_m^s}\right) + x_0^L\left(\frac{V_m^L}{V_m^s} - 1\right)}{x_0^L\frac{V_m^L}{V_m^s}(1 - k_{part})}\right] = -\ln\left(\frac{\frac{\lambda}{2} - y}{\frac{\lambda}{2}}\right) \quad (11.17)$$

Equation (11.17) can be transformed into:

$$x^L = \frac{x^s}{k_{part}} = x_0^L \left[\left(1 - \frac{2y}{\lambda}\right)^{-\left(1 - k_{part}\frac{V_m^L}{V_m^s}\right)} - \frac{\frac{V_m^L}{V_m^s} - 1}{1 - k_{part}\frac{V_m^L}{V_m^s}} \right] \quad (11.18a)$$

or

$$x^L = \frac{x^s}{k_{part}} = x_0^L \left[(1-f)^{-\left(1 - k\frac{V_m^L}{V_m^s}\right)} - \frac{\frac{V_m^L}{V_m^s} - 1}{1 - k_{part}\frac{V_m^L}{V_m^s}} \right] \quad (11.18b)$$

Equation (11.18b) is valid for case IIa, i.e. when the diffusion of the alloying element in the solid phase is low and the supplied melt has a constant composition. It describes the concentration of the alloying element within the volume element during solidification as a function of the solidification fraction f.

Case IIb
Assumptions:

1. The diffusion of the alloying element in the solid phase is high.
2. The added melt has a constant composition $x^{L'} = x_0^L$.

The material balance will be the same as in case Ib and we can use Equation (11.9):

$$\frac{x^s y A}{V_m^s} + \frac{x^L \left(\frac{\lambda}{2} - y\right) A}{V_m^L} + x^{L'}\left(\frac{1}{V_m^s} - \frac{1}{V_m^L}\right)dyA = \frac{(x^s + dx^s)(y + dy)A}{V_m^s} + \frac{(x^L + dx^L)\left[\frac{\lambda}{2} - (y + dy)\right]A}{V_m^L} \quad (11.9)$$

After inserting $x^{L'} = x_0^L$ and reduction we obtain, if all terms of the type $dxdy$ are neglected,

$$\left(\frac{V_m^L}{V_m^s} - 1\right)\frac{x_0^L}{V_m^L}dy + \left(\frac{x^L}{V_m^L} - \frac{x^s}{V_m^s}\right)dy = \frac{ydx^s}{V_m^s} + \frac{\left(\frac{\lambda}{2} - y\right)dx^L}{V_m^L} \quad (11.19)$$

Inserting $x^s = s \cdot x^L$ and multiplication with V_m^L give:

$$\left(yk_{part}\frac{V_m^L}{V_m^s} + \frac{\lambda}{2} - y\right)dx^L = \left[x^L\left(1 - k_{part}\frac{V_m^L}{V_m^s}\right) + x_0^L\left(\frac{V_m^L}{V_m^s} - 1\right)\right]dy \quad (11.20)$$

Equation (11.20) is integrated between the limits x_0^L to x^L and 0 to y:

$$\int_{x_0^L}^{x^L} \frac{dx^L}{x^L\left(1 - k_{part}\frac{V_m^L}{V_m^s}\right) + x_0^L\left(\frac{V_m^L}{V_m^s} - 1\right)} = \int_0^y \frac{dy}{\frac{\lambda}{2} - y\left(1 - k_{part}\frac{V_m^L}{V_m^s}\right)} \quad (11.21)$$

and we have:

$$\frac{1}{1 - k_{part}\frac{V_m^L}{V_m^s}}\ln\left[\frac{x^L\left(1 - k_{part}\frac{V_m^L}{V_m^s}\right) + x_0^L\left(\frac{V_m^L}{V_m^s} - 1\right)}{x_0^L\frac{V_m^L}{V_m^s}(1 - k_{part})}\right] = \frac{1}{-\left(1 - k_{part}\frac{V_m^L}{V_m^s}\right)}\ln\left[\frac{\frac{\lambda}{2} - y\left(1 - k_{part}\frac{V_m^L}{V_m^s}\right)}{\frac{\lambda}{2}}\right]$$

or

$$\frac{x^L\left(1 - k_{part}\frac{V_m^L}{V_m^s}\right) + x_0^L\left(\frac{V_m^L}{V_m^s} - 1\right)}{x_0^L\frac{V_m^L}{V_m^s}(1 - k_{part})} = \left[1 - \frac{2y}{\lambda}\left(1 - k_{part}\frac{V_m^L}{V_m^s}\right)\right]^{-1} \quad (11.22)$$

which can be transformed into:

$$x^L = \frac{x^s}{k_{part}} = x_0^L\left\{\left[1 - \frac{2y}{\lambda}\left(1 - k_{part}\frac{V_m^L}{V_m^s}\right)\right]^{-1} - \frac{\frac{V_m^L}{V_m^s} - 1}{(1 - k_{part})\frac{V_m^L}{V_m^s}}\right\} \quad (11.23a)$$

or

$$x^L = \frac{x^s}{k_{part}} = x_0^L\left\{\left[1 - f\left(1 - k_{part}\frac{V_m^L}{V_m^s}\right)\right]^{-1} - \frac{\frac{V_m^L}{V_m^s} - 1}{(1 - k_{part})\frac{V_m^L}{V_m^s}}\right\} \quad (11.23b)$$

Equation (11.23b) is valid in case IIb, i.e. when the diffusion rate of the alloying element in the solid phase is high and the supplied melt has a constant composition. It describes the concentration of the alloying element in the volume element during solidification as a function of the solidification fraction f.

Example 11.1
During continuous casting of steel an increased carbon concentration in the centre is frequently observed (see Figure 11.1 on page 370). This segregation results when melt is supplied from above according to the diagram shown in order to compensate for the solidification and cooling shrinkage.

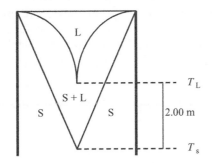

$$T_L - T_s = 100\ ^\circ C$$

A certain sort of steel has a carbon concentration of 0.80 wt-% and a solidification shrinkage of $\beta = 0.0275$. The two-phase region has a length of 2.00 m and the temperature difference between its upper and lower parts is 100 °C. It is reasonable to assume that the dendrite arm distance is about 100 μm.

(a) Calculate the maximum carbon concentration at the centre if the cooling shrinkage is neglected and only the solidification shrinkage is taken into consideration. This is reasonable at ingot casting. The answer should be given in weight-%.
(b) Does the central carbon concentration increase or decrease if the cooling shrinkage of the completely solidified material is also taken into consideration? This is necessary during continuous casting. Qualitative arguments, not quantitative calculations, are required in the answer.

Solution:
(a) The average vertical temperature gradient over the whole length interval $= 100/2.0 = 50.0\ \mathrm{K/m}$.

Since the temperature difference between two adjacent dendrite arms is very small, the error is minor if we assume that the melt from above has the same composition as the melt within a volume element with a vertical length of $\lambda/2$.

The solidification shrinkage results in inverse segregation. The reasoning we had about inverse segregation during ingot casting with a feeder can also be applied here for calculation of the carbon concentration in the small vertical volume element with length $\lambda/2$. Since carbon has a high diffusion rate in steel, case Ib is valid.

Consequently, we have $x^L = x^{L'}$ and can apply Equation (11.14b) for calculation of the carbon concentration in the volume element when completely solidified:

$$x^L = \frac{x^s}{k_{part}} = x_0^L \left[1 - f\left(1 - k_{part} \frac{V_m^L}{V_m^s} \right) \right]^{\frac{(1-k_{part})\frac{V_m^L}{V_m^s}}{1-k_{part}\frac{V_m^L}{V_m^s}}} \quad (1')$$

The carbon concentration in the centre is calculated from Equation (1') when all the material has solidified, i.e. when $f = 1$.

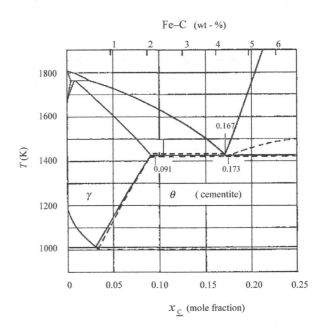

Fe–C (wt - %)

From the phase diagram for the Fe–C system we obtain:

$$k_{part} = \frac{x^s}{x^L} = \frac{0.091}{0.173} = 0.526$$

We also have:

$$\frac{V_m^L}{V_m^s} = \frac{\dfrac{M_{Fe}}{\rho^L}}{\dfrac{M_{Fe}}{\rho^s}} = \frac{\rho^s}{\rho^L} = \frac{1}{1-\beta} = \frac{1}{1-0.0275} = 1.03$$

where $\beta =$ the solidification shrinkage (see Chapter 10, page 329).

In the calculation of the concentration we have to use mole fraction or atom-%. The carbon concentration, given in weight-%, has to be recalculated into atom-% (Chapter 9, page 257):

$$x_0^L = \frac{\dfrac{c_0^L}{M_C}}{\dfrac{c_0^L}{M_C} + \dfrac{1-c_0^L}{M_{Fe}}} = \frac{\dfrac{0.80}{12}}{\dfrac{0.80}{12} + \dfrac{99.2}{55.85}} = 0.0362 = 3.62\ \text{at-\%}$$

Inserting these values in Equation (1′), we obtain:

$$
x^s = k_{part} x_0^L \left[1 - f \left(1 - k_{part} \frac{V_m^L}{V_m^s} \right) \right]^{-\frac{(1-k_{part})\frac{V_m^L}{V_m^s}}{1-k_{part}\frac{V_m^L}{V_m^s}}}
$$

or

$$
x^s = 0.526 \times 0.0362[1 - (1 - 0.526 \times 1.03)]^{-\frac{(1-0.526)1.03}{1-0.526\times1.03}}
$$
$$
= 0.0366
$$

The calculated carbon concentration is given in mole fraction and has to be recalculated into weight-% (Chapter 9, page 257):

$$
c^s = \frac{M_C x^s}{M_{Fe} - (M_{Fe} - M_C)x^s} \times 100
$$
$$
= \frac{12 \times 0.0366 \times 100}{55.85 - (55.85 - 12) \times 0.0366} = 0.81 \text{ wt-%}
$$

We could have performed the calculations in weight-% from the beginning but then Equation (11.14b) is no longer valid and a new material balance has to be set up. This requires more effort than the unit transformations of the concentrations. In Section 11.9.3 we will use weight-% throughout.

(b) When the cooling shrinkage is taken into account, another term has to be added in the material balance because the volume of the cooling shrinkage is filled with melt from outside.

A fourth term has to be added in the material balance on the left-hand side of Equation 11.9. The cooling shrinkage requires melt from outside. Consequently, more alloying element is supplied from outside, which results in an increase in the centre concentration.

Answer:

(a) The carbon concentration at the centre, due to the solidification, is calculated as 0.81 wt-%. The influence of the cooling shrinkage is neglected in this case.

(b) When both solidification and cooling shrinkage are considered, the calculated carbon concentration in the centre will be higher.

In reality, the macrosegregation in the centre in Example 11.1b is much larger than the answer in Example 11.1a, indicates. *Both* the solidification shrinkage *and* the cooling shrinkage must be considered during continuous casting. Macrosegregation as a result of cooling shrinkage during continuous casting will be treated in Section 11.5.

11.2.2 Degree of Macrosegregation

The general conclusion of the last section is that a volume element due to macrosegregation will contain *more* alloying element than the amount due to microsegregation only.

The increase in the concentration of the alloying element is called the *degree of macrosegregation* and is defined by the relation

$$
\Delta x = \overline{x^s} - x_0^L \tag{11.24}
$$

where

$\overline{x^s}$ = average concentration of the alloying element in the volume element after complete solidification

x_0^L = concentration of the alloying element in the initial melt.

The average concentration of the alloying element can be calculated from the relation

$$
\overline{x^s} = \frac{\int_0^{\lambda/2} x^s dy}{\int_0^{\lambda/2} dy} \tag{11.25}
$$

Equation (11.24) is the general expression for the degree of macrosegregation. Equation (11.25) shows that one has to know x^s as a function of y to be able to calculate the degree of macrosegregation.

The average concentration of the alloying element for the four cases Ia, Ib, IIa and IIb is discussed below and listed in the box on page 377. When $\overline{x^s}$ has been calculated, the degree of macrosegregation can easily be obtained with the aid of Equation (11.24).

It is important to note that the *degree of macrosegregation* is measured in atom-% or weight-% whereas the *degree of microsegregation* (Chapter 7, page 201) is a dimensionless ratio.

Calculation of the Average Concentration $\overline{x^s}$ of the Alloying Element in the Solid Phase

Cases Ia and IIa

In these cases the diffusion of the alloying element is very low and *Scheil's equation* is valid (Figure 11.9). When

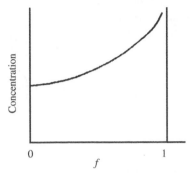

Figure 11.9 Scheil's equation: $x^s = k_{part} x_0^L (1-f)^{-(1-k_{part})}$

shrinkages is taken into consideration, we will use the equation:

$$x^L = \frac{x^s}{k_{part}} = x_0^L \left(1 - \frac{2y}{\lambda}\right)^{-(1-k_{part})\frac{V_m^L}{V_m^s}} \quad (11.8a)$$

When x^s is plotted as a function of the solidification fraction f a bent curve is obtained. Integration is necessary to obtain $\overline{x^s}$ with the aid of Equation (11.25).

The shape of Equation (11.18a) shows that integration is necessary also in case IIa (see page 374).

Cases Ib and IIb

In these cases, the alloying element diffuses easily through the solid phase. The concentration of the alloying element in the solid phase is the same *everywhere in the solid phase* and the *lever rule* [Equation (7.15), page 189] is valid:

$$x^L = \frac{x^s}{k_{part}} = \frac{x_0^L}{1 - f(1 - k_{part})} \quad \text{lever rule} \quad (11.26)$$

However, Equation (11.26) is valid only if the solidification shrinkage is neglected and cannot be used in this case. Instead, we use the statement above and conclude that the average concentration $\overline{x^s}$ must be equal to the value of x^s when the solidification is complete, i.e. $f = 1$:

$$\overline{x^s} = x^s \quad (11.27)$$

The average value in cases Ib and IIb can therefore be obtained directly without integration. The $\overline{x^s}$ values are obtained when $f = 1$ is inserted into Equations (11.14b) and (11.23b).

Example 11.2

During a casting process, an Al–Cu melt with 2.50 at-% Cu is cast into the mould and into the feeder. The melt in the casting and the feeder solidifies at the same rate. The ratio V_m^L/V_m^s between the molar volumes of the alloy is 1.04.

According to the phase diagram for the system Al–Cu, the solid Al phase has a solubility of 2.50 at-% Cu at the eutectic temperature. The eutectic melt has the composition 17.3 at-% Cu.

Case Ia:

Low diffusion in the solid and $x^L = x^{L'}$.

$$\overline{x^s} = \frac{\int_0^{\lambda/2} x^s dy}{\int_0^{\lambda/2} dy} = \frac{\int_0^{\lambda/2} k_{part} x_0^L \left(1 - \frac{2y}{\lambda}\right)^{-(1-k_{part})\frac{V_m^L}{V_m^s}} dy}{\frac{\lambda}{2}} \quad (11.28)$$

Case Ib:

High diffusion in the solid and $x^L = x^{L'}$.

$$\overline{x^s} = k_{part} x_0^L \left(k_{part} \frac{V_m^L}{V_m^s}\right)^{-\frac{(1-k_{part})\frac{V_m^L}{V_m^s}}{1-k_{part}\frac{V_m^L}{V_m^s}}} \quad (11.29)$$

Case IIa:

Low diffusion in the solid and $x^{L'} = x_0^L$.

$$\overline{x^s} = \frac{\int_0^{\lambda/2} k_{part} x_0^L \left[\left(1 - \frac{2y}{\lambda}\right)^{-(1-k_{part})\frac{V_m^L}{V_m^s}} - \frac{\frac{V_m^L}{V_m^s} - 1}{1 - k_{part}\frac{V_m^L}{V_m^s}}\right] dy}{\frac{\lambda}{2}} \quad (11.30)$$

Case IIb:

High diffusion in the solid and $x^{L'} = x_0^L$.

$$\overline{x^s} = k_{part} x_0^L \left[\left(k_{part} \frac{V_m^L}{V_m^s}\right)^{-1} - \frac{\frac{V_m^L}{V_m^s} - 1}{(1 - k_{part})\frac{V_m^L}{V_m^s}}\right] \quad (11.31)$$

(a) What fraction of the alloy solidifies with eutectic structure?

(b) Calculate the average composition of the alloy when it has solidified completely.

(c) Calculate the degree of macrosegregation.

Solution:

Since the molar volumes cannot be regarded as equal, macrosegregation is to be expected. Since the melt in the feeder and the casting solidify at an equal rate, case I is valid.

To be able to use the equations, we must know the value of the partition coefficient k_{part}. It is calculated in the same

way as in Example 7.1 (page 187):

$$k_{part} = \frac{x^s}{x^L} = \frac{2.50}{17.3} = 0.1445 \qquad (1')$$

(a) It is reasonable to assume that the melt left when the temperature has decreased to the eutectic temperature will solidify with eutectic structure. Hence the problem can be solved by calculation of the degree of solidification $f = 2y/\lambda$ when the melt has reached the eutectic concentration. We apply Equation (11.8a on page 372), which is valid for case Ia:

$$x^L = x_0^L \left(1 - \frac{2y}{\lambda} \right)^{-\left(1-k_{part}\right)\frac{V_m^L}{V_m^s}} \qquad (2')$$

or

$$0.173 = 0.0250 \times (1 - f_E)^{-(1-01445)\times 1.04} \qquad (3')$$

where the eutectic fraction $f_E = 2y_E/\lambda$.

When the eutectic point is reached, the solidification front is situated at $y = y_E$. The remaining melt solidifies with eutectic structure. The fraction which solidifies with eutectic structure is obtained from Equation (3'):

$$1 - f_E = 1 - 2y_E/\lambda = \left(\frac{0.173}{0.0250} \right)^{\frac{-1}{0.8555\times 1.04}} = 0.114 \quad (4')$$

(b) Equation (11.25 on page 376) can *not* be applied without further notice in this case. The composition of the melt follows Equation (11.8a) (page 372) up to $y = y_E$ and is subsequently constant and equal to x_E. The average concentration of Cu becomes:

$$\frac{x_{cu}^s}{V_m^s} \frac{\lambda}{2} = \frac{1}{V_m^s} \int_0^{y_E} x^s dy + \frac{x_E \left(\frac{\lambda}{2} - y_E \right)}{V_m^E} \qquad (5')$$

We assume that $V_m^s \approx V_m^E$ and apply Equation (11.8a) to Equation (5'):

$$\overline{x_{Cu}^s} = \frac{2}{\lambda} \int_0^{y_E} k_{part} x_0^L \left(1 - \frac{2y}{\lambda} \right)^{-(1-k_{part})\frac{V_m^L}{V_m^s}} dy + x_E \left(y - \frac{2y_E}{\lambda} \right) \qquad (6')$$

After integration we obtain:

$$\overline{x_{Cu}^s} = \frac{k_{part} x_0^L}{1 - (1 - k_{part}) \cdot \frac{V_m^L}{V_m^s}} \left[1 - \left(1 - \frac{2y_E}{\lambda} \right)^{1-(1-k_{part})\frac{V_m^L}{V_m^s}} \right] + x_E \left(y - \frac{2y_E}{\lambda} \right) \qquad (7')$$

Inserting the values gives:

$$\overline{x_{Cu}^s} = \frac{0.1445 \times 0.0250}{1 - 0.8555 \times 1.04} \left[1 - 0.114^{(1-0.8555\times 1.04)} \right] + 0.173 \times 0.114 = 0.0267$$

(c) The degree of macrosegregation is defined as:

$$\overline{x_{Cu}^s} - x_0^L = 0.027 - 0.025 = 0.002 = 0.2 \text{ at-}\%$$

Answer:

(a) Approximately 11.4 % solidifies with eutectic structure.
(b) The average Cu concentration in the solidified alloy is 2.7 at-%.
(c) The macrosegregation is approximately 0.2 at-%.

11.3 MACROSEGREGATION DURING UNIDIRECTIONAL SOLIDIFICATION

In the early 1960s, Flemings and co-workers performed a great number of solidification experiments such as that in Example 11.2. The aim was to study and analyse the macrosegregation in alloys due to solidification shrinkage during unidirectional solidification (Chapter 6, page 159).

In one of their experiments, an Al alloy with 4.6 wt-% Cu was solidified. Their apparatus is shown in Figure 11.10. After the solidification, the composition of the alloy was examined and recorded together with the positions of the test samples. The measured Cu concentrations were plotted as a function of the distance from the cooled surface. The result is shown in Figure 11.11, which indicates that the Cu concentration close to the cooled surface is 4–5 % higher than the average value in the interior of the alloy.

Below we discuss qualitatively the distribution of alloying elements in this type of experiment and try to explain the appearance of macrosegregation close to the surface and in the interior of a continuously cast strand. Macrosegregation is a severe problem in continuous casting.

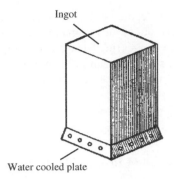

Ingot

Water cooled plate

Figure 11.10   Unidirectional solidification of an Al–4.6 wt-% Cu alloy ingot. Reproduced by permission of The Minerals, Metals & Materials Society.

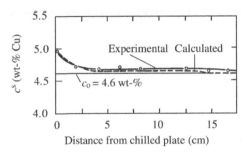

Distance from chilled plate (cm)

Figure 11.11   Cu concentration in the solid ingot as a function of the distance from the chilled plate.

11.3.1   Macrosegregation at Constant Surface Temperature

Owing to solidification shrinkage, melt from the interior part of the melt will be sucked into the melt–solid phase two-phase region. This transport of melt occurs within the whole extension of the two-phase region.

Consider a volume element according to Figure 11.12 (a). Its lower surface has an initial position which coincides with the solidification front. The volume element remains at rest during unidirectional solidification and the solidification front gradually moves upwards [Figures 11.12 (a)–(d)].

The more the melt 'invades' the volume element, the more the concentration of the alloying element in the melt will increase within the interdendritic regions since solid phase precipitates continuously. The result is that the concentration of the alloying element increases gradually within the two-phase region. It is highest in the lowest parts of the volume element. The melt which is fed to the lower parts of the two-phase region is strongly enriched. Hence the concentration will be maximum at the cooled surface and decreases inwards, in agreement with known observations (Figure 11.11).

At the very top of the specimen, melt is left to be sucked into the two-phase region. Pores are formed between the dendrite arms and the measured concentration at the very

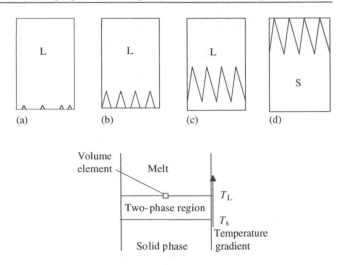

(e) Unidirectional solidification

Figure 11.12   (a)–(d) Position of the two-phase region as a function of time during a unidirectional solidification process. (e) Position of the volume element.

top corresponds to the concentration at the centres of the dendrite arms.

Exactly how the concentration of the alloying element will vary depends on the cooling conditions and the concentration in the added melt and is closely related to the phase diagram of the alloy (Figure 11.13).

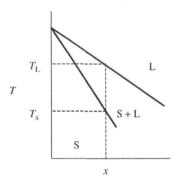

Figure 11.13   Principle sketch of the phase diagram of a binary alloy. It can be seen that the width of the solidification interval $T_L - T_S$ depends on the concentration x of the alloying element.

The magnitude of the macrosegregation depends entirely on the cooling conditions. If the cooling is designed in such a way that the two-phase width is *large* and *increases* during the solidification, the melt must pass a longer and longer distance through the dendritic network. The result is that the melt cools on its way, solid phase is precipitated and the alloying element accumulates in the remaining melt. The wider the two-phase region is, the larger will be the accumulation and the macrosegregation.

On the other hand, if the cooling is such that the two-phase region is *narrow* and its width *decreases* during the solidification process, the accumulation of the alloying element in the remaining melt is moderate. The narrower the two-phase region is, the smaller will be the accumulation of alloying elements and the macrosegregation.

11.3.2 Macrosegregation at Variable Surface Temperature

The deviations from the average composition in the interior of a strand from the surface and inwards are often moderate during normal solidification conditions. However, the variation of the carbon concentration in a steel alloy is not as simple and regular as is described in Figure 11.11. It shows a wave pattern as indicated in Figure 11.14.

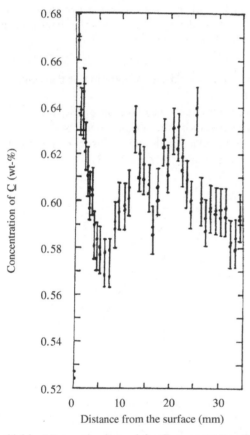

Figure 11.14 Measured values of the C concentration from the surface and inwards towards the centre of a strand with a cross-sectional area 160×160 mm^2. Reproduced by permission of Bo Rosberg.

During continuous casting, there are strong temperature changes both at the surface and at the centre of the strand. This is illustrated in Figure 10.44 (page 338).

With a violent temperature increase or decrease of the temperature of the surface, the solidified shell will expand or shrink owing to the cooling shrinkage. Melt is sucked into or squeezed out of the two-phase region for this reason, which causes changes in the average composition in the interior of the strand, as can be seen in Figure 11.14. This topic will be analysed more carefully in the next section.

11.4 INVERSE MACROSEGREGATION

11.4.1 Definition of Inverse Macrosegregation

Macrosegregation was primarily observed and investigated during casting of large ingots. Different types of concentration deviations from the average composition were observed.

Positive macrosegregation means that the concentration of the alloying element in a region is *higher* than the average concentration. If the local concentration is *lower* than the average concentration the macrosegregation is said to be *negative*.

The early observations showed that the surface and the region close to the surface have positive macrosegregation (Figures 11.11 and 11.14). This phenomenon was called *inverse macrosegregation*, probably because one had expected the opposite sign of the macrosegregation which was presumed to compensate for the strongly positive macrosegregation at the centre (Figure 11.1 on page 370).

11.4.2 Mathematical Model of Inverse Macrosegregation

Scheil was the first to examine the mechanism behind the so-called inverse macrosegregation, during the 1940s. Later Flemings and co-workers treated the phenomenon in more detail (page 378).

The mathematical model which was used in Section 11.2 to treat inverse segregation in continuously cast materials is greatly simplified compared with the reality to be described. The variation of the densities of the melt and the solid phase with temperature was not considered, for example.

In this section, we will introduce a more sophisticated one-dimensional model, which was developed by Fredriksson and Rogberg. Rogberg used the model to calculate the inverse segregation in high-carbon steel. By computer calculations he was able to simulate the conditions during continuous casting and study how density variations, different cooling conditions, and expansion and shrinkage of the surface influence the macrosegregation of carbon in high-carbon steel.

A simple analytical solution of the differential equation of the model is given below in order to illustrate the principle of the calculations. The model is general and valid for both ingots and strands during continuous casting. It is used below to explain the increased C concentration and the decrease close to the surface of the strand (Figures 11.1 and 11.14).

Mathematical Model of Inverse Segregation

Figure 11.15 shows a unidirectionally solidified sample, which is part of a strand. When the strand moves downwards in the casting machine, the solidification front moves simultaneously towards the centre of the strand with a velocity u. The width of the two-phase region is designated L.

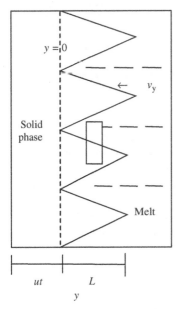

Figure 11.15 Continuously cast strand, which moves with a velocity u through a two-phase region. Reproduced by permission of Bo Rosberg.

We consider a small volume element in the shape of a thin slice in the two-phase region. Its thickness is dy and its cross-sectional area is A. By dividing the area into two parts A_L and A_s, we can schematically describe the fractions of melt and solid phase in the two-phase region. The volume element must be large enough to give a reliable value of the solidification fraction f but at the same time small enough to be treated as an infinitesimal quantity. The volume element is situated in the two-phase region and does not follow the solidification front in its motion inwards.

The reason for the segregation is that melt is sucked into the dendrite network during the solidification process. The composition of the entering melt determines the concentration of the alloying element in the volume element.

During continuous casting, the cooling conditions vary considerably. This means that the velocity of the solidification front varies and that the thickness of the solidifying shell grows and sometimes even shrinks as a consequence of the temperature variations.

We make the following five assumptions in order to set up a material balance (concentrations of alloying elements in weight-%) for the volume element:

1. The solid phase stays in equilibrium with the melt in the volume element. The equilibrium is described by the relation:

$$\frac{c^s}{c^L} = k_{\text{part}} \qquad (11.32)$$

where k is the partition coefficient.

2. Mass is supplied and removed from the volume element only by a flow of melt.

3. No pores are formed within the volume element.

4. The volume of the element is a function of temperature and is changed by change in the lateral area A_L (Figure 11.16). The changes cause a flow of melt into or out of the volume element.

5. The dendrite arm distance is very small compared with the width L of the two-phase region. Consequently, the solidification fraction or the fraction of melt can be expressed with the aid of a continuous function.

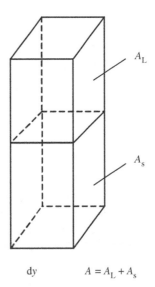

Figure 11.16 Enlargement of the volume element in Figure 11.15.

The model includes consideration to the variation of the densities of the solid phase and the melt with temperature and allows an increase and decrease of the two-phase area by variation of A_L.

We choose a coordinate system with the origin of the coordinates at the solidification front and the y-axis in the direction of its motion (Figure 11.15). The mass of the volume element can then be written as:

$$dm = A_s \rho_s dy + A_L \rho_L dy + A_L \rho_L (-v_y) dt \qquad (11.33)$$

where

$A_{s,\,L}$ = lateral areas of the solid phase and the melt, respectively, in the volume element

$\rho_{s,\,L}$ = densities of the solid phase and the melt, respectively

v_y = velocity of the melt; it moves in the negative y direction and v_y is therefore negative.

The first two terms in Equation (11.33) correspond to the mass within the volume element. The third term represents the mass contribution due to the inflow during the time interval dt. The velocity v_y of the melt is negative and the minus sign is added to describe how the inflow of melt occurs in the direction of the *negative y*-axis. The contribution of melt is *positive*.

The mass of the volume element is obviously a function of two variables y and t. This can be expressed mathematically as a total differential:

$$\mathrm{d}m = \frac{\partial m}{\partial y}\mathrm{d}y + \frac{\partial m}{\partial t}\mathrm{d}t \qquad (11.34)$$

Equation (11.34) can be identified with Equation (11.33) and we have:

$$\frac{\partial m}{\partial y} = A_s\rho_s + A_L\rho_L \qquad (11.35)$$

$$-\frac{\partial}{\partial y}(A_L c^L \rho_L v_y) = c^L \rho_L \frac{\partial A_L}{\partial t} + A_L \rho_L \frac{\partial c^L}{\partial t} + A_L c^L \frac{\partial \rho_L}{\partial t} + k_{\text{part}}\, c^L \rho_s\left(\frac{\partial A}{\partial t} - \frac{\partial A_L}{\partial t}\right) + k_{\text{part}}\, c^L (A - A_L)\frac{\partial \rho_s}{\partial t} + k_{\text{part}}\,\rho_s(A - A_L)\frac{\partial c^L}{\partial t} \qquad (11.42)$$

and

$$\frac{\partial m}{\partial t} = A_L\rho_L(-v_y) \qquad (11.36)$$

Equations (11.35) and (11.36) depend on each other. We form the mixed second derivatives by partial derivatization of Equation (11.35) with respect to t and Equation (11.36) with respect to y. The partial derivatives are equal since the

result does not depend on the order of derivatization.

$$-\frac{\partial}{\partial y}(A_L\rho_L v_y) = \frac{\partial}{\partial t}(A_s\rho_s + A_L\rho_L) \qquad (11.37)$$

Analogously, we obtain from a corresponding total material balance of the alloying element:

$$-\frac{\partial}{\partial y}(A_L c^L \rho_L v_y) = \frac{\partial}{\partial t}(A_s c^s \rho_s + A_L c^L \rho_L) \qquad (11.38)$$

In addition we have by definition:

$$A = A_s + A_L \qquad (11.39)$$

and

$$\frac{\partial A}{\partial t} = \frac{\partial A_s}{\partial t} + \frac{\partial A_L}{\partial t} \qquad (11.40)$$

Partial derivatization of the relation $c^s = k_{\text{part}} c^L$ with respect to time gives:

$$\frac{\partial c^s}{\partial t} = k_{\text{part}}\frac{\partial c^L}{\partial t} \qquad (11.41)$$

We have six equations [(11.32) and (11.37)–(11.41)] and want to find a differential equation that contains c^L and its partial time derivative.

The right-hand side of Equation (11.38) is developed and A_s and c_s and their partial time derivatives are eliminated with the aid of Equations (11.32) and (11.39)–(11.41). Equation (11.38) can then be written as:

We develop the left-hand side of Equation (11.42) and obtain:

$$-\frac{\partial}{\partial y}(A_L c^L \rho_L v_y) = -c^L \frac{\partial}{\partial y}(A_L \rho_L v_y) - A_L \rho_L v_y \frac{\partial c^L}{\partial y} \qquad (11.43)$$

We develop the right-hand side of Equation (11.37) and eliminate A_s and $\partial A_s/\partial t$ with the aid of Equations (11.39) and (11.40). Equation (11.37) can then be written as:

$$-\frac{\partial}{\partial y}(A_L\rho_L v_y) = (A - A_L)\frac{\partial \rho_s}{\partial t} + \rho_s\left(\frac{\partial A}{\partial t} - \frac{\partial A_L}{\partial t}\right) + \rho_L \frac{\partial A_L}{\partial t} + A_L \frac{\partial \rho_L}{\partial t} \qquad (11.44)$$

The right-hand side of Equation (11.44) is substituted into Equation (11.43) and the new expression is inserted into Equation (11.42). After these manipulations, Equation (11.42) becomes:

$$[A_L\rho_L + k_{\text{part}}\,\rho_s(A - A_L)]\frac{\partial c^L}{\partial t} + c^L \rho_s(k_{\text{part}} - 1)\left(\frac{\partial A}{\partial t} - \frac{\partial A_L}{\partial t}\right) + c^L(A - A_L)(k_{\text{part}} - 1)\frac{\partial \rho_s}{\partial t} = -A_L\rho_L v_y \frac{\partial c^L}{\partial y} \qquad (11.45)$$

Equation (11.45) is the desired and fundamental differential equation. Hopefully ρ_s, ρ_L, and A are known functions of time and temperature. In order to solve Equation (11.45), we must know either A_L or c^L. If A_L is known, c^L can be calculated and vice versa.

The flow rate v_y of the melt is obtained by integration of Equation (11.37):

$$v_y = \frac{-1}{A_L(y)\rho_L(y)} \int_{y_L}^{y} \frac{\partial}{\partial t}[A_s(y)\rho_s(y) + A_L(y)\rho_L(y)]dy \quad (11.46)$$

The lower integration limit corresponds to the first solidification front where the temperature equals the liquidus temperature T_L.

11.5 MACROSEGREGATION DURING CONTINUOUS CASTING

Macrosegregation is important during continuous casting and cannot be avoided. The dominant metal cast by continuous casting is steel in terms of volume of production. For this reason, we will discuss macrosegregation at continuous casting of steel below, but the statements and outlines are valid for other metals also.

We will apply the theory of inverse macrosegregation, given in Section 11.4.2. Figure 11.14 shows the carbon distribution at the surface and inwards in a strand after continuous casting. As was mentioned on page 379, the cooling conditions strongly influence the degree of macrosegregation. The surface temperature of the strand changes with

the cooling conditions (see Chapters 5 and 10). This changes the solidification rate and also the strain or compression of the shell. Both result in volume changes and influence the macrosegregation.

It is also observed that the carbon concentration on the surface is very high and may amount to up to 10–15 %. An important parameter when the high macrosegregation at the surface is to be explained is the ratio of the densities of the melt and the solid phase. We will restrict the discussion to the simplest case when ρ_L and ρ_s are both constant. In more complicated cases, and unfortunately also in reality, they vary with the temperature and the composition of the alloy.

The advantage of the simplification is that there is an analytical solution to the differential equations if the densities are constant. In more complicated cases, only numerical solutions are available.

11.5.1 Simple Application of the Model of Inverse Macrosegregation at the Surface of a Strand

In certain simple cases there is a relatively simple analytical solution to the differential Equation (11.45). We will try to calculate the macrosegregation at the surface of a steel strand as a function of the constant densities of the melt and the solid phase.

Unfortunably, available data for steel densities at various carbon concentrations are incomplete. It is customary to assume that the solidification shrinkage in steel, which solidifies as austenite, is 4 % and we use this value in our calculations unless some other value is given in the text.

Macrosegregation at the Strand Surface in the Case of Constant Densities

It is of interest to find out how the macrosegregation varies quantitatively at various values of ρ_s and ρ_L and to compare the theoretical values with experimental data.

- Figure 11.15 shows that the sucked melt has to stop at the surface (solidification front), i.e. $v_y = 0$.
- In addition, we assume that the area A of the volume element is constant.
- We introduce the designation g for relative volume. According to Figure 11.16, the fractions of liquid and solid phases can then be written as:

$$g_L = A_L/A \qquad \text{and} \qquad g_s = A_s/A \quad (11.47)$$

When these values and designations are inserted into Equation (11.45), we obtain:

$$[g_L\rho_L + k_{part}\rho_s(1 - g_L)]\frac{\partial c^L}{\partial t} - c^L\rho_s(k_{part} - 1)\frac{\partial g_L}{\partial t} + c^L(1 - g_L)(k_{part} - 1)\frac{\partial \rho_s}{\partial t} = -g_L\rho_L v_y\frac{\partial c^L}{\partial y} = 0 \quad (11.48)$$

This equation is valid at the surface where $v_y = 0$. It will be the basis for the examination of the solution below.

If we divide Equation (11.45) by ρ_s, the new equation contains only the ratios ρ_L/ρ_s and $(\partial \rho_s/\partial t)/\rho_s$. In this case, the time derivative is zero and therefore the only parameter is the constant ratio ρ_L/ρ_s The value of the partition constant $k_{part} = 0.33$ will be used in the calculations.

Calculation of the Macrosegregation as a Function of the Initial Carbon Concentration at Constant Densities
In this case both $v_y = 0$ and $\partial \rho_s/\partial t = 0$ at the surface and Equation (11.48) contains only three terms.

$$[g_L\rho_L + k_{part}\rho_s(1 - g_L)]\frac{\partial c^L}{\partial t} - c^L\rho_s(k_{part} - 1)x^2\frac{\partial g_L}{\partial t} = 0$$

$$(11.49)$$

We want to solve c^L as a function of g_L. Equation (11.49) is separable and we obtain:

$$\int_{c_0^L}^{c^L} \frac{dc^L}{c^L} = \int_1^{g_L} \rho_s(k_{part} - 1) \frac{dg_L}{g_L(\rho_L - k_{part}\rho_s) + k_{part}\rho_s}$$

or

$$\ln\left(\frac{c^L}{c_0^L}\right) = \frac{\rho_s(k_{part} - 1)}{\rho_L - k_{part}\rho_s} \ln\left[\frac{g_L(\rho_L - k_{part}\rho_s) + k_{part}\rho_s}{\rho_L}\right]$$

The solution of c^L can be used to calculate c^s:

$$c^s = k_{part}\,c^L = k_{part}c_0^L\left[\frac{\rho_L}{g_L(\rho_L - k_{part}\rho_s) + k_{part}\rho_s}\right]^{\frac{1-k_{part}}{\frac{\rho_L}{\rho_s} - k_{part}}} \tag{11.50}$$

The degree of macrosegregation can then be written as (page 376):

$$\Delta c = c^s - c_0^L = k_{part}c_0^L\left[\frac{\frac{\rho_L}{\rho_s}}{g_L\left(\frac{\rho_L}{\rho_s} - k_{part}\rho_s\right) + k_{part}}\right]^{\frac{1-k_{part}}{\frac{\rho_L}{\rho_s} - k_{part}}} - c_0^L \tag{11.51}$$

Equation (11.51) is the relation between the degree of macrosegregation and the initial carbon concentration in the melt in the case of constant densities. Δc is plotted as a function of c_0^L in Figure 11.17 for some different values of the constant ratio ρ_L/ρ_s. Figure 11.17 shows that the differences in degree of macrosegregation amount to only a few hundredths of a weight-% when the initial carbon concentration is increased by a factor of three.

The theory of inverse segregation shows that the macrosegregation at the surface of a strand depends not only

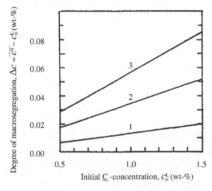

Figure 11.17 Theoretical degree of macrosegregation close to the surface as a function of the initial carbon concentration during continuous casting of steel alloys. ρ_L and ρ_s are assumed to be constant. Curve 1, $\rho_L/\rho_s = 0.98$; curve 2, $\rho_L/\rho_s = 0.95$; curve 3, $\rho_L/\rho_s = 0.92$. Reproduced by permission of Bo Rosberg.

on the solidification shrinkage but also on the density variations in the solid phase. Equation (11.45) can be solved numerically in more complicated cases with densities, which depend on temperature and time.

It is a puzzling fact that the calculated values of the inverse macrosegregation at the surface are in all cases 10–15 % lower than the observed values. One explanation for the high experimental carbon concentration at the surface of the strand during continuous casting of steel may be that the surface probably is strained, owing to an increase of the temperature. Strain during unidirectional solidification has been discussed briefly in Section 10.7.3 (page 346) and is illustrated in Figure 10.58 (page 348). The influence of strain on the carbon concentration will be estimated below.

Calculation of the Macrosegregation When the Strand Surface is Strained

If the surface is strained Equation (11.50) is no longer valid. The initial carbon concentration c_0^L has to be replaced by the expression:

$$\overline{c_{0\,strain}^L} = \frac{\overline{c_0^L} + \varepsilon\dfrac{c^L}{k_{part}}}{1 + \varepsilon} \tag{11.52}$$

where

$\overline{c_{0\,strain}^L}$ = average carbon concentration in the melt at the given solidification fraction of the strained surface

$\overline{c_0^L}$ = average carbon concentration in the melt at the given solidification fraction of the strain-free surface

c^L = carbon concentration of the sucked melt in the strained volume

ε = strain due = $\Delta l/l$.

For calculation of the average carbon concentration in the case of strain, we consider an unstrained volume element filled with melt with the average carbon concentration $\overline{c_0^L}$ at a given temperature. When the temperature increases, the surface of the volume element becomes strained and its average carbon concentration increases to $\overline{c_{0\,strain}^L}$.

In order to calculate $\overline{c_{0\,strain}^L}$, we must know ε and the concentration c^L as functions of temperature. The strain ε is related to the thermal expansion coefficient of the melt and the temperature change and can be derived from these quantities; c^L is obtained with the aid of the phase diagram when the temperature is known.

When the values of ε and c^L have been derived, $\overline{c_{0strain}^L}$ can be calculated with the aid of Equation (11.52) and the strained macrosegregation is obtained with the aid of the relation:

$$\Delta c_{strain} = c^s - \overline{c_{0\,strain}^L} \tag{11.53}$$

The total macrosegregation is obtained from Equation (11.51) if we add the effect of strain:

$$\Delta c_{\text{total}} = \Delta c + \Delta c_{\text{strain}} \qquad (11.54)$$

Comparison between calculated and experimental values of the degree of macrosegregation shows that satisfactory agreement is achieved when the contribution of the strain is added.

11.5.2 Macrosegregation in the Interior of a Strand

Experimental values of the macrosegregation in the interior of a strand are shown in Figure 11.14 on page 380. It is important to apply the theory of inverse segregation to the interior of the strand and compare the results of the simulation calculations with the experimental values.

Calculation of the macrosegregation of continuously cast billets requires a simultaneous solution of:

1. the differential equation of the macrosegregation model:
2. the general law of thermal conduction including boundary conditions in combination with
3. the equilibrium conditions in the phase diagram of the alloy.

The *first* Equation is identical with Equation (11.45). With the aid of this equation, the macrosegregation can be derived.

The *second* equation [Fourier's second law, Equation (4.10) (page 62)] controls the heat transport and the temperature field and thus the solidification rate of the alloy.

The *third* condition which has to be fulfilled cannot be described mathematically but represents the equilibrium conditions which have to be fulfilled. The concentrations of the alloying elements in the solid and liquid phases must be in equilibrium with each other. They are functions of temperature.

The equations form an equation system. The solution of this system provides the desired information about the solidification process and the simultaneous macrosegregation formation.

The theoretical calculations supply the following information:

- the macrosegregation as a function of the distance from the cooled surface
- the temperature as a function of position and time
- information about macrosegregation under various conditions, for example the influence of cooling conditions and expansion/shrinkage of the strand on the macrosegregation.

The last point will be discussed below.

11.5.3 Influence of Cooling Conditions and Strand Expansion/Shrinkage on Macrosegregation in the Interior of a Strand

Theoretical calculations with proper assumptions can be performed to simulate different cooling conditions and expansion or shrinkage of the strand.

Influence of Cooling Conditions

To simulate different cooling conditions, three different cooling arrangements were chosen during the cooling of a strand. The initial carbon concentration was 1.0 wt-% in all cases.

The model was a slab with thickness 100 mm, which was assumed to have been continuously cast. The cooling conditions were simulated with the aid of different heat transfer coefficients during the one-dimensional heat flow.

1. $h = 500$ J/m^2 K during the whole solidification process; $\rho_L/\rho_s = 0.95$.
2. $h = 1500$ J/m^2 K during the first 15 s and then 500 J/m^2 K; $\rho_L/\rho_s = 0.95$.
3. $h = 500$ J/m^2 K during the first 45 s and then 1500 J/m^2 K; $\rho_L/\rho_s = 0.95$.

The results of the calculations are seen in Figures 11.18–11.20.

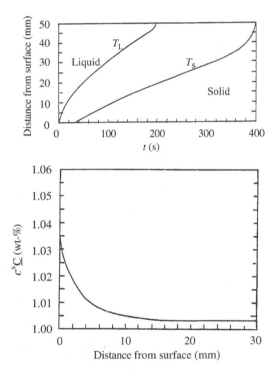

Figure 11.18 Case 1. The upper figure shows the motion of the solidification fronts. The lower figure shows the resultant $\underline{C}$ concentration as a function of the distance from the surface. Reproduced by permission of Bo Rosberg.

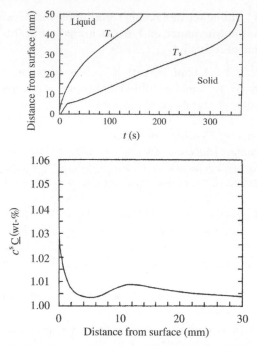

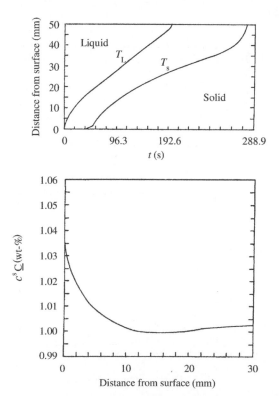

Figure 11.19 Case 2. The upper figure shows the motion of the solidification fronts. The lower figure shows the resultant $\underline{C}$ concentration as a function of the distance from the surface. Reproduced by permission of Bo Rosberg.

Figure 11.20 Case 3. The upper figure shows the motion of the solidification fronts. The lower figure shows the resultant $\underline{C}$ concentration as a function of the distance from the surface. Reproduced by permission of Bo Rosberg.

A comparison between the three cases shows that the influence of varying cooling conditions on the solidification rate is small in the present case and considerably smaller than shown in Figure 11.14. The influence cannot be neglected in the general case.

Influence of Strand Expansion and Strand Shrinkage

In simulation calculations to show the effect of a changing strand area, the cooling conditions were kept constant and the area A changed linearly from A to $1.01A$ with time. Three different time intervals were studied. The heat transfer coefficient h was $1000 \ \mathrm{J/m^2 \ K}$ and the density ratio ρ_L/ρ_s was 0.95. The result is illustrated in Figure 11.21.

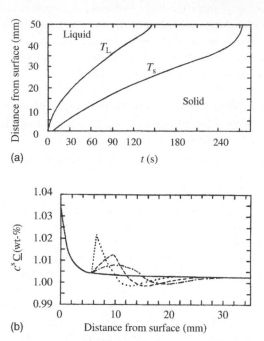

Figure 11.21 (a) Motion of the solidification fronts, i.e. the positions of the solidification fronts marked by T_L (solidification starts) and T_s (complete solidification) as functions of time. (b) The $\underline{C}$ concentration as a function of the distance from the surface for three different expansion time intervals of the strand: 30–33, 30–50 and 30–70 s. Reproduced by permission of Bo Rosberg.

The conclusion is that an expanding or a shrinking strand will in this case cause minor concentration variations in the solid phase during the solidification process. However, the effect increases strongly if the length strain coefficient is >0.01.

11.6 CENTRE SEGREGATION DURING CONTINUOUS CASTING

The inverse macrosegregations at the surface and the interior of a strand during continuous casting are small, as was found above. The opposite is true for centre segregation.

Centre segregation and pipe formation are the two most serious problems in continuous casting. Macrosegregation during continuous casting is closely related to pipe formation because the origin of the two phenomena is the same. Pipe formation during continuous casting has been treated in Chapter 10 and we will refer to the description given there.

The yield of slabs, billets, and blooms is much larger in continuous casting than in ingot casting. For this reason, the continuous casting method is of great economic interest. The part of the strand which contains a pipe has to be discarded and remelted, which decreases the yield. It is therefore important to reduce pipe formation as much as possible.

It is even more important to reduce macrosegregation than pipe formation. It is well known that it is more difficult to cast alloys with high concentrations of alloying elements than alloys with low concentrations. The former are more inclined than the latter to form external and internal cracks and macrosegregation in the centre.

The problem mentioned above restricts the suitability of the continuous casting method. In order to be able to tackle the problems with pipe formation and centre segregation during continuous casting, one has to acquire a thorough knowledge of the phenomena. In addition, it is necessary to examine the causes of their appearances, find a useful theory, and use it to establish what precautions should be taken to minimize pipe formation and macrosegregation during continuous casting.

11.6.1 Centre Segregation

At the beginning of the 1980s, Fredriksson, Rogberg and Raihle performed an extensive experimental investigation of pipe formation and macrosegregation during continuous casting of steel strands with a cross-section of 10×10 cm (Figure 11.22). The study was particularly focused on the influences of casting rate and cooling conditions on pipe formation and macrosegregation.

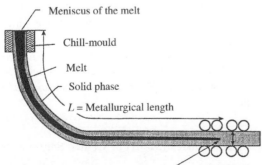

Figure 11.22 Schematic diagram of the machine for continuous casting which was used in the experiments. The secondary cooling was divided into several zones. Reproduced with permission from the American Society for Metals (ASM).

Experimental Method

The casting rate could be varied and rates between 2.5 and 3.5 m/min were used, depending on the steel quality. The level of the melt in the chill-mould was kept constant by automatic control.

The Machine had a spray-water ring immediately below the mould and two additonal zones of water cooling further down. The cooling intensity could be controlled partly by the water pressure and partly by varying the cross-sections of the nozzles and their distances from each other. In order to keep a constant surface temperature, the cooling intensity was gradually decreased along the strand.

Numerous experiments with varying carbon concentrations were performed. The macrostructure of the cast material in each experiment was analysed with the aid of cross-section samples taken at steady-state positions. The samples were etched with hydrochloric acid at 70 °C and examined carefully.

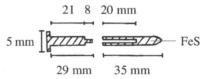

Figure 11.23 Schematic diagram of an FeS nail. The cavity of the front part is filled with FeS and then the two parts are pushed together. Reproduced with permission from the Scandanavian Journal of Metallurgy.

In order to map the flow of the melt in the central parts of the strand, injections of FeS into the strand during the casting was used. The position of the injected FeS could be observed after complete solidification with the aid of the so-called 'sulfur print' method. FeS was injected into the strand with the aid of a 'nail'. An example of the design of such a nail is given in Figure 11.23.

Experimental Results

Numerous experiments with various carbon concentrations and casting rates at varying cooling rates of the strand were performed. The macrosegregation turned out to depend to a great extent on the cooling conditions. The pipe formation and the macrosegregation were fairly independent of the $\underline{C}$ concentration. Consequently, the results of two experiments can be compared even if the $\underline{C}$ concentrations were different in the two cases.

Along the central axis, pipe-like cavities were often formed (Figure 11.24). The cavities were similar to the pores observed at the bottom of the big pipes formed at the end of casting operations and discussed in Section 10.6.2 (see, for example, pages 336–337). The pores are formed by the same mechanism, which will be discussed below.

If a nail is shot into the strand a pipe is immediately formed below it. Figures 11.25 and 11.26 show the result of such an experiment.

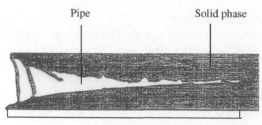

Pipe   Solid phase

Pipe depth = 0.8 m

Figure 11.24 Typical pipe formed at the end of continuous casting. The picture was drawn after a reprint of a sulfur print. Reproduced with permission from the American Society for Metals.

With the aid of a sulfur print, it is possible to map the flow of melt in the central region. The solidification and cooling shrinkage in the centre results in a flow of melt downwards, if the strand is supposed to be vertical, and fills the cavity as long as the channel is open.

Figure 11.25 shows a situation when an FeS nail has been adapted in the strand just before the two solidification fronts meet (see Chapter 10, page 338). The nail has blocked the further flow of melt downwards before the melt in the centre has solidified. A bridge has been formed and below it a pipe.

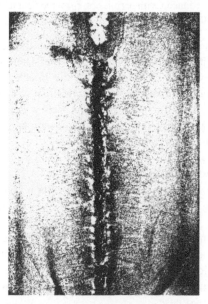

Figure 11.25 Sulfur print above and around a pipe. Reproduced with permission from the Scandanavian Journal of Metallurgy, Blackwell.

Melt has flowed from the region above the bridge through the dendritic network in the two-phase region. Its passage has formed a U-shaped pattern, as shown in Figure 11.26 below level A–A. The regions close to the bridge and at the centre immediately above the bridge consist of material with a $\underline{C}$ concentration lower than the average (*negative*

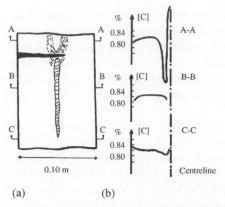

0.10 m

(a)   (b)

Figure 11.26 (a) Principal sketch of the structure in Figure 11.25. (b) The $\underline{C}$ concentration at various levels relative to the nail as a function of the distance from the strand surface. Reproduced with permission from the Scandanavian Journal of Metallurgy, Blackwell.

segregation). Some 3 cm above the bridge (level A–A) the central $\underline{C}$ concentration is much higher than the average (strongly *positive* segregation); 10 cm below the nail, at the end of the pipe, the segregation at the centre is weakly positive. At a large distance below the pipe, the $\underline{C}$ concentration has approximately the same distribution as that 3 cm above the nail (A–A). This is the normal segregation pattern in the strand.

In connection with the appearance of centre segregation, so called V-segregates are also formed. With the aid of a sulfur print two different types of V-segregates have been observed, simple V-segregates and closely packed V-segregates, characteristic of the flow of the melt.

Closely packed V-segregates (Figure 11.27), which form an angle of ~45° with the central axis, appear when there is a wide equiaxed zone with many small, randomly orientated crystals. This type of structure appears at the end of the continuous casting when the temperature in the centre has decreased strongly (Chapter 10, page 338).

If the thickness of the equiaxed zone decreases in favour of the columnar zone, the distance between the V-segregates increases and they can be characterized as simple V-segregates. Below this type of segregates a pipe often appears (Figures 11.25 and 11.28).

The centre segregation of the strand and the pipe formation depend to a great extent on the cooling conditions and also on the casting rate. We shall come back to this later in this chapter.

11.6.2 Mathematical Model of Centre Segregation

Model of Pipe Formation

In Section 10.6.2, we treated the theory of pipe formation during continuous casting. We start with the equations that we derived there and initially repeat some important concepts and designations.

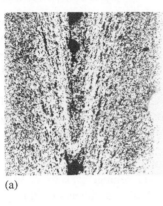

(a)

(b)

(c)

Figure 11.27 (a) Sulfur print of closely packed V-segregates. Length cutting of the strand. (b) Schematic diagram of closely packed V-segregates. Length cutting of the strand. (c) Schematic diagram of closely packed V-segregates. Cross-section of the strand. Reproduced with permission from the Scandanavian Journal of Metallurgy, Blackwell.

A material balance gives the relationship:

$$\Delta V_{cool} = \Delta V_{surface} + \Delta V_{centre} \qquad (11.55)$$

where
ΔV_{cool} = volume change of the strand due to thermal expansion or contraction
$\Delta V_{surface}$ = volume change of the strand due to thermal expansion or contraction caused by the change of surface temperature
ΔV_{centre} = volume change in the central region due to thermal contraction.

(a)

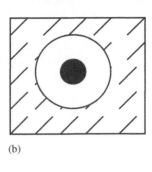

(b)

Figure 11.28 Schematic diagram of a simple V-segregate along the strand. (a) Length cutting through the strand; (b) cross-section of the strand. Reproduced with permission from the Scandanavian Journal of Metallurgy, Blackwell.

Since the temperature variations in the length direction are much smaller than the temperature changes as a function of time perpendicular to the length direction, we can neglect the length contraction and consider only the volume changes caused by changes of the cross-section area. In this case the volume changes can be expressed with the aid of area changes (Figure 11.29) and we obtain:

$$\Delta A_{cool} = \Delta A_{surface} + \Delta A_{centre} \qquad (11.56)$$

The three different areas in Equation (11.56) were calculated in Chapter 10 on pages 339–341. For a square strand

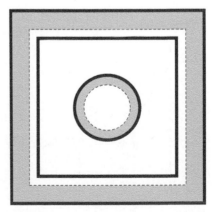

Figure 11.29 Grey square = $\Delta A_{surface}$; grey circular ring = ΔA_{centre}; area between the black (continuous) squares = ΔA_{cool}.

they are [see Equations (10.49), (10.51), and (10.52)]:

$$\Delta A_{\text{surface}} = \int_0^{\frac{x_0}{2}} 2\alpha_l \times 4(x_0 - 2x)\,dx\, \frac{\frac{x_0}{2} - x}{\frac{x_0}{2}} \Delta T_{\text{surface}}$$

$$= \frac{4}{3}\,\alpha_l\, x_0^2\, \Delta T_{\text{surface}} \qquad (11.57)$$

$$\Delta A_{\text{centre}} = \int_0^{\frac{x_0}{2}} 2\alpha_l \times 4(x_0 - 2x)\,dx\, \frac{x}{\frac{x_0}{2}} \Delta T_{\text{centre}}$$

$$= \frac{2}{3}\,\alpha_l\, x_0^2\, \Delta T_{\text{centre}} \qquad (11.58)$$

$$\Delta A_{\text{cool}} = \Delta A_{\text{centre}} + \Delta A_{\text{surface}}$$

$$= \frac{2}{3}\,\alpha_l\, x_0^2\, \Delta T_{\text{centre}} - \frac{4}{3}\alpha_l\, x_0^2\, \Delta T_{\text{surface}} \qquad (11.59)$$

where

α_l = length strain coefficient

x_0 = width of the strand = side of the square strand

$\Delta T_{\text{surface}}$ = temperature change at the surface of the strand

ΔT_{centre} = temperature change at the centre of the strand.

Mathematical Model of Centre Segregation
Figures 11.25 and 11.26 illustrate the formation of a bridge when a nail is shot into the strand. Above the nail, melt flows downwards through the two-phase region to fill the cavity caused by the centre shrinkage. A simple V-segregate structure, illustrated in Figure 11.28, is formed in a similar

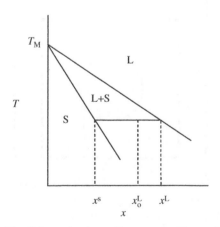

Figure 11.30 Schematic phase diagram of a binary alloy.

way when a bridge of equiaxed crystals appears between the two solidification fronts. This material bridge prevents the continued flow of melt downwards. Part of the melt which passes the bridge solidifies. Material with lower $\underline{C}$ concentration than the average solidifies at the bridge and the melt accumulates carbon during its further passage downwards according to the phase diagram (Figure 11.30).

Figure 11.27 (a) shows closely packed V-segregates, which are typical of a structure consisting of equiaxed crystals. The V-shaped segregates start at a certain distance from the centre. Nail experiments on this type of structure show that the melt flows from the colder and outer region to a warmer and inner region. The melt transport occurs in the channels of the V-segregates.

This type of material transport is caused by thermal stress. When the temperature falls strongly in the central region, the two-phase region will be compressed and melt is forced inwards towards the centre. These thermal forces are assumed to have a direction corresponding to an angle of 45° with the central axis. The melt is therefore forced to form oblique channels through the two-phase region, in agreement with the law of least resistance.

The melt, which moves downwards and fills the central pore, is in equilibrium with the solid phase. It will change its composition by freezing or melting, depending on its flow direction when it flows from a warmer to a colder region or from a colder to a warmer part of the two-phase region.

When these conditions are known it is possible to calculate the average $\underline{C}$-concentration in the central region.

Consider a volume element in the two-phase region with volume V, equal to the area A multiplied by the height h of the two-phase region (Figure 11.31). Then the average $\underline{C}$ concentration can be written as:

$$\bar{c} = c_T^{\text{L}}\, \frac{k_{\text{part}}\, \rho_s\, A_s + \rho_{\text{L}}\, A_{\text{L}}}{\rho_s\, A_s + \rho_{\text{L}}\, A_{\text{L}}} \qquad (11.60)$$

where

ρ_s = density of the solid phase

A_s = area of the solid phase in the volume element

k_{part} = partition coefficient

ρ_{L} = density of the melt

A_{L} = area of the melt in the volume element

c_T^{L} = equilibrium concentration of the melt in equilibrium with the solid phase at temperature T.

A change dA of the area element at constant temperature will result in a change in its average concentration since melt will flow in or out to compensate for the volume change. The

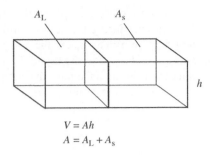

$$V = Ah$$
$$A = A_L + A_s$$

Figure 11.31 Volume element.

new average concentration of the volume element can be written:

$$\bar{c} + \bar{d}c = c_T^L \frac{k_{part}\,\rho_s A_s + \rho_L(A_L + dA)}{\rho_s A_s + \rho_L(A_L + dA)} \qquad (11.61)$$

By subtracting Equation (11.60) from Equation (11.61), we obtain an expression for the centre segregation, i.e. the concentration change in the volume element due to the area change dA:

$$d\bar{c} = c_T^L \frac{\rho_L \rho_s A_s (1 - k_{part}) dA}{(\rho_s A_s + \rho_L A_L)[\rho_s A_s + \rho_L(A_L + dA)]} \qquad (11.62)$$

We use the earlier introduced g-functions to express A_L and A_s as fractions of A (Section 11.5.1; page 383):

$$A_s = g_s A \qquad (11.63)$$
$$A_L = g_L A \qquad (11.64)$$

Inserting these expressions of A_L and A_s into Equation (11.62) gives:

$$d\bar{c} = c_T^L \frac{\rho_L \rho_s g_s (1 - k_{part})}{(\rho_s g_s + \rho_L g_L)\left[\rho_s g_s + \rho_L\left(g_L + \dfrac{dA}{A}\right)\right]} \frac{dA}{A}$$

$$\qquad (11.65)$$

This is the equation which is the basis of numerical calculations of the centre segregation. The terms g_s and g_L are functions of the temperature T. These functions are given below without derivation.

$$g_s = 1 - g_L = \frac{T_L - T + \dfrac{2}{\pi}(T_s - T_L)\left\{1 - \cos\left[\dfrac{\pi(T - T_L)}{2(T_s - T_L)}\right]\right\}}{(T_L - T_s)\left(1 - \dfrac{2}{\pi}\right)}$$

$$\qquad (11.66)$$

where

g_s = fraction of solid phase
g_L = fraction of melt
T_L = liquidus temperature

T_s = solidus temperature
T = temperature of the volume element.

The centre segregation is particularly temperature dependent. All quantities on the right-hand side of Equation (11.65), except possibly k_{part}, change with temperature. The latter is time dependent and is determined by solving a differential equation of thermal conduction simultaneously. Numerical stepwise calculations are necessary. By combining Equation (11.65) with expressions which describe A and dA, it is possible to calculate the macrosegregation in the central region of the strand.

dA is chosen to be equal to dA_{centre}. In order to obtain a concrete expression for A we have to choose a realistic initial value of A in the numerical calculations. A reasonable value of the area of macrosegregation at the centre of a strand with dimensions 10×10 cm is, for example, $A = 2.5 \times 10^{-5}\,\text{m}^2$, which is the area of a circle with radius 2.8 mm.

Equations (11.65) and (11.66) show that the centre segregation is strongly influenced by the temperatures and the cooling rates at the centre and at the surface of the strand, respectively. Since the temperature at the surface of the strand can be affected, a possibility opens up of reducing the centre segregation. This topic will be discussed below.

11.6.3 Methods to Eliminate Centre Segregation

As is apparent from above, centre segregation during continuous casting is highly dependent of the shrinkage in the central region during solidification and cooling. Since it is desirable to avoid both centre segregation and pipe formation, it is important to find methods which reduce this shrinkage to a minimum during the solidification process. There are two main methods, mechanical influence and thermal influence.

Mechanical Influence during the Solidification Process
Experiments have shown that it is possible to eliminate centre segregation by exposing the strand to an external pressure during the solidification process.

Experience shows that centre segregation is very sensitive to *how* and *where* the pressure is adapted. It is also important not to interrupt the pressure treatment before the strand has solidified completely.

Figure 11.32 gives an example of the strong effect which is obtained when the strand is exposed to an extra external pressure during the final part of the solidification process.

Thermal Influence during the Solidification Process
The most common method to decrease the risk of centreline segregation for slabs is the mechanical method discussed above. It is called *mechanical soft reduction*. This method is not easy to use for billets. Instead, the so-called *thermal soft reduction* process is used, i.e. the secondary cooling of the strand is designed in an optimal way.

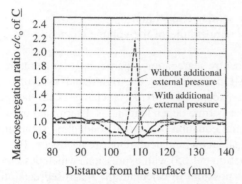

Figure 11.32 Segregation ratio of carbon in the middle part of a slab with and without external pressure during the final part of the solidification. Reproduced with permission from the Scandanavian Journal of Metallurgy, Blackwell.

It is evident from the equations:

$$d\bar{c} = c_T^L \frac{\rho_L \rho_s g_s (1 - k_{part})}{(\rho_s g_s + \rho_L g_L)\left[\rho_s g_s + \rho_L \left(g_L + \dfrac{dA}{A}\right)\right]} \frac{dA}{A}$$

(11.65)

and (11.58) on page 390 that the macrosegregation at the centre depends strongly on the temperature change at the centre of the strand. However, ΔA_{centre}, $\Delta A_{surface}$, and A_{cool} are not independent of each other. If we assume that the temperature changes at the surface and at the centre vary linearly between the centre and the surface, as we have done in Chapter 10 and in Equation (11.66), we obtain the net shrinkage of the strand surface with the aid of Equation (11.59).

If we introduce the factor G (Chapter 10, page 341), we obtain:

$$\Delta A_{cool} = \frac{2}{3} \alpha_l x_0 (1 - 2G) \Delta T_{centre}$$

(11.67)

where

$$\alpha_l = \text{length strain coefficient}$$
$$x_0 = \text{cross-section side of the strand}$$
$$dT_{surface}/dt = \text{cooling rate at the surface of the strand}$$
$$dT_{centre}/dt = \text{cooling rate at the centre of the strand}$$
$$G = (dT_{surface}/dt)/(dT_{centre}/dt).$$

It can be seen from Equation (11.67) that it is the G-factor, the ratio of the cooling rate at the surface to that at the centre, which determines the sign of ΔA_{cool}. The expression has a physical meaning only if it is positive. If $G > 0.5$ the centre segregation becomes negative, no pipe is formed and no macrosegregation appears. If both pipe formation and centre segregation are to be avoided, the following condition is therefore extremely important:

The secondary cooling ought to be designed in such a way that it fulfils the condition $G > 0.5$ during the whole solidification process in the centre.

Figure 11.33 gives a typical example of the temperature distribution of the strand during the solidification process. Before the solidification fronts meet, the temperature at the centre is constant and equal to the liquidus temperature. When the solidification fronts have met, the supply of solidification heat ceases and the temperature in the interior falls rapidly.

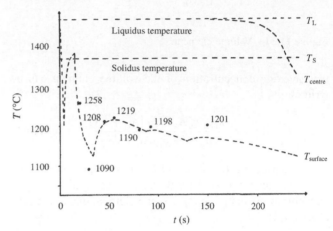

Figure 11.33 Temperature at the centre and at the middle of a side in a square strand during continuous casting as a function of time.

The temperature at the surface of the strand can be influenced by the design of the secondary cooling. The secondary cooling should be designed in such a way that the condition $G > 0.5$ is fulfilled everywhere. In practice, this condition is not so easy to fulfil. It is evident from Figure 11.33 that the risk of centre segregation appears during the last part of the solidification process when the solidification fronts meet and the temperature in the centre starts to decrease rapidly.

Fredriksson and Raihle found, by using the condition for pipe formation and centre segregation, that it is possible to minimize the pipe formation and completely avoid centre segregation (the macrosegregation ratio $c_{centre}/c_0^L = 1$), by applying *extra cooling* at the end of the solidification process. The result of the calculations is plotted in Figure 11.34. Two alternatives have been simulated, with strong and soft secondary cooling. The position of the extra cooling was varied. The cooling started at the positions given in Figure 11.34 and lasted until the solidification was complete.

It is desirable that the centre segregation ratio $c_{centre}/c_0^L = 1$ (dashed horizontal line in Figure 11.34).

It can be concluded from Figure 11.34 that:

- With both soft and strong cooling the centre segregation can be influenced considerably by extra cooling.
- The contraction of the strand surface can be so large that the centre segregation $(c - c_0^L)$ becomes negative, which gives a ratio $c/c_0^L < 1$.

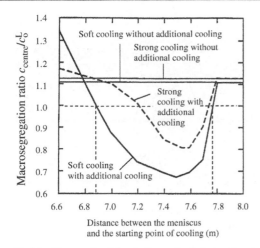

Figure 11.34   The result of simulation calculations of centre segregation during continuous casting as a function of position of the extra cooling. The two horizontal continuous lines show the macrosegregation with strong extra cooling (upper line) and soft extra cooling (lower line). Reproduced by permission of ASM International.

- Soft cooling without extra cooling gives a better result than strong cooling without extra cooling.
- If the extra cooling starts before the optimal position, the effect may be the opposite – one may obtain a ratio that is much greater than 1.

With the result in mind that the model gave good agreement with experimental results in the case of pipe formation (Chapter 10, page 342) it is reasonable to assume that the method with extra cooling also works in the case of centre segregation. This has been confirmed by experiments on an industrial scale.

11.7   FRECKLES

Inside solidified ingots and on their surfaces there are in certain cases vertical channels where the material has a different composition to the surroundings. The phenomenon is called *freckles*.

Examination with an etching technique has shown that the tracks may contain an excess of eutectic material, particles from secondary phases, for example pores and small, randomly oriented crystals. Freckles are a form of macrosegregation and have a pencil-like appearance. On the surface the pencils look like dots. This explains the term freckles.

Freckles are casting defects which appear in certain cases during remelting processes with an electric arc or during electroslag refining. They may also be formed during processes involving unidirectional solidification. Figure 11.35 shows a typical example, which we will return to later in this section.

The origin of freckles was unknown for a long time. After model experiments with the system $NH_4Cl–H_2O$, the

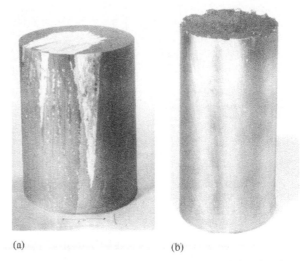

(a)   (b)

Figure 11.35   (a) Freckles in an ingot with unidirectional structure of the alloy Ni–20 at-% Al. (b) An ingot with unidirectional structure of the alloy Ni–13 at-% Ta. No freckles. Reproduced by permission of ASM International.

development of a theory for freckles and tests of the latter on metallic systems, the explanation given below is generally accepted.

11.7.1   Model Experiments with the System NH₄Cl–H₂O

By model experiments with the system $NH_4Cl–H_2O$, which solidifies with a dendritic structure like most metals, it has been possible to study the solidification process in detail and even film it since both the solution and solid NH_4Cl and ice have the great advantage of being transparent. The parameters have been varied, one at a time, and the changes caused by each parameter separately have been studied. In this way, basic knowledge has been achieved about the phenomenon of freckles.

Copley and co-workers used a solution containing 30 wt-% NH_4Cl. A great number of experiments were performed. A sketch of the apparatus and the phase diagram of the system are given in Figure 11.36 (a) and (b). Different phases and regions with different composition could be distinguished visually and be photographed thanks to differences in refractive index. The differences in refractive index may be caused by differences in density or temperature.

Jet Streams and Freckles

In a solidifying 'ingot' of $NH_4Cl–H_2O$, three different regions can be identified: an upper liquid zone, a two-phase region with dendrites, liquid and a solid phase in contact with the chilled Cu plate.

At a certain distance from the Cu plate, there are distinct traces with an unmistakeable resemblance to the freckles that sometimes occur in metals, for example in nickel-base

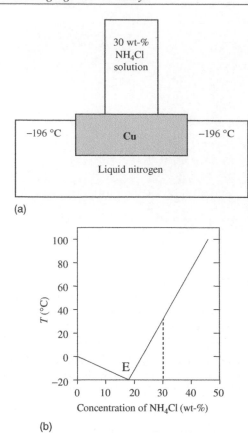

(a)

(b)

Figure 11.36 (a) Experimental equipment for study of the solidification process of the NH$_4$Cl–H$_2$O system. A glass or plastic mould placed on and in good thermal contact with a Cu plate, chilled with liquid nitrogen (-196 °C). (b) Phase diagram of the NH$_4$Cl–H$_2$O system. Reproduced by permission of ASM International.

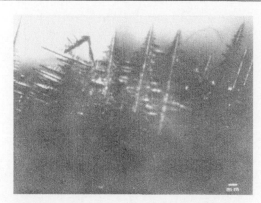

Figure 11.37 Dendrite fragments. Reproduced by permission of ASM International.

Figure 11.38 Typical freckles traces in NH$_4$Cl–H$_2$O at the external surface of the glass. Reproduced by permission of ASM International.

alloys. It has been possible to observe that these traces are formed by strong so-called *jets*, streams directed upwards in the two-phase region. These jets break the dendritic network into fragments on their way upwards (Figure 11.37). In this way, a great number of new nuclei are formed, which are the sources of new dendrites. Since the jet streams originate from the bottom of the two-phase zone, it is reasonable to assume that the streams have eutectic temperature and composition.

The cross-section of the jet stream extends over four to nine dendrite arms. Larger particles remain in the trace and form grains with a random orientation. The smaller particles (secondary dendrite arms) are carried away with the jet stream and remelt in warmer surroundings. This leads to macrosegregation, i.e. a different composition of the stream compared with the new surroundings. Substances with positive segregation ($k_{part} < 1$) accumulate in the jet stream, whereas it is depleted of substances with negative segregation ($k_{part} > 1$), because the composition of the liquid is controlled by microsegregation.

Along the external surface of the glass mould, the typical structure of these traces can be observed (Figure 11.38). It

can also be seen that numerous dendrite arms have grown in the vicinity of the traces. This indicates that the liquid in the jet streams is colder than the surrounding liquid. The eruptions of jet streams upwards also show that the density in the stream is lower than the density of the surrounding liquid.

Special experiments designed to map the influence of various parameters on the traces gave the following results:

- Freckle channels or just freckles appear only at a certain distance from the chilled surface.
- If the cross-sectional area of the two-phase region is *planar*, the freckles become *evenly distributed* there.
- If the cross-sectional area of the two-phase region is *convex*, freckles appear only in the interior of the mould. If the cross-section area of the two-phase region is *concave*, freckles appear only at the external surface of the mould.
- If the mould is inclined in such a way that the two-phase region forms an angle with the vertical line, this does not influence the vertical direction of the jet streams. They remain parallel to the gravitation force [Figure 11.39 (a) and (b)].

- The formation of jets increases when the temperature gradient and the solidification rate of the dendrites decrease.
- Since the jet streams carry fragments of the dendrite network they accumulate alloying elements with positive segregation ($k_{part} < 1$) and are depleted of alloying elements with negative segregation ($k_{part} > 1$).

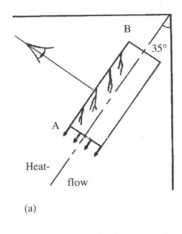

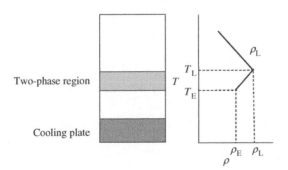

(b)

Figure 11.39 (a) Freckles lines along the upper part of a cylinder-shaped ingot of the alloy Mar-M200. The cylinder has been inclined 35° relative to the vertical line during the solidification process. (b) The appearance of a cylinder surface with freckles. The cylinder has solidified in an inclined position. Reproduced with permission from the American Society for Metals (ASM).

Driving Force of Jet Streams

All liquids and melts with normal length expansion coefficients have densities which decrease with increasing temperature (Figure 11.40 and Table 11.1).

However, the temperature gradient is not the only cause of the inverted density of the two-phase region in the

Figure 11.40 Inverted density in NH_4Cl solution. Index E = eutectic. Reproduced by permission of the American Society for Metals.

TABLE 11.1 Definition of normal and inverse density.

Temperature distribution	T_1 low T_2 high
Normal density	$\rho_1 > \rho_2$
Inverse density	$\rho_1 < \rho_2$

solidifying NH_4Cl solution. A concentration gradient, which cooperates with the temperature gradient, also contributes to the inverted density. This is illustrated in Figure 11.41. The inverted density leads to convection, which is the origin of the jet streams, which cause freckles.

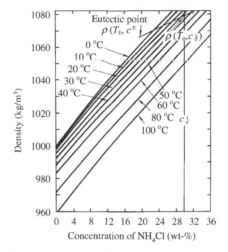

Figure 11.41 Density of NH_4Cl solution as a function of concentration with temperature as a parameter. Reproduced by permission of ASM International.

11.7.2 Causes of Freckles in Metals

A comparison with water-based systems of the type $NH_4Cl–H_2O$ and metallic systems shows that the origins of the convection, which causes jet streams and freckles, are the same in both cases. However, there are considerable differences between the values of the material constants and the magnitude of the density inversion, which is the driving force of the jet streams.

Concentration-driven Convection in Metal Systems

The existence of concentration-driven convection in metallic systems depends on two factors, the magnitude of the density inversion and the transport processes. They are determined by material constants such as dendritic growth rate, diffusion in the melt and the solid phase, density of the melt and the solid phase, and the temperature conditions.

The density inversion of the two-phase region is possible in:

- metallic systems with *lighter* elements with a positive or *normal* segregation ($k_{part} < 1$);
- metallic systems with *heavier* elements, which have a *negative* or *inverted* segregation ($k_{part} > 1$).

Only these two alternatives lead to the inverted density distribution, which is a necessary condition for the appearance of freckles.

By a *light* element we mean an alloying element which has *lower* density than the basic element of the alloy. Similarly, a *heavy* element is an alloying element with *higher* density than that of the basic element.

By *positive segregation* we mean that the melt with decreasing temperature attains the concentration of the alloying element, which is *higher* than the initial one (Figure 11.42). *Negative segregation* means that the melt with decreasing temperature attains the concentration of the alloying element, which is *lower* than the initial one (Figure 11.42).

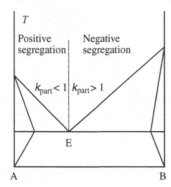

Figure 11.42 Regions of positive and negative segregation in a binary phase diagram.

The determination of the magnitude of the density inversion in metallic systems was done from measurements of the density as a function of concentration of alloying element and temperature of molten Zn–Al alloys. The results showed that the density change of the melt in the two-phase region due to concentration changes was much larger than density changes due to temperature changes. Calculations also showed that the density inversion of the metallic system exceeded the corresponding one of the system NH_4Cl–H_2O by a factor of >10.

The main difference between water-based systems and metallic systems concerning transport processes is the magnitude of the thermal diffusion, which is about 100 times larger in metallic systems than in water-based systems.

Since high density inversion and high diffusion both favour convection, there are good opportunities to explain the appearance of freckles during unidirectional solidification and during remelting processes with arcs in vacuum.

Example 11.3
Explain with the aid of the phase diagrams of the two alloys why freckles appear during casting of the alloy Ni–20 at-% Al but not at all during casting of the alloy Ni–13 at-% Ta. This is verified by Figures 11.35 (a) and (b) on page 393.

Solution and Answer:

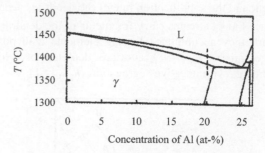

Part of the phase diagram of the Ni–Al system.

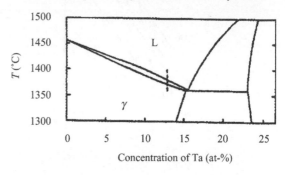

Part of the phase diagram of the Ni–Ta system.

The top phase diagram shows that aluminium has *normal* or *positive* segregation. Because Al is a *light* element, compared with Ni, the conditions for the appearance of freckles are fulfilled. The bottom phase diagram shows that tantalum also has a *normal* segregation. However, since Ta is a *heavy* element compared with Ni, freckles cannot appear.

11.7.3 Mathematical Model of Macrosegregation at Freckles

Freckles appear by concentration-driven convection in the two-phase region and consist of channels with a different composition to the surroundings. By introducing a simple mathematical model for freckles, we will try to derive the conditions for the appearance of this casting defect.

Raleigh predicted in 1916 the condition for a stationary flow pattern of the type given in Figure 11.43. He found that the condition for this type of flow is that the product of the Grashof number and the Prandtl number must be ≥ 1700:

$$Gr \times Pr \geq 1700 \qquad (11.68)$$

We apply this condition to find the conditions for the appearance of freckles. The Grashof and Prandtl numbers are defined as:

$$Gr = \frac{g\beta L^3 \Delta T}{v_{kin}^2} \qquad (11.69)$$

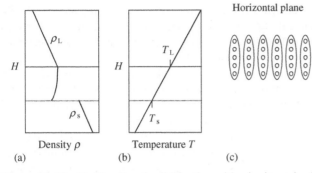

Figure 11.43 Unidirectional solidification with a horizontal solidification front. (a) Density as a function of the height H. (b) Temperature as a function of the height H. (c) Convection pattern seen from above around the freckles in the two-phase region.

$$Pr = \frac{v_{kin}}{\alpha} \qquad (11.70)$$

where

g = acceleration due to gravity
β = volume strain coefficient (K^{-1})
L = characteristic length
v_{kin} = kinematic viscosity coefficient (η/ρ)
α = thermal diffusqutivity ($k/\rho c_p$) (Chapter 4, page 62)
T_F = temperature at the solidification front
T_s = solidus temperature of the melt.

The convection occurs in this case in the two-phase region. Hence it is reasonable to choose the width of the two-phase region as the characteristic length L.

ΔT is the temperature difference between the temperature of the solidification front T_F and the solidus temperature T_s. In this case we can approximate the temperature of the solidification front by the liquidus temperature T_L.

Instead of the Prandtl number, we will in this case use the Smidt number because we want to describe the influence of the concentrations of the alloying elements on the density:

$$S = \frac{v_{kin}}{D} \qquad (11.71)$$

where v_{kin} is the kinematic viscosity coefficient and D the diffusion coefficient.

We obtain the length of the two-phase region by dividing the temperature difference by the temperature gradient G:

$$L = \frac{T_L - T_s}{G} \qquad (11.72)$$

The final condition for freckle formation is:

$$\frac{K_p \, g \, \beta (T_L - T_s)^4}{v_{kin} D G^3} \geq 1700 \qquad (11.73a)$$

By introduction of $[-(1/\rho)(\Delta\rho/(T_L - T_s)]$ instead of β (see Example 11.4, page 398), this expression can be transformed into:

$$-\frac{K_p \, g \, \Delta\rho(T_L - T_s)^3}{\rho \, v_{kin} D G^3} \geq 1700 \qquad (11.73b)$$

where

K_p = permeability coefficient
$\Delta\rho$ = $\rho_{top} - \rho_{bottom}$ = density difference of the melt at the top and bottom of the melt in the two-phase region
G = temperature gradient in the two-phase region.

The permeability coefficient K_p concerns the friction between the dendrites and the melt. The permeability coefficient is a measure of the ability of the melt to penetrate the dendritic network. The *higher* this coefficient is, the *more easily* the melt will flow into and out from the two-phase region. K_p varies between 0 (no penetration) and 1 (total penetration).

As the phase diagram of a system offers a relationship between temperature and concentration, it is possible to obtain a relationship between the temperature gradient G and the concentrations c^L and c^E instead of Equation (11.73b) between G and T. The former relationship is derived in Example 11.4 (page 398). It can be used to find a useful condition for freckle formation or avoiding freckles.

$$G \leq \left[\frac{g \, m_L^3 \left(m_L \frac{\partial\rho}{\partial T} + \frac{\partial\rho}{\partial c} \right)}{\rho_L \, v_{kin} D \times 1700} \right]^{\frac{1}{3}} (c^E - c^L)^{\frac{4}{3}} \qquad (11.74)$$

Condition for Avoiding Freckles

Freckles cause the quality of castings to deteriorate and represent a potential risk of external and internal crack formation. It is therefore desirable to prevent their formation. In alloys with a risk of freckles, it is advisable to perform calculations of the kind illustrated in Example 11.4 and design the casting process in such a way that this casting defect is prevented.

Equation (11.73b) is the basic condition for freckles. The liquidus and solidus temperatures are related to the concentration of the alloying element with the aid of the phase diagram for the alloy in question and the condition (11.73b) can be replaced by Equation (11.74) between the concentration of the alloying element and the temperature gradient. As an example, the curve for the Pb–Sn system given in Figure 11.44 (a) was calculated with the aid of the phase diagram in Figure 11.44 b and Equation (11.74).

For each value of the concentration of the alloying element Sn [Figure 11.44 (a)] we can read the minimum value

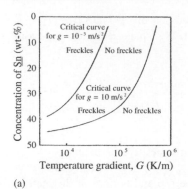

(a)

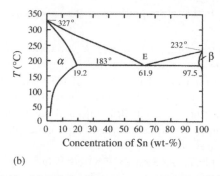

(b)

Figure 11.44 (a) Relation between concentration of alloying element and temperature gradient for Pb–Sn alloys. The limit between macrosegregation and no macrosegregation depends on the value of the acceleration due to gravity, g. The curve is drawn for two different g values. (b) Phase diagram of the Pb–Sn system.

G_{cr} of the temperature gradient which can be allowed if freckle formation is to be avoided. If the sign of equality is used in Equation (11.74), the critical temperature gradient G_{cr} is obtained.

Figure 11.44 a and Equation (11.74) show that if freckles are to be *avoided*, the temperature gradient G must *exceed* the critical value G_{cr}:

$$G > G_{cr} \Rightarrow \text{no freckles} \qquad (11.74)$$

The temperature gradient should in reality exceed the critical value with a good margin for safety reasons.

Figure 11.44 (a) also shows that a a *high* value of g *increases* the critical value G_{cr} and the risk of freckle formation. This is important in the case of centrifugal casting. On the other hand, a *low* value of g *reduces* the risk of freckle formation as the critical curve is displaced to the left in Figure 11.44 (a). This situation is discussed in Section 11.8.1 (page 400).

Example 11.4
Examine under what solidification conditions the risk of freckle formation arises for Pb–Sn alloys by calculating the critical temperature gradient as a function of the Sn concentration for some different Pb–Sn alloys. Perform the calculations for alloys with initial Sn concentrations of 20, 30, 40, 50 and 60 wt-%.

Give the result (a) as a table and (b) as plot of the Sn concentration as a function of the critical temperature gradient. Indicate in the figure when there *is* a risk and where there is *no* risk of freckle formation.

The phase diagram of the Pb–Sn system is given in Figure 11.44 (b). The diffusion constant is assumed to be 3.0×10^{-9} m^2/s and $K_p = 1$. Remaining material constants have to be taken from standard reference tables. $g = 9.8$ m/s^2.

Solution:
The basis of our calculations is Equation (11.73a):

$$\frac{K_p\, g\beta\, (T_L - T_S)^4}{v_{kin}\, D G^3} \geq 1700 \qquad (1')$$

To be able to apply this equation, we have to express the quantities we miss in other known or available quantities:

$$V = V_0 + \beta V\, \Delta T \qquad \text{or} \qquad \Delta V = \beta V\, \Delta T \qquad (2')$$

which can be written as:

$$\beta = \frac{1}{V} \frac{\Delta V}{\Delta T} \qquad (3')$$

We also know that:

$$V\rho = \text{constant} \qquad (4')$$

Logarithmic differentiation of Equation (4') gives:

$$\frac{\Delta V}{V} = -\frac{\Delta \rho}{\rho} \qquad (5')$$

With the aid of Equations (3') and (5'), we obtain:

$$\beta = \frac{1}{V} \frac{\Delta V}{\Delta T} = -\frac{1}{\rho} \frac{\Delta \rho}{\Delta T} \qquad (6')$$

In addition, we have with the lowest possible $T_s = T_E$:

$$L = \frac{T_L - T_s}{\dfrac{dT}{dy}} \approx \frac{T_L - T_E}{G} = \frac{m_L\, (c^L - c^E)}{G} \qquad (7')$$

where
$m_L = dT/dc = (T_L - T_E)/(c^L - c^E) = $ the slope of the liquidus line
$G = dT/dy = $ the temperature gradient.

Inserting Equation (7') for L into Equation (11.73b) we obtain for $K_p = 1$ and $T_L - T_s \approx T_L - T_E$:

$$\frac{-g\, \Delta\rho}{\rho v_{kin}\, D} \frac{(T_L - T_E)^3}{G^3} \geq 1700$$

or

$$\frac{-g\,\Delta\rho}{\rho v_{\text{kin}}\,D}\,\frac{m_L^3\,(c^L - c^E)^3}{G^3} \geq 1700 \qquad (8')$$

The density ρ is a function of two variables, the temperature T and the concentration c. We assume that ρ, within the temperature and concentration regions in question, can be written as:

$$\rho = \rho_L + \frac{\partial\rho}{\partial T}\,(T - T_L) + \frac{\partial\rho}{\partial c}\,(c - c^L) \qquad (9')$$

Application of Equation (9') for $T = T_E$ gives:

$$\Delta\rho = \rho_E - \rho_L = \frac{\partial\rho}{\partial T}\,(T_E - T_L) + \frac{\partial\rho}{\partial c}\,(c^E - c^L) \qquad (10')$$

Equation (13') is the condition for macrosegregation freckles. We solve G from Equation (13'):

$$G \leq \left[\frac{g\,m_L^3\left(m_L\,\dfrac{\partial\rho}{\partial T} + \dfrac{\partial\rho}{\partial c}\right)}{\rho_L v_{\text{kin}}\,D \times 1700}\right]^{\frac{1}{3}}\,(c^E - c^L)^{\frac{4}{3}} \qquad (14')$$

This is the condition for the formation of freckles. The critical value G_{cr} can be derived from Equation (14') when an equality sign is used.

From Equation (14') we can calculate G_{cr} as a function of $c^E - c^L$ by inserting the values $g = 9.8 \text{ m/s}^2$, $D = 3.0 \times 10^{-9} \text{ m}^2/\text{s}$, $m_L = -2.33 \text{ K/wt-}\% \text{ Sn}$ (derived from the phase diagram), and $c^E - 61.9$ wt-% Sn (read from the phase diagram).

The partial derivatives of the density are calculated with the aid of tables of ρ as a function of temperature and Sn concentration. The calculations are given in the table shown below.

c^L	$\left(\dfrac{\partial\rho}{\partial T}\right)_{c^L,T_L}$	$\left(\dfrac{\partial\rho}{\partial c}\right)_{c^E,T_E}$	$c^E - c^L$	ρ_L	v_{kin}	G_{cr}
(wt-% Sn)	(10^3 kg/m^3 K)	(10^3 kg/m^3 wt-%)	(wt-% Sn)	(10^3 kg/m^3)	(10^{-4} m^2/s)	(10^2 K/m)
20	-10.48	-46.8	41.9	9.67	2.73	152
30	-10.11	-41.1	31.9	9.24	2.81	1836
40	-9.71	-37.1	21.9	8.85	2.92	4218
50	-8.76	-35.2	11.9	8.49	3.06	7205
60	-8.69	-31.8	1.9	8.16	3.26	10728
From text	Table values	Table values	Calculated	Table values	Table values	Calculated from Equation (14')

Inserting $T_E - T_L = m_L\,(c^E - c^L)$ into Equation (10'), we obtain:

$$\Delta\rho = \frac{\partial\rho}{\partial T}\,m_L\,(c^E - c^L) + \frac{\partial\rho}{\partial c}\,(c^E - c^L) \qquad (11')$$

This value of $\Delta\rho$ is inserted into Equation (8'):

$$\frac{-g\left(m_L\,\dfrac{\partial\rho}{\partial T} + \dfrac{\partial\rho}{\partial c}\right)(c^E - c^L)}{\rho v_{\text{kin}}\,D}\,\frac{m_L^3\,(c^L - c^E)^3}{G^3} \geq 1700 \qquad (12')$$

As a value of ρ we choose ρ_L, since it gives the lowest possible value of the left-hand side of Equation (12'):

$$\frac{g\,m_L^3\left(m_L\,\dfrac{\partial\rho}{\partial T} + \dfrac{\partial\rho}{\partial c}\right)(c^E - c^L)^4}{\rho_L v_{\text{kin}}\,DG^3} \geq 1700 \qquad (13')$$

Answer:

(a) The desired table is shown on the next page.

(b) The desired diagram is shown below. On the right-hand side of the curve there is no risk of freckles.

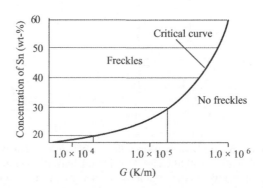

$[c_{Sn}]$ (wt-%)	G_{cr} (K/m)
20	1.5×10^4
30	1.8×10^5
40	4.2×10^5
50	7.2×10^5
60	1.1×10^6

The temperature gradient G_{cr} is determined by the temperature distribution during the solidification process. The latter also determines the solidification rate. The larger G is, the larger will be the solidification rate and the smaller the risk of freckles.

11.8 MACROSEGREGATION DURING HORIZONTAL SOLIDIFICATION

Figure 11.45 (a) shows the density distribution of an alloy which solidifies by unidirectional solidification, perpendicularly to the gravity field. The density of the solid phase increases linearly with decreasing temperature and is higher than the density of the melt. We have assumed that the density of the melt decreases within the two-phase region and has a minimum close to the solid phase. Inside the two-phase region the density of the melt can be either higher or lower than that outside, which we have seen in Section

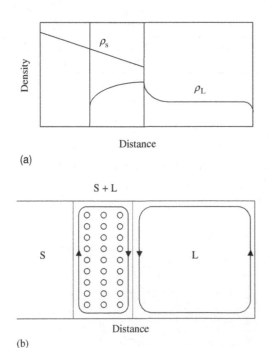

(a)

(b)

Figure 11.45   (a) Density distribution in melt and solid phase during unidirectional solidification. (b) Natural convection in the melt and inside the two-phase region for the density distribution in (a).

11.7 above. This density distribution causes natural convection in the melt and within the two-phase region according to the pattern illustrated in Figure 11.45 (b).

The velocity of the melt inside the two-phase region is certainly very low but the natural convection around the two-phase region has the consequence that some melt is carried away from the mushy zone. The melt in the two-phase region has a higher concentration of the alloying element than the melt outside, owing to microsegregation. Hence the natural convection leads to a gradual increase in the concentration of the alloying element in the bulk melt.

11.8.1   Influence of Gravitation on Macrosegregation during Horizontal Solidification

Fredriksson and co-workers performed a series of experiments with two Pb–Sn alloys on Earth with normal gravitation and under microgravity in space. The sizes of the samples were $100 \times 100 \times 25$ mm.

The two Pb–Sn alloys were allowed to solidify by unidirectional solidification. The directions of solidification and the temperature gradient were perpendicular to the gravity field. The aim of the experiment was to study the influence of gravitation on macrosegregation during horizontal solidification.

One of the alloys had a high concentration of Pb (Pb–15 wt-% Sn) and a density of the melt in the two-phase region which was *lower* than the initial density. The other alloy had a high concentration of Sn (Sn–10 wt-% Pb) and a density of the melt in the two-phase region which was *higher* than the initial density. In both cases, the flow patterns were simulated by computer calculations and the expected macrosegregation was calculated. The experimental results showed good agreement with the computer calculations, as described in the captions for Figures 11.46 and 11.47.

Space Experiments
The corresponding experiments on solidification of Pb–Sn alloys in space in the absence of gravitation were performed at the beginning of the 1990s. By reduction or elimination of the gravitational force the convection was strongly reduced and macrosegregation could be avoided.

The numerical calculations show that the convection in the interdendritic regions is different in the two cases and controlled by the density effect of Sn and Pb on the melt. By changing the orientation of the gravity force the convection can be reduced in the Sn–10 wt-% Pb alloy by solidification from bottom to top with the gravity vector and the solidification direction in opposite directions.

For the Pb–Sn alloy, no orientation of the solidification direction can eliminate the convection. In the case when the solidification direction is oriented opposite to the gravity vector, macrosegregation, i.e. freckles, will be formed. The

Case I: Pb–15 wt-% Sn

The interdendritic melt had a *lower* density than the initial melt.

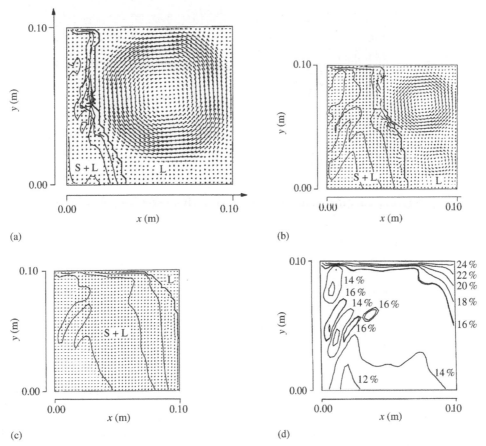

(a)

(b)

(c)

(d)

Figure 11.46 (a)–(c) Unidirectional solidification of the alloy Pb–15 wt-% Sn. Calculated flow pattern and calculated solidification process. The curves show the positions of the different phases 12, 24, and 36 min after the start of the casting process. (d) Concentration of alloying element (Sn) in the different parts of the casting after complete solidification. When the densities in the two-phase region are *lower* than in the melt, a *high positive* macrosegregation in the *upper* part of the casting is obtained. The % figures represent Sn concentrations. Reproduced by permission of ASM International.

only way to eliminate the convection is to reduce the gravity force to a minimum. This can be done in space.

11.9 MACROSEGREGATIONS IN STEEL INGOTS

In steel ingots, many different types of macrosegregations appear. Figure 11.48 shows a survey of macrostructures and types of macrosegregations of general occurrence.

Positive segregation means that the concentration of the alloying element exceeds the average concentration. On *negative segregation* there is instead a local lack of the alloying element.

Figure 11.48 shows that there is an increased concentration of the alloying element at the top and more pure material at the bottom of the ingot. The left part of the figure shows that the ingot has a zone of equiaxed crystals and that this zone is located at the bottom of the ingot. As we

shall see in Section 11.9.1, positive and negative macrosegregations have their origins in sedimentation of free crystals in the melt during the solidification process.

The right part of Figure 11.48 shows that so-called A-segregates occur. They have the same shape as freckles and can be described as pencil-like channels, filled with an alloy, which contain high concentrations of the alloying elements. In the centre of the ingot there are in addition V-segregates. These are of the same shapes and types that are present in continuously cast strands. The V-segregates will be discussed in more detail in Section 11.9.3.

11.9.1 Macrosegregation during Sedimentation

Small crystals keep floating in a melt owing to the thermal motion, mainly by convection, which is considerable in a metal melt. The crystals grow dendritically by continuous solidification of melt. When a crystal has reached a certain

Case II: Sn–10 wt-% Pb
The interdendritic melt had a *higher* density than the initial melt.

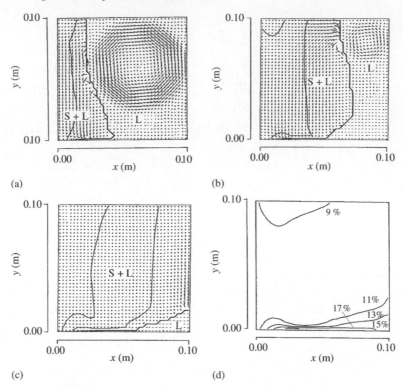

(a)

(b)

(c)

(d)

Figure 11.47 (a)–(c) Unidirectional solidification of the alloy Sn–10 % Pb. Calculated flow pattern and calculated solidification process. The curves show the positions of the different phases (a) 3, (b) 6, and (c) 9 min after the start of the casting process. (d) Concentration of alloying element (Pb) in the different parts of the casting after complete solidification. When the densities in the two-phase region are *lower* than in the melt, a *high positive* macrosegregation in the *lower* part of the casting is obtained. The % figures represent Pb concentrations. Reproduced by permission of ASM International.

critical size, the thermal motion can no longer compensate for the gravitation force minus the buoyancy force and the crystal sinks towards the bottom or *sediments.*

The growing crystal that sediments has not yet solidified completely but contains melt of interdendritic composition.

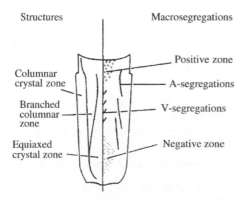

Figure 11.48 Survey of the structures and types of macrosegregations which often occur large ingots. Reproduced with permission from the Scandanavian Journal of Metallurgy, Blackwell.

The melt outside the crystal has a different and lower concentration of the alloying element ('bulk liquid'). When the crystal sediments, it passes regions with lower alloy composition and the interdendritic melt carried by the crystal is partly rinsed away and replaced by melt from the surroundings. The melt within the crystal acquires an intermediate composition between the concentration of the surrounding melt and the high concentration of the melt in the interdendritic regions inside the crystal.

The exchanged interdendritic melt stays in the upper parts of the ingot and crystals while partly more pure (lower concentration of alloying elements) melt sinks towards the bottom. The result is *positive* segregation in the *upper* part of the ingot and *negative* segregation in the *lower* parts.

Simple Mathematical Model of Sedimentation Segregation

Some Definitions
Macrosegregation, due to sedimentation of free crystals, has been treated theoretically. The free crystals contain

interdendritic melt and carry additional bulk melt when they sediment. In this way one part of the crystals is accompanied by W parts of bulk liquid, which become embedded in the sediment zone. The average concentration of the alloying element in the sediment zone can be calculated from the relation:

$$c^+ = \frac{c^s + W c^L}{1 + W} \qquad (11.75)$$

where

c^s = concentration of the alloying element in the crystals

c^L = concentration of the alloying element in the bulk melt

c^+ = average concentration of the alloying element in the sediment zone.

The partition coefficient is defined as c^s/c^L. The segregation due to sedimentation can be described by an effective partition coefficient k_{part}^+. It can be calculated by dividing Equation (11.75) by c^L:

$$k_{part}^+ = \frac{c^+}{c^L} = \frac{k_{part} + W}{1 + W} \qquad (11.76)$$

Calculation of the Relative Height of the Sediment Zone
A possible and simplified mechanism for the formation of the equiaxed zone is

1. Free crystals nucleate, grow in the liquid at a given undercooling, and settle simultaneously.
2. No more crystals form in the melt during the settling time.

The concentration of the alloying element in the melt can be calculated with the aid of the lever rule [Equation (7.15), page 189]:

$$c^L = \frac{c_0^L}{1 - f \left(1 - k_{part}\right)} \qquad (11.77)$$

where

c_0^L = initial concentration of the alloying element

f = total volume fraction of solid phase in the melt at the time of sedimentation.

The volume fraction of solid phase can be calculated from Equation (11.77). The result is:

$$f = \frac{c^L - c_0^L}{c^L \left(1 - k_{part}\right)} \qquad (11.78)$$

The volume fraction of crystals including their associated melt is a factor $(1 + W)$ times larger than f:

$$f_{crystal} = f (1 + W) \qquad (11.79)$$

According to Figure 11.49, the volume fraction of crystals, including their associated melt, can also be written as:

$$f_{crystal} = \frac{h}{H} \qquad (11.80)$$

where H and h are the heights before and after sedimentation, respectively.

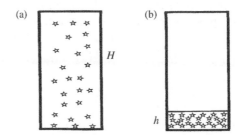

Figure 11.49 (a) Equiaxed crystals in the melt before sedimentation; (b) equiaxed crystals, including their associated melt, after sedimentation. Reproduced with permission from the Scandanavian Journal of Metallurgy, Blackwell.

Equations (11.79) and (11.80) give:

$$f = \frac{\dfrac{h}{H}}{1 + W} \qquad (11.81)$$

Example 11.5
Find expressions for c^L and c^+ in terms of c_0^L, k_{part}, and W.

Solution:
If we replace f in Equation (11.77) by Equation (11.81), we obtain:

$$c^L = \frac{c_0^L}{1 - f\left(1 - k_{part}\right)} = \frac{c_0^L}{1 - \left(1 - k_{part}\right)\dfrac{\frac{h}{H}}{1 + W}} = \frac{c_0^L (1 + W)}{(1 + W) - \left(1 - k_{part}\right)\dfrac{h}{H}}$$

This expression for c^L in combination with Equation (11.75) gives:

$$c^+ = \frac{c^s + W c^L}{1 + W} = \frac{k_{part} c^L + W c^L}{1 + W} = \frac{k_{part} + W}{1 + W} c^L$$

or

$$c^+ = \frac{k_{part} + W}{1 + W} c^L = \frac{k_{part} + W}{1 + W} \frac{c_0^L (1 + W)}{(1 + W) - \left(1 - k_{part}\right)\dfrac{h}{H}}$$

$$= \frac{c_0^L \left(k_{part} + W\right)}{(1 + W) - \left(1 - k_{part}\right)\dfrac{h}{H}}$$

Answer:

$$c^{L} = \frac{c_0^{L}(1 + W)}{(1 + W) - (1 - k_{part})\frac{h}{H}}$$

and

$$c^{+} = \frac{c_0^{L}(k_{part} + W)}{(1 + W) - (1 - k_{part})\frac{h}{H}}$$

Calculation of the Sedimentation Segregation
The theory predicts that the average concentration of the alloying element is a simple function of the relative height of the sediment zone. We can write:

$$c^{+}(h) = c^{+}(0)(1 - h/H)^{(k_{part}^{+} - 1)} \qquad (11.82)$$

which gives the sedimentation segregation:

$$c^{+}(h) - c_0^{L} = c^{+}(0)(1 - h/H)^{(k_{part}^{+} - 1)} - c_0^{L} \qquad (11.83)$$

Equation (11.83) in combination with Equation (11.76) describes the macrosegregation due to sedimentation of equiaxed crystals if the relative height of the sediment zone, the partition constant, and *W* are known.

A small sediment zone *h* gives a large negative macrosegregation at the bottom of the ingot, but the contribution to the positive macrosegregation at the top will be small.

Influence of Crystal Size on Sedimentation Segregation
Another reason for the fact that the average composition of the sedimented crystals deviates from the initial composition may be that, during the sedimentation, they have grown at a lower undercooling than the surrounding crystals.

It has been found that in ingots with a fine-grained equiaxed crystal zone the tendency for negative segregation at the bottom is larger than that in ingots with an equiaxed zone of large crystals. When there is a large number of free crystals in the melt, they grow with a lower undercooling than if there are fewer and larger crystals. This also facilitates the exchange of melt in the interdendritic regions because the melt flows more easily through wide than narrow channels.

11.9.2 A-segregations

In Section 11.7, we found that the channels in castings, which are called freckles, are caused by natural convection due to temperature and especially concentration gradients in the two-phase region during the solidification process in an ingot.

At the vertical solidification front in an ingot, the type of convection appears that was mentioned in the preceding section [Figure 11.45(b)]. This flow may, in analogy with freckle formation, result in channel formation. These channels are called *A-segregations*. A-segregations are thus caused by natural convection in the melt during the solidification process, due to the density variations, and result in macrosegregation in the material.

A-segregations occur in large ingots and appear generally in the upper third of the ingot up to the upper surface. They are sometimes called 'ghost lines'. A typical example is shown in Figure 11.50.

Centreline

Figure 11.50 Sulfur print of A-segregations in a 1.7 ton ingot. Reproduced with permission from the Scandanavian Journal of Metallurgy, Blackwell.

The circular, nearly vertical channels are formed when melt with a high concentration of the alloying element is carried to a region with a higher temperature than its own and a melt with lower alloy composition. The addition of melt from outside leads to a change in the composition in the region, which causes the liquidus temperature to decrease and some crystals to melt. The melt, which flows through the dendritic network, finds its way to where the flow resistance is smallest.

When the melt flows from a colder to a warmer region, part of the already formed dendritic network melts and channels are formed according to the same principles which are valid for freckle formation.

In the same way as described in Figure 11.46, we obtain positive segregation at the top in this case also. In large ingots the positive segregation at the top is mainly caused by convection in the A-segregation channels. This can be described quantitatively in a relatively simple way.

Simple Mathematical Model for A-segregations

The convection in the A-segregation channels is driven by the density differences in the melt. The convection pattern appears in principle as illustrated in Figure 11.51. The bent curves in Figure 11.51 are two streamlines, which illustrate the flow of the melt in the two-phase region. The channel formation starts at the transition from a colder to a warmer region. These positions P are marked in Figure 11.51.

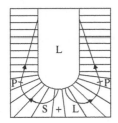

Figure 11.51 Convection pattern in a solidifying ingot.

The directions of the channels coincide with the directions of the flow lines. The channels 'intersect' the vertical solidification front under a certain angle and give elliptical intersections with the front surface (Figure 11.52).

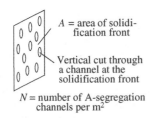

Figure 11.52 The A-segregation channels are not perfectly vertical. Consequently, they have intersections with a vertical plane.

The density difference and the number and sizes of the A-segregations determines the segregation pattern, i.e. the size of the macrosegregation.

We set up a material balance of the melt, which flows through the A-segregation channels:

$$\frac{\mathrm{d}c}{\mathrm{d}t}V \qquad = \qquad v_A A_A (c_A - c) \qquad (11.84)$$

Received amount of	Supplied amount of
alloying element	alloying element
per unit time	per unit time

where

c = concentration of alloying element in the melt
V = volume of the melt
c_A = concentration of the melt in the A-channels

v_A = average flux of the melt in the A-channels
A_A = total cross-sectional areas of the A-channels
t = time.

We have to find expressions for A_A, v_A, and V to be able to solve Equation (11.84).

A_A can be connected with the area of the solidification front (see Figure 11.52) with the aid of the relation:

$$A_A = A N \pi r_A^2 \qquad (11.85)$$

where

A = area of the solidification front
N = number of A-segregation channels per unit area
r_A = radius of an A-channel (Figure 11.53).

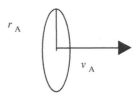

Figure 11.53 Flux of melt in an A-channel.

The average flux v_A can be calculated with the aid of the Hagen–Poiseuille law, which is valid for a flowing liquid in a tube [Equation (3.20), page 48]:

$$v_A = \frac{r_A^2 \Delta p}{8\eta l} = \frac{r_A^2 \Delta \rho g l}{8\eta l} = \frac{r_A^2 \Delta \rho g}{8\eta} \qquad (11.86)$$

where

Δp = pressure difference between the two ends of a tube or channel
$\Delta \rho$ = density difference within the two-phase region
η = viscosity coefficient of the melt
l = length of the tube (channel).

By combining equations (11.84)–(11.86), the concentration change of the molten phase as a function of time can be calculated. The final equation becomes:

$$\frac{\mathrm{d}c}{\mathrm{d}t} = Bc\frac{A}{V} \qquad (11.87)$$

where B is a summarized constant. After integration, we obtain:

$$c = c_0^L \exp\left(B \int_0^t \frac{A}{V}\mathrm{d}t\right) \qquad (11.88)$$

where c_0^L is the initial concentration of the melt.

It should be emphasized that both A and V are functions of time and therefore must stay inside the integral sign in Equation (11.88). If we make assumptions about the motion of the solidification front during the solidification, it is possible to find expressions for A and V as functions of time during the solidification process, and then perform approximate calculations of the macrosegregation at various heights from the bottom in an ingot.

Figure 11.54 gives an example of such a calculation of the macrosegregation and a comparison with experimentally measured values. It can be seen that the macrosegregation is considerable. At the top of the ingot the concentration of the alloying elements is twice as large as in the interior of the ingot. The positive segregation at the top is also influenced, although to a minor extent, by the sedimentation of crystals (see Section 11.9.1, page 401) within the ingot.

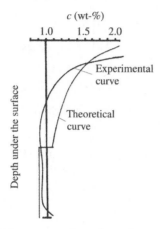

Figure 11.54 Macrosegregation along the central axis in an ingot. Reproduced with permission from the Scandanavian Journal of Metallurgy, Blackwell.

11.9.3 V-segregations

In connection with the treatment of centre segregation during continuous casting of strands, we briefly discussed V-segregations (Section 11.6, pages 388–389). It is apparent from Figure 11.48 that they also occur in ingots.

With the aid of radioactive tracers Kohn studied how the solidification front moved within a 3.5-ton ingot during the solidification process. The result of the experiment is illustrated in Figure 11.55. It can be seen that:

- A layer of solid/liquid material is built up more rapidly from the bottom of the ingot than from the sides.
- During the solidification process, a strong lowering or dip of material in the centre will sometimes occur within the time interval 34–50 min.

A very thick zone with a crystal framework and an interdendritic melt, with strong accumulation of alloying elements, is formed at the bottom.

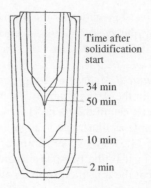

Figure 11.55 Position of the solidification front during the solidification process in a 3.5-ton ingot as a function of time. Reproduced with permission from The Metal Society, Institute of Materials, Minerals & Mining.

An explanation of the settling at the centre during the solidification process may be that the freshly formed crystal framework cannot withstand the ferrostatic pressure but becomes compressed.

Hultgren has given a more likely explanation for settling. He claims that the vacuum region, caused by the solidification shrinkage in the lower parts of the ingot, cannot be filled by melt from above because the distance is long and the resistance from the dendrite arms is high. The pressure difference which arises between the different parts of the ingot results in a general settling of the crystals in the middle zone.

This settling causes either deformation of the crystal framework in the middle zone or framework cracks along regions which already have achieved considerable stability. These cracks are filled more or less completely by melt, which has accumulated alloying elements and impurities. Since the pressure which causes the settling has a maximum at the centre and decreases radially towards the periphery of the ingot, the cracked parts show a V-shape and have a 'stripy' segregation profile.

The V-segregation pattern is influenced by the shape of the ingot mould. Experience shows that a shorter and wider mould gives less V-segregations than a longer and more narrow mould. An ingot with a large conicity gives fewer V-segregations than an ingot with a small conicity.

SUMMARY

■ Macrosegregation

Local deviations from the average concentration of the alloying elements are called macrosegregation. The most common causes of macrosegregation are solidification and cooling shrinkage.

Degree of Macrosegregation

$$\Delta x = \overline{x^s} - x_0^L \quad \text{or} \quad \Delta c = \overline{c^s} - c_0^L$$

■ Centre Segregation during Continuous Casting

Centre segregation is caused, like pipe formation, by solidification and cooling shrinkage. Melt flows through a two-phase region and fills the cavity, which is caused by the shrinkage at the centre.

Simple and closely packed V-segregations appear in the region above and below bridges across the pipes.

Centre segregation and pipe formation at continuous casting can be minimized by:

- reduction of the casting rate during the final stage of the casting;
- application of an external pressure on the strand at an optimal distance from the mould;
- no interruption of the pressure treatment until the strand has solidified completely;
- design of the secondary cooling in an optimal way; additional cooling from a certain optimal position is favourable.

■ Freckles

Freckles are casting defects which appear in certain cases during ingot casting and remelting processes with arcs under vacuum or during electroslag remelting.

Freckles are a type of macrosegregation. They have a pencil-like appearance, which looks like dots in the cross-section of the material.

Freckles are caused by jet streams, which appear during concentration-driven convection. A condition for this is density inversion in the melt. The formation of freckles has been studied in detail for the transparent system $NH_4Cl–H_2O$.

The existence of concentration-driven convection in metallic systems depends on two factors, the magnitude of the *density inversion* and the *transport processes*, which are controlled by material constants such as dendritic growth, diffusion and densities in melt and solid phase, and temperature conditions.

Density inversion in the two-phase region is possible in alloy systems if:

- the *lighter* alloying element (compared to the matrix element) has a *positive segregation*;
- the heavier alloying element (compared to the matrix element) has a *negative segregation*.

The condition for the occurrence of freckles is:

$$\frac{K_p g \beta (T_L - T_S)^4}{v_{kin} DG^3} \geq 1700$$

or alternatively:

$$1700 \leq \frac{K_p g \Delta\rho (T_L - T_S)^3}{\rho v_{kin} DG^3}$$

The casting process must be designed in such a way that freckles are avoided. If a temperature gradient G, which is large enough, is used freckles cannot appear:

$$G > G_{cr} \quad \Rightarrow \quad \text{no freckles}$$

■ Macrosegregation during Unidirectional Solidification

When an alloy solidifies by horizontal unidirectional solidification, the density distribution causes natural convection in the melt and the two-phase region, which leads to macrosegregation in the material.

Within the two-phase region, the density of the melt can be either higher or lower than the density of the melt outside.

The gravity constant influences the size of the macrosegregation strongly. Space experiments show that a decreased gravitation force reduces the convection strongly and macrosegregations of this type can be avoided.

■ Macrosegregation in Ingots

When crystals in an ingot sediment, there is an exchange of interdendritic melt included inside the crystals and the surrounding bulk melt. The exchanged interdendritic melt remains in the upper parts of the ingot and crystals with partly more pure melt sink to the bottom. The result is positive segregation in the upper parts of the ingot and negative segregation in the lower parts.

Sedimentation Segregations

Sedimentation of equiaxed crystals results in an uneven distribution of the alloying elements. Each crystal carries a certain amount of the melt when it sediments.

The average concentration of the alloying element in the sediment zone is:

$$c^+ = \frac{c^s + Wc^L}{1 + W}$$

The effective partition coefficient k_{part}^+ is:

$$k_{part}^+ = \frac{c^+}{c^L} = \frac{k_{part} + W}{1 + W}$$

The concentrations of the bulk melt c^L and c^+ can be expressed in terms of c_0^L, k_{part}, and W.

Sedimentation Segregation

$$c^+(h) - c_0^L = c^+(0)\left(1 - h/H\right)^{(k^+ - 1)} - c_0^L$$

A-segregations

A-segregations are circular, approximately vertical, channels with deviating composition relative to the surroundings. They are reminiscent of freckles.

A-segregations are caused by the natural convection in the melt and the two-phase region during the solidification process. The convection leads to remelting of already formed crystals.

A-segregations may occur in large ingots, preferably in the upper third of the ingot, and cause considerable macrosegregation there.

V-segregations

When ingots solidify, a layer of solid/liquid material is built up more rapidly from the bottom of the ingot than from the sides. A large lowering or dip in the material is formed at the centre. This settling causes either deformation of the crystal framework in the middle zone or framework cracks along regions which have already achieved considerable stability. The cracks have a maximum at the centre and decrease radially towards the periphery of the ingot. They show a V-shape and have a 'stripy' segregation profile.

EXERCISES

11.1 (a) Calculate the carbon concentration at the surface of a continuously cast slab that contains 1.0 wt-% carbon. The partition coefficient of the cast alloy is 0.33. The ratio of the liquid and solid densities $\rho_L/\rho_s = 0.96$.

Hint B5

(b) In addition, calculate the average carbon concentration at the surface of the strand if it is strained 3 % when 70 % of the surface has solidified.

Hint B83

11.2 A steel sample with 0.50 wt-% carbon is kept in a temperature gradient. Part of the sample is liquid and the rest is solid. Transfer of heat from the hot side to the cold side keeps the temperature gradient constant. The partition coefficient of the alloy is 0.33. The ratio of the liquid and solid densities $\rho_L/\rho_s = 0.96$. The sample is exposed to strain.

(a) Calculate the average carbon concentration in the liquid/solid region as a function of the fraction of liquid and the total strain.

Hint B27

(b) Plot the function in (a) for the two strain levels 10 and 50 % in a diagram.

Hint B111

11.3 (a) Give some different suggestions of practical solutions to avoid centre segregation in continuously cast materials.

Hint B250

(b) Discuss also if there are differences in this respect between slabs and billets.

Hint B122

11.4 In a steel works experiments with two different casting practices were made to study the effect on centre segregation during continuous casting. In one case strong cooling and a high excess temperature (compared with the melting point of the alloy) were used. In the other case soft cooling and a low excess temperature (above the melting point) were used. The results of the two experiments are illustrated in Figure 11.33 on page 392 and those shown here.

Soft cooling and low excess temperature

Strong cooling and high excess temperature

Explain the difference in structure and centre segregation in the two cases.

Hint B213

11.5 A steel company produces ball-bearing steel which contains 1.0 wt-% C and 1.0 % Cr by ESR electroslag remelting. Sometimes macrosegregations occur in the ingots. The macrosegregations are related to freckles, which are formed during the remelting process.

(a) Describe the formation process of the macrosegregations (description, mathematical model, importance of the temperature gradient for formation of macrosegregation).

Hint B42

(b) Once, an ESR-refined ingot of ball-bearing steel showed 93 freckles/m^2 at the top after refining. The remelting rate was 1.5 m/h and the height of the ingot was 5.0 m. The depth of the molten pool was 0.15 m and the initial $\underline{C}$ concentration of its melt was 1.0 wt-%. The $\underline{C}$ concentration in the freckle channels was 1.5 wt-% and they had a diameter of 5.0 mm.

Materials constants of ball-bearing steel

$\eta_{melt} = 0.0067 \ N\,s/m^2$
Density difference of the melt within the two-phase region:
$\rho_L^{top} - \rho_L^{bottom} \approx 100 \ kg/m^3$

The density of the steel melt depends on its carbon concentration. To simplify the problem you may assume that the density difference within the two-phase region is constant during the ESR process. Estimate the carbon concentration of the melt in the molten pool as a function of the growing height H of the refined ingot. Set up a differential equation which gives [$\underline{C}$] as a function of time. Solve the equation and plot [$\underline{C}$] as a function of the growing height H of the ingot.

Hint B316

(c) Calculate the carbon concentration of the remelted and solidified ingot as a function of its growing height H. Plot the function in the diagram (b). What happens when the molten pool solidifies?

Hint B390

11.6 There are several types of macrosegregations in steel ingots, among others positive segregation at the top and negative segregation at the bottom of the ingot.

(a) How do these macrosegregations arise?

Hint B305

(b) Is it possible to affect them by change of casting temperature, composition and/or casting practice?

Hint B178

11.7 During casting of steel in ordinary cast iron moulds, it has been found that A-segregations are often obtained:

(a) in large ingots

Hint B14

(b) if the mould has the wider end up

Hint B288

(c) at high Si concentrations.

Hint B205

Explain these observations.

11.8 During squeeze casting of a two-step stairs profile, a difference in composition between the thin and the thick parts of the product is observed. The casting process can be described as follows.

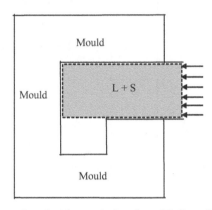

Reproduced with permission from McGraw-Hill.

A rectangular rod made of an Al–10 wt-% Mg alloy is heated to 550 °C and immediately moved into a so-called 'preform' (the grey volume in the upper figure). The preform is then squeezed in one step to the final shape of the casting.

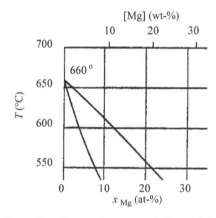

(a) Use the phase diagram of the Al–Mg system shown in the lower figure and explain the formation and distribution of macrosegregation during the process.

Hint B6

(b) Estimate roughly the maximum and minimum <u>Mg</u> concentratrations which appear at different parts of the cast product.

Hint B80

11.9 A steel company casts steel alloys with a carbon concentration of 0.50 wt-% into ingots with a height of 1.2 m. It is of interest to know if the macrosegregation due to sedimentation of free crystals increases with the height of the sediment crystal zone or not. A theoretical calculation may give the answer.

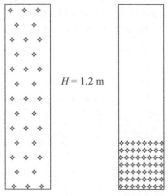

$H = 1.2$ m

Reproduced with permission from McGraw-Hill.

When free crystals sediment in a steel melt, each crystal is accompanied by a certain volume of bulk liquid, equal

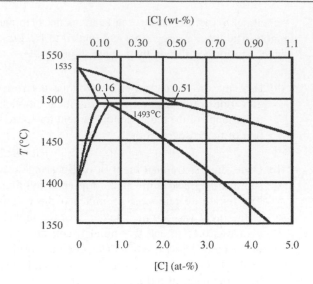

to W times its own volume. Part of the phase diagram of the Fe–C binary system is shown in the figure.

(a) Calculate the average carbon concentration in the sediment zone of an ingot with a height of 1.2 m. Assume that the crystals have grown at an undercooling of 8 °C. The initial carbon concentration of the melt is 0.50 wt-%.

Hint B48

(b) Plot the height of the sediment layer as a function of the degree of macrosegregation for the three alternatives $W = 0.5$, 1.0, and 2.0.

Hint B24

Answers to Exercises

The answers below are short in many cases. More extensive information is given in the Guide to Exercises, which is a supplement to the main book. SI units are used unless some other units are clearly declared.

CHAPTER 3

3.1 Investment casting (large series) or Shaw process (small series).

3.2 The difference between the upper and lower diameters is of the order of 0.5 mm, which can be neglected. A straight sprue can be used. Diameter $= 13$ mm.

3.3 ~ 50 s.

3.4 $v_3 = 0.010\sqrt{h}$.

3.5 $v_{\text{cast}} = 0.54$ m/min; $\quad d_{\text{outlet}} = 33$ mm.

3.6 13 mm.

3.7a Increased cooling capacity of the mould $\Rightarrow$ low L_f.
Increased surface tension $\Rightarrow$ low L_f.
Increased viscosity $\Rightarrow$ low L_f.
Wide solidification interval of the alloy $\Rightarrow$ low L_f.

3.7b Pure metals and eutectic alloys $\Rightarrow$ high L_f.
Increased temperature $\Rightarrow$ low viscosity $\Rightarrow$ high L_f.
Formation of dendrites prevents the flow strongly $\Rightarrow$ low L_f.
Low thermal conductivity and high heat of fusion increases L_f.
Laminar flow gives a larger fluidity length than turbulent flow.

3.8 $\Delta T = T - T_L = \left(\dfrac{6\varepsilon\sigma_B}{\rho_L c_p^L R} t + T_i^{-3} \right)^{-\frac{1}{3}} - T_L$
where $\Delta T \geq 0$.

3.9a $L_f = v t_f = \text{constant} \times \dfrac{v \rho_L c_p^L}{h} \times \ln \dfrac{T_L + (\Delta T)_i - T_0}{T_L - T_0}$
where

$\text{constant} = \dfrac{a \sin\alpha}{4(1 + \sin\frac{\alpha}{2})} = 0.0011$.

Materials Processing during Casting H. Fredriksson and U. Åkerlind
Copyright © 2006 John Wiley & Sons, Ltd.

3.9b $L_f = 5.4$ cm.

3.10a $R_{\text{Al}} = \dfrac{\sigma\sqrt{2}}{\rho_{\text{Al}} g h} = \dfrac{8.00 \times 10^{-5}}{h}$

3.10b $R_{\text{Fe}} = \dfrac{R_{\text{Al}}}{8.7}$
An Fe corner is 'sharper' than an Al corner.

CHAPTER 4

4.1 10.5 min ($\lambda = 0.90$).

4.2 11 min ($\lambda = 0.79$).

4.3a 1.5 h.

4.3b 1 h. The Fe ingot solidifies more rapidly than the Al ingot owing to a lower heat of fusion and higher driving force.

4.4

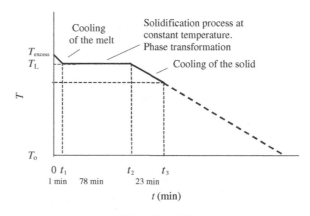

Exercise 4.4

4.5 40 min.

4.6 $t_{\text{sand}} = 13$ s; $t_{\text{metal}} = 5$ s. The production capacity will be more than doubled in the latter case.

4.7 $t = 77 z$, where z is the distance from the top; $t_{\text{total}} \approx 8$ s.

4.8a $t = \dfrac{\rho(-\Delta H)}{(T_{\text{melt}} - T_0)} \left(\dfrac{R^2}{4k} + \dfrac{R}{2h} \right)$

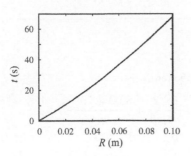

Exercise 4.8a

4.8b $\left| \dfrac{dy_L}{dt} \right| = \dfrac{T_{\text{melt}} - T_0}{\rho(-\Delta H)} \times \dfrac{1}{\dfrac{R_0 - y_L}{2k} + \dfrac{1}{2h}}$

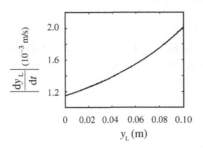

Exercise 4.8b

4.9a Figure 4.17 all cases. Figure 4.27 if $\text{Nu} = \dfrac{hy_L}{k} \ll 1$.

4.9b See figures.

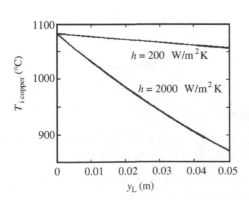

Exercise 4.9b Copper

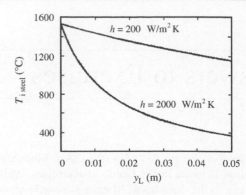

Exercise 4.9b Steel

4.10a Region I linear growth law. Region II parabolic growth law.

4.10b $\sim 1.5 \times 10^2 \, \text{W/m}^2 \, \text{K}$ in region I.

4.11a Read $T_{i\,\text{metal}}$ and y_L values for each curve in the exercise figure and calculate the h value for each curve. The result shows that h decreases with increasing y_L values.

4.11b The y_L-t curve shown is a plot of the corresponding y_L and t values for each curve in the exercise figure.

The discontinuity of the curve is due to air gap formation. The solidification rate is the derivative of the curve. The rate increases significantly at the end owing to decreasing area of the solidification front.

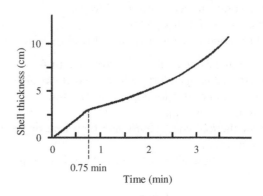

Exercise 4.11b

CHAPTER 5

5.1a $\sim 6\%$.

5.1b $1.5 \times 10^2 \, \text{MJ}$.

5.1c 15 cm (unrealistic). An insulating pore is often formed below the solidified shell.

5.1d Very little radiation is emitted provided that the upper surface is properly insulated. It solidifies *later* than the centre of the ingot.

5.2a $y = \dfrac{k_s}{h_1}\left(\dfrac{h_1}{h_2}\dfrac{T_L - T_0}{\Delta T} - 1\right)$ where ΔT = excess temperature.

5.2b $\sim 400\,°\text{C}$.

5.3 Convection in the melt. See *Hint A246* in the Guide to Exercises for details.

5.4a The oil or casting powder film thickness decreases with increasing distance from the top and the heat flux increases with distance from the top.

5.4b The film has evaporated. Good direct contact between metal and mould.

5.4c The heat flux decreases owing to air gap formation and increasing thickness of the shell.

5.5a 21 min.

5.5b 0.94 m/min.

5.6 1.3, 0.71, 0.54, 0.57 kW/m², respectively.

5.7 The solidification fronts meet ~ 0.5 mm from the upper surface of the strip.

5.8 $-\dfrac{dT}{dt} = \dfrac{h}{c_p \rho d}\left(T_{\text{strip}} - T_0\right) \approx 5.1 \times 10^3\,\text{K/s}$.

5.9 $v_{\max} = \dfrac{9.4 \times 10^{-3}\,\text{m}^2/\text{min}}{d}$.

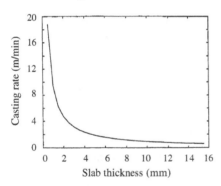

Exercise 5.9

5.10a 0.10 m/s.

5.10b 0.10 m/s.

5.11 $v_{\max} = \dfrac{64 \times 10^{-3}\,\text{m}^2/\text{min}}{d}$.

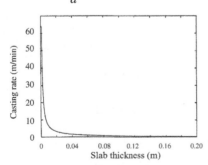

Exercise 5.11

5.12a Reasonable.

5.12b 0.80 m.

5.12c 6.1×10^3 K/s.

5.12d No turbulence, atmosphere with oxygen, and rapid solidification.

5.12e $t = \dfrac{\rho(-\Delta H)}{2\sigma_B \varepsilon\left(T_L^4 - T_0^4\right)} R = 2.0 \times 10^3 R$.

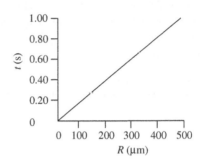

Exercise 5.12e

CHAPTER 6

6.1 >22 g.

6.2 $\lambda_{\text{den}} = 4.7 \times 10^{-4}\sqrt{y_L}$.

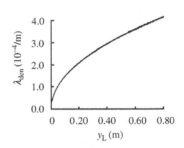

Exercise 6.2

6.3 $\lambda_{\text{den}} = \dfrac{6.5 \times 10^{-5}}{\sqrt{p}}$ m (p in atm).

6.4 See the figure.

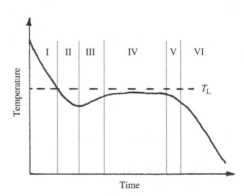

Exercise 6.4

I: Cooling of the melt. The surface crystal zone and later the columnar crystal zone are formed at the outer surface (see Figure 6.38, page 163).

II: Equiaxed crystals are formed ahead of the growing dendrites.

III: Equiaxed crystal zone replaces the columnar zone.

IV: Steady state. Equiaxed crystals are formed in the central part of the ingot.

V: Decreasing area of the solidification front gives less solidification heat. The ingot solidifies completely.

VI: The solid casting cools.

6.5 Basic equations give:

$$\frac{dr}{dt} = \frac{1}{6} C \frac{t^{-\frac{5}{6}}}{N^{\frac{1}{3}}}.$$

6.5a Combination with $v_{growth}\, \lambda_{den}^2 = $ constant shows that λ_{den} increases with N.

6.5b Combination with $v_{growth} = \mu(T_E - T)^n$ shows that the undercooling and the risk of cementite precipitation decrease when N is increased. Reasons for inoculation: (1) reduced risk of white solidification; (2) better mechanical properties.

6.6 Nuclei can probably be formed but not grow in the melt ahead of the solidification front because the melt is not undercooled.

6.7 $h < 0.6 \,\mathrm{kW/m^2\,K}$.

6.8a Convection causes crystal multiplication, remelting of crystal bases, and fast decrease of the super heat of the melt.

6.8b $v_{crystal} = 12 \times 10^2/(N + 3 \times 10^6)\,\mathrm{m/s}$.

6.8c $4 \times 10^{-4}\,\mathrm{m/s}$. $v_{crystal}$ is always smaller than this value. If the melt is inoculated, N increases strongly and the free crystals stop the dendrite growth and the transition from columnar to equiaxed zone occurs.

6.9a $-\dfrac{dT}{dt} = \dfrac{0.13}{L^2}.$

6.9b $L = 4.7\,\mathrm{cm}$. (Grey cast iron for $L > 4.7\,\mathrm{cm}$.)

6.10 $h = \dfrac{\rho(-\Delta H) \times 10^{-11}}{r_0(T_L - T_0)} \dfrac{r_0 - y}{\lambda_{den}^2}$

or

$$h = 0.21 \times \frac{65 \times 10^{-6} - y}{\lambda_{den}^2}$$

h decreases with increasing distance from the nozzle.

CHAPTER 7

7.1 $2.9\,\%$.

7.2 $12\,\%$.

7.3a $S = 4.2$, 1.5, and 1.0, respectively.

7.3b The segregation decreases when the diffusion rate of the alloying element in the solid increases.

7.4 $S = \left(\dfrac{B}{1+B}\right)^{-(1-k_{part})} = \left(\dfrac{0.53y^{1/3}}{1 + 0.53y^{1/3}}\right)^{-0.27}$

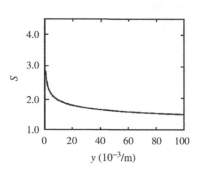

Exercise 7.4

7.5 $S_{Ni} = 1.5$, $S_{Cr} = 1.6$, and $S_{Mo} = 3.1$.

7.6 $x_{\underline{V}}^0 = 8.3 \times 10^{-3}[0.4 + 0.6(1 - f_E)](1 - f_E)^{0.6}$

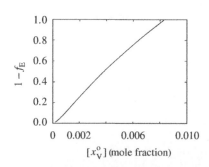

Exercise 7.6

7.7 $[\underline{Si}] \leq 0.4$ wt-%; $[\underline{Cu}] \leq 2$ wt-%.

7.8a $c_{\underline{P}}^{0L}/(7.2\,\mathrm{wt}\text{-}\%)$.

7.8b $1.4\,\%$.

7.9a

$c^0_{\underline{C}}$ (wt-%)	$1 - f$	$c^L_{\underline{Cr}} = 1.5(1-f)^{-0.27}$ (wt-%)	$S_{\underline{Cr}} = \dfrac{c^s_{max}}{c^s_{min}} = \dfrac{c^L_{\underline{Cr}}}{1.5 \times 0.73}$
1.5–2.0	0.0100	5.2	4.8
2.1	0.0332	3.8	3.4
2.2	0.0710	3.1	2.8
2.3	0.114	2.7	2.5
2.4	0.158	2.5	2.3
2.5	0.203	2.3	2.1
2.6	0.249	2.2	2.0
2.7	0.295	2.1	1.9
2.8	0.341	2.0	1.8

For experimental reasons the values of $(1-f) < 1\%$ in the table are all replaced by 0.01. Equations for determination of $(1-f)$ are:

$$c^L_{\underline{Cr}} = 68.8 - 16c^L_{\underline{C}}$$

$$c^L_{\underline{C}} = \frac{c^s_{\underline{C}}}{k_{part\,\underline{C}}} = \frac{c^0_{\underline{C}}}{k_{part\,\underline{C}} + (1-f)(1 - k_{part\,\underline{C}})}$$

$$c^L_{\underline{Cr}} = \frac{c^s_{\underline{Cr}}}{k_{part\,\underline{Cr}}} = c^0_{\underline{Cr}}(1-f)^{-\left(1 - k_{part\,\underline{Cr}}\right)}$$

7.9b $\quad S_{\underline{Cr}} = \dfrac{c^s_{max}}{c^s_{min}} = \dfrac{c^L_{\underline{Cr}}}{c^0_{\underline{Cr}} k_{part\,\underline{Cr}}}$

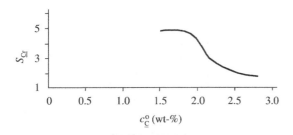

Exercise 7.9b

Back-diffusion cannot be neglected in alloys with low carbon concentrations. No such calculations are required here, however.

CHAPTER 8

8.1 $\quad t = 0.058 \dfrac{\lambda^2_{eut}}{D}$.

8.2 Dissolution time below T_E is $\sim 3.3\,\text{h}$ and above T_E $\sim 6.4\,\text{h}$. The answer is no.

8.3 $\quad t = 10 \ln\left(\dfrac{1}{1 - 2.0 f^L_0}\right)$.

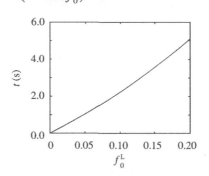

Exercise 8.3

8.4a Below T_E: the higher temperature the better. The concentration difference increases with T, which promotes homogenization. D increases with T, which also promotes homogenization.

Above T_E: the concentration difference decreases with T, which prolongs the homogenization. D increases with T but the effect does not dominate over the decrease in concentration difference.

8.4b The best choice of temperature is just below T_E, in this case for the sample 540 °C.

8.5a $t_{dis} = 10\,\text{h}$. Convection shortens the time, however.

8.5b $t_{hom} = 11\,\text{h}$.

8.6 1 h.

8.7 $\quad t_{dis} = -\dfrac{C_2 + C_3 y}{\pi^2 D} \ln\left[1 - \dfrac{f^0_0 \cdot \left(x^0 - x^m\right)}{x^{a/0} - x^m}\right]$

where

$$\lambda^2_{den} = C_1 \frac{\rho(-\Delta H)}{T_L - T_0} \frac{k + hy}{hk} = C_2 + C_3 y$$

The thinner the casting is, the shorter t_{dis} will be. D increases with temperature. The higher the temperature is, the shorter will be t_{dis}.

8.8 3.8 h.

8.9 12 min.

8.10 2.1 days.

CHAPTER 9

9.1 Solidification pores with spiny surfaces are formed if [G] < 10 ppm.

Round pores with smooth surfaces are formed if [G] > 20 ppm.

Elongated pores with smooth surfaces are formed if the gas concentration fulfils the condition 10 ppm < [G] < 20 ppm.

9.2a v_{growth} and [G]. Low solidification rate and high gas concentration promote the pore growth.

9.2b $v_{pore} = \dfrac{db}{dt} = -\dfrac{RT}{p}\dfrac{D_{\underline{H}}}{V_m}\dfrac{dx_{\underline{H}}}{dy} = 0.006\,\text{m/s}.$

The pore is assumed to be a cylinder with radius r and length b. The pore escapes to be trapped at the solidification front if $v_{pore} > v_{growth}$.

9.3 2.7×10^{-6} wt-%.

9.4a $[\underline{C}] = 0.12$ wt-% and $p_{CO} \approx 10$ atm.

9.4b $p_{CO} \approx 20$ atm.

9.5 $c_{\underline{O}}^0 = \dfrac{0.0019}{0.050}\left[1 - \left(\dfrac{c_{\underline{H}}^0}{0.0025}\right)^2\right]$

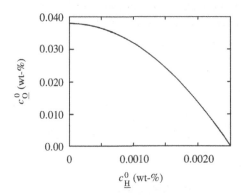

Exercise 9.5

9.6 4.0 wt-%. For safety add 4.5 wt-% Mn.

9.7a At *low* $\underline{C}$-concentrations, a *monotectic* reaction occurs, which gives δ-phase, which later is transformed into γ-phase and liquid MnS. The latter solidifies as MnS colonies of type I or II.

At *high* $\underline{C}$ concentrations, Mn and S are concentrated in the remaining melt when the steel solidifies as γ-phase. Solid MnS of type III or IV starts to precipitate by a *eutectic* reaction.

9.7b Ti reacts much more easily with C than with S. It is therefore useless to add Ti with the intention of obtaining TiS instead of MnS. The added Ti forms TiC and not TiS if the $\underline{C}$ concentration in the steel is high.

9.8 0.90.

9.9 0.05 wt-% is much more than sufficient. Only $[Ce] \geq 2 \times 10^{-7}$ wt-% is required.

9.10a The particle moves upwards with the velocity:

$$v = \frac{2g}{9\eta}(\rho_{melt} - \rho_{particle})r^2$$

9.10b 1×10^{-6} and 1×10^{-4} m/s.

9.10c $L_1 \approx 0.06t$ (mm, min) and $L_{10} \approx 6t$ (mm, min).

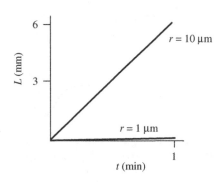

Exercise 9.10c

9.10d $r > 2\,\mu\text{m}$ $(v_{particle} > v_{growth})$.

CHAPTER 10

10.1a $h_{feeder} > 6$ cm, for safety ~ 8 cm ($\beta = 0.04$).

10.1b $h_{feeder} > 15$ cm, for safety ~ 18 cm. The insulated feeder requires less material and is to be preferred.

10.2 Bronze is no alternative as h_f becomes negative, owing to the high CFR value of this alloy. The only possible material choice is brass.

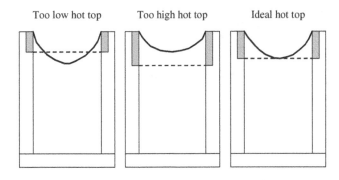

Exercise 10.2

Two feeders are chosen to avoid pipes in the cylinder. For safety their heights have to be $\sim 40\,\text{cm}$. The optimal positions of the feeders are shown in the figure.

10.3 A too low hot top gives a pipe and waste of material. A too high hot top means waste of material.

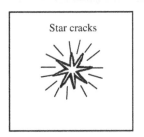

Exercise 10.3

10.4a A rough value and too low value of the height of the hot top is $6\,\text{cm}$ ($\beta = 0.04$).

10.4b More accurate calculations give the value $\sim 8\,\text{cm}$. For safety, choose $\sim 10\,\text{cm}$.

10.4c No.

10.5 $5.3\,\text{cm}$, for safety $\sim 8\,\text{cm}$.

10.6 $\overline{\sigma_x} = \dfrac{2}{3} E\alpha\Delta T_0 - E\alpha\Delta T_0\left(1 - \dfrac{y^2}{c^2}\right)$

10.7a The sudden temperature increase at the surface of the strand leads to expansion forces at the surface and tensile forces in the solidified region. The material is brittle in the region between the brittle/ductile transition temperature and the solidus temperature. The material is particularly sensitive to crack formation in the vicinity of the two-phase region in the interior of the strand. The risk of longitudinal midway cracks is considerable.

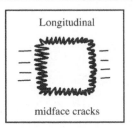

Exercise 10.7a

10.7b The cooling rates differ between the surface and the centre and all the way in the intermediate space. The stresses are strongest in the material at a temperature between the transition temperature (brittle/ductile zones) and the solidus temperature. This region is weak, i.e. especially sensitive to cracks. Star-like cracks are formed at the centre.

Exercise 10.7b

10.8 $y_{\text{L}} = -121\delta + \sqrt{(121\delta)^2 + 1.63 \times 10^{-3}}$

$T_{\text{i}} = \dfrac{1350}{1 + 8.28 \times 10^{-3}\dfrac{y_{\text{L}}}{\delta}} + 100$

δ (m)	y_{L} (m)	T_{i} (°C)
10^{-5}	0.00392	140
10^{-4}	0.00300	487
10^{-3}	0.00656	1380

10.9 The strain is equal to 2.6×10^{-3}. The thermal stress is $\sim 5 \times 10^7\,\text{N/m}^2$. The strain leads to very high stresses in the material, which do not correspond to equilibrium. Either the material becomes deformed plastically, e.g. bent, or it breaks.

10.10a Temperature inside the shell:

$T = 1260 + 7.0 \times 10^3 y$

The temperature decrease is:

$\Delta T(y) = (T_{\text{ss}} - T_{\text{so}})\left(1 - \dfrac{y}{s}\right)$

or

$$\Delta T(y) = 220\left(1 - \frac{y}{0.030}\right){}^\circ\mathrm{C}$$

10.10b Thermal strain:

$$\varepsilon^T(y) = \alpha(T_{ss} - T_{so})\left(\frac{1}{2} - \frac{y}{s}\right)$$

The equation

$$\varepsilon^T(y) = 5.0 \times 10^{-5} \times 220\left(\frac{1}{2} - \frac{y}{0.030}\right)$$

is plotted in the figure.

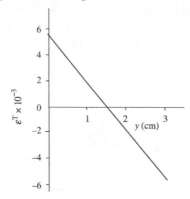

Exercise 10.10b Strain

Stress in the length direction:

$$\sigma^T(y) = -E\varepsilon^T(y)$$

or

$$\sigma^T(y) = -E\alpha(T_{ss} - T_{so})\left(\frac{1}{2} - \frac{y}{s}\right)$$

The equation

$$\sigma^T(y) = -5.0 \times 10^6 \times 220\left(\frac{1}{2} - \frac{y}{0.030}\right)$$

is plotted in the figure.

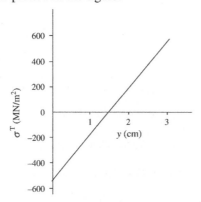

Exercise 10.10b Stress

10.10c Elimination of *y* between the stress equation above and the relation between *T* and *y* in Exercise 10.10a gives $\sigma(T)$ below.

Stress as a function of temperature:

$$\sigma^T(T) = -5.0 \times 10^6 \times 220\left(\frac{1}{2} - \frac{T - 1260}{210}\right)$$

is plotted in the figure.

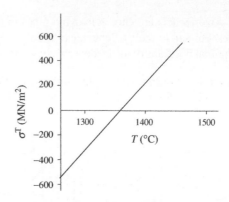

Exercise 10.10c Stress

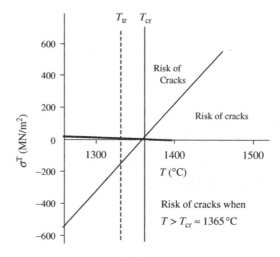

Exercise 10.10c Cracks

Risk of cracks: the line $\sigma(T)$ in the Exercise 10.10c Stress figure intersects the rupture limit at the critical temperature T_{cr}.

The rupture limit is found with the aid of the rupture limit versus temperature diagram in Chapter 10 (page 355), and is plotted in the Exercise 10.10c Cracks figure. This figure shows that there is risk of cracks when $T > T_{cr} \approx 1365\,{}^\circ\mathrm{C}$.

CHAPTER 11

11.a 1.03 wt-%.

11.b 1.15 wt-%.

11.2a

$$\overline{c_{stran}^0} = \frac{\frac{(1-g_L)k + \frac{\rho_L}{\rho_s}g_L}{1-g_L+\frac{\rho_L}{\rho_s}g_L} + \frac{\varepsilon}{k}}{1+\varepsilon} \frac{c_0^L}{k+g_L\cdot(1-k)}$$

11.2b

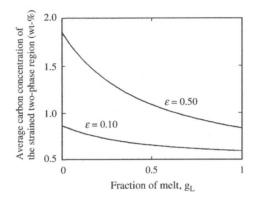

Exercise 11.2b

11.3a Mechanical reduction (external pressure).
Thermal reduction (cooling) with the condition:

$$G = \frac{dT_{surface}}{dt} \bigg/ \frac{dT_{centre}}{dt} \geq 0.5$$

11.3b Mechanical reduction in a slab caster. Thermal reduction in a billet caster.

11.4 Strong cooling and high excess temperature give no pipe, no centresegregation, and simple V-segregations. Soft cooling and low excess temperature give a pipe, centresegregation, and closely packed V-segregations.

11.5a Freckles arise by concentration-driven convection, which generates jet streams through the two-phase region. With *positive* segregation this is possible if $\rho_{alloying\ element} < \rho_{matrix}$. With negative segregation, the opposite condition is valid.

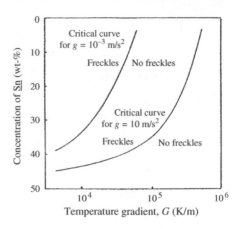

Exercise 11.5a

Conditions for freckle formation:
1. A stationary flow pattern of the melt.
2. A temperature gradient G smaller than the critical temperature gradient G_{cr}, which is given by the relation:

$$1700 = \frac{K_p g \Delta\rho (T_L - T_s)^3}{\rho v_{kin} D G_{cr}^3}$$

where $\Delta\rho = \rho^{top} - \rho^{bottom}$.
No freckles if $G > G_{cr}$.

11.5b The differential equation for determining the desired function is:

$$\frac{dc_{melt}}{dt} l_{pool} = \frac{r_F^2 \Delta\rho g}{8\eta} N\pi r_F^2 (c^F - c_{melt}) + v_{growth}c_0^L - v_{growth}c_{melt}$$

The solution is of the type:

$$c_{melt} = \frac{a - (a-b)e^{-bt}}{b}$$

where $a = 0.00486$ and $b = 0.00417$.

The carbon concentration in the molten pool as a function of the growing height of the refined ingot can be written as:

$$c_{melt} = 1.16 - 0.16e^{-10H}$$

where $0 \leq H \leq 5.0\,\text{m}$.

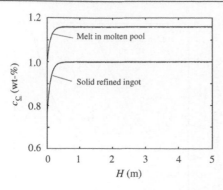

Exercise 11.5b

11.5c The carbon concentration in the refined ingot as a function of its growing height can be written as:

$$c_{\text{solid}} = c_{\text{melt}}\left(1 + \frac{r_F^4 \Delta\rho g}{8\eta}\frac{N\pi}{v_{\text{growth}}}\right) - c^F \frac{r_F^4 \Delta\rho g}{8\eta}\frac{N\pi}{v_{\text{growth}}}$$

or

$$c_{\text{solid}} = 1.00 - 0.25e^{-10H}$$

The molten pool cools from all sides and the conditions for freckles are no longer present. It solidifies as a normal ingot. The average carbon concentration of the solidified pool is hence equal to 1.16 wt-%. This concentration is illustrated in the figure by the thick line for the interval $5.0\,\text{m} < H < 5.15\,\text{m}$.

The first and the last parts of the remelted ingot have a lower and a higher $\underline{C}$ concentration, respectively, than the average. They have to be removed, which reduces the yield.

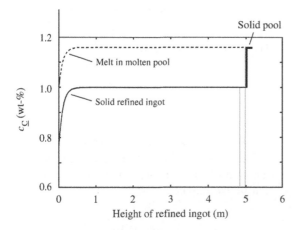

Exercise 11.5c

11.6a Small crystals, floating in the melt owing to natural convection, grow continuously. At a certain size convection can no longer compensate for gravitation and the crystals sediment. They are surrounded by interdendritic melt. Owing to the crystal growth, the interdendritic melt has a higher concentration of the alloying elements than bulk melt. Owing to exchange of melt during the sedimentation, the interdendritic melt of the bottom crystals has a lower concentration of the alloying elements than the interdendritic melt around the top crystals. This gives positive segregation at the top and negative segregation at the bottom.

11.6b
1. Decrease in the volume of the equiaxed zone by increase in the temperature of the melt before casting.
2. Heating of the upper surface of the melt by use of very effective exothermic hot top materials in order to reduce the natural convection.
3. Choice of suitable alloying elements in order to achieve a constant density of the melt in the interdendritic regions.

11.7a The interdendritic melt tries to move upwards through the dendritic network of the crystals, owing to a lower density than that of the bulk melt. The average concentration of alloying elements increases and results in a decrease in the melting point. Remelting occurs at the passage and round channels of different composition (A-segregations) are formed.

11.7b When the interdendrite melt moves upwards, the width of the dendrite arms decreases gradually, owing to remelting and the increasing width of the mould. Hence the passage of melt becomes easier and the formation of A-segregations will be facilitated.

11.7c The light Si atoms, which occur with a higher concentration in the interdendrite melt than elsewhere, give a particularly low density to the melt. As the density difference is the driving force of convection, the formation of A-segregations is facilitated for ingots with high Si concentrations, compared with other ingots.

11.8a The phase diagram of the Al–Mg system and the known average composition and temperature of the alloy show that the grey volume is a two-phase region.

The applied pressure squeezes a good deal of the interdendritic melt (23 %) out of the dendrite network in the grey volume as the melt flows away owing to the applied pressure. It is reasonable to assume that the melt fills the empty thin part of the future cast product.

11.8b 7 wt-% <u>Mg</u> in the thick part of the casting.
20 % <u>Mg</u> in the thin part of the casting.

11.9a 0.31 wt-% ($W = 0.5$). 0.36-% ($W = 1.0$).
0.42 wt-% ($W = 2.0$).

11.9b 0.16 m ($W = 0.5$). 0.22 m ($W = 1.0$).
0.33 m ($W = 2.0$).

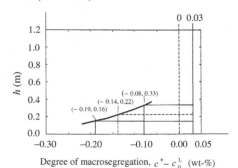

Exercise 11.9b

The macrosegregation increases with increasing height of the sediment layer.

Index

Note: Page numbers in *italics* refer to figures and those in **bold** refer to tables.

Materials Processing During Casting   H. Fredriksson and U. Åkerlind
Copyright © 2006 John Wiley & Sons, Ltd.

With kind thanks to Geraldine Begley for creation of this index.

Printed and bound by CPI Group (UK) Ltd, Croydon, CR0 4YY

Printed and bound by CPI Group (UK) Ltd, Croydon, CR0 4YY
06/10/2021
03085904-0004